Coral Reefs of the World

Volume 21

Series Editors

Coral Reefs of the World is a series presenting the status of knowledge of the world's coral reefs authored by leading scientists. The volumes are organized according to political or regional oceanographic boundaries. Emphasis is put on providing authoritative overviews of biology and geology, explaining the origins and peculiarities of coral reefs in each region. The information is so organized that it is up to date and can be used as a general reference and entry-point for further study. The series will cover all recent and many of the fossil coral reefs of the world.

Prospective authors and/or editors should consult the Series Editors B.M. Riegl and R.E. Dodge for more details. Any comments or suggestions for future volumes are welcomed:

Dr. Bernhard M. Riegl/Dr. Richard E. DodgeNova Southeastern UniversityDania Beach, FL 33004USA

e-mail: rieglb@nova.edu and dodge@nova.edu

Serge Planes
Editor

The Future of Coral Reefs

Evidence from Research

Springer

Editor

Serge Planes
UAR 3278 CRIOBE
PSL Research University
CNRS-EPHE-UPVD
Perpignan, France

ISSN 2213-719X ISSN 2213-7203 (electronic)
Coral Reefs of the World
ISBN 978-3-031-98583-6 ISBN 978-3-031-98584-3 (eBook)
https://doi.org/10.1007/978-3-031-98584-3

Cover Image: With kind permission from A. Rosenfeld.

This Springer imprint is published by the registered company Springer Nature Switzerland AG
The registered company address is: Gewerbestrasse 11, 6330 Cham, Switzerland

Foreword—The State of Coral Reefs Research in a Changing World

Coral reefs are viewed among the most captivating and vital ecosystems on Earth. Often considered as the "rainforests of the sea," these underwater landscapes harbor an extraordinary diversity of life—from microscopic plankton to iconic reef fishes and apex predators. They play critical roles in coastal protection, nutrient cycling, and carbon sequestration, and they provide indispensable services to human communities through fisheries, tourism, and cultural heritage (Moberg and Folke 1999; Spalding et al. 2001). Yet, in the face of rapid global change, coral reefs are confronting challenges of unprecedented scale.

Recent research has vividly illustrated that rising sea temperatures and ocean acidification are fundamentally altering the physical and biological fabric of reef ecosystems (Hughes et al. 2022; Baker et al. 2020). However, mass bleaching events triggered by anomalously high sea surface temperatures (van Hooidonk et al. 2016) are now occurring with increasing frequency and severity making mass bleaching events as the most critical for the future of coral reefs. These events disrupt the critical symbiosis between corals and their photosynthetic algae, often leading to widespread mortality and a significant loss of reef complexity (van Woesik et al. 2021). Numerous studies across the Indo-Pacific and Caribbean have shown declines in live coral cover of up to 50% in just a few decades (Donner et al. 2020; Madin et al. 2022, Souter et al. 2021), with long-lasting repercussions for biodiversity and ecosystem functioning (van Hooidonk et al. 2014).

In parallel, ocean acidification—stemming from increased atmospheric CO_2—impairs the ability of corals to secrete and maintain their calcium carbonate skeletons. This weakening of coral structures not only undermines the stability of individual colonies but also compromises the resilience of entire reef frameworks, making them more susceptible to erosion and physical disturbances (Altieri et al. 2020; Fabricius et al. 2020). Local stressors, including nutrient runoff, sedimentation from coastal development, and unsustainable fishing practices, further amplify these global pressures. The cumulative effects of these stressors lead to profound shifts in reef community composition and function, ultimately reducing the capacity of these ecosystems to recover from disturbances (Ferrario et al. 2020; Pochon et al. 2021).

Recent technological advances—such as high-resolution remote sensing, environmental DNA (eDNA) monitoring, and innovative geochemical analyses—are now providing us with unprecedented insights into these rapid changes. These tools allow researchers to capture fine-scale variations in reef health and track shifts in biodiversity with remarkable precision (Thomsen and Willerslev 2020; Pochon et al. 2021). Such innovations have not only enriched our understanding of coral reef dynamics but have also underscored the urgent need for integrated conservation strategies.

This book is conceived as a synthesis of cutting-edge research deployed by the Laboratory of Excellence CORAIL network—a dynamic consortium of French scientists and institutions dedicated to advancing coral reef science through interdisciplinary collaboration. LabEx CORAIL has been at the forefront of addressing the multifaceted challenges facing coral reefs, bringing together expertise from molecular biology, ecology, oceanography, geochemistry, and archaeology. The network's commitment to collaborative research and innovation has laid the groundwork for many of the studies presented in this volume. Notably, our own work (Planes

2022) has contributed to a deeper understanding of reef dynamics and resilience strategies under rapidly changing environmental conditions.

A Collaborative Vision for Reef Research

The chapters assembled in this book represent the fruits of research efforts supported and inspired by LabEx CORAIL during 15 years. While our work is rooted in addressing urgent ecological and environmental challenges, it is equally driven by a deep-seated curiosity about the natural world. At LabEx CORAIL, scientists have leveraged a wide array of tools and methodologies—from traditional field surveys to high-throughput molecular techniques—to unravel the complex interplay of factors that influence reef resilience and decline. This integrative approach is critical for understanding the multifactorial impacts of global climate change and local anthropogenic pressures on coral reefs. One of the strengths of LabEx CORAIL is that its laboratories are spread across the French Overseas Territories, giving access to coral reefs in three oceans and thus providing a global view of reef functioning.

For instance, studies within our network have revealed how some coral species are capable of modifying their symbiotic relationships and metabolic pathways in response to thermal stress, suggesting potential adaptive mechanisms that could bolster reef resilience (Hughes et al. 2022; Baird et al. 2021). Other research efforts have focused on reconstructing historical reef conditions through high-resolution geochemical analyses of coral skeletons, providing essential context for interpreting current trends and forecasting future trajectories (Thomsen and Willerslev 2020; Pochon et al. 2021).

The LabEx CORAIL has championed the integration of local ecological knowledge and traditional practices with state-of-the-art scientific research. This synergy has proven invaluable in developing community-based management strategies that are both culturally relevant and scientifically sound (Berkes 2020; Altieri et al. 2020). The commitment to an inclusive research paradigm, one that values both modern technologies and traditional wisdom, is a hallmark of the LabEx CORAIL approach and is reflected throughout the chapters of this volume.

A Multifaceted Overview of Coral Reef Science

This volume weaves together 11 interrelated chapters that collectively form a comprehensive panorama of coral reef science and management. Rather than presenting discrete, isolated studies, the text unfolds as a continuous narrative where each chapter informs and builds upon the others, illustrating the dynamic interplay of processes that shape these intricate ecosystems.

The journey begins with a holistic synthesis of reef biodiversity, presented in a chapter that integrates genetic, molecular, and community-level data to provide a comprehensive snapshot of reef health. This cross-scale perspective is essential for guiding conservation strategies, as it demonstrates how biodiversity metrics can be used to assess ecosystem integrity and inform management decisions. The integration of such diverse data streams echoes the interdisciplinary approach that is the hallmark of this volume (Pellerin et al. 2025).

The holistic synthesis of reef biodiversity links then to an exploration of trophic ecology and energy fluxes, which lays the foundation by examining the intricate network of energy flow that sustains reefs. Here, researchers detail how both in situ production—driven by symbiotic algae and benthic microalgae—and external nutrient inputs fuel the rapid recycling of organic matter (Briand et al. 2025). This understanding of nutrient cycling is critical, as it provides context for later discussions on how energy dynamics influence coral physiology and adaptive responses.

Building on these ecological fundamentals, the narrative transitions to the study of calcification and environmental archives. In this section, researchers demonstrate how coral skeletons serve as natural recorders of past environmental conditions by using geochemical proxies—such as isotopic ratios and elemental compositions—to reconstruct historical climate variability (Canesi et al. 2025). The ability to trace past conditions not only illuminates the origins of current vulnerabilities but also interacts with the emerging understanding of bleaching dynamics and stress responses. For instance, by linking calcification rates with environmental stress, the chapter underscores how historical baselines can help predict future reef trajectories.

Complementing these discussions is an exploration of coral heterotrophy and nutritional plasticity. Recognizing that many corals rely not solely on photosynthetic energy but also on heterotrophic feeding, this section examines how flexible nutritional strategies help corals endure periods of thermal stress and bleaching. The chapter reveals the intimate connection between energy acquisition and reproductive success, reinforcing the narrative of adaptability that permeates the volume (Meunier et al. 2025).

As the book progresses, attention shifts to the critical processes of coral bleaching and symbiosis (Dube et al. 2025). Here, the discussion centers on the physiological and molecular mechanisms that underpin the breakdown of the coral–algae partnership during thermal stress. The loss of symbiotic algae triggers bleaching events that lead to widespread mortality and destabilize reef communities. This section not only reveals the underlying biochemical pathways but also sets the stage for understanding the recovery mechanisms detailed later. Indeed, adaptive responses—such as shifts in symbiont communities—are essential for coral survival and are examined further in discussions of coral reproduction and restoration.

The narrative then moves to the mechanisms governing coral reproduction and restoration, emphasizing how both sexual and asexual reproduction contribute to reef regeneration after disturbances. This discussion is deeply interwoven with the earlier exploration of energy dynamics and symbiotic resilience, highlighting how genetic diversity and phenotypic plasticity can enhance recovery. Innovative restoration techniques, including larval enhancement and assisted gene flow, are presented as promising approaches to scale up reef regeneration efforts, linking back to the adaptive strategies discussed in the bleaching chapter (Fellous et al. 2025).

A significant biological challenge to reef health is posed by outbreaks of the crown-of-thorns sea star. In an innovative chapter, emerging technologies such as environmental DNA biosensors and machine-learning algorithms are introduced as effective tools for monitoring and managing these destructive predators. The discussion of these novel approaches not only illustrates the application of advanced molecular methods but also reinforces the integrative theme of the volume by linking biological threats with broader ecological processes (Gorman et al. 2025).

While the chapters addressing climate-related stressors form a core part of the volume, the text also broadens its scope to include emerging chemical threats. The discussion on contaminants of emerging concern illustrates how microplastics, persistent organic pollutants, and trace metals interfere with coral physiology and disrupt fundamental nutrient cycles. The insights here are directly connected to the earlier chapters on trophic ecology, as contaminant interactions can compromise energy flow and exacerbate latent vulnerabilities within the reef ecosystem (Gardon et al. 2025).

The next chapter underpins the concept of latent vulnerability in social-ecological systems, where the interplay between immediate stressors and long-term pressures (such as overfishing or nutrient imbalances) further refines our understanding of reef resilience (Brun et al. 2025).

Adding another layer of complexity, the volume addresses reef health from the perspective of associated fauna, specifically focusing on fish parasites. The diversity of metazoan parasites found in reef fishes is presented as a sensitive indicator of ecosystem imbalance, with shifts in parasite communities providing early warning signals of deteriorating reef conditions. These findings dovetail with broader biodiversity assessments and serve to remind us of the intricate connections between host organisms and their environments (Pavloudi et al. 2025).

Finally, the narrative comes full circle with a chapter that bridges natural and human sciences by examining the long history of human–reef interactions through island archaeology. By reconstructing ancient marine socio-ecosystems, this section provides valuable insights into how coastal communities have historically adapted to and managed reef environments. This long-term perspective not only enriches our understanding of current adaptive strategies but also highlights the enduring relationship between humans and coral reefs—a relationship that underscores many of the resilience mechanisms discussed throughout the volume (Conte et al. 2025).

Each chapter in this volume is not an isolated contribution but rather a dynamic part of an integrated narrative. The recurring themes—energy flow, environmental archives, adaptive strategies, and human interactions—are revisited across chapters, reinforcing their interdependence and collectively enriching our understanding of coral reef ecosystems. Through these interconnections, the book offers a multidimensional view of coral reef science, one that is as comprehensive as it is collaborative.

By weaving together diverse methodological approaches and areas of expertise, this volume underscores the complexity and urgency of coral reef conservation. From trophic ecology to archaeological insights, each chapter adds a distinct perspective to the ongoing conversation about how best to protect and restore these invaluable ecosystems. Ultimately, the synergistic insights presented here aim to guide researchers, policymakers, and communities toward strategies that ensure coral reefs continue to thrive in an ever-changing world.

Throughout the continuous narrative of this volume, recurring themes—such as energy flow, environmental archives, adaptive strategies, and human interactions—are interwoven to reinforce their interdependence and collectively enhance our understanding of coral reef ecosystems. The synergistic insights presented here, drawn from a diverse array of methodological approaches and scientific disciplines, provide a multidimensional view of reef science. This integrated narrative reflects the collaborative spirit of the research deployed by LabEx CORAIL, which has been instrumental in fostering the interdisciplinary approaches that underpin the chapters in this volume. Together, these contributions guide us toward strategies that not only aim to protect and restore coral reefs but also ensure that these extraordinary ecosystems continue to thrive in a world of constant change.

Bridging Science and Society

The research presented in this volume is not just an academic exercise—it is a call to action. The challenges facing coral reefs today demand that we translate scientific insights into practical, community-based conservation strategies. Advanced techniques such as high-resolution geochemical analyses, remote sensing, and eDNA monitoring are now being combined with traditional field studies and archaeological investigations to develop a comprehensive understanding of reef dynamics. This integrated approach is vital for devising effective management measures that can mitigate the rapid environmental changes threatening these ecosystems.

LabEx CORAIL has played an instrumental role in fostering this integrative research ethos. By uniting researchers from diverse disciplines and encouraging collaboration between modern science and traditional knowledge, the network has helped create a robust foundation for innovative reef conservation strategies. The contributions from LabEx CORAIL, though one element among many in this volume, highlight the value of interdisciplinary research and serve as a testimony to what can be achieved when expertise and innovation converge.

Looking to the Future

While the challenges facing coral reefs are formidable, the work compiled in this volume offers a hopeful vision for the future. It demonstrates that coral reefs possess an inherent capacity for adaptation and recovery—provided that the stressors are sufficiently mitigated and that conservation efforts are effectively implemented. However, realizing this potential will require sustained, coordinated action across scientific, governmental, and community sectors.

The chapters in this volume collectively offer a roadmap for advancing coral reef science and conservation. They elucidate the complex interplay of factors that govern reef health—from energy fluxes and nutrient cycling to the subtle shifts in symbiotic relationships and reproductive strategies—and they provide innovative solutions for monitoring and managing these dynamic ecosystems. As we move further into the Anthropocene, our ability to understand and protect coral reefs will hinge on the integration of advanced scientific techniques, interdisciplinary collaboration, and the active participation of local communities.

In closing, it is our hope that this book serves as both an inspiration and a resource for scientists, conservationists, policymakers, and all those who are committed to the future of our oceans. Coral reefs are dynamic, living systems whose health is inextricably linked to the well-being of our planet and its people. By harnessing the power of interdisciplinary research—exemplified by the work of LabEx CORAIL—and by translating these insights into actionable conservation strategies, we can work together to ensure that these extraordinary ecosystems continue to flourish for generations to come.

Serge Planes
UAR3278 CRIOBE, PSL Research University, CNRS-EPHE-UPVD
Perpignan, France

Emilie Boissin
Laboratoire d'Excellence "Corail"
Perpignan, France

References

Altieri AH, Harrison SB, Seemann J et al (2020) Tropical dead zones and mass mortalities on coral reefs. Proc Natl Acad Sci 117(14):8280–8286

Baird AH, Guest JR, Willis BL (2021) Systematic and biogeographical patterns in the reproductive biology of scleractinian corals. Mar Biol 168(1):3

Baker AC, Glynn PW, Riegl B (2020) Climate change and coral reef bleaching: an ecological assessment of long-term impacts, recovery trends, and future outlook. Estuar Coast Shelf Sci 238:106815

Berkes F (2020) Sacred ecology: traditional ecological knowledge and resource management. Routledge

Conte E, Traversat G, Rurua VA, Béarez P, Galzin R, Molle G (2025) Living among the reefs: insights and perspectives from Island archaeology in the Pacific. In Coral Reef Resilience in a Changing World

Donner SD, Skirving WJ, Little CM et al (2020) Global assessment of coral bleaching and required rates of adaptation under climate change. Glob Chang Biol 26(5):2661–2671

Fabricius KE et al (2020) Impacts of ocean acidification on coral reef ecosystems: a meta-analysis. Glob Chang Biol 26(3):1630–1644

Fellous A, Hédouin L, Putnam H (2025) Evaluating the role of natural and human assisted coral reproduction to maintain and recover disturbed reefs. In Coral Reef Resilience in a Changing World

Ferrario F, Beck MW, Storlazzi CD et al (2020) The effectiveness of coral reefs for coastal hazard risk reduction and adaptation. Nat Commun 11:3794

Gorman LM, Byrne M, Beldade R, et al (2025) The use of new biologically based technological development in the management of crown-of-thorns sea stars. In Coral Reef Resilience in a Changing World

Hughes TP, Anderson KD, Connolly SR et al (2022) Spatial and temporal patterns of mass bleaching of corals in the Anthropocene. Science 377(6605):193–197

Madin JS, van Woesik R et al (2022) Coral reef degradation in the Anthropocene: the shifting baseline syndrome and new paradigms for resilience. Front Mar Sci 9:843132

Meunier V, Ferrier-Pagès C, Dromard CR, Houlbrèque F (2025) Coral heterotrophy in an era of climate change. In Coral Reef Resilience in a Changing World

Pavloudi C, Boxshall GA, Bray RA et al (2025) The biodiversity of metazoan parasites of coral reef fishes of the Central and Eastern Indo-Pacific. In Coral Reef Resilience in a Changing World

Pellerin F, Espel D, Ballard J et al (2025) Snapshots of coral reef biodiversity. In Coral Reef Resilience in a Changing World

Pochon X et al (2021) Environmental DNA (eDNA) monitoring of marine biodiversity: current progress and future prospects. Front Mar Sci 8:1–12

Souter D, Planes S, Wicquart J, Logan M, Obura D, Staub F (eds) (2021) Status of coral reefs of the world: 2020 report. Global Coral Reef Monitoring Network (GCRMN) and International Coral Reef Initiative (ICRI)

Spalding MD, Ravilious C, Green EP (2001) World atlas of coral reefs. University of California Press

Thomsen PF, Willerslev E (2020) Environmental DNA—an emerging tool in conservation for monitoring past and present biodiversity. Biol Conserv 248:108784

van Hooidonk R, Maynard JA, Planes S (2014) Opposite latitudinal gradients in projected ocean acidification and bleaching impacts on coral reefs. Glob Chang Biol 20:103–112

van Hooidonk R, Maynard J, Tamelander J, Gove J, Ahmadia G, Rymundo L, Williams G, Heron SF, Planes S (2016) Local-scale projections of coral reef futures and implications of the Paris Agreement. Sci Rep 6:39666

van Woesik R et al (2021) Recurrent coral bleaching in the northern Gulf of Mexico and implications for resilience. Mar Ecol Prog Ser 672:29–38

Contents

1 Trophodynamics in Coral Reefs: Current Knowledge and Research Perspectives

Marine J. Briand, Charlotte R. Dromard, Mireille Harmelin-Vivien, Valeriano Parravicini, and Yves Letourneur

Abstract

Coral reef trophodynamics is extremely complex and still poorly understood, while investigated for decades. Recent advances on organic matter sources, fluxes, trophic niches of species, and spatiotemporal variations of food webs are first summarized. Then, tools used in trophodynamics research are briefly documented, from traditional pioneer methods of stomach content analysis to more recent biochemical methods (stable isotopes, fatty acids, biochemical composition, metabolomics, DNA metabarcoding, etc.) and modeling approaches. Eventually, future challenges and perspectives are presented addressing the diverse scenarios of coral reef trophic functioning response in front of the ongoing changes in global and local stressors.

Keywords

Trophic relationships · Biochemical tracers · Stable isotopes · DNA metabarcoding · Modeling

M. J. Briand · V. Parravicini
CRIOBE, PSL Research University, USR 3278, EPHE-CNRS-UPVD, LabEx « Corail », Université de Perpignan, Avenue Paul Alduy, Perpignan Cedex, France
e-mail: valeriano.parravicini@ephe.psl.eu

C. R. Dromard
UMR 8067 BOREA, Univ Antilles, MNHN, SU, UCN, CNRS, IRD, Pointe-à-Pitre, Guadeloupe, France
e-mail: Charlotte.Dromard@univ-antilles.fr

M. Harmelin-Vivien
Aix Marseille Université, Université de Toulon, CNRS, IRD, MIO, UM 110, Campus Luminy, Marseille, France
e-mail: mireille.harmelin@mio.osupytheas.fr

Y. Letourneur (✉)
UMR 250 ENTROPIE, EMR 9001 SantEco, UR, IRD, CNRS, IFREMER, UNC, LabEx « Corail », Université de La Nouvelle-Calédonie, Nouméa Cedex, New Caledonia
e-mail: yves.letourneur@unc.nc

1.1 Context of Trophic Ecology in Coral Reefs

Coral reefs are characterized by poorly resolved innumerable and nonrandom linkages across an intricate network of ecological interactions (Jordano 2016). Accounting for such complexity is critical to define energetic pathways and, ultimately, ecosystem functioning, productivity, and mechanisms behind the maintenance of biodiversity (Tilman et al. 2014). Indeed, a food web perspective may emerge as the most appropriate for verifying and assessing the integrated effects of multiple pressures on natural ecosystems, as internal food web mechanisms can buffer or even reverse the effects exerted at the population or species scale (Gray et al. 2014). Many aspects of the trophodynamic processes have already been investigated in coral reefs over the last decennies and are summarized below (Fig. 1.1).

1.1.1 Main Questionings in the Trophic Studies

1.1.1.1 Major Organic Matter Resources and Fluxes

Understanding the mechanisms that maintain ecosystem productivity requires a definition of the major sources of nutrients and energy that fuel the entire coral reef ecosystem (Letourneur et al. 2013; Robinson et al. 2024). The characterization of these energy flows is still complicated due to a high diversity of organic matter (OM) sources available to primary consumers in these systems. Coral reef food webs rely both on an allochthonous oceanic production supported by phytoplankton and on autochthonous benthic primary producers, that is, coral symbionts, macroalgae, algal turf, microphytobenthos, crustose coralline algae, and seagrasses (Dromard et al. 2013; Briand et al. 2015, 2016; Fey et al. 2020). Moreover, both benthic and pelagic potential food sources, plus the inputs from the terrestrial realm, may be important indirect contributors to the OM pools, that is,

S. Planes (ed.), *The Future of Coral Reefs*, Coral Reefs of the World 21, https://doi.org/10.1007/978-3-031-98584-3_1

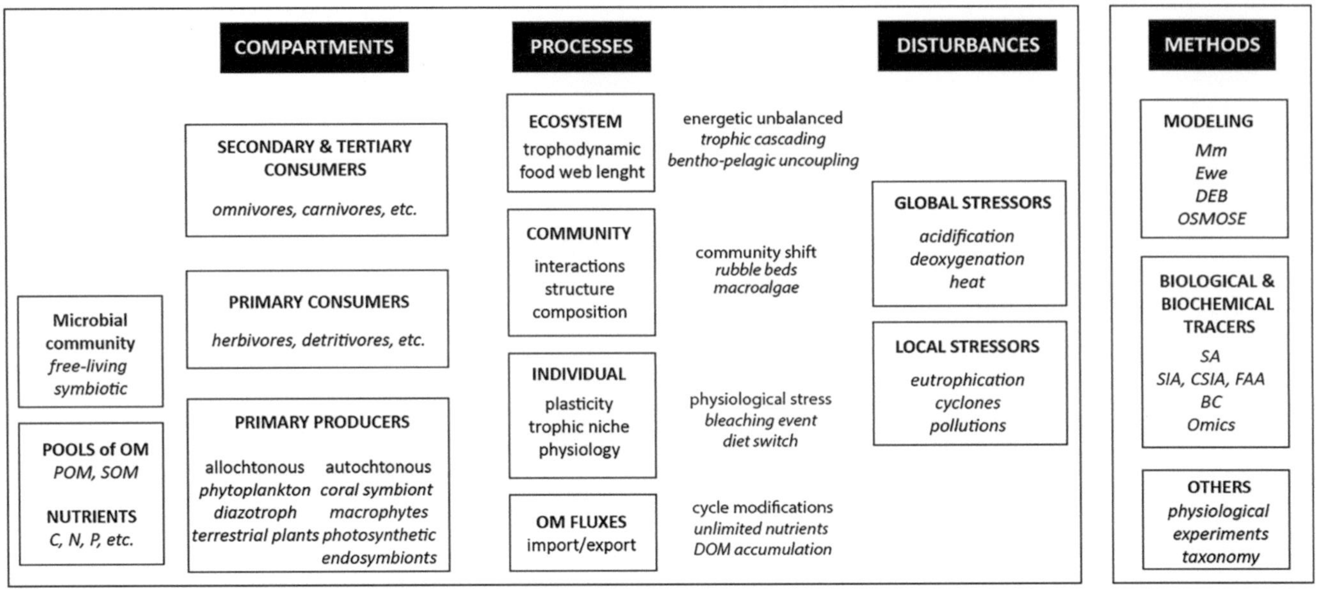

Fig. 1.1 Summary scheme of the main compartments, processes, disturbances, and methods (abbreviations in the text) involved in trophodynamics research for the last decades and in the future

detritus, particulate, and sedimentary OM (Crossman et al. 2001; Wyatt et al. 2013).

The rapid recycling of OM is a characteristic of coral reef functioning, sustaining high productivity in relatively oligotrophic environments. The roles of different organisms and pathways having a key influence on nutrient biogeochemistry and fluxes were thus pointed out. For example, sponges are both filter feeders and hosts to abundant symbiont communities, and have a major role in the transfer of dissolved organic matter (DOM) to higher trophic levels in the benthos, for example, sponge-associated and free-living detritivores (Rix et al. 2017). Several species recycle carbon by directly assimilating dissolved organic carbon (DOC) into biomass (McMurray et al. 2016) or converting DOC exudates released by corals and algae into bioavailable particulate organic carbon (POC) and cellular detritus through the sponge loop pathway (De Goeij et al. 2013). In addition, sponges are important sink and/or sources of dissolved inorganic nitrogen, phosphate, and silicate through their metabolism and the inextricable activity of their associated microbial communities (Pawlik and McMurray 2020). Indeed, microorganisms also efficiently recycle the dissolved organic matter (DOM) released by sponges and other primary producers through the microbial loop (Haas et al. 2013; Silveira et al. 2017). Recent studies even highlighted the significant impact of DOM components like metabolites (e.g., tryptophan, phenylalanine, caffeine) produced by sponges on the success of microbial communities and organisms associated with microbial symbionts (Fiore et al. 2017). Finally, planktonic diazotrophs fix and reduce atmospheric N_2 and release part of this nitrogen into bioavailable ammonium (NH_4^+) in seawater. Abundant in coral lagoon waters, these prokaryotes are providing sufficient nitrogen stocks for the development of the planktonic food web in oligotrophic waters (Bonnet et al. 2016).

By consuming these sources, primary consumers are initiating tranfers of OM into the food webs. Their food preferences are therefore decisive and can be explained by several factors; for instance, mature forms of macroalgae are more or less consumed by herbivores according to their palatability, that being a function of the algal morphology (e.g., soft or calcareous), physiological strategies (e.g., synthesis of repellent molecules), and nutritional quality (Dromard et al. 2017 and included references). All other trophic network interactions also lead to important fluxes of energy, and many studies reinforce the evidence of a tight bentho-pelagic coupling that might become an increasingly important characteristic in future coral reefs (Fey et al. 2020; Skinner et al. 2021).

1.1.1.2 Trophic Niche and Plasticity of Species

It remains difficult to clearly capture dynamic energetic processes on coral reefs (Harmelin-Vivien 2002; Briand et al. 2016), and understanding species' roles and interactions requires an accurate assessment of their trophic status and niche. Thus, the trophic position is an important concept typically used to calculate food chain length, degree of omnivory or compound bioaccumulation into food webs (Post 2002). The trophic niche, issued from the ecological niche theory applied to feeding processes, is also largely employed; its size and specialization result from complex interactions between biological traits and local constraints, generating difficulties to disentangle the respective effects of each characteristic (Futuyma and Moreno 1988). This balance between different traits and characteristics may lead to different responses and thus influence feeding plasticity in response to changes of resource availability (Letourneur et al. 2017). Therefore, it is critical and urgent to better assess how and through what mechanisms species or functional groups can adapt and cope with changing environmental conditions.

Identifying interspecific diet and niches specializations is critical (Cybulski et al. 2022 and included references). For instance, the diverse range of feeding behaviors and adaptive morphologies employed by invertivores on coral reefs is a clear indication of the widely underestimated role of cryptofauna in the reef trophodynamics (Glynn et al. 2011). However, it has been challenging to delineate such complex behaviors and interactions, and a paucity of empirical information often leads to inconsistent definitions of trophic guilds based on expert opinions (Parravicini et al. 2020). Dietary variation of individuals within species is also widespread (Wyatt et al. 2019), so characterizing intraspecializations within a larger population is essential for delineating a species' trophic niche. Besides, reef organisms may change their resource use with ontogeny (Cummings et al. 2010), and alter their resource use according to what is available by plasticity process (Fey et al. 2021a).

1.1.1.3 Spatial and Temporal Variations of Food Webs

Due to changes in environmental conditions and in the structure of benthic and pelagic primary producers, OM resources fluctuate within and between reefs (Johnson et al. 1995). In one hand, variations along spatial scales have shown influence of numerous environmental factors, such as the distance from the coast and the terrestrial or oceanic exposure, or the connectivity between adjacent habitats (Briand et al. 2015). On the other hand, short- as long-term temporal changes have been less investigated, and, for example, few data were recorded about the effects of nycthemeral variations on reef communities (de Lira et al. 2014). As a result, relevancy of isotopic spatiotemporal comparisons is limited unless isotopic baselines are standardized between sampling periods or locations (Tamelander et al. 2009), and therefore understanding such variations associated with trophic structure is often complicated.

1.1.2 Tools Used in Trophodynamic Research

1.1.2.1 Traditional Methods

From the 1950s, in parallel to direct observation of foraging, stomach analysis (SA) is one of the first and the most applied methods for qualitative and quantitative evaluation of consumers' diet (Hyslop 1980). It concerns the analysis of food items in all portions of the digestive tract (mostly the stomach), through (1) a sampling step using nonlethal (i.e., regurgitation) or, most often, lethal methods (i.e., extraction of the digestive tract), and (2) an identification step requiring visual and/or DNA-based techniques (Da Silveira et al. 2020 and included references). SA provides punctual information on ingested food (not assimilated one), at individual to community levels, revealing intra- and interspecific relationships strong enough to schemed coherent trophic functioning (Harmelin-Vivien 1981). However, SA also goes with high sampling costs and laborious laboratory work, which is reliant on potential degradation and/or fragmentation of samples, implying a possible nonidentification of the food item, and does not allow the determination of the origins of consumed inorganic matter.

1.1.2.2 Biochemical Tracers

From the 1980s, biochemical tracers were used to reveal detailed and integrated diet information, with accuracy and high resolution (Nielsen et al. 2018). The synthesis of body tissues is related to the chemical elements present in consumed foods, so both should be isotopically related. Stable isotope analysis (SIA) of carbon ($\delta^{13}C$), nitrogen ($\delta^{15}N$), and sulfur ($\delta^{34}S$) has then emerged as a powerful method for revealing "time-integrated assimilated food" information and for giving insight into their trophic ecology. Specifically, there is a historically accepted change in bulk isotopic ratio values at each trophic level (i.e., ~1‰ for $\delta^{13}C$ and ~ 3‰ for $\delta^{15}N$; Post 2002, McCutchan Jr et al. 2003), widely used to estimate the basal sources of carbon and nitrogen in a particular food chain, prey composition, and trophic position (Michener and Kaufman 2007). Such assumptions may have led to biased results in the past, as since values of trophic fractionation factors have been largely reevaluated (Letourneur et al. 2024); they are generally higher between primary producers and herbivores than between consumers of medium to high levels, and vary substantially among species (Wyatt et al. 2010). Thanks to improvements in analytical techniques, plus given the much larger range of variation in resource sulfur signatures and the low consumer-diet shift (McCutchan Jr et al. 2003), the $\delta^{34}S$ ratio is increasingly being used as a third tracer to disentangle sources and trophic level identification in food webs (Skinner et al. 2022). Despite many assets, the SIA tool needs to rely on isotopically distinct food sources for accurate interpretations, a condition that is often not observed in the field (Da Silveira et al. 2020).

After that, a refinement of the SIA method called compound-specific isotope analysis (CSIA) has been successfully applied in trophic research. While SIA evaluates the "bulk" isotopic concentration of a given biological sample, CSIA evaluates isotopic signatures of specific individual macronutrients such as amino acids and lipids (e.g., $\delta^{15}N$ of glutamic acid or phenylalanine, $\delta^{13}C$ of fatty acids; Twining et al. 2020). The conservative and predictable assimilation patterns obtained in consumers allow for a more accurate identification of the food source (McMahon et al. 2016), and provide specific information on both the assimilated baseline and the length of food chains (Fey et al. 2021a).

Fatty acids analysis (FAA) is also a conservative method providing integrated qualitative information regarding the diet of consumers in short- and long-term approaches (Iverson et al. 2004; Iverson 2009). FAs are carbon-rich compounds that are ubiquitous and greatly diversified in most organisms (25–70 unique molecules). Although some are biosynthesized de novo by specific organisms at the base of the food chain (e.g., bacteria, diatoms, macroalgae, vascular plants), most FAs are acquired through diet and are poorly metabolized by the consumers (Dalsgaard et al. 2003). Owing to their predictable assimilation patterns and their multivariate profiles, these lipid biomarkers are useful tool to distinguish potential dietary items at a low identification level (Iverson et al. 2004). Data on the presence, amount, and ratios of different FAs can also be used to infer food consumption and energetic mobilizations along the consumers' life cycles (Dalsgaard et al. 2003; Fey et al. 2021a). As CSIA, the FAA is an expensive and time-consuming method, and profiles can change dramatically due to environmental conditions (e.g., temperature, light, nutrients; Da Silveira et al. 2020).

At last, the analysis of the biochemical composition (BC) of the OM as the sum of lipids, proteins, and soluble carbohydrates concentrations has been used to provide information on the energetic value of food material potentially available to consumers (Krogdahl et al. 2005).

The rise of omics approaches in recent years provides molecular-level insights into food relationships and biogeochemical cycling. High-resolution *metabolomics* techniques were used to probe the composition of OM (e.g., DOM) including metabolites relevant to the growth of many organisms (Fiore et al. 2017; Wegley Kelly et al. 2022). Untargeted or discovery-based analysis allows for a semiquantitative profile-view of low-molecular-weight molecules within a sample, while the targeted analysis provides a quantitative comparison of a small subset of metabolites (Patti et al. 2012; Kido Soule et al. 2015). DNA metabarcoding (DNAm) methods can capture individual-level dietary variations, improving trophodynamic assessments, but they are limited to the gut contents of the animal at the time of sampling (Casey et al. 2019).

Each method has its own objectives, sampling and analysis procedures, data resolution, spatial and temporal coverage, results interpretation, and limitations. The combination of these approaches, including stomach content analysis, surpasses most of these limitations and is therefore greatly recommended (Da Silveira et al. 2020; Robinson et al. 2024). A joint perspective that can be applied to different ecological levels (population, community, and ecosystem), providing valuable diet information on intra- (e.g., ontogenetic, reproductive, migration) and interspecific relationships, clarifying the origin of food resources explored, spatial-temporal patterns, and trophic-web structuration (Matley et al. 2018).

1.1.2.3 Modeling Approaches

Biochemical methods are not able to estimate the amounts of consumed foods without modeling (Iverson et al. 2004). Consequently, in the last 20 years, several analytical methods called "Bayesian models" have been proposed to convert stable isotope data into robust derived metrics. Mixing models (MM) allow (i) the estimation of the relative contribution of isotopically different OM sources in a mixture and the reconstruction of animal diets from isotopic ratios and/or multiple chemical tracers (e.g., *"SIAR"*, then "SIMMR" and "MIXSIAR" models in R; Stock et al. 2018), (ii) the calculation of consumer trophic position at the population level (*"tRophicPosition"* in R; Quezada-Romegialli et al. 2018), (iii) the measurements of the isotopic niche space of populations and communities ["SIBER" model in R; Jackson et al. 2011, or also MCPs (minimum convex polygon); Fey et al. 2021b], and (iv) the evaluation of entire food webs ("IsoWeb" model; Kayoda et al. 2012). Needless to say, that outputs must be carefully interpreted, in the light of strengths and weaknesses of each model, and that current stomach-content analysis and in situ direct observations should guide model inputs (Twining et al. 2020).

Trophic ecology also provides basic information for structuring complex ecotrophic models. *ECOPATH* has been developed to estimate mean annual biomass, production, and consumption for components of an ecosystem (Polovina 1984). The software and techniques have been improved to include methods of comparing ecosystems using Ecological Network Analysis (Christensen and Pauly 1992), to model dynamic changes using *ECOSIM* (Walters et al. 1997), to model spatial changes using *ECOSPACE* (Walters et al. 1999), and to model dynamics using the spatial-temporal framework and the habitat foraging capacity model (Christensen et al. 2014). Is born out a combined *Ecopath with Ecosim (EwE)* software package, widely applied for modeling marine ecosystems and able to address many of the questions asked by managers on marine policy issues (Christensen and Walters 2004). For example, *ECOTRACER* included in EwE simulate and analyze the transport of contaminants through coral reefs (e.g., methylmercury or radiocesium, Walters and Christensen 2018). However, still to date, easy access with limited quality control contributes to the low utilization of EwE models for management applications.

Other models less used in coral environment are nevertheless of interest. Models of energetics called dynamic energy budget or "DEB" claims to provide a comprehensive description of the metabolic organization and rates underlying the physiological functions of an organism throughout its life cycle in a fluctuating environment (Kooijman 2010). The predator-prey model *OSMOSE* is a two-dimensional, individual-based, and multispecies model explicitly representing major processes in the life cycle of high trophic level (HTL) groups of fish and invertebrate species (Grüss et al. 2015). Such models have already been useful to predict the effects of global change, and better understand trophic dynamics, environmental stressors effects, species geographical patterns, bio-production optimization, or management of exploited resources (Lesser 2013).

1.2 Future Challenges and Prospectives

Pristine areas are no more (Pandolfi et al. 2003), and even under best-case emissions trajectories, coral reefs will likely continue to be transformed by the global anthropogenic climate change (Hoey et al. 2016) and face escalating local human threats which cause severe destruction (i.e., coastal development, overfishing, pollution, and eutrophication; Zaneveld et al. 2016). Due to the complexity of coral reef ecology plus local and regional differences, understanding and identifying effective research and management practices to address corals' present and future fate is therefore challenging (Wilkins et al. 2021). In particular, among the coral regions studied by the French research community, the Caribbean and the French Polynesia should be prioritized for inclusion in international conservation programs (Guan et al. 2020). Therefore, multiple scientific challenges, economic progress, as well as a surge in technological innovations will be of importance for the future of coral reefs. We have chosen to focus on several critical scenarios here, for which trophic research might provide answers and solutions (Fig. 1.1).

1.2.1 Scientific Key Lever for Future Researches

1.2.1.1 Changes in OM Fluxes

The pool of nutrients can be altered by several drivers of natural and anthropogenic origin (D'Angelo and Wiedenmann 2014), and changes in OM fluxes are a crucial problematic, not only to assess food web functioning and the resilience of coral reefs, but also to anticipate unexpected cascading effects and undesirable ecological surprises. In the context of climate change, marine heat waves are becoming more intense and frequent (Hughes et al. 2018), and water acidification is increasing energy requirements of calcifying organisms, in particular scleractinian corals. As well, anthropogenic nutrient enrichment in reef waters, mostly of nitrogen and phosphorus compounds, induces direct and indirect negative effects on coral physiological performances and ecosystem functioning (D'Angelo and Wiedenmann 2014). Because of the combined effect of heat, CO_2, and nutrient stress, coral reefs are experiencing increasingly frequent and devastating bleaching events (Hughes et al. 2017).

The global warming disrupts the key symbiosis interaction between corals and their endosymbiotic dinoflagellates and therefore decreases coral nitrogen acquisition capacity (Godinot et al. 2011). Thus, the ability of coral reefs to cope with future ocean conditions will depend on their capacity to fulfil their metabolic requirements (Meunier et al. 2021). More physiological studies on coral nutrition are therefore crucial to understand the feeding strategies of these polytrophic and opportunistic feeders faced with future life conditions, that is, via capture of plankton and organic particles by polyps and/or via translocation of photosynthetic products from their endosymbiotic algae (Duprey et al. 2016). An interesting track to resist and recover from stressing episodes is the store of energy reserves and/or the switch from an autotrophic to a heterotrophic diet. Selective feeding on picoplankton-sized unicellular cyanobacteria (*Synechococcus* spp.) and prochlorophytes (*Prochlorococcus* sp.) is a significant rich food N-input, and benefit from increasing feeding rates on zooplankton has been less studied (Meunier et al. 2021). Besides, both the activity and geographical distribution of diazotrophs, for example, *Trischodesmium* spp., will likely increase with future rising sea surface temperature (Ani et al. 2023). They could be an increasingly external alternative nutrient source for bleached corals (Meunier et al. 2019), reducing calcification sensitivity to heat stress and high pCO_2 conditions by a surplus of bioavailable N (Meunier et al. 2022). Such increasing atmospheric N_2 inputs into planktonic trophic webs may have positive consequences on the functioning of coral reefs. Thus, reef systems characterized by high planktonic diazotroph dynamic fluctuations in terms of abundance, community composition, and activity could be more resistant to climate change, as already observed in different regions, i.e. New Caledonia, Papua New Guinea, the Australian Great Barrier Reef, Hawaii, the Caribbean Sea, and the Red Sea (Luo et al. 2012; Bonnet et al. 2017). With corals decline, OM fluxes could also be disturbed by other organisms implicated into nutrient cycles. For instance, certain sponges may increase in abundance and in some cases, become functionally dominant on coral reefs (Bell et al. 2018). However, sponges are also negatively affected under future climate change scenarios with impacts on their symbionts and higher mortality rates (Fan et al. 2013). A thermal stress coupled with extreme acidification and deoxygenation may cause metabolic shifts and disturb nutrient cycling, by inducing a reduction or a break in the loop pathway and therefore in the particulate and detritus production, and by reducing photosynthetic functioning of associated photosymbionts (Maggioni et al. 2023).

Ongoing climate changes may also have an important effect by modifying the exchanges of OM at the interface between ocean and land (Liénart et al. 2018). For instance, higher rainfalls due to more recurrent and intense cyclonic events (Kossin 2018) might generate an increase of terrigenous river runoffs and sediment resuspension (Devlin et al. 2001). Even considerably distant from reefs, increase in dissolved nutrient level might have secondary direct combined effects that need to be furthered. Indeed, in conditions of sporadic events, nutrients are rapidly used by primary producers to fuel planktonic and benthic food webs (Delesalle et al. 1993). But in case of too recurrent events and nutrient excess, induced phytoplankton blooms and subsequent post-bloom changes produce increases of decomposing OM, bacterial load, and altered oxygen levels (D'Angelo and Wiedenmann 2014), which could ripple through the entire food web. The escalating risk of common mechanisms resulting from such an unbalanced ecosystem and subsequent consequences is then likely the overgrowth by successive, conspicuous blooms of benthic macroalgae (Harmelin-Vivien 1994), the prevalence and severity of coral diseases (Voss and Richardson 2006), but also the outbreaks of coral eaters and bioeroders. Top-down control processes can reduce the impact of negative indirect effects of elevated nutrient levels for example, but involve abundant and diversified assemblages of reef consumers, as grazers of benthic algae, detritivores, or predators of corallivores (Hoey and Bellwood 2009).

In other words, several scientific levers are still largely unknown under future scenarios and are of high priority for next trophic investigations. Among them, consequences of quantitative and qualitative changes of the planktonic communities on the entire reef system, impacts of modifications in organism physiology and community ecology on the quality and quantity of OM fluxes within reefs, results on the pelagic-benthic coupling or food web dynamics across both fine and broad spatial and temporal scales. Besides, trophic ecology might help in knowledge-based optimization of management of multiple combined stress that becomes crucial for coral reef conservation (Brodie et al. 2012). As eutrophication driven by nitrogen (and other nutrients) is a major management challenge in many coastal settings (Bonsdorff 2021), quantifying the nitrogen contributed by diazotroph such as *Trichodesmium* spp. is a fundamental knowledge gap to fulfill. A supply of planktonic diazotrophs on the reefs could be included among innovative management approaches to improve the resilience of coral reefs with high conservation potential, during and after bleaching episodes (Hoey et al. 2016). Management strategies should aim for sustaining top-down control processes as well as reducing the nutrient influx in seawater, and future assessment will also strongly involve the optimal utilization of water quality bioindicators (Fabricius et al. 2012).

1.2.1.2 Becoming of Reef Communities

Trophic interactions underpin many of the processes that drive coral reef community structure and the key ecological services these ecosystems provide (Brandl et al. 2019). Many marine food webs are shaped by a small number of strong species interactions that have large effects on a community, that is, involving keystone and foundation species, plus species nested among many weaker interactions (Paine 1992). Consequently, trophic cascades tend to be particularly strong in marine systems (Shurin et al. 2002). When these impactful species interactions are altered by environmental conditions, they may act as ecological leverage points in the community that can magnify the effects of environmental change, for example, shifts in competitive dominance, intensification of predation/herbivory, increases in the frequency and severity of disease/parasite outbreaks, and disruption of facilitation/mutualism (Kroeker and Sanford 2022). Thus, ongoing fluctuations in ocean temperature, dissolved oxygen, and pH can impact community members differently, leading to shifts in the strength and outcome of species interactions and community dynamics. Indeed, environmental changes are translated into community changes via multiple organismal physiologies, that is, different physiological optima and environmental tolerances of species (Somero et al. 2017). In order to improve our ability to predict the emergent effects of ocean change, and allow marine scientists and managers to anticipate which ecosystems are especially vulnerable, trophic ecology can help a further exploration of leverage points in marine communities by way of a continued integration of physiology and ecology (Kroeker and Sanford 2022).

To cope with increased energetic demands imposed by ocean warming, deoxygenation, and ocean acidification, heterotrophs may, for instance, reallocate energy and/or increase rates of consumption; both responses can have substantial consequences for species interactions, including altering growth rates and competitive outcomes, and changing the intensity of herbivory or predation (Kroeker and Sanford 2022). Future research should therefore explore the presence of potential sublethal effects of habitat degradation on higher-order consumers, such as lower nutritional condition (Hempson et al. 2018). A given resource-consumer interaction might also vary with environmental changes (Gaylord et al. 2019). Thus, understanding how a species' dietary niche width relates to its functional role within an ecosystem is critical, given that increasing complexity of species interactions may serve to mitigate the effect of biodiversity loss on ecosystem function in future scenarios (Bregman et al.

2015). Indeed, omnivores play an important role in dampening potential trophic cascades (Bruno and O'Connor 2005) and would be expected to have a greater potential of short-term adaptive response to changes in habitat degradation, whereas specialist species would be more vulnerable (Letourneur et al. 2017). However, species-level analyses may, therefore, mask individual differences in dietary variation and limit the ability to identify more complex energy flows and trophodynamics within a specific study area or differences in species' ecological roles. So, determining the degree of trophic plasticity within species will help understand how they may cope with environmental change (Cybulski et al. 2022). The more deeply the trophic structure of the communities will be identified, the best the consequences of reef modifications on food chain characteristics, stability, and resilience will be understood.

Local and global pressures are leading to substantial benthic alterations in coral reefs (De'ath et al. 2012). For example, the primary impacts of ocean acidification include reductions in structural complexity, skeleton density and coral recruitment, as well as increases in bioeroders (Fabricius et al. 2017). Current habitat degradation and extensive coral mortality suggest that shifts to algal- and/or rubble-dominated states may be more prevalent in a future coral reefs system (Wolff et al. 2018). Habitat degradation is thought to be a major driver of trophic structure of ecosystems (Álvarez-Filip et al. 2013), and, for example, the generation of rubble beds can have lasting effects on structural habitat complexity, species interactions, food webs, and reef communities (Wolfe et al. 2021 and references therein). Though considered as seemingly barren, featureless habitats, rubble beds actually form complex microhabitats that can host a great density and diversity of predominantly cryptic life (Enochs and Manzello 2012); a suite of taxa from microbes and biofilms, sessile and encrusting organisms, motile cryptofauna, and cryptobenthic or juvenile fishes. Despite their rising prevalence in the coral reef benthos, less is known about the biology and ecology of rubble beds, and empirical data on direct transfers of biomass and energy from the cryptobenthos to higher-order fishes are scarce (Wolfe et al. 2020). Yet, these data are required to more holistically parameterize and predict how bottom-up processes may enhance the system productivity as reefs degrade. Besides, rubble generation sustaining or even enhancing a cryptic biodiversity initially depends on framework production by corals (Wolfe et al. 2021). Thus, the potential benefits of reef erosion on rubble-dwelling taxa and ecological functioning may be impaired over time, as the production of corals—and by extension potential for rubble generation—declines (Enochs and Manzello 2012). All these key ecological points require greater attention with the integration of more uniform and standardized trophic approaches to define and characterize degraded habitats and their associated community to improve the ability to monitor, predict, and manage these ecosystems into the future (Wolfe et al. 2021).

1.2.2 Next Steps in Means and Actions

1.2.2.1 Method Enhancements and Innovative Techniques

In the field of biochemical tracer (SIA, CSIA, FA, etc.), generating more data across a wider range of sources to get suitable baselines is a future priority. Also, lowering costs and increasing laboratory availability are necessary to combine approaches and answer macroevolutionary questions (Twining et al. 2020 and included references). By default, marine food web studies should employ a tri-isotope approach at least, this way additional examples of three-dimensional SIBER that explore cross-guild relationships should come out (Cybulski et al. 2022). However, more research into the effect of tissue lipid levels on $S^{34}S$ is sorely needed. Also, 70% of coral reef food web SIA studies focus solely on a single tissue type, mostly muscles (Skinner et al. 2022); although measuring stable isotope ratios across multiple consumer tissues can provide important information about the temporal dynamics of resource use, other tissues like liver or gonads are still less often utilized. Also, studies are beginning to perform SIA on fish eye lenses allowing reconstruction of an individual's trophic and habitat histories (Bell-Tilcock et al. 2021). Finally, hydrogen isotope (^2H) values have demonstrated a typically larger isotopic separation between diet sources and consumers than the more traditional C and N isotopes (e.g., 40‰ vs. 2‰ between aquatic and terrestrial plants), because of the greater relative mass differences associated with such a light isotope (Macko et al. 1983). Ecological research on stable hydrogen isotope ratios (δ^2H or δD) of fatty acids is still in its infancy, so future studies need to address basic questions (Twining et al. 2020), including the spatiotemporal variability of different sources, the trophic discrimination from one trophic level to the next, and the analytical limitations. Also, must be taken into account changes in hydrogen isotopes of environmental waters that can consequently affect δ^2H values in consumers, including the compound-specific level.

Robust application of the different biomarkers faces the same difficulties, with a lack of information concerning trophic discrimination and most studies limited to qualitative applications. For example, with the FA, researchers have only superficial acknowledgment of the uncertainty associated with the trophic transfer and storage in consumers of these bioactive molecules with diverse structures and functions (Iverson 2009), and these biotracers are rarely used to determine predator–prey interactions which form the basis of ecosystem models (Galloway and Budge 2020 and included references). Yet, when genetic analyses are paired with empirical feeding studies, it is possible to link demonstrated synthesis of particular molecules, including fatty acids, with upregulation of specific genes (Kabeya et al. 2018). So, accounting for lipid trophic modification in consumers through project-specific feeding trials or by measurement and estimation of general dietary "calibration coefficients' is critical for fatty acid biomarker applications, especially quantitative ones (Jardine et al. 2020). Besides, despite a dazzling contribution in recent years, even omics methods cannot fully resolve the complex nature of some sources, for example, DOM, particularly the labile fraction within the low-molecular-weight component (i.e., metabolites; Wegley Kelly et al. 2022). It is likely due to both the limited representation of marine metabolites in current libraries and data quality that is compromised by sample complexity. Furthermore, measuring ecologically relevant changes in the concentrations of specific compounds often exceeds instrument precision (Fiore et al. 2017). Therefore, progress in both hardware advances of high-resolution tandem mass spectrometers and computational tools are still needed to semiquantitatively detect and characterize rare and diverse compounds (e.g., Petras et al. 2017; Ernst et al. 2019; Nothias et al. 2020).

As well, several cutting-edge methods are not yet widely used in ecological studies, but hold the potential to address some of the limitations of current isotopic techniques. Then, intramolecular isotope pattern studies are emerging as another promising tool to contribute to the development of new techniques using CSIA to study trophic ecology. Within a molecule's isotopic "fine structure," isotopomers are similar in isotopic composition but differ in the position of isotopes within the molecule and are preferentially used to investigate consumers' metabolic pathways by revealing enzyme activities and the nature of reordering reactions (Wiechert et al. 2013). Isotopologues, as molecules that differ in their isotope composition and profiles, offer more information on carbon fluxes and dietary routing of metabolites (Postle and Hunt 2009). However, although they offer finer-scale resolutions than CSIA for either $\delta^{13}C$ or $\delta^{2}H$, isotopomer and isotopologue labeling studies have yet to be applied in studies of marine trophic ecology (Twining et al. 2020).

1.2.2.2 Beyond the Prism of Trophic Ecology

The use of biochemical tracers requires bridging three distinct fields, that is, ecology, physiology, and chemistry. One fundamental limitation in the field at this point is that the experts may not understand the nuances of the other disciplines, therefore the path forward will involve interdisciplinary collaboration (Galloway and Budge 2020; Robinson et al. 2024). Indeed, there is the need to break traditional barriers across disciplines (e.g., biochemistry, microbiology, and ecology) and subdisciplines (e.g., limnology, ornithology, and soil ecology) in order to better understand trophic interactions and metabolism at the scale of whole food webs. Also, the need to cross new methodological frontiers, such as routinely combining empirical ecophysiological studies with food web analyses, to achieve deeper insight in how dietary tracers move through consumers, within food webs and across complex ecosystems (Twining et al. 2020). Thus, determining many physiological process and drivers of consumer trophic ecology represents an area ripe for further investigation (e.g., ingestion vs. excretion, thermal regulation, metabolic rates). Besides, despite a decline in interest in this discipline, taxonomy is still essential, and will be even more in future, to acknowledge and characterize ongoing massive changing in community composition. New feeding experimental works on the uptake and transfer in individual organisms are also required for the development and refinement of biomarker models that can accurately estimate animal diets, and to advance the field and make meaningful use of these tools at the scale of populations or ecosystems (Galloway and Budge 2020). Developments in modeling will definitely require acquisition of more and recent data, the use of multimodel approaches, and could rapidly evolve shortly thanks to progress in the mathematics of algorithms and the contributions of artificial intelligence (Xu et al. 2021).

Trophic ecology can offer improvements and solutions to problems that hamper coral restoration, and empirically testing methods of incorporating trophic interactions in restoration designs to assess their costs, effectiveness, and utility under different scenarios is an important priority (Ladd and Shantz 2020). Yet, despite rapidly growing interest, only 15% of restoration publications considered trophic interactions, highlighting a clear mismatch between the fundamental role of trophic ecology on coral reefs and its consideration in restoration efforts. So, the dire future of the planet's coral reefs will require the best working scientific collaborations to resolve the problems mentioned above.

References

Álvarez-Filip L, Carricart-Ganivet JP, Horta-Puga G, Iglesias-Prieto R (2013) Shifts in coral assemblage composition do not ensure persistence of reef functionality. Sci Rep 3:1–5

Ani CJ, Smithers SG, Lewis S, Baird M, Robson B (2023) eReefs modelling suggests *Trichodesmium* may be a major nitrogen source in the Great Barrier Reef. Estuar Coast Shelf Sci 285:108306

Bell JJ, Rovellini A, Davy SK, Taylor MW, Fulton EA, Dunn MR, Bennett HM, Kandler NM, Luter HM, Webster NS (2018) Climate change alterations to ecosystem dominance: how might sponge-dominated reefs function? Ecology 99(9):1920–1931

Bell-Tilcock M, Jeffres CA, Rypel AL, Sommer TR, Katz JVE, Whitman G, Johnson RC (2021) Advancing diet reconstruction in fish eye lenses. Methods Ecol Evol 12:449–457

Bonnet S, Berthelot H, Turk-Kubo K, Fawcett S, Rahav E, L'Helguen S, Berman-Frank I (2016) Dynamics of N^2 fixation and fate of diazotroph-derived nitrogen in a low-nutrient, low-chlorophyll ecosystem: results from the VAHINE mesocosm experiment (New Caledonia). Biogeosciences 13:2653–2673

Bonnet S, Caffin M, Berthelot H, Moutin T (2017) Hot spot of N_2 fixation in the western tropical South Pacific pleads for a spatial decoupling between N_2 fixation and denitrification. Proc Natl Acad Sci USA 114:E2800–E2801

Bonsdorff E (2021) Eutrophication: Early warning signals, ecosystem-level and societal responses, and ways forward. Ambio 50(4):753–758

Brandl SJ, Rasher DB, Côté IM, Casey JM, Darling ES, Lefcheck JS, Duffy JE (2019) Coral reef ecosystem functioning: eight core processes and the role of biodiversity. Front Ecol Environ 17(8):445–454

Bregman TP, Lees AC, Seddon N, MacGregor HE, Darski B, Aleixo A, Bonsall MB, Tobias JA (2015) Species interactions regulate the collapse of biodiversity and ecosystem function in tropical forest fragments. Ecology 96(10):2692–2704

Briand MJ, Bonnet X, Goiran C, Guillou G, Letourneur Y (2015) Major sources of organic matter in a complex coral reef lagoon: Identification from isotopic signatures ($\delta^{13}C$ and $\delta^{15}N$). PLoS One 10:e0131555

Briand MJ, Bonnet X, Guillou G, Letourneur Y (2016) Complex food webs in highly diversified coral reefs: Insights from $\delta^{13}C$ and $\delta^{15}N$ stable isotopes. Food Webs 8:12–22

Brodie JE, Kroon F, Schaffelke B, Wolanski E, Lewis S, Devlin M, Bohnet I, Bainbridge Z, Waterhouse J, Davis A (2012) Terrestrial pollutant runoff to the Great Barrier Reef: an update of issues, priorities and management responses. Mar Pollut Bull 65(4–9):81–100

Bruno JF, O'Connor MI (2005) Cascading effects of predator diversity and omnivory in a marine food web. Ecol Lett 8:1048–1056

Casey JM, Meyer CP, Morat F, Brandl SJ, Planes S, Parravicini V (2019) Reconstructing hyperdiverse food webs: Gut content metabarcoding as a tool to disentangle trophic interactions on coral reefs. Methods Ecol Evol 10:1157–1170

Christensen V, Pauly D (1992) ECOPATH II—a software for balancing steady-state ecosystem models and calculating network characteristics. Ecol modell 61(3–4):169–185

Christensen V, Walters CJ (2004) Ecopath with Ecosim: methods, capabilities and limitations. Ecol Modell 172(2–4):109–139

Christensen V, Coll M, Steenbeek J, Buszowski J, Chagaris D, Walters CJ (2014) Representing variable habitat quality in a spatial food web model. Ecosystems 17:1397–1412

Crossman DJ, Choat HJ, Clements KD, Hardy T, McConochie J (2001) Detritus as food for grazing fishes on coral reefs. Limnol Oceanogr 46(7):1596–1605

Cummings DO, Booth DJ, Lee RW, Simpson SJ, Pile AJ (2010) Ontogenetic diet shifts in the reef fish *Pseudanthias rubrizonatus* from isolated populations on the North-West Shelf of Australia. Mar Ecol Progr Ser 419:211–222

Cybulski JD, Skinner C, Wan Z, Wong CKM, Toonen RJ, Gaither MR, Soong K, Wyatt ASJ, Baker DM (2022) Improving stable isotope assessments of inter- and intra-species variation in coral reef fish trophic strategies. Ecol Evol 12(9):e9221

D'Angelo C, Wiedenmann J (2014) Impacts of nutrient enrichment on coral reefs: new perspectives and implications for coastal management and reef survival. Curr Opin Environ Sustain 7:82–93

Da Silveira EL, Semmar N, Cartes JE, Tuset VM, Lombarte A, Ballester ELC, Vaz-dos-Santos AM (2020) Methods for trophic ecology assessment in fishes: a critical review of stomach analyses. Rev Fish Sci Aquac 28:71–106

Dalsgaard J, St. John M, Kattner G, Muller-Navarra D, Hagen W (2003) Fatty acid trophic markers in the pelagic marine environment. Adv Mar Biol 46:225–340

De Goeij JM, Van Oevelen D, Vermeij MJA, Osinga R, Middelburg JJ, De Goeij AFPM, Admiraal W (2013) Surviving in a marine desert: the sponge loop retains resources within coral reefs. Science 342(6154):108–110

de Lira SMA, de Teixeira IÁ, de Lima CDM, de Santos GS, Leitão SN, Schwamborn R (2014) Spatial and nycthemeral distribution of the zooneuston off Fernando de Noronha, Brazil. Braz J Oceanogr 62:35–45

De'ath G, Fabricius KE, Sweatman H, Puotinen M (2012) The 27-year decline of coral cover on the Great Barrier Reef and its causes. Proc Natl Acad Sci USA 109:17995–17999

Delesalle B, Pichon M, Frankignoulle M, Gattuso JP (1993) Effects of a cyclone on coral reef phytoplankton biomass, primary production and composition (Moorea Island, French Polynesia). J Plankton Res 15(12):1413–1423

Devlin M, Waterhouse J, Brodie J (2001) Community and connectivity: summary of a community-based monitoring program set up to assess the movement of nutrients and sediments into the Great Barrier Reef during high flow events. Water Sci Technol 43(9):121–131

Dromard CR, Bouchon-Navaro Y, Cordonnier S, Fontaine MF, Verlaque M, Harmelin-Vivien M, Bouchon C (2013) Resource use of two damselfishes, *Stegastes planifrons* and *Stegastes adustus*, on Guadeloupean reefs (Lesser Antilles): Inference from stomach content and stable isotope analysis. J Exp Mar Biol Ecol 440:116–125

Dromard CR, Bouchon-Navaro Y, Harmelin-Vivien M, Bouchon C (2017) The nutritional quality of non-calcified macroalgae in Guadeloupe (Lesser Antilles) evaluated by their biochemical composition. Gulf Caribb Res 28(1):GCFI1–GCFI6

Duprey NN, Yasuhara M, Baker DM (2016) Reefs of tomorrow: eutrophication reduces coral biodiversity in an urbanized seascape. Glob Chang Biol 22:3550–3565

Enochs IC, Manzello DP (2012) Responses of cryptofaunal species richness and trophic potential to coral reef habitat degradation. Diversity 4:94–104

Ernst M, Kang KB, Caraballo-Rodríguez AM, Nothias LF, Wandy J, Chen C, Wang M, Rogers S, Marnix MH, Dorrestein PC, Van Der Hooft JJ (2019) MolNetEnhancer: Enhanced molecular networks by integrating metabolome mining and annotation tools. Metabolites 9(7):144

Fabricius KE, Cooper TF, Humphrey C, Uthicke S, De'ath G, Davidson J, LeGrand H, Thompson A, Schaffelke B (2012) A bioindicator system for water quality on inshore coral reefs of the Great Barrier Reef. Mar Pollut Bull 65(4–9):320–332

Fabricius KE, Noonan SHC, Abrego D, Harrington L, De'ath G (2017) Low recruitment due to altered settlement substrata as primary constraint for coral communities under ocean acidification. Proc R Soc B: Biol Sci 284(1862):20171536

Fan L, Liu M, Simister R, Webster NS, Thomas T (2013) Marine microbial symbiosis heats up: the phylogenetic and functional response of a sponge holobiont to thermal stress. ISME J 7(5):991–1002

Fey P, Parravicini V, Lebreton B, Meziane T, Galzin R, Zubia M, Bănaru D, Letourneur Y (2020) Sources of organic matter in an atypical phytoplankton rich coral ecosystem, Marquesas Islands: composition and properties. Mar Biol 167:1–13

Fey P, Parravicini V, Bănaru D, Dierking J, Galzin R, Lebreton B, Meziane T, Polunin NVC, Zubia M, Letourneur Y (2021a) Multitrophic markers illuminate the understanding of the functioning of a remote, low coral cover Marquesan coral reef food web. Sci Rep 11:20950

Fey P, Letourneur Y, Bonnabel S (2021b) The α-minimum convex polygon as a relevant tool for isotopic niche statistics. Ecol Indic 130:108048

Fiore CL, Freeman CJ, Kujawinski EB (2017) Sponge exhalent seawater contains a unique chemical profile of dissolved organic matter. Peer J 5:e2870

Futuyma DJ, Moreno G (1988) The evolution of ecological specialization. Annu Rev Ecol and Syst 19:207–233

Galloway AWE, Budge SM (2020) The critical importance of experimentation in biomarker-based trophic ecology. Philos Trans R Soc B: Biol Sci 375:20190638
Gaylord B, Barclay KM, Jellison BM, Jurgens LJ, Ninokawa AT, Rivest EB, Leighton LR (2019) Ocean change within shoreline communities: from biomechanics to behaviour and beyond. Conserv Physiol 7:coz077
Glynn PW, Colley SB, Guzman HM, Enochs IC, Cortés J, Maté JL, Feingold JS (2011) Reef coral reproduction in the eastern Pacific: Costa Rica, Panamá, and the Galápagos Islands (Ecuador). VI. Agariciidae, *Pavona clavus*. Mar Biol 158:1601–1617
Godinot C, Houlbrèque F, Grover R, Ferrier-Pagès C (2011) Coral uptake of inorganic phosphorus and nitrogen negatively affected by simultaneous changes in temperature and pH. PLoS One 6:1–10
Gray C, Baird DJ, Baumgartner S, Jacob U, Jenkins GB, O'Gorman EJ, Lu X, Ma A, Pocock MJO, Schuwirth N (2014) Ecological networks: the missing links in biomonitoring science. J Appl Ecol 51:1444e1449
Grüss A, Schirripa MJ, Chagaris D, Drexler M, Simons J, Verley P, Shin YJ, Karnauskas M, Oliveros-Ramos R, Ainsworth CH (2015) Evaluation of the trophic structure of the West Florida Shelf in the 2000s using the ecosystem model OSMOSE. J Mar Syst 144:30–47
Guan Y, Hohn S, Wild C, Merico A (2020) Vulnerability of global coral reef habitat suitability to ocean warming, acidification and eutrophication. Glob Chang Biol 26:5646–5660
Haas AF, Nelson CE, Rohwer F, Wegley Kelly L, Quistad SD, Carlson CA, Leichter JJ, Hatay M, Smith JE (2013) Influence of coral and algal exudates on microbially mediated reef metabolism. Peer J 1:e108
Harmelin-Vivien ML (1981) Trophic relationships of reef fishes in Tulear (Madagascar). Oceanol Acta 4(3):365–374
Harmelin-Vivien ML (1994) The effects of storms and cyclones on coral reefs: a review. J Coast Res Special Issues 12:211–231
Harmelin-Vivien ML (2002) Energetics and fish diversity on coral reefs. In: Sale PF (ed) Coral reef fishes. Dynamics and diversity in a complex ecosystem. Acad Press, San Diego, pp 265–274
Hempson TN, Graham NAJ, McNeil MA, Bodin N, Wilson SK (2018) Regime shifts shorten food chains for mesopredators with potential sublethal effects. Funct Ecol 32(3):820–830
Hoey AS, Bellwood DR (2009) Limited functional redundancy in a high diversity system: single species dominates key ecological process on coral reefs. Ecosystems 12:1316–1328
Hoey A, Howells E, Johansen J, Hobbs JP, Messmer V, McCowan D, Wilson S, Pratchett M (2016) Recent advances in understanding the effects of climate change on coral reefs. Diversity 8:12
Hughes TP, Kerry JT, Álvarez-Noriega M, Álvarez-Romero JG, Anderson KD, Baird AH, Babcock RC, Beger M, Bellwood DR, Berkelmans R, Bridge TC, Butler IR, Byrne M, Cantin NE, Comeau S, Connolly SR, Cumming GS, Dalton SJ, Diaz-Pulido G, Eakin CM, Figueira WF, Gilmour JP, Harrison HB, Heron SF, Hoey AS, Hobbs JPA, Hoogenboom MO, Kennedy EV, Kuo CY, Lough JM, Lowe RJ, Liu G, McCulloch MT, Malcolm HA, McWilliam MJ, Pandolfi JM, Pears RJ, Pratchett MS, Schoepf V, Simpson T, Skirving WJ, Sommer B, Torda G, Wachenfeld DR, Willis BL, Wilson SK (2017) Global warming and recurrent mass bleaching of corals. Nature 543(7645):373–377
Hughes TP, Kerry JT, Baird AH, Connolly SR, Dietzel A, Eakin CM, Heron SF, Hoey AS, Hoogenboom MO, Liu G, McWilliam MJ, Pears RJ, Pratchett MS, Skirving WJ, Stella JS, Torda G (2018) Global warming transforms coral reef assemblages. Nature 556:492–496
Hyslop EJ (1980) Stomach contents analysis – a review of methods and their application. J Fish Biol 17(4):411–429
Iverson SJ (2009) Tracing aquatic food webs using fatty acids: from qualitative indicators to quantitative determination. In: Arts MT, Brett MT, Kainz MJ (eds) Lipids in aquatic ecosystems. Dordrecht (the Netherlands). Springer, pp 281–307
Iverson SJ, Field C, Bowen WD, Blanchard W (2004) Quantitative fatty acid signature analysis: a new method of estimating predator diets. Ecol Monogr 74(2):211–235
Jackson AL, Inger R, Parnell AC, Bearhop S (2011) Comparing isotopic niche widths among and within communities: SIBER e stable Isotope Bayesian Ellipses in R. J Anim Ecol 80:595e602
Jardine TD, Galloway AWE, Kainz MJ (2020) Unlocking the power of fatty acids as dietary tracers and metabolic signals in fishes and aquatic invertebrates. Philos Trans R Soc B 375:20190639
Johnson C, Klumpp D, Field J, Bradbury R (1995) Carbon flux on coral reefs: effects of large shifts in community structure. Mar Ecol Progr Ser 126:123–143
Jordano P (2016) Chasing Ecological Interactions. PLoS Biol 14:e1002559
Kabeya N, Fonseca MM, Ferrier DEK, Navarro JC, Bay LK, Francis DS, Tocher DR, Castro LFC, Monroig Ó (2018) Genes for de novo biosynthesis of omega-3 polyunsaturated fatty acids are widespread in animals. Sci Adv 4:eaar6849
Kayoda T, Osada Y, Takimoto G (2012) IsoWeb: A Bayesian isotope mixing model for diet analysis of the whole food web. PLoS One 7:e41057
Kido Soule MC, Longnecker K, Johnson WM, Kujawinski EB (2015) Environmental metabolomics: analytical strategies. Mar Chem 177:1–62
Kooijman S (2010) Dynamic energy budget theory for metabolic organisation. Cambridge University Press, Cambridge
Kossin JP (2018) A global slowdown of tropical-cyclone translation speed. Nature 558(7708):104–107
Kroeker KJ, Sanford E (2022) Ecological leverage points: species interactions amplify the physiological effects of global environmental change in the ocean. Ann Rev Mar Sc 14:75–103
Krogdahl Å, Hemre GI, Mommsen TP (2005) Carbohydrates in fish nutrition: digestion and absorption in postlarval stages. Aquac Nut 11(2):103–122
Ladd MC, Shantz AA (2020) Trophic interactions in coral reef restoration: a review. Food Webs 24:e00149
Lesser MP (2013) Using energetic budgets to assess the effects of environmental stress on corals: are we measuring the right things? Coral Reefs 32(1):25–33
Letourneur Y, Lison de Loma T, Richard P, Harmelin-Vivien ML, Cresson P, Banaru D, Fontaine MF, Gref T, Planes S (2013) Identifying carbon sources and trophic position of coral reef fishes using diet and stable isotope ($\delta^{15}N$ and $\delta^{13}C$) analyses in two contrasted bays in Moorea, French Polynesia. Coral Reefs 32:1091–1102
Letourneur Y, Briand MJ, Graham NAJ (2017) Coral reef degradation alters the isotopic niche of reef fishes. Mar Biol 164:12
Letourneur Y, Fey P, Dierking J, Galzin R, Parravicini V (2024) Challenging trophic position assessments in complex ecosystems: Calculation method, choice of baseline, trophic enrichment factors, season and feeding guild do matter: a case study from Marquesas Islands coral reefs. Ecol Evol 14:e11620
Liénart C, Savoye N, David V, Ramond P, Rodriguez Tressa P, Hanquiez V, Marieu V, Aubert F, Aubin S, Bichon S, Boinet C, Bourasseau L, Bozec Y, Bréret M, Breton E, Caparros J, Cariou J, Claquin P, Conan P, Corre AM, Costes L, Crouvoisier M, Del Amo Y, Derriennic H, Dindinaud F, Duran R, Durozier M, Devesa J, Ferreira S, Feunteun E, Garcia N, Geslin S, Grossteffan E, Gueux A, Guillaudeau J, Guillou G, Jolly O, Lachaussée N, Lafont M, Lagadec V, Lamoureux J, Lauga B, Lebreton B, Lecuyer E, Lehodey JP, Leroux C, L'Helguenm S, Mac E, Maria E, Mousseau L, Nowaczyk A, Pineau P, Petit F, Pujo-Pay M, Raimbault P, Rimmelin-Maury P, Rouaud V, Sauriaud PG, Sultan E, Susperregui E (2018) Dynamics of particulate organic matter composition in coastal systems: forcing of spatio-temporal variability at multi-systems scale. Progr Oceanogr 162:271–289

Luo YW, Doney SC, Anderson LA, Benavides M, Berman-Frank I, Bode A, Bonnet S, Boström KH, Böttjer D, Capone DG, Carpenter EJ, Chen YL, Church MJ, Dore JE, Falcón LI, Fernández A, Foster RA, Furuya K, Gómez F, Gundersen K, Hynes AM, Karl DM, Kitajima S, Langlois RJ, LaRoche J, Letelier RM, Marañón E, McGillicuddy DJ Jr, Moisander PH, Moore CM, Mouriño-Carballido B, Mulholland MR, Needoba JA, Orcutt KM, Poulton AJ, Rahav E, Raimbault P, Rees AP, Riemann L, Shiozaki T, Subramaniam A, Tyrrell T, Turk-Kubo KA, Varela M, Villareal TA, Webb EA, White AE, Wu J, Zehr JP (2012) Database of diazotrophs in global ocean: abundance, biomass and nitrogen fixation rates. Earth Syst Sci Data 4:47–73

Macko SA, Estep ML, Lee WY (1983) Stable hydrogen isotope analysis of food webs on laboratory and field populations of marine amphipods. J Exp Mar Biol Ecol 72:243–249

Maggioni F, Bell JJ, Pujo-Pay M, Shaffer M, Cerrano C, Lemonnier H, Letourneur Y, Rodolfo-Metalpa R (2023) Sponge organic matter recycling: Reduced detritus production under extreme environmental conditions. Mar Pollut Bull 190:114869

Matley JK, Maes GE, Devloo-Delva F, Huerlimann R, Chua G, Tobin AJ, Fisk AT, Simpfendorfer CA, Heupel MR (2018) Integrating complementary methods to improve diet analysis in fishery-targeted species. Ecol Evol 8(18):9503–9515

McCutchan JH Jr, Lewis WM Jr, Kendall C, McGrath CC (2003) Variation in trophic shift for stable isotope ratios of carbon, nitrogen, and sulfur. Oikos 102(2):378–390

McMahon KW, Thorrold SR, Houghton LA, Berumen ML (2016) Tracing carbon flow through coral reef food webs using a compound specific stable isotope approach. Oecologia 180:809–821

McMurray SE, Johnson ZI, Hunt DE, Pawlik JR, Finelli CM (2016) Selective feeding by the giant barrel sponge enhances foraging efficiency. Limnol Oceanogr 61(4):1271–1286

Meunier V, Bonnet S, Pernice M, Benavides M, Lorrain A, Grosso O, Lambert C, Houlbrèque F (2019) Bleaching forces coral's heterotrophy on diazotrophs and *Synechococcus*. ISME J. 13:2882–2886

Meunier V, Geissler L, Bonnet S, Rädecker N, Perna G, Grosso O, Lambert C, Rodolfo-Metalpa R, Voolstra CR, Houlbrèque F (2021) Microbes support enhanced nitrogen requirements of coral holobionts in a high CO_2 environment. Mol Ecol 30:5888–5899

Meunier V, Bonnet S, Camps M, Benavides M, Dubosc J, Rodolfo-Metalpa R, Houlbrèque F (2022) Ingestion of diazotrophs makes corals more resistant to heat stress. Biomolecules 12:537

Michener RH, Kaufman L (2007) Stable isotope ratios as tracers in marine food webs: an update. Stable Isot Ecol Environ Sci 2:238–282

Nielsen JM, Clare EL, Hayden B, Brett MT, Kratina P (2018) Diet tracing in ecology: method comparison and selection. Methods Ecol Evol 9(2):278–291

Nothias LF, Petras D, Schmid R, Dührkop K, Rainer J, Sarvepalli A, Protsyuk I, Ernst M, Tsugawa H, Fleischauer M, Aicheler F, Aksenov AA, Alka O, Allard PM, Barsch A, Cachet X, Caraballo-Rodriguez AM, Da Silva RR, Dang T, Garg N, Gauglitz JM, Gurevich A, Isaac G, Jarmusch AK, Kameník Z, Kang KB, Kessler N, Koester I, Korf A, Le Gouellec A, Ludwig M, Martin CH, McCall LI, McSayles J, Meyer SW, Mohimani H, Morsy M, Moyne O, Neumann S, Neuweger H, Nguyen NH, Nothias-Esposito M, Paolini J, Phelan VV, Pluskal T, Quinn RA, Rogers S, Shrestha B, Tripathi A, van der Hooft JJJ, Vargas F, Weldon KC, Witting M, Yang H, Zhang Z, Zubeil F, Kohlbacher O, Böcker S, Alexandrov T, Bandeira N, Wang M, Dorrestein PC (2020) Feature-based molecular networking in the GNPS analysis environment. Nat Methods 17(9):905–908

Paine RT (1992) Food-web analysis through field measurement of per capita interaction strength. Nature 355:73–75

Pandolfi JM, Bradbury RH, Sala E, Hughes TP, Bjorndal KA, Cooke RG, McArdle D, McClenachan L, Newman MJH, Paredes G, Warner RR, Jackson JB (2003) Global trajectories of the long-term decline of coral reef ecosystems. Science 301(5635):955–958

Parravicini V, Casey JM, Schiettekatte NMD, Brandl SJ, Pozas-Schacre C, Carlot J, Edgar GJ, Graham NAJ, Harmelin-Vivien M, Kulbicki M, Strona G, Stuart-Smith RD (2020) Delineating reef fish trophic guilds with global gut content data synthesis and phylogeny. PLoS Biol 18:e3000702

Patti GJ, Yanes O, Siuzdak G (2012) Innovation: Metabolomics: the apogee of the omics trilogy. Nat Rev Mol Cell Biol 13:263–269

Pawlik JR, McMurray SE (2020) The emerging ecological and biogeochemical importance of sponges on coral reefs. Ann Rev Mar Sci 12:315–337

Petras D, Koester I, Da Silva R, Stephens BM, Haas AF, Nelson CE, Wegley Kelly L, Aluwihare LI, Dorrestein PC (2017) High-resolution liquid chromatography tandem mass spectrometry enables large scale molecular characterization of dissolved organic matter. Front Mar Sci 4:405

Polovina JJ (1984) Model of a coral reef ecosystem: I. The ECOPATH model and its application to French Frigate Shoals. Coral Reefs 3:1–11

Post DM (2002) Using stable isotopes to estimate trophic position: models, methods, and assumptions. Ecology 83:703–710

Postle AD, Hunt AN (2009) Dynamic lipidomics with stable isotope labelling. J Chromatogr B 877:2716–2721

Quezada-Romegialli C, Jackson AL, Hayden B, Kahilainen KK, Lopes C, Harrod C (2018) Trophic position, an R package for the Bayesian estimation of trophic position from consumer stable isotope ratios. Methods Ecol Evol 9:1592–1599

Rix L, de Goeij JM, van Oevelen D, Struck U, Al-Horani FA, Wild C, Naumann MS (2017) Differential recycling of coral and algal dissolved organic matter via the sponge loop. Funct Ecol 31(3):778–789

Robinson JP, Benkwitt CE, Maire E, Morais R, Schiettekatte NM, Skinner C, Brandl SJ (2024) Quantifying energy and nutrient fluxes in coral reef food webs. Trends Ecol Evol 39(5):467–478

Shurin JB, Borer ET, Seabloom EW, Anderson K, Blanchette CA, Broitman B, Cooper SD, Halpern BS (2002) A cross-ecosystem comparison of the strength of trophic cascades. Ecol Lett 5:785–791

Silveira CB, Cavalcanti GS, Walter JM, Silva-Lima AW, Dinsdale EA, Bourne DG, Thompson CC, Thompson FL (2017) Microbial processes driving coral reef organic carbon flow. FEMS Microbiol Rev 41(4):575–595

Skinner C, Mill AC, Fox MD, Newman SP, Zhu Y, Kuhl A, Polunin NVC (2021) Offshore pelagic subsidies dominate carbon inputs to coral reef predators. Sci Adv 7:eabf3792

Skinner C, Pei YD, Morimoto N, Miyajima T, Wyatt ASJ (2022) Stable isotopes elucidate body-size and seasonal fluctuations in the feeding strategies of planktivorous fishes across a semi-enclosed tropical embayment. Front Ecol Evol 10:942968

Somero GN, Lockwood BL, Tomanek L (2017) Biochemical adaptation: response to environmental challenges from life's origins to the Anthropocene. Sinauer Ass Inc Publ, Sunderland

Stock BC, Jackson AL, Ward EJ, Parnell AC, Phillips DL, Semmens BX (2018) Analyzing mixing systems using a new generation of Bayesian tracer mixing models. Peer J 6:e5096

Tamelander T, Kivimäe C, Bellerby RGJ, Renaud PE, Kristiansen S (2009) Base-line variations in stable isotope values in an Arctic marine ecosystem: effects of carbon and nitrogen uptake by phytoplankton. Hydrobiol 630:63–73

Tilman D, Isbell F, Cowles JM (2014) Biodiversity and ecosystem functioning. Ann Rev Ecol Evol Syst 45:471–493

Twining CW, Taipale SJ, Ruess L, Bec A, Martin-Creuzburg D, Kainz MJ (2020) Stable isotopes of fatty acids: current and future perspectives for advancing trophic ecology. Philos Trans R Soc B 375:20190641

Voss JD, Richardson LL (2006) Nutrient enrichment enhances black band disease progression in corals. Coral Reefs 25(4):569–576

Walters WJ, Christensen V (2018) Ecotracer: analyzing concentration of contaminants and radioisotopes in an aquatic spatial-dynamic food web model. J Environ Radioact 181:118–127

Walters C, Christensen V, Pauly D (1997) Structuring dynamic models of exploited ecosystems from trophic mass-balance assessments. Rev Fish Biol Fish 7(2):139

Walters C, Pauly D, Christensen V (1999) Ecospace: prediction of mesoscale spatial patterns in trophic relationships of exploited ecosystems, with emphasis on the impacts of marine protected areas. Ecosystems 2:539–554

Wegley Kelly L, Nelson CE, Petras D, Koester I, Quinlan ZA, Arts MG, Nothiasc LF, Comstockg J, Whitee BM, Hopmansf EC, van Duylf FC, Carlsong CA, Aluwiharea LI, Dorresteinc PC, Haas AF (2022) Distinguishing the molecular diversity, nutrient content, and energetic potential of exometabolomes produced by macroalgae and reef-building corals. Proc Nationl Acad Sci 119(5):e2110283119

Wiechert W, Nöh K, Weitzel M (2013) Metabolic isotopomer labeling systems. Part III: Path tracing. Math Biosci 244:1–12

Wilkins KW, Rosa-Marín A, Cziesielski M, Hughes H, Love C, Nowakowski C (2021) Short-and long-term visions for protecting coral reefs. Limnol Oceanogr Bull 30(1):13–15

Wolfe K, Anthony K, Babcock R, Bay L, Bourne D, Burrows D, Byrne M, Deaker D, Diaz-Pulido G, Frade P, Gonzalez-Rivero M, Hoey A, Hoogenboom M, McCormick M, Ortiz JC, Razak T, Richardson A, Roff G, Sheppard-Brennand H, Stella J, Thompson A, Watson SA, Webster N, Audas D, Beeden R, Carver J, Cowlishaw M, Dyer M, Groves P, Horne D, Thiault L, Vains J, Wachenfeld D, Weekers D, Williams G, Mumby PJ (2020) Priority species to support the functional integrity of coral reefs. Oceanogr Mar Biol, Annu Rev 58:179–318

Wolfe K, Kenyon TM, Mumby PJ (2021) The biology and ecology of coral rubble and implications for the future of coral reefs. Coral Reefs 40:1769–1806

Wolff NH, Mumby PJ, Devlin M, Anthony KRN (2018) Vulnerability of the Great Barrier Reef to climate change and local pressures. Glob Change Biol 24:1978–1991

Wyatt ASJ, Waite AM, Humphries S (2010) Variability in isotope discrimination factors in coral reef fishes: Implications for diet and food web reconstruction. PLoS One 5:e13682

Wyatt ASJ, Lowe RJ, Humphries S, Waite AM (2013) Particulate nutrient fluxes over a fringing coral reef: source–sink dynamics inferred from carbon to nitrogen ratios and stable isotopes. Limnol Oceanogr 58:409–427

Wyatt ASJ, Matsumoto R, Chikaraishi Y, Miyairi Y, Yokoyama Y, Sato K, Ohkouchi N, Nagata T (2019) Enhancing insights into foraging specialization in the world's largest fish using a multi-tissue, multi-isotope approach. Ecol Monogr 89:e01339

Xu Y, Liu X, Cao X, Huang C, Liu E, Qian S, Liu X, Wu Y, Dong F, Qiu CW, Qiu J, Hua K, Su W, Wu J, Xu H, Han Y, Fu C, Yin Z, Liu M, Roepman R, Dietmann S, Virta M, Kengara F, Zhang Z, Zhang L, Zhao T, Dai J, Yang J, Lan L, Luo M, Liu Z, An T, Zhang B, He X, Cong S, Liu X, Zhang W, Lewis JP, Tiedje JM, Wang Q, An Z, Wang F, Zhang L, Huang T, Lu C, Cai Z, Wang F, Zhang J (2021) Artificial intelligence: a powerful paradigm for scientific research. Innovation 2(4):100179

Zaneveld JR, Burkepile DE, Shantz AA, Pritchard CE, McMinds R, Payet JP, Welsh R, Correa AMS, Lemoine NP, Rosales S, Fuchs C, Maynard JA, Thurber RV (2016) Overfishing and nutrient pollution interact with temperature to disrupt coral reefs down to microbial scales. Nat Commun 7(May):1–12

2 Disentangling Current and Latent Dimensions of Vulnerability in Social-Ecological Systems

Victor Brun, Cecile Brugere, and Joachim Claudet

Abstract

The effects of global environmental change, already tangible worldwide, are projected to increase in the future, putting at further risk ecosystems and societies. Vulnerability assessments have become widely used to study how social-ecological systems are exposed and resilient to climate change or cumulative stressors. Many of these assessments, however, implicitly add up entangled current and latent dimensions of vulnerability. In this chapter, we argue that conceptualizing the difference between current and latent, as well as present and future dimensions of vulnerability, would benefit vulnerability assessments as time represents both a barrier and a lever to adaptation. Using published literature, we assess the links between past, present, and future vulnerability, propose a definition of current and latent dimensions of vulnerability and then discuss practical methods and indicators to assess them.

Keywords

Adaptation · Climate change · Human well-being · Resilience · Vulnerability assessment

V. Brun (✉)
UAR3278 CRIOBE, PSL Université Paris, CNRS-EPHE-UPVD, Maison de l'Océan 195, Paris, France

Laboratoire d'Excellence CORAIL, Moorea, French Polynesia

Ocean & Climate Platform, Paris, France
e-mail: vbrun@ocean-climate.org

C. Brugere
Soulfish Research & Consultancy, York, UK

J. Claudet
National Center for Scientific Research, PSL Université Paris, CRIOBE, CNRS-EPHE-UPVD, Maison de l'Océan 195, Paris, France

Laboratoire d'Excellence CORAIL, Moorea, French Polynesia
e-mail: joachim.claudet@cnrs.fr

2.1 Introduction

People are dependent on the ecosystems they inhabit, use, shape, and ultimately threaten, hence making vulnerable fundamental elements of their well-being (Brondízio et al. 2019). These ecosystems, landscapes, species, or genetic resources underpin human health and represent the cornerstone of entire cultures and ways of life. The ubiquitous decline in nature calls for immediate transformative actions to preserve its contribution to human well-being (Díaz et al. 2019). As trade-offs exist between environmental conservation and human health, win-win solutions should be found to ensure the interventions dedicated to improve the status of environmental features do not impede the ability of people to benefit from them (Hopkins 2021). Climate change is a major driver of decline in nature's contributions to people (Brondízio et al. 2019), and directly threatens human societies through climate hazards such as cyclones and sea-level rise (IPCC 2022).

Vulnerability assessments (VAs), rooted primarily in the development, risk-hazard and resilience research tradition, are valuable tools to study and plan the sustainability of social-ecological systems (Adger 2006; Bennett et al. 2016; Depietri 2020; Thiault et al. 2021). While there are many acceptances of vulnerability (Brugère and Young 2015; Eakin and Luers 2006), in its most common definition, vulnerability represents the degree to which a system is exposed and expected to cope with one or several stressors; it is conceptually and practically linked to (a lack of) resilience (Adger 2006; Füssel 2007). Applied to social-ecological systems (SESs), VAs use indicators to identify how changes in environmental conditions both result from socio-economic drivers such as poverty or inadequate infrastructure and in turn threaten communities, households or individuals. In this chapter, we will conceptualize vulnerability as composed of three dimensions, following Thiault et al. (2021): (i) exposure (i.e., the degree to which a system is subject to a stressor), (ii) sensitivity (i.e., the conditions making a system prone to change in the short term when facing exposure), and

S. Planes (ed.), *The Future of Coral Reefs*, Coral Reefs of the World 21, https://doi.org/10.1007/978-3-031-98584-3_2

(iii) adaptive capacity (i.e., the latent ability to cope and recover from the impacts of a stressor, linked to the system's resilience). This stems from an earlier definition of the Intergovernmental Panel on Climate Change (IPCC 2022), and has been used in many studies and decision-making materials. In a review encompassing 4432 papers, Kuhlicke et al. (2023) found out that about 90% of them lacked of a theoretical underpinning and emphasized the importance of theory in vulnerability research. VAs have the potential to produce materials relevant for decision-making in order to reduce exposure, sensitivity, and enhance the adaptive capacity of SESs in the face of global change (Thiault et al. 2021). Capturing heterogeneity in social and ecological contexts, drivers, and capacities is instrumental for VAs which only hold an explanatory potential when they tease out what distinguishes SESs' vulnerability from one another: whether (i) it is systems, or places within a system that are differently vulnerable (i.e., spatial heterogeneity); (ii) it is the source (exposure, sensitivity, and adaptive capacity) of their vulnerability, and/or (iii) it is time that makes them more or less vulnerable (i.e., temporal heterogeneity). Spatial heterogeneities of vulnerability usually constitute the main focus of social-ecological VAs, and while focusing on a particular scale also explore the links with other scales (Adger 2006; Turner et al. 2003). Built at the global level, VAs are used to compare countries in order to direct adaptation funds and other adaptation measures (Brooks et al. 2005; Comte et al. 2018; Joakim et al. 2015; Klein and Möhner 2011).

Measuring vulnerability can only be relevant when characterizing how "a specified system [is vulnerable] to a specific range of hazards" (Brooks 2003, p. 3) to compare vulnerability across space, time, and contexts. Ecosystems, as much as social institutions, are subject to dramatic and constant change in the face of natural and anthropogenic drivers. Past drivers of change are shaping and informing present vulnerability, and at any moment, present drivers are further influencing its future (Bennett et al. 2016; Duvat et al. 2017; Fazey et al. 2016; Terti et al. 2015). Measuring vulnerability and planning adaptation should explicitly account for temporal dynamics. Yet VAs often fall short of including time, and many assessments do not study changes in vulnerability, nor project future vulnerability trajectories (Jurgilevich et al. 2017). Many VAs promote a static understanding of human-environment interactions (Ford et al. 2018). When dynamics are included, they are not necessarily properly and explicitly accounted for: for instance, they only consider future changes in exposure and integrate them in a present and static indicator. While present adaptive capacity can shape the ability of a society to cope with future threats, this adaptive capacity also has chances to evolve over time. Only considering changes in one particular component of vulnerability, a procedure often justified by lack of data, makes vulnerability projections blind to the potential evolution of its other components. Moreover, mixing and compressing present and future indicators jeopardizes a temporally explicit understanding of change, and the opportunity to answer the following question: when is change expected to come, and how can a timely adaptation be planned?

Some authors have pointed out that vulnerability has latent dimensions, in particular adaptive capacity (Cinner et al. 2018; Whitney et al. 2017). The resources a society is able to mobilize in order to face a potential risk are only effectively mobilized when the risk realizes. Current dimensions of vulnerability are to be opposed to latent ones, and those dimensions together constitute the vulnerability of a social-ecological system at any point in time. In this chapter, we therefore propose a distinction between *past, present,* and *future* vulnerability, as well as *current* and *latent* dimensions of vulnerability to better design VAs and inform adaptation actions. We then review practical methods and indicators used in the literature to assess current and latent dimensions of vulnerability.

2.2 Current and Latent Dimensions of Vulnerability: Past, Present, and Future

2.2.1 Current and Latent Dimensions of Vulnerability

Assessments of vulnerability to climate change typically focus on future changes in climate patterns and their potential to affect human systems. The main way of doing so is by only including the future of dimensions for which data is readily available. Typically, it is future exposure that is included, for example as a change in sea-surface temperature (Blasiak et al. 2017; Jara et al. 2020), or future exposure to droughts (Brown et al. 2016). Future exposure, and present sensitivity and adaptive capacity are then aggregated in a unique indicator. This temporal disconnection must be noted, as sensitivity and adaptive capacity are also subject to a high level of change, both in the case of ecological and social dimensions. In relation to this idea, many programs designed to lower vulnerability are meant to increase adaptive capacity over time rather than decreasing exposure (Smit and Wandel 2006).

Here, we propose that the projected evolutions in vulnerability dimensions are latent, as opposed to current dimensions that are measurable at present. This slightly differs from the usual acceptation of adaptive capacity being latent (Cinner et al. 2018), as in our definition adaptive capacity can both have current features, resources, and capacities that can be mobilized now, and latent features, that is, resources and capacities that could be mobilized in the future. We add and emphasize that other dimensions of vulnerability (exposure and sensitivity) also have latent dimensions that can be projected. An illustration could be the vulnerability to sea-level rise that a coastal community is facing. Current exposure could be the current average sea level and the risk it poses if a storm was to come; latent exposure could be sea-level rise projections. Current sensitivity, at the community scale, could be measured by the number of houses in exposed areas and the population it represents; latent sensitivity could be linked to projected changes in demography. Finally, current adaptive capacity could be measured by the average wealth or availability in insurance schemes this community can access now, whereas latent adaptive capacity could be linked to long-term development planning and potential future factors of resilience. While two different communities, "A" and "B," could have a similar level of current vulnerability at present, if the community "A" had lower projected sea-level rise, a projected decrease in exposed coastal population, and strong development policy planning, it would have a lower level of latent vulnerability and would therefore be less vulnerable at this present time.

2.2.2 Past, Present, and Future Vulnerability

Current and latent dimensions of vulnerability result in a particular vulnerability at a given point in time, past, present and future (Fig. 2.1). At present, this vulnerability derives from past trajectories, and is influenced by future pathways that should be the target of adaptation planning.

To understand past vulnerability and connect it to the present, authors have focused on the study of "trajectories of exposure and vulnerability" (Duvat et al. 2017; Magnan et al. 2012; Magnan and Duvat 2016). Duvat et al. (2017) define these trajectories as "storylines of driving factors and processes that have influenced past and present territorial system exposure and vulnerability to impacts associated with climate variability and change." Recent past changes in vulnerability are useful to understand the present trajectory on which the system is engaged (Thiault et al. 2018a). When facing change, another important feature of SESs' resilience is their social-ecological memory. The literature proposes several definitions of the concept: while some authors insist on the collective forms of knowledge linking people and ecosystems engaging populations in sustainability patterns of resource use (Barthel et al. 2010), others insist on the effect of past disturbances on present social-ecological resilience (Folke 2006; von Heland and Nykvist 2012). The latter is more tightly linked to the concept of ecological memory, which describes how disturbance legacies drive longer-term ecosystem processes. In forest ecosystems, for example, past disturbances have been proved to affect their present resilience (Johnstone et al. 2016; Seidl et al. 2014), as well as other ecosystem dynamics and characteristics such as landscape pattern (Peterson 2002) and tree growth and mortality (Bond-Lamberty et al. 2014). Similarly, the extent of bleaching on coral reef ecosystems has also been proved to be dependent on recent bleaching events in the Great Barrier Reef. Hughes et al. (2019) demonstrated that despite a much higher exposure to heat stress in 2017 compared to 2016, the bleaching response of corals was diminished.

2.2.3 Vulnerability Profiles and Implications for Adaptation

Scenarios of latent vulnerability, once built, should already describe different adaptation options. Their results should also, in an iterative way, further guide decision-making on adaptation by displaying different profiles of social-ecological vulnerability. Adding latent states of vulnerability to current ones can display four kinds of situations for different SESs (Fig. 2.2):

1. Lowest concern (low current vulnerability and low latent vulnerability): Systems in this quadrant will be characterized by low exposure to the threats studied, while showing a relatively high resilience to these threats. Likewise, in the future, the latent states of their vulnerability components are projected to be relatively low. These SESs, which can be labeled as "lowest concern," have to be kept on this path of low vulnerability. Lower concern does, however, not mean lower attention, for two reasons. First, unpredictable evolutions and nonlinear changes could occur and lead to a higher state of vulnerability. Second, these systems must be given attention to understand the underlying reasons for their higher preparedness and potentially replicating these conditions to other contexts of higher vulnerability. Such a system could be represented by a wealthier coastal community having infrastructure well-designed to face the impacts of climate change, while having developed rigorous future adaptation policies made to decrease future sensitivity and increase future adaptive capacity.
2. High current vulnerability (and low latent vulnerability): Systems in this quadrant are characterized by relatively higher levels of current stressors and lower resilience to their effects. Their latent states of vulnerability are not predicted to be much higher. In this situation, adaptation actions must focus on reducing present stressors while increasing social-ecological resilience. Attention should nonetheless be kept on the future, as changes could still occur and lead to higher vulnerability. The drivers of latent adaptive capacity should also be a focus to lead adaptation measures in other contexts. An example of such system would be an impoverished coastal commu-

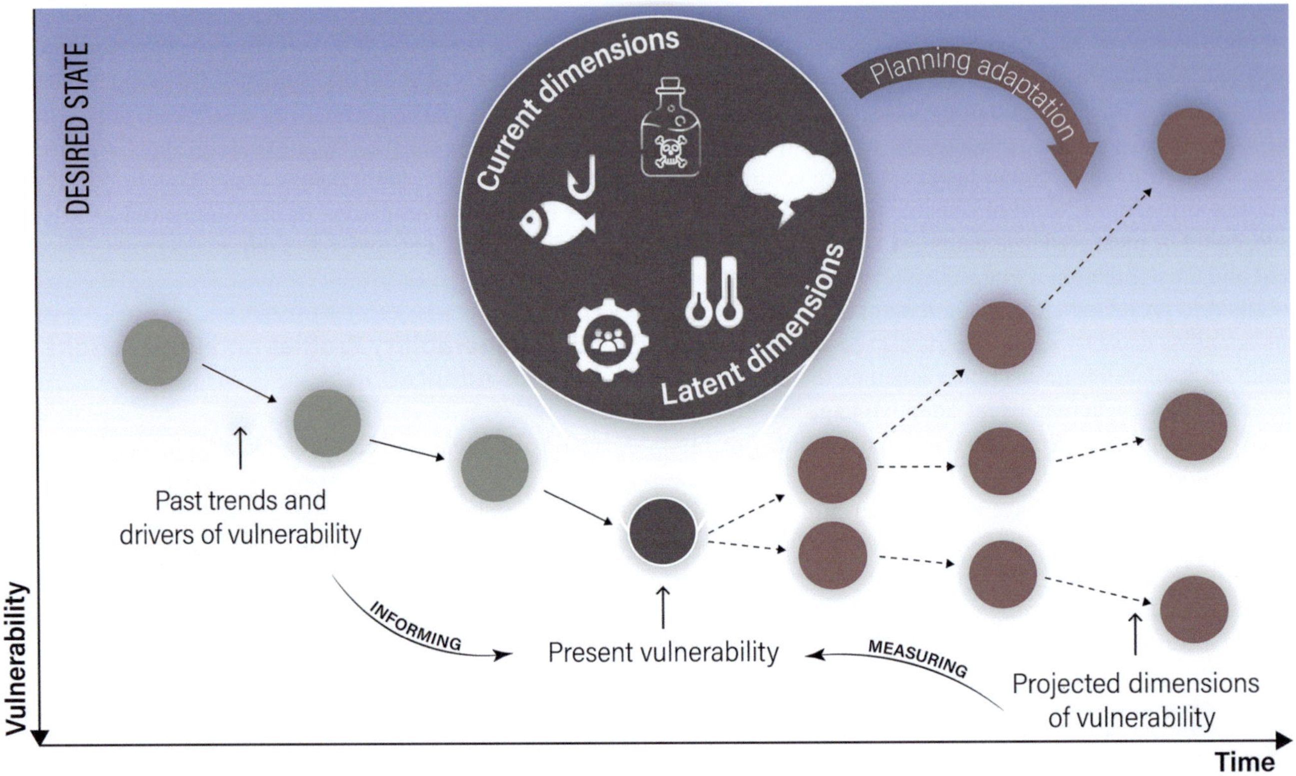

Fig. 2.1 Past, present, and future vulnerability of a social-ecological system are integrated in current and latent dimensions of vulnerability. Past trends in vulnerability can inform present vulnerability. Projections of the evolution of the vulnerability's components (i.e., exposure, sensitivity, and adaptive capacity), which can be proposed in different contrasting scenarios, are used to measure the latent dimensions of present vulnerability. This measure of present vulnerability can be used to plan adaptation and transformation actions and act recursively on the projected dimensions

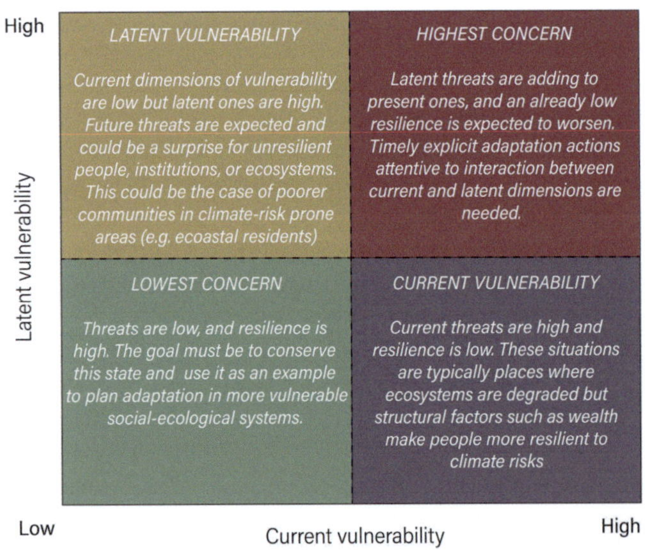

Fig. 2.2 Four different vulnerability profiles of a social-ecological system based on its current and latent dimension of vulnerability

nity with little climate readiness, making this community at risk now. With a high sensitivity and low adaptive capacity, climate impacts in the present or near future would already be threatening.

3. High latent vulnerability (and high latent vulnerability): Examples of systems in this quadrant could be a coastal zone that few people inhabit, causing both few threats to the ecosystem and a low cumulated social dependence on ecosystem services. Yet, some coastal development could be planned in this area, leading, for instance, to the development of touristic infrastructures in low-lying areas, with low resilience and high projected exposure to climate change. Latent states of vulnerability would then be very high, and invisible at the moment. The solution must be timely adaptation planning, while including potential future threats in decision-making processes over time, where relevant.
4. Highest concern (high current vulnerability and high latent vulnerability): When current vulnerability is high,

and latent social-ecological states are projected to lead to even higher vulnerability, attention must be given to both reduce present factors of vulnerability while looking at the future, and curbing threats while promoting resilience. Systems in this quadrant could be represented by a densely populated area, where pressures on marine ecosystems are high, social dependence is high, and adaptive capacity is low due to low incomes. Coral reef fisheries can represent an eloquent example where both current and latent vulnerability are very high as threats are high and expected to disproportionately increase in the future putting people and nature at an even higher level of risk (Hughes et al. 2017). In this case, adaptation planning must relate present stressors to latent ones to establish a more holistic understanding of vulnerability patterns.

Table 2.1 Review of coral reef social-ecological system vulnerability approaches and indicators contrasted as current and latent dimensions

	Current	Latent
Exposure	Sea surface temperature (Cinner et al. 2012) Fishing effort (Thiault et al. 2018b)	Socioeconomic drivers including population, infrastructure, tourism (Nicholls et al. 2011) Climate forecasts for future disasters (Maina et al. 2016) Future sea-level rise (Brown et al. 2018)
Sensitivity	Current dependence on coral reefs (Licuanan et al. 2015; Pendleton et al. 2016)	Future coastal population (Neumann et al. 2015)
Adaptive capacity	Adaptive capacity to coastal disasters (Adger et al. 2005; Krishnan et al. 2019) Adaptive capacity of marine fisheries to climate change (Blasiak et al. 2017)	Future resilience of energy systems (Kishita et al. 2017) Future coral reef resilience (Delevaux et al. 2018)

2.3 Tools to Measure the Current and Latent Vulnerability of Social-Ecological Systems

Understanding the vulnerability of a SES according to which past drivers of change it resulted from already provides more valuable information than a static VA; both regarding the root causes of vulnerability, and the actual impacts we can expect from the set of stressors the SES was exposed to. Once these past dimensions have been explored and the present state of vulnerability is thus better understood, the next step is to assess future vulnerability trends and drivers. Adapting to future threats necessitates considering the future possible states of a SESs (e.g., in terms of demography, socioeconomics, governance, large-scale or small-scale ecological components, ways of valuing and using ecosystems). Latent states of vulnerability must be studied as depending on different pathways including the effects of projected changes in exposure, but also changes in sensitivity and adaptive capacity. This section reviews the literature on VAs to discuss practical ways in which latent vulnerability dimensions can be assessed using four complementary approaches: (i) climate projections, (ii) socioeconomic pathways, (iii) participatory methods, and (iv) modeling. We then provide examples of indicators and methods used in previous studies to assess both current and latent states of vulnerability (Table 2.1).

2.3.1 Climate Projections

Climate projections extracted from climatic models are commonly used in VAs in order to measure future exposure related to changes in temperature and weather patterns, extreme events, or sea-level rise (Allison et al. 2009; Johnson et al. 2016; Weißhuhn et al. 2018). IPCC-class models certainly remain the most widely used models for these assessments (Stock et al. 2011). This data, related to past and present observations of climate variables, is informative on what constitutes, in time, the different exposures of SESs to climate change. They do not, however, inform on potential changes in sensitivity and adaptive capacity, and should therefore be complemented with other methods.

2.3.2 Socioeconomic Pathways

Socioeconomic scenarios produced by the IPCC, or "social system pathways" (SSP) can be used to build scenarios focusing on the social dimensions of vulnerability. They predict the main orientations or paths a society can take throughout time, and can be derived to local scale or sectoral assessments (Maury et al. 2017). Different profiles of systems can be distinguished depending on whether they emphasize sustainability and cooperation, versus noncooperative behaviors and policies, and economic development. These large trends may have the inconvenient to be broad. Yet, they include the possibility to categorize very different dimensions in a restricted number of representative scenarios. In order to precise these trends, demographic projections can also be used. Changes in demography are a very important driver of change in vulnerability—though some highly populated area can display high adaptive capacity—and authors have produced assessments of changes in coastal population in order to capture the subsequent future states of vulnerability, in particular through changes in exposure (Neumann et al. 2015).

2.3.3 Participatory Methods

Through exchanges with stakeholders and experts, participatory approaches remain a widely used tool in order to identify possible futures of SES. Visioning methods and forecasting scenarios, for example, aim at questioning what desirable future can lie ahead, and how to reach it (Kishita et al. 2017; Robinson 1990; Svenfelt et al. 2011; Wiek and Iwaniec 2014). Pre-built elements of scenarios, projections, as well as maps, can be used to guide discussions and produce scenarios that examine a range of "possible futures" under different framing conditions. Expert elicitation also belongs to this category of approaches, and has been used in the field of cumulative impact studies, and interactions between sustainable development goals (SDGs) (Ban et al. 2014; Singh et al. 2017, 2018). Scenarios of "cascading vulnerability" have also been built using interviews of stakeholders in a participatory approach, in order to assess potential futures of groundwater management and potential vulnerability pathways (Parisi et al. 2018). The related concept of predicting "cascading disasters" (Pescaroli and Alexander 2016) has also been approached with participative methods.

2.3.4 Modeling Vulnerability

Latent states of vulnerability can be assessed using models incorporating "policy-relevant scenarios" (Carlsen et al. 2016). Like SSP approaches, these models aim at providing restricted sets of possible futures to guide decision-making on adaptation. Participatory approaches can also feed models in order to assess social-ecological dynamics and include temporal elements (Leenhardt et al. 2017). Social-ecological models are able to make interact different dimensions, for example, by displaying the links between fishery yields, tourism and climate change effects (Wabnitz et al. 2018). Predictive models of present vulnerability (Tilley et al. 2018) could also be used to feed predictions of latent states of vulnerability. These quantitative approaches are extremely important when considering the large prevalence of quantitative approaches to measure vulnerability.

2.4 Conclusion

Human agency and decision-making, in a context of uncertainty, both appear as major shapers of tomorrow's vulnerability. Assessing latent vulnerability—the threats to be faced in the future depending on our actions in the present—constitutes a way to better guide adaptation measures in time. As the future will never be predicted in an accurate manner, negotiating scenarios and pathways while keeping our attention on the observed impacts of climate change and other stressors thus becomes our best chance to avoid deleterious futures.

References

Adger WN (2006) Vulnerability. Global Environ Change 16:268–281. https://doi.org/10.1016/j.gloenvcha.2006.02.006

Adger WN, Hughes TP, Folke C, Carpenter SR, Rockström J (2005) Social-Ecological Resilience to Coastal Disasters. Science 309:1036–1039. https://doi.org/10.1126/science.1112122

Allison EH, Perry AL, Badjeck MC, Neil Adger W, Brown K, Conway D, Halls AS, Pilling GM, Reynolds JD, Andrew NL, Dulvy NK (2009) Vulnerability of national economies to the impacts of climate change on fisheries. Fish Fish 10:173–196. https://doi.org/10.1111/j.1467-2979.2008.00310.x

Ban SS, Pressey RL, Graham NAJ (2014) Assessing interactions of multiple stressors when data are limited: a Bayesian belief network applied to coral reefs. Global Environ Change 27:64–72. https://doi.org/10.1016/j.gloenvcha.2014.04.018

Barthel S, Folke C, Colding J (2010) Social-ecological memory in urban gardens-retaining the capacity for management of ecosystem services. Global Environ. Change 20:255–265. https://doi.org/10.1016/j.gloenvcha.2010.01.001

Bennett NJ, Blythe J, Tyler S, Ban NC (2016) Communities and change in the anthropocene: understanding social-ecological vulnerability and planning adaptations to multiple interacting exposures. Reg Environ Chang 16:907–926. https://doi.org/10.1007/s10113-015-0839-5

Blasiak R, Spijkers J, Tokunaga K, Pittman J, Yagi N, Österblom H (2017) Climate change and marine fisheries: least developed countries top global index of vulnerability. PLoS One 12:e0179632. https://doi.org/10.1371/journal.pone.0179632

Bond-Lamberty B, Rocha AV, Calvin K, Holmes B, Wang C, Goulden ML (2014) Disturbance legacies and climate jointly drive tree growth and mortality in an intensively studied boreal forest. Glob Change Biol 20:216–227. https://doi.org/10.1111/gcb.12404

Brondízio ES, Settele J, Díaz S, Ngo HT (eds) (2019) The global assessment report of the intergovernmental science-policy platform on biodiversity and ecosystem services. Intergovernmental Science-Policy Platform on Biodiversity and Ecosystem Services (IPBES), Bonn

Brooks N (2003) Vulnerability, risk and adaptation: a conceptual framework. Tyndall Centre for Climate Change Research

Brooks N, Adger WN, Kelly PM (2005) The determinants of vulnerability and adaptive capacity at the national level and the implications for adaptation. Global Environ Change 15:151–163. https://doi.org/10.1016/j.gloenvcha.2004.12.006

Brown JR, Kluck D, McNutt C, Hayes M (2016) Assessing drought vulnerability using a socioecological framework. Rangelands 38:162–168. https://doi.org/10.1016/j.rala.2016.06.007

Brown JM, Morrissey K, Knight P, Prime TD, Almeida LP, Masselink G, Bird CO, Dodds D, Plater AJ (2018) A coastal vulnerability assessment for planning climate resilient infrastructure. Ocean Coast Manag 163:101–112. https://doi.org/10.1016/j.ocecoaman.2018.06.007

Brugère CD, Young C (2015) Assessing climate change vulnerability in fisheries and aquaculture: available methodologies and their relevance for the sector. FAO, Rome

Carlsen H, Lempert R, Wikman-Svahn P, Schweizer V (2016) Choosing small sets of policy-relevant scenarios by combining vulnerability and diversity approaches. Environ Model Softw 84:155–164. https://doi.org/10.1016/j.envsoft.2016.06.011

Cinner JE, McClanahan TR, Graham NAJ, Daw TM, Maina J, Stead SM, Wamukota A, Brown K, Bodin Ö (2012) Vulnerability of coastal communities to key impacts of climate change on coral reef fisheries. Global Environ. Change 22:12–20. https://doi.org/10.1016/j.gloenvcha.2011.09.018

Cinner JE, Adger WN, Allison EH, Barnes ML, Brown K, Cohen PJ, Gelcich S, Hicks CC, Hughes TP, Lau J, Marshall NA, Morrison TH (2018) Building adaptive capacity to climate change in tropical coastal communities. Nat Clim Chang 8:117–123. https://doi.org/10.1038/s41558-017-0065-x

Comte A, Pendleton LH, Bailly D, Quillérou E (2018) Conceptual advances on global scale assessments of vulnerability : informing investments for coastal populations at risk of climate change. Mar Policy 1–10:391. https://doi.org/10.1016/j.marpol.2018.10.038

Delevaux JMS, Jupiter SD, Stamoulis KA, Bremer LL, Wenger AS, Dacks R, Garrod P, Falinski KA, Ticktin T (2018) Scenario planning with linked land-sea models inform where forest conservation actions will promote coral reef resilience. Sci Rep 8:12465. https://doi.org/10.1038/s41598-018-29951-0

Depietri Y (2020) The social–ecological dimension of vulnerability and risk to natural hazards. Sustain Sci 15:587–604. https://doi.org/10.1007/s11625-019-00710-y

Díaz S, Settele J, Brondízio ES, Ngo HT, Agard J, Arneth A, Balvanera P, Brauman KA, Butchart SHM, Chan KMA, Garibaldi LA, Ichii K, Liu J, Subramanian SM, Midgley GF, Miloslavich P, Molnár Z, Obura D, Pfaff A, Polasky S, Purvis A, Razzaque J, Reyers B, Chowdhury RR, Shin Y-J, Visseren-Hamakers I, Willis KJ, Zayas CN (2019) Pervasive human-driven decline of life on earth points to the need for transformative change. Science 366:eaax3100. https://doi.org/10.1126/science.aax3100

Duvat VKE, Magnan AK, Wise RM, Hay JE, Fazey I, Hinkel J, Stojanovic T, Yamano H, Ballu V (2017) Trajectories of exposure and vulnerability of small islands to climate change. WIREs Clim Change 8:e478. https://doi.org/10.1002/wcc.478

Eakin H, Luers AL (2006) Assessing the vulnerability of social-environmental systems. Annu Rev Environ Resour 31:365–394. https://doi.org/10.1146/annurev.energy.30.050504.144352

Fazey I, Wise RM, Lyon C, Câmpeanu C, Moug P, Davies TE (2016) Past and future adaptation pathways. Clim Dev 8:26–44. https://doi.org/10.1080/17565529.2014.989192

Folke C (2006) Resilience: the emergence of a perspective for social-ecological systems analyses. Global Environ Change 16:253–267. https://doi.org/10.1016/j.gloenvcha.2006.04.002

Ford JD, Pearce T, McDowell G, Berrang-Ford L, Sayles JS, Belfer E (2018) Vulnerability and its discontents: the past, present, and future of climate change vulnerability research. Clim Chang 151:189–203. https://doi.org/10.1007/s10584-018-2304-1

Füssel H-M (2007) Vulnerability: a generally applicable conceptual framework for climate change research. Global Environ Change 17:155–167. https://doi.org/10.1016/j.gloenvcha.2006.05.002

Hopkins SR (2021) How to identify win–win interventions that benefit human health and conservation. *Nat Sustain* 4:298–304

Hughes TP, Barnes ML, Bellwood DR, Cinner JE, Cumming GS, Jackson JBC, Kleypas J, Van De Leemput IA, Lough JM, Morrison TH, Palumbi SR, Van Nes EH, Scheffer M (2017) Coral reefs in the Anthropocene. Nature 546:82–90. https://doi.org/10.1038/nature22901

Hughes TP, Kerry JT, Connolly SR, Baird AH, Eakin CM, Heron SF, Hoey AS, Hoogenboom MO, Jacobson M, Liu G, Pratchett MS, Skirving W, Torda G (2019) Ecological memory modifies the cumulative impact of recurrent climate extremes. Nat Clim Chang 9:40–43. https://doi.org/10.1038/s41558-018-0351-2

IPCC (2022) Summary for policymakers. In: Pörtner H-O, Roberts DC, Tignor M, Poloczanska ES, Mintenbeck K, Alegría A, Craig M, Langsdorf S, Löschke S, Möller V, Okem A, Rama B (eds) Climate change 2022: impacts, adaptation, and vulnerability. contribution of working group II to the sixth assessment report of the intergovernmental panel on climate change. Cambridge University Press, Cambridge/New York, NY, pp 3–33. https://doi.org/10.1017/9781009325844.001

Jara HJ, Tam J, Reguero BG, Ganoza F, Castillo G, Romero CY, Gévaudan M, Sánchez AA (2020) Current and future socio-ecological vulnerability and adaptation of artisanal fisheries communities in Peru, the case of the Huaura province. Mar Policy 119:104003. https://doi.org/10.1016/j.marpol.2020.104003

Joakim EP, Mortsch L, Oulahen G (2015) Using vulnerability and resilience concepts to advance climate change adaptation. Environ Hazards 14:137–155. https://doi.org/10.1080/17477891.2014.1003777

Johnson JE, Welch DJ, Maynard JA, Bell JD, Pecl G, Robins J, Saunders T (2016) Assessing and reducing vulnerability to climate change: moving from theory to practical decision-support. Mar Policy 74:220–229. https://doi.org/10.1016/j.marpol.2016.09.024

Johnstone JF, Allen CD, Franklin JF, Frelich LE, Harvey BJ, Higuera PE, Mack MC, Meentemeyer RK, Metz MR, Perry GLW, Schoennagel T, Turner MG (2016) Changing disturbance regimes, ecological memory, and forest resilience. Front Ecol Environ 14:369–378. https://doi.org/10.1002/fee.1311

Jurgilevich A, Räsänen A, Groundstroem F, Juhola S (2017) A systematic review of dynamics in climate risk and vulnerability assessments. Environ Res Lett 12:013002. https://doi.org/10.1088/1748-9326/aa5508

Kishita Y, McLellan BC, Giurco D, Aoki K, Yoshizawa G, Handoh IC (2017) Designing backcasting scenarios for resilient energy futures. Technol Forecast. Soc. Change 124:114–125. https://doi.org/10.1016/j.techfore.2017.02.001

Klein RJT, Möhner A (2011) The political dimension of vulnerability: implications for the green climate fund. IDS Bull 42:15–22. https://doi.org/10.1111/j.1759-5436.2011.00218.x

Krishnan P, Ananthan PS, Purvaja R, Joyson Joe Jeevamani J, Amali Infantina J, Srinivasa Rao C, Anand A, Mahendra RS, Sekar I, Kareemulla K, Biswas A, Kalpana Sastry R, Ramesh R (2019) Framework for mapping the drivers of coastal vulnerability and spatial decision making for climate-change adaptation: a case study from Maharashtra, India. Ambio 48:192–212. https://doi.org/10.1007/s13280-018-1061-8

Kuhlicke C, Madruga De Brito M, Bartkowski B, Botzen W, Doğulu C, Han S, Hudson P, Nuray Karanci A, Klassert CJ, Otto D, Scolobig A, Moreno Soares T, Rufat S (2023) Spinning in circles? A systematic review on the role of theory in social vulnerability, resilience and adaptation research. Global Environ Change 80:102672. https://doi.org/10.1016/j.gloenvcha.2023.102672

Leenhardt P, Stelzenmüller V, Pascal N, Probst WN, Aubanel A, Bambridge T, Charles M, Clua E, Féral F, Quinquis B, Salvat B, Claudet J (2017) Exploring social-ecological dynamics of a coral reef resource system using participatory modeling and empirical data. Mar Policy 78:90–97. https://doi.org/10.1016/j.marpol.2017.01.014

Licuanan WY, Samson MS, Mamauag SS, David LT, Borja-del Rosario R, Quibilan MCC, Siringan FP, Sta Maria MYY, España NB, Villanoy CL, Geronimo RC, Cabrera OC, Martinez RJS, Aliño PM (2015) I-C-SEA change: a participatory tool for rapid assessment of vulnerability of tropical coastal communities to climate change impacts. Ambio 44:718–736. https://doi.org/10.1007/s13280-015-0652-x

Magnan A, Duvat VKE (2016) Trajectoires de vulnérabilité et adaptation au changement climatique à la Réunion. IDDRI 53:395. https://doi.org/10.1177/0022146512469014

Magnan A, Duvat V, Garnier E (2012) Reconstituer les « trajectoires de vulnérabilité » pour penser différemment l'adaptation au changement climatique. Nat Sci Soc 20:82–91. https://doi.org/10.1051/nss/2012008

Maina J, Kithiia J, Cinner J, Neale E, Noble S, Charles D, Watson JEM (2016) Integrating social–ecological vulnerability assessments with climate forecasts to improve local climate adaptation planning for coral reef fisheries in Papua New Guinea. Reg Environ Chang 16:881–891. https://doi.org/10.1007/s10113-015-0807-0

Maury O, Campling L, Arrizabalaga H, Aumont O, Bopp L, Merino G, Squires D, Cheung W, Goujon M, Guivarch C, Lefort S, Marsac F, Monteagudo P, Murtugudde R, Österblom H, Pulvenis JF, Ye Y, van Ruijven BJ (2017) From shared socio-economic pathways (SSPs) to oceanic system pathways (OSPs): building policy-relevant scenarios for global oceanic ecosystems and fisheries. Global Environ. Change 45:203–216. https://doi.org/10.1016/j.gloenvcha.2017.06.007

Neumann B, Vafeidis AT, Zimmermann J, Nicholls RJ (2015) Future coastal population growth and exposure to sea-level rise and coastal flooding – A global assessment. PLoS One 10:e0118571. https://doi.org/10.1371/journal.pone.0118571

Nicholls RJ, Woodroffe CD, Burkett V, Hay J, Wong PP, Nurse L (2011) Scenarios for coastal vulnerability assessment, in: treatise on estuarine and coastal science. Elsevier, pp 289–303. https://doi.org/10.1016/B978-0-12-374711-2.01217-1

Parisi A, Monno V, Fidelibus MDD (2018) Cascading vulnerability scenarios in the management of groundwater depletion and salinization in semi-arid areas. Int J Disaster Risk Reduct 30:292–305. https://doi.org/10.1016/j.ijdrr.2018.03.004

Pendleton L, Comte A, Langdon C, Ekstrom JA, Cooley SR, Suatoni L, Beck MW, Brander LM, Burke L, Cinner JE, Doherty C, Edwards PET, Gledhill D, Jiang L-Q, Van Hooidonk RJ, Teh L, Waldbusser GG, Ritter J (2016) Coral reefs and people in a high-CO2 world: where can science make a difference to people? PLoS One 11:e0164699. https://doi.org/10.1371/journal.pone.0164699

Pescaroli G, Alexander D (2016) Critical infrastructure, panarchies and the vulnerability paths of cascading disasters. Nat Hazards 82:175–192. https://doi.org/10.1007/s11069-016-2186-3

Peterson GD (2002) Contagious disturbance, ecological memory, and the emergence of landscape pattern. Ecosystems 5:329–338. https://doi.org/10.1007/s10021-001-0077-1

Robinson JB (1990) Futures under glass. A recipe for people who hate to predict. Futures 22:820–842. https://doi.org/10.1016/0016-3287(90)90018-D

Seidl R, Rammer W, Spies TA (2014) Disturbance legacies increase the resilienceof forest ecosystem structure, composition, and functioning. Ecol Appl 24:2063. https://doi.org/10.1017/CBO9781107415324.004

Singh GG, Sinner J, Ellis J, Kandlikar M, Halpern BS, Satterfield T, Chan KMA (2017) Mechanisms and risk of cumulative impacts to coastal ecosystem services: an expert elicitation approach. J Environ Manag 199:229–241. https://doi.org/10.1016/j.jenvman.2017.05.032

Singh GG, Cisneros-Montemayor AM, Swartz W, Cheung W, Guy JA, Kenny TA, McOwen CJ, Asch R, Geffert JL, Wabnitz CCC, Sumaila R, Hanich Q, Ota Y (2018) A rapid assessment of co-benefits and trade-offs among sustainable development goals. Mar Policy 93:223–231. https://doi.org/10.1016/j.marpol.2017.05.030

Smit B, Wandel J (2006) Adaptation, adaptive capacity and vulnerability. Global Environ. Change 16:282–292. https://doi.org/10.1016/j.gloenvcha.2006.03.008

Stock CA, Alexander MA, Bond NA, Brander KM, Cheung WWL, Curchitser EN, Delworth TL, Dunne JP, Griffies SM, Haltuch MA, Hare JA, Hollowed AB, Lehodey P, Levin SA, Link JS, Rose KA, Rykaczewski RR, Sarmiento JL, Stouffer RJ, Schwing FB, Vecchi GA, Werner FE (2011) On the use of IPCC-class models to assess the impact of climate on living marine resources. Prog Oceanogr 88:1–27. https://doi.org/10.1016/j.pocean.2010.09.001

Svenfelt Å, Engström R, Svane Ö (2011) Decreasing energy use in buildings by 50% by 2050 — a backcasting study using stakeholder groups. Technol Forecast Soc Change 78:785–796. https://doi.org/10.1016/j.techfore.2010.09.005

Terti G, Ruin I, Anquetin S, Gourley JJ (2015) Dynamic vulnerability factors for impact-based flash flood prediction. Nat Hazards 79:1481–1497. https://doi.org/10.1007/s11069-015-1910-8

Thiault L, Marshall P, Gelcich S, Collin A, Chlous F, Claudet J (2018a) Space and time matter in social-ecological vulnerability assessments. Mar Policy 88:213–221. https://doi.org/10.1016/j.marpol.2017.11.027

Thiault L, Marshall P, Gelcich S, Collin A, Chlous F, Claudet J (2018b) Mapping social-ecological vulnerability to inform local decision making: mapping social-ecological vulnerability. Conserv Biol 32:447–456. https://doi.org/10.1111/cobi.12989

Thiault L, Jupiter SD, Johnson JE, Cinner JE, Jarvis RM, Heron SF, Maina JM, Marshall NA, Marshall PA, Claudet J (2021) Harnessing the potential of vulnerability assessments for managing social-ecological systems. Ecol Soc 26:art 1. https://doi.org/10.5751/ES-12167-260201

Tilley A, López-Angarita J, Herrón P, Espinosa S, Angarita JL, Box S (2018) Predicting vulnerability to management changes in data-limited, small-scale fisheries. Mar Policy 94:39–45. https://doi.org/10.1016/j.marpol.2018.04.013

Turner BL, Kasperson RE, Matson PA, McCarthy JJ, Corell RW, Christensen L, Eckley N, Kasperson JX, Luers A, Martello ML, Polsky C, Pulsipher A, Schiller A (2003) A framework for vulnerability analysis in sustainability science. Proc Natl Acad Sci 100:8074–8079. https://doi.org/10.1073/pnas.1231335100

von Heland J, Nykvist B (2012) Social-ecological memories as a source of general and specific resilience. Ecol. Soc. 19:12

Wabnitz CCC, Cisneros-Montemayor AM, Hanich Q, Ota Y (2018) Ecotourism, climate change and reef fish consumption in Palau: benefits, trade-offs and adaptation strategies. Mar Policy 88:323–332. https://doi.org/10.1016/j.marpol.2017.07.022

Weißhuhn P, Müller F, Wiggering H (2018) Ecosystem vulnerability review: proposal of an interdisciplinary ecosystem assessment approach. Environ Manag 61:904. https://doi.org/10.1007/s00267-018-1023-8

Whitney CK, Bennett NJ, Ban NC, Allison EH, Armitage D, Blythe JL, Burt JM, Cheung W, Finkbeiner EM, Kaplan-Hallam M, Perry I, Turner NJ, Yumagulova L (2017) Adaptive capacity: from assessment to action in coastal social-ecological systems. Ecol Soc 22:art22. https://doi.org/10.5751/ES-09325-220222

Wiek A, Iwaniec D (2014) Quality criteria for visions and visioning in sustainability science. Sustain Sci 9:497–512. https://doi.org/10.1007/s11625-013-0208-6

Scleractinian Coral Skeletons as Natural Archives of Global Change: A Cross-Disciplinary Investigation

3

Marine Canesi, Éric Douville, Chloé Brahmi, Laetitia Hédouin, Serge Planes, Paolo Montagna, and Stéphanie Reynaud

Abstract

Scleractinian coral skeletons serve as valuable environmental archives, offering key insights into past climate and ocean variability. Understanding the mechanisms underlying coral calcification is essential for reconstructing accurate paleoenvironmental records. However, calcification is a complex process regulated by both biological and environmental factors. This study explores the interplay between coral skeletal geochemistry and physiological processes, focusing on species-specific growth patterns, biomineralization dynamics, and the impact of diagenesis. We highlight the use of elemental ratios (e.g., Mg/Ca, Sr/Ca, Li/Mg) as proxies for environmental reconstructions while addressing biases introduced by skeletal heterogeneity, bioerosion, and secondary mineral precipitation. Additionally, we discuss how physiological traits, such as symbiosis, nutrient uptake, and respiration, influence calcification and may affect the preservation of geochemical archives over time. The integration of high-resolution geochemical analyses with biological insights is of great importance for refining paleoenvironmental proxies and enhancing the accuracy of climate reconstructions. This multidisciplinary approach provides a broader perspective on coral biomineralization, a critical issue for assessing reef resilience in a changing ocean.

Keywords

Calcification · Geochemical proxies · Biomineralization · Coral physiology · Scleractinian corals · Global change

M. Canesi (✉) · L. Hédouin · S. Planes
Laboratoire d'Excellence CORAIL, Perpignan, France

USR 3278 CRIOBE, PSL Université Paris, EPHE-UPVD-CNRS, Papetoai, Mo'orea, French Polynesia
e-mail: laetitia.hedouin@criobe.pf; planes@univ-perp.fr

É. Douville
Laboratoire des Sciences du Climat et de l'Environnement, LSCE/IPSL, CEA-CNRS-UVSQ, Université Paris-Saclay, Gif-sur-Yvette, France
e-mail: eric.Douville@lsce.ipsl.fr

C. Brahmi
Université de la Polynésie française, ILM, Ifremer, IRD, UMR SECOPOL, Tahiti, French Polynesia
e-mail: chloe.brahmi@upf.pf

P. Montagna
Istituto di Scienze Polari (ISP), Consiglio Nazionale delle Ricerche (CNR), Bologna, Italy
e-mail: paolo.montagna@cnr.it

S. Reynaud
Centre Scientifique de Monaco, Monaco, Monaco

LIA ROPSE, Laboratoire International Associé Université Côte d'Azur – Centre Scientifique de Monaco, Monaco, Monaco
e-mail: sreynaud@centrescientifique.mc

3.1 Massive Scleractinian Corals as Natural Archives

Scleractinian corals, also known as "hard" or "stony" corals comprising many symbiotic and reef-building species, are some of the most important organisms in low latitude due to their ability to construct the framework that supports the diverse array of species found in these habitats (Spalding 2001; Porter and Tougas 2001). Among them, massive corals are distributed throughout tropical and subtropical warm shallow waters around the world and are characterized by their potentially large size (a few meters), slow growth (3–20 mm.yr.$^{-1}$), massive band structures, and dome-shaped appearance (Kleypas et al. 1999; Fig. 3.1). Massive corals are of particular scientific interest because of their remarkable longevity, which can extend over several centuries, and their ability to continuously record environmental history in their skeletons as they grow. In addition, their fossil form of a few centimeter/meter in length serves as a natural archive for the past (using drills that provide cores of skeleton) with the potential to provide invaluable insights into historical climate and environmental variations and changes in the marine environment at high resolution (season, year, decade) and over long (from a few years/decades to centuries) and

S. Planes (ed.), *The Future of Coral Reefs*, Coral Reefs of the World 21, https://doi.org/10.1007/978-3-031-98584-3_3

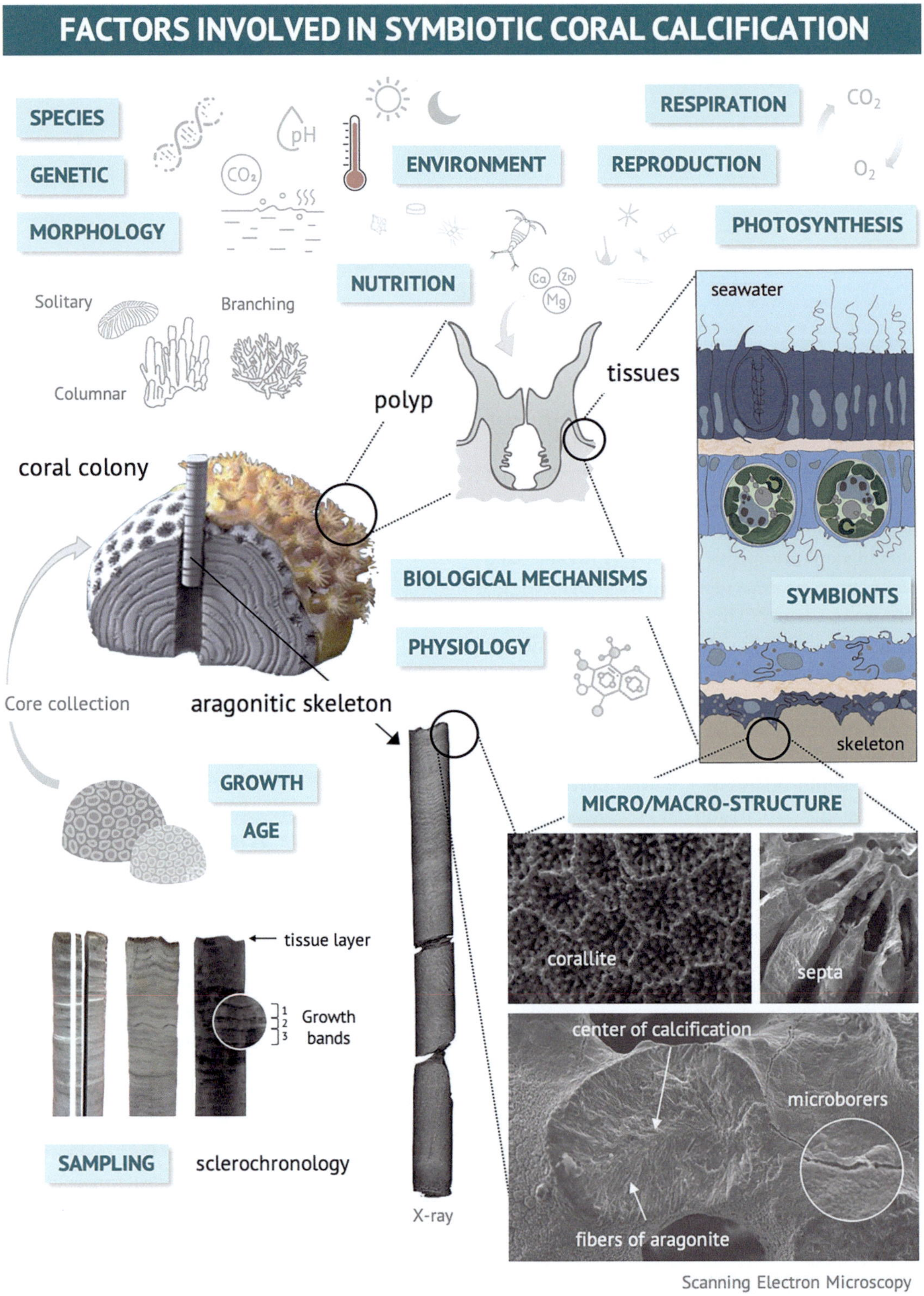

Fig. 3.1 Factors involved in coral calcification: from nanometric to macrometric scale

various periods (Anthropocene/Holocene, last climate cycles, Quaternary, and beyond). Among hundreds of scleractinian coral species, massive corals such as *Porites*, *Orbicella*, *Diploastrea, Siderastrea,* and *Montastraea* represent a wide range of long-life genera offering a high potential for decennial/centennial paleo-reconstructions for shallow waters of tropical areas. Among these genera studied since the end of the 1970s (Goreau 1977; Emiliani et al. 1978), *Porites* is the most extensively studied, as it is one of the most common and widely distributed coral genera in tropical areas (Thompson 2021). Other genera such as *Orbicella, Siderastrea, and Montastraea*, endemic massive corals of the Caribbean region, have also been the focus of numerous studies aimed at reconstructing past marine changes for the tropical West Atlantic region (Goreau 1977; Emiliani et al. 1978; DeLong et al. 2011; Flannery et al. 2017, 2018; Paterne et al. 2023). More recently, in the last two decades, *Diploastrea* has been implicated in Indo-Pacific paleoclimate reconstructions due to its attributes as a slow-growing (3–5 $mm.yr.^{-1}$), long-lived genus (>3 centuries), and its colony size of a few meters potentially allows for longer temporal records than *Porites* for the same length of sampling (Watanabe et al. 2003; Bagnato et al. 2004; Wu et al. 2018).

The continuous secretion of the calcium carbonate skeleton of massive corals, under favorable conditions, presents a unique opportunity for them to serve as natural archives of climate and environmental changes. This is because, during the aragonitic skeleton formation, the incorporation and partitioning of major (Ca, C, O), minor (Sr, Mg), trace (e.g., U, Li, B, Ba, REE) elements and stable isotopes (e.g., ^{11}B, ^{13}C, ^{18}O, ^{15}N) are strongly sensitive to various environmental factors, including the physicochemical properties of the surrounding seawater (e.g., SST, salinity, pH, and carbonate chemistry), light intensity, the availability of food and nutrients (Barnes and Lough 1996; Quinn and Sampson 2002), as well as biotic factors such as symbiotic activities (e.g., the transfer of photosynthates from the symbiotic microalgae to coral). This characteristic has captured the interest of marine biologists and paleoclimatologists, as coral skeletons offer continuous, high-resolution records spanning from seasonal to millennial timescales, filling gaps in instrumental records, which are typically limited to last decades (Lough 2010; Lee et al. 2014; Saha et al. 2016; etc.). Coral geochemistry has thus emerged as a powerful tool for reconstructing past oceanic conditions through the analysis of key climate or environmental proxies. Due to a range of thermodynamic, kinetic, crystallographic, and biological processes, including Rayleigh fractionation, enzymatic element transport, crystal growth rate, and ionic substitution in aragonite lattice, the isotopic and elemental signatures of coral skeletons reflect, at a certain length scale, the properties of the surrounding seawater at the time of skeletal accretion. For example, several geochemical proxies, such as $\delta^{18}O$, Sr/Ca, Mg/Ca, Li/Mg, and U/Ca, have been developed and applied to reconstruct sea surface temperatures (Corrège 2006; Montagna et al. 2014; Fowell et al. 2016; DeCarlo et al. 2016; Cuny-Guirriec et al. 2019). $\delta^{13}C$ reflects the balance between photosynthesis by symbiotic microalgae and coral respiration, while also being influenced by dissolved inorganic carbon in seawater, variations in primary productivity, and anthropogenic inputs such as the Suess effect (e.g., Watanabe et al. 2002). Initially used to indicate seawater pH and carbonate chemistry, $\delta^{11}B$ and B/Ca are now more frequently employed to reconstruct the internal properties of corals during calcification (see below). Similarly, $\delta^{15}N$ serves as a proxy for nutrient sources and human-induced impacts like eutrophication and wastewater discharge and is recognized as a key indicator of oxygen-deficient zones (Duprey et al. 2024). Additionally, Ba/Ca and Cd/Ca ratios reveal signals from riverine inputs, phytoplankton blooms, or upwelling, whereas Mn/Ca, REE/Ca, and heavy metal (e.g., Pb) concentrations indicate continental influences or industrial contamination. Although integrating all these geochemical proxies is beyond the current scope, numerous studies have advanced our ability to constrain environmental changes over the Quaternary using skeletal geochemistry. Moreover, each seasonal skeletal growth band, or scleroband, forms an environmental record (Fig. 3.1) that provides essential insights into historical growth parameters, such as linear extension, skeletal density, and calcification rate (Lough and Cooper 2011), thereby underpinning models for predicting future coral growth.

Geochemical and physiological observations are crucial for understanding the ocean's historical response to climate variability and for assessing long-term patterns in coral growth and calcification. However, promising geochemical proxies, such as those used to reconstruct sea surface temperature (SST) via Sr/Ca and Li/Mg ratios, are limited by biological (also referred to as « vital effect ») and diagenetic effects, resulting in uncertainties of approximately <1 °C (Cuny-Guirriec et al. 2019; Canesi et al. 2024b). Recent studies combining proxies (e.g., Sr/Ca–Li/Mg or Sr-U) show the potential to reduce these uncertainties by mitigating growth rate effects (Fowell et al. 2016; DeCarlo et al. 2016; Canesi et al. 2024b), but further research is needed to achieve precision within a few tenths of a degree and thereby enhance the reliability of coral-based SST records for climate modeling. Similarly, boron isotopes and B/Ca ratios were originally used to reconstruct seawater pH and carbonate chemistry (Hemming and Hanson 1992; Hönisch et al. 2004; Reynaud et al. 2004; Douville et al. 2010; Wu et al. 2018), have been reevaluated as proxies for the carbonate chemistry of the coral calcifying fluid (McCulloch et al. 2012; DeCarlo et al. 2018; Canesi et al. 2024a), though numerous assumptions remain to be validated. Addressing these challenges through further investigations w ill help establish, for each coral genus or species, robust relationships among environmental factors, internal calcifying fluid properties, and coral calcification and growth.

3.2 Exploring the Surrounding Environmental Properties

Coral calcification, by precipitating calcium carbonate, creates the primary framework that supports the entire reef ecosystem. However, the effects of climate and environmental changes, pollution, and habitat degradation recently exacerbated during the Anthropocene pose significant threats to corals. Ocean warming and acidification are the two major stressors of global change due to massive CO_2 emissions into the atmosphere. This increases the oceanic CO_2 uptake that has been shown to significantly impact coral calcification (Kleypas et al. 2001; Dove et al. 2020; Cornwall et al. 2021, 2022). In this context of global change, the rate of coral calcification may decline due to the decrease of the aragonite saturation state (Ω_{arag}) and pH in seawater, making it more difficult for scleractinian corals to maintain proper conditions at the mineralization site to form their skeletal structures (Gattuso et al. 1998; Mollica et al. 2018). While several recent studies revealed that warm seawater can be beneficial to the calcification of some massive corals, but also, that each coral genus may have its response in terms of calcification (e.g., Canesi et al. 2023), the concomitant decrease in seawater pH reflecting ocean acidification induced by an increase in dissolved CO_2 concentration may also reduce the rate of calcification as it limits the availability of carbonate ions (CO_3^{2-}), which consequently affects the upregulation of the aragonite saturation state $(\Omega_{arag})_{cf}$ in the calcifying fluid that corals pilot to build their skeleton (Canesi et al. 2024a). These factors may lead to a weakening of tropical coral reefs in the coming decades/centuries, resulting in a reduction in coral diversity. This will have serious implications for the health of coral reef ecosystems, the socioeconomic activities associated with these marine ecosystems, and, in the longer term, it may also contribute to the erosion of tropical coastlines (Hoegh-Guldberg 2011; Descombes et al. 2015; Eddy et al. 2021). Understanding the response of coral calcification to ongoing environmental change is a widely studied topic worldwide, intending to elucidate the physiological mechanisms by which each coral species may be resilient or not to future global changes. Understanding these responses is critical for predicting reef resilience and informing effective conservation strategies.

In addition to seawater temperature and carbonate chemistry changes, it is important to recognize that several other natural environmental variables, regional or local, may have a significant impact on both the calcification process and coral growth. For example, biotic and/or abiotic factors, such as nutrient availability (i.e., food, nitrates, phosphates), dissolved oxygen, and light (photosynthetically active radiation—PAR) may also influence coral growth (Mw 1995; Rodolfo-Metalpa et al. 2008). For example, blue light, independent of photosynthesis, significantly enhances calcification in *Porites lutea* and *Acropora variabilis* corals, challenging the traditional understanding of light-enhanced calcification (LEC), a mechanism that relies on photosynthetic CO_2 uptake to raise pH at the mineralization site and promote calcification. The influence of light quality (including the light spectrum), and the presence of UVA and UVB radiation on coral-microalgae symbiosis, and calcification, needs to be further investigated. In addition, a mesocosm study demonstrated the positive effects of temperature and light on the photophysiology and calcification of *Acropora nasuta* corals (Ross et al. 2022). It has been also demonstrated that feeding enhances coral growth rate and increases calcification rates (Reynaud et al. 2002; Houlbrèque et al. 2003). Moreover, research conducted on *Acropora millepora* corals in aquariums showed the beneficial effect of oxygenation on calcification rates (Wijgerde et al. 2014), which is consistent with previous studies on various scleractinian corals (Colombo-Pallotta et al. 2010; Wijgerde et al. 2012). In recent decades, additional stressors on coral development have emerged as a result of demographic changes and human activities, particularly in tropical coastal areas. Notable among these anthropogenic pressures are the effects of urbanization and coastal development, which affect coral reproduction and calcification through light effects (Levy et al. 2007). Substantial releases of dissolved pollutants, including heavy/toxic metals and nutrients, which contribute to eutrophication (Fabricius 2011), as well as increasing inputs of various solid materials (including plastics), constitute an additional threat to coral growth (Hall et al. 2015; Lamb et al. 2018). The effects of these increasing pressures on coral calcification, which have just been raised, are at present poorly assessed and require future investigation.

In general, most studies on coral calcification are conducted in mesocosms or aquaria under controlled conditions, where the effects of one or two parameters, or their combination, are investigated. However, in the natural environment, a large number of parameters vary simultaneously, such as carbonate chemistry, temperature, oxygen level, light, and/or productivity during seasons, but are not considered in such studies on live corals due to the lack of local time series, the availability of complete oceanographic data, or technical constraints to acquire all the necessary information. It is known, for example, that the $\delta^{18}O$ value of the skeleton is influenced by the seawater temperature. However, it is also influenced by sea level, salinity, and light intensity (Reynaud-Vaganay et al. 2001) through photosynthetic activity and feeding (Reynaud et al. 2002). Under these terms, assessing the accuracy of any seawater temperature proxy, such as Sr/Ca, Li/Mg, or the $\delta^{18}O$ ratio in corals, for reconstructing a single environmental factor (i.e., SST), without disentangling the precise and local effects of other environmental or biological drivers, becomes challenging.

In fact, for paleoclimatologists, geochemistry and biology offer essential tools for reconstructing historical environmental changes and identifying long-term trends (a few decades or centuries) in coral growth and calcification. However, to assess the precise effects of environmental properties on coral calcification, it is necessary to select, from databases or in situ measurements, the seawater properties that are most representative and closest to the medium in which the corals form their skeleton (a few centimeters, meters, or kilometers from the colonies studied). Accurate and complete descriptions of the study site from which coral cores are collected are critical, as they can contribute significantly to the uncertainty of proxies, both in their calibration and in their application. Therefore, ensuring the reliability of current and past data, and databases remains a key challenge in using corals as reliable natural archives.

Although data loggers now enable continuous monitoring of pH and dissolved oxygen, direct measurement and long-term monitoring of oxygen levels, pH, carbonate chemistry, and nutrient concentrations in seawater remain resource-intensive and logistically complex, particularly due to limitations in data acquisition frequency and duration. In contrast, parameters such as temperature, salinity, light, and chlorophyll a can be simultaneously and continuously monitored in real time using advanced autonomous sensors. Dedicated efforts are essential to accelerate the development of efficient, user-friendly tools for rapid and accurate assessment of oxygen, pH, DIC, and nutrient levels in seawater (Hasumoto et al. 2006), given their substantial impact on the calcification process. Numerous marine databases such as PANGEA, Copernicus, and GLODAP provide in situ or interpolated measured seawater data available to the scientific community but, mostly for open ocean (Mishima and Nakamura 2009; Koubarakis et al. 2018; Lauvset et al. 2021). One of the main challenges in improving global change impact studies using corals is the limited availability of high-resolution time series of variables, over a few decades or centuries, close to coral habitats, and for many contrasting tropical environments. Today, many SST records from recent decades or centuries in tropical areas have been published or synthesized, as there is clear evidence and model for global warming of the shallow ocean at low latitudes (Tierney et al. 2015). In contrast, past changes in seawater carbonate chemistry or other environmental variables are so far been poorly documented or even nonexistent for many tropical regions. Overcoming technical limitations by standardizing simultaneous measurements of parameters such as PAR, nutrients, oxygen, and others, in addition to commonly collected data, is key. Integrating marine data from diverse sites across seasons and environments enhances the accuracy of reef ecosystem analysis and strengthens coral skeleton geochemical assessments. High-resolution satellite data on SST, light, carbonate chemistry, and primary productivity from the past two decades offer a valuable complement to the limited tropical monitoring stations. Combined with long-term records from oceanographic cruises and monitoring stations, these datasets will enable a more detailed investigation of key environmental drivers influencing coral geochemistry and calcification. It is also important to note that variables such as nutrients and oxygen are known to be correlated with primary environmental factors such as temperature or light (Takahashi et al. 1993; Badran 2001). These correlations may mask or bias the singular effect of temperature or the other variable on coral geochemistry or growth. The interdependence of temperature, carbonate chemistry, oxygen, nutrients, and light can obscure the interpretation of coral geochemical analyses. Isolating the effects of individual environmental factors allows accurate quantification of coral responses to specific stressors and assessment of their adaptation or acclimation to changing conditions. Advanced statistical models have been developed to disentangle these influences, particularly in *Porites* colonies, enabling differentiation between the long-term effects of seawater temperature and ocean acidification on coral calcification (D'Olivo et al. 2019; Guo et al. 2020). D'Olivo et al. (2019) estimated that ~82% of the decline in pH_{cf} is driven by ocean acidification, with ~17% attributed to the long-term warming from 1939 to 2013. However, these models, applied primarily to *Porites* colonies in the Great Barrier Reef, require validation across other massive coral genera in diverse tropical regions. Moreover, they do not systematically account for secondary factors such as deoxygenation or eutrophication (Mollica et al. 2018; D'Olivo et al. 2019). To improve precision and interpretation, decoupling models must integrate all major drivers of calcification beyond ocean warming and acidification. Controlled experiments offer a way forward by simultaneously monitoring key environmental variables (e.g., SST, light, carbonate chemistry, food, pollutants), analyzing geochemical tracers in coral skeletons (e.g., trace elements, stable isotopes), and assessing coral growth and calcification rates. Addressing the interdependence of environmental factors requires culture-based, multivariate dynamic experiments exposing multiple coral species to varying conditions in natural (non-semi-controlled) settings. Additionally, emerging climate models, statistical tools, and AI-driven approaches hold promise for refining the decoupling of individual environmental influences (e.g., Clampitt 2023).

3.3 Diagenesis and Structural Issues/ Skeletal Heterogeneity

The preservation of the aragonitic coral skeleton is shaped by biotic and abiotic factors, with diagenetic processes and bioerosion playing critical roles. Among alterations due to diagenetic processes, the replacement of pristine aragonite by calcite, and secondary precipitations may be observed. Indeed, precipitation of cement such as abiogenic calcite and aragonite may occur in the skeleton pore spaces. Moreover, secondary precipitations (e.g., high magnesium calcite, secondary aragonite, brucite) may fill galleries formed into the skeleton by endolith microorganisms (e.g., filamentous microalgae, fungi, cyanobacteria) (Nothdurft et al. 2007; Webb et al. 2009). These mineral cements/precipitations exhibit distinct geochemical signatures (e.g., Sr/Ca, Mg/Ca) that can affect coral-based environmental reconstructions (Sarmento et al. 2020). In parallel, biological erosion and dissolution further alter coral skeletons, maintaining a delicate balance between calcification (bioaccretion) and erosion/dissolution (destruction), a key role in determining the prospects of coral reefs (Tribollet and Golubic 2011; Alvarado et al. 2017). While coral skeletons grow through calcification, they face persistent erosion over time, primarily from bioerosion agents which systematically remove calcium carbonate structures. Bioerosion is due to different biological activities of organisms varying in size, ranging from microborers like algae, fungi, and cyanobacteria to larger macroborers species such as sponges, polychaete worms, sipunculids, mollusks, crustaceans, echinoids, and grazing species like parrotfish, acanthurids, and pufferfish (Glynn and Manzello 2015). Increasing seawater temperatures and acidification may have exacerbated microborer activities, creating favorable conditions for these erosive organisms to thrive (Schönberg et al. 2017).

These diagenesis processes can obscure or alter the original environmental signals recorded by geochemical signatures in coral skeletons, particularly in older or fossilized colonies (Quinn and Taylor 2006; Allison et al. 2007; Nothdurft and Webb 2009). The selection of high-quality, well-preserved aragonite coral cores, which have not undergone significant early or post-diagenesis is therefore crucial for accurate paleoclimate studies.

Various parameters can be assessed within corals to evaluate their preservation status, and thus increase the reliability of result interpretation in paleoclimatology or impact studies. One of these is the calcite/aragonite ratio in the aragonite matrix, which can significantly alter the chemical composition of corals and introduce geochemical bias into paleoreconstructions. As an example of early diagenesis, Lazareth et al. (2016) highlighted the influence of calcite on Sr/Ca or Li/Mg ratios in *Porites* coral skeletons, which affects the SST reconstructions by about +0.14 °C and +0.26 °C per 1% calcite, respectively. Calcite presence on boron isotopes ($\delta^{11}B$) has also been observed and consequently may strongly bias ocean pH reconstructions.

In the process of calcification, the extracellular organic matrix secreted by the calicoblastic cells (Puverel et al. 2005), and composed of macromolecules (e.g., proteins, glycoproteins, polysaccharides, lipids), plays a crucial role in regulating the growth and spatiotemporal arrangement of aragonite crystals that form coral skeleton (Allemand et al. 1998). A part of these organic matrix compounds is integrated into the skeleton (i.e., intraskeletal organic matrix). Looking at the micrometric and sub-micrometric levels, a heterogeneity of structural and geochemical composition of coral skeletal is observed. Several studies have demonstrated that, at the ultrastructural length scale, two components form the coral skeleton, that is, centers of calcification (COC) and fibers of aragonite (FA) (also referred to as rapid accretion deposits and thickening deposits, respectively; Stolarski 2003), which are characterized by different nanostructural organization (Stolarski and Mazur 2005; Cuif and Dauphin 2005; Benzerara et al. 2011).

Microanalytical analyses of elemental (e.g., Mg/Ca, Sr/Ca, Li/Mg) and isotopic ratios reveal significant geochemical heterogeneities between COCs and FA in symbiotic corals, including massive *Porites* (Meibom et al. 2007; Rollion-Bard and Blamart 2015). In several coral species, COCs exhibit higher Mg/Ca ratios than FA. Similarly, Li/Mg ratios differ between these skeletal components. Mg enrichment is often associated with rapid skeletal growth, occurring not only within COCs but also at larger spatial scales. For instance, dissepiments, which form at rates exceeding 50 μm/day, show higher Mg concentrations compared to more slowly forming structures such as septa and spines in *Pocillopora damicornis* (Brahmi et al. 2012). The relative architecture between COC and FA is also influenced by external physicochemical processes, even if biomineralization is attributed to the biological activity of the coral (Sugiura et al. 2021). As observed by Nothdurft and Webb (2009), both macro- and microstructural features of coral skeletons, such as polyp size, columella, and septothecal wall, are not only genus-specific but also species-specific. These skeletal traits serve as key taxonomic criteria for coral classification (Cuif et al. 2003). These differences are also notable between *Porites* and other massive corals such as *Siderastrea, Monstratrea, or Diploastrea*, challenging the use of a single generalized growth model as the effects of macro- and microstructures on reconstructions. Consequently, the skeletal structures of massive corals pose a challenge to the universal use of geochemical proxies and can be a source of uncertainty in environmental reconstructions due to the effects of microstructures on element or stable isotope partitioning (Rollion-Bard et al. 2003, 2011). Further studies of the relative abundances of FA and COC in the coral skeleton and their effects on geochemistry through direct microstructural observations are recommended to better understand coral skeletogenesis and obtain accurate geochemical data (Montagna and Douville 2020).

Additionally, skeleton carbonate chemistry shows large heterogeneity across the macrostructure of a coral polyp (i.e., columella vs. inter-polyp walls). This macrostructure is specific to each coral genus (e.g., *Porites* vs *Diploastrea)* and has to be considered during sampling steps before all pertinent interpretations of geochemical studies (Chalk et al. 2021), and limits generally the temporal resolution of reconstructions using corals at a few months. Heterogeneity within annual growth bands compromises the accuracy of universal geochemical proxies (Allison and Finch 2009), demonstrating the lack of accurate, structure-independent universal geochemical proxies. To improve reconstructions, multi-axis sampling within a single colony is essential to model elemental and isotopic incorporation in three dimensions. Sampling resolution from nano- to centimeter scale mitigates spatial and temporal biases, enhancing measurement reliability. Advancing analytical techniques and implementing rigorous archive quality verification are critical to refining high-resolution coral-based reconstructions. Therefore, the use of nondestructive methods for mapping (micro)structures or local calcification/density heterogeneity is required, such as systematic quantification of FA/COC ratios (proportions, dimensions) using SEM or Raman (e.g., Stolarski 2003), calcite/aragonite ratios using micro X-ray diffraction analysis (μXRD), or key trace element quantities (B, Sr, U, Mg, Li) and stable isotopes using lasers coupled with new generations of inductively coupled plasma mass spectrometers (ICPMS) or micro-/nanoscale secondary ion mass spectrometry (SIMS), would enable targeted and reliable sampling. Recently, the application of tools such as machine learning has optimized the analysis of a large number of SEM images (Reznikov et al. 2020; Alaguarda et al. 2022). X-ray micro- or nano-tomographies are innovative techniques that have recently provided insights into the grain morphology and intra-particle pores of carbonate sand particles (Zhou et al. 2022a), as well as the zooid wall layers of a bryozoan skeleton (Pakhnevich 2021).

Overall, the dynamics of biomineralization in corals can benefit from innovative techniques or microsensors applied to coral skeletons, enabling the mapping or modeling of skeletal structures at various scales, and exploring correlations with environmental (e.g., dissolved oxygen, nutrients, turbidity) and physiological (e.g., calcification rate, photosynthesis, respiration) variables. Microtomography, in particular, facilitates detailed three-dimensional studies of internal biomineralized structures, offering valuable insights as observed in coralline algae (Torrano-Silva et al. 2015). Its nondestructive nature can also improve coral sampling methods and generate new geochemical proxies to reduce diagenetic bias or calcification effects in paleoclimate reconstructions.

3.4 Coral Growth, Vital, and Age-Related Effects

Recent attention has been focused on addressing the inherent ambiguities associated with coral paleothermometers that arise from the biomineralization processes themselves. These ambiguities include the vital effect (i.e., biological effect), first described by Urey et al. (1951) and Erez (1978), and the growth-rate-related kinetic effect, highlighted by McConnaughey (1989) and Cohen and McConnaughey (2003). In particular, coral growth rates have been found to influence the oxygen and carbon isotopic ratio ($\delta^{18}O$, $\delta^{13}C$) (McConnaughey 1989; Cohen and McConnaughey 2003; Felis et al. 2003; Maier et al. 2004), as well as elemental ratios such as Sr/Ca (de Villiers et al. 1995; Hayashi et al. 2013), Mg/Ca (Inoue et al. 2007), and U/Ca (DeCarlo et al. 2015). More recently, the development of multi-proxy calibrations, based on the combination of Sr/Ca with Li/Mg in a unique Sr/Ca-Li/Mg proxy (±0.3–1.2 °C; Fowell et al. 2016; D'Olivo et al. 2018; Cuny-Guirriec et al. 2019; Zinke et al. 2019) or Sr/Ca with U/Ca leading to the recent Sr-U approach (±0.5–0.93 °C; DeCarlo et al. 2016; Alpert et al. 2017; Ross et al. 2019; Canesi et al. 2024b), has improved the reliability of using these proxies as paleothermometers, limiting the effects of coral growth on reconstructions. Recent studies have substantially reduced temperature uncertainties by a factor of about two (± 0.6 °C), reducing the impact of vital effects on coral-derived paleoreconstructions, highlighting the importance of developing and applying multi-proxy approaches in the future. However, further research is needed to identify the limitations of these universally applicable calibrations before their use can be expanded.

Moreover, inter- and intra-colonial variability in coral geochemistry can pose challenges for paleoclimate reconstructions (Stephans et al. 2004). Inter-colonial variability refers to differences in geochemical signatures between different colonies of the same species living in the same area and period, while intra-colonial variability refers to variation within a single colony (as described above). These variations can be caused by differences in local environmental conditions (i.e., depth or light incidence) or by biological factors such as coral size, age, reproduction, gender, genetics, growth rate, or axis or photosynthesis. To improve the reliability of paleoclimate records, replication across multiple colonies (ideally three per site) is recommended (Lough 2004). However, this approach is both costly and time-consuming and may lead to excessive sampling of live coral colonies.

In addition, coral density and extension rate provide valuable insights into environmental conditions, as these parameters are known to correlate with factors such as temperature, light, and nutrient availability (Reed et al. 2021). Recent calibration studies have established a correlation between seawater temperature and the shades of gray of X-ray images, which reflect the density of the coral skeleton. This has led to the development of a temperature-density calibration with mean absolute errors ranging from 0.73 to 0.99 °C (Bautista et al. 2021). These values are comparable to the mean absolute errors (0.5 to 0.93 °C) reported for other proxies such as Sr-U (DeCarlo et al. 2016; Alpert et al. 2017; Ross et al. 2019). Canesi et al. (2021) examined the effect of a density anomaly (Δdensity = 0.47 g·cm^{-3}) on SST reconstructions, resulting in differences in ΔSST ranging from 1.5 °C for Sr-U to 5.1 °C for Sr/Ca. This density anomaly also affected the elemental (ΔB/Ca) and isotopic ($\Delta\delta^{11}$B) boron composition of the *Porites* corals, resulting in variations in the reconstructed pH_{cf} of about 0.07 units (Canesi et al. 2021). This study clearly shows that the removal of biases in reconstructions due to geochemical coral data linked to calcification processes is still a challenge.

Additional other factors such as bleaching events can lead to discontinuous growth records and alter the geochemical composition of coral skeletons (Suzuki et al. 2003; Cantin and Lough 2014; Wang et al. 2021). For example, skeletal δ^{13}C decreased immediately after bleaching in *M. capitata*, but not in *P. compressa* and skeletal δ^{18}O of both species failed to record the warming event (Rodrigues and Grottoli 2006). Another study showed that short-term coral bleaching is not recorded by skeletal boron isotopes (Schoepf et al. 2014), which highlights the need to evaluate the link between bleaching and other proxies. Isotopic micro-profiling may hold great potential for detecting signatures of previous bleaching events by pinpointing gaps in coral growth (Cantin and Lough 2014) and thus, could be used as a proxy for coral bleaching (and therefore calcification and growth). To tackle these issues, advanced techniques like micro-computed tomography can be employed to examine the three-dimensional architecture of coral skeletons and precisely measure variations in density and extension at microscale, for example, during abrupt events of coral bleaching. Understanding how coral growth dynamics, age-related processes, and calcification interact is essential for revealing hidden geochemical signatures within coral skeletons, providing valuable insights into short- and long-term environmental changes affecting coral reef ecosystems.

3.5 Physiological Mechanisms Behind Internal Regulation

Scleractinian coral calcification is a complex biomineralization process involving chemical elements and organic matrix compounds, resulting in the precipitation of calcium carbonate ($CaCO_3$) into aragonite crystals, forming a highly organized lattice structure known as the coral exoskeleton (Fig. 3.1). The calcification reaction involves calcium ions (Ca^{2+}) and dissolved inorganic carbon (DIC) species, which exist in three different chemical forms (i.e., CO_2, HCO_3^-, CO_3^{2-}), depending mainly on seawater pH. The formation of calcium carbonate crystals takes place in the calcification fluid (cf), at the interface of the calicoblastic layer of the polyp and the skeleton (Tambutté et al. 2011; Fig. 3.1). Since ions used for calcification ultimately come from the external surrounding seawater, they have to cross epithelial layers to reach the calcifying site. In addition to the tissue layers, corals also produce a mucus layer which ions must cross, and which may play an important role in ion transport, given that it is enriched in calcium (Clode and Marshall 2002). Reactions involving CO_2 and HCO_3^- generate protons (H^+), which, if not expelled from the cf., would hinder the formation of $CaCO_3$. To prevent a decrease in the pH of the calcification fluid (pH_{cf}), the Ca^{2+}-ATPase pump transports Ca^{2+} into the cf. while removing protons (McConnaughey 1989; Cohen and McConnaughey 2003; Zoccola et al. 2004; Venn et al. 2013). This chemical regulation, occurring within the calicoblastic cells (Muscatine et al. 1997; Tambutté et al. 2007), increases the concentration of carbonate ions (CO_3^{2-}) and maintains an aragonite saturation state (Ω_{arag}) up to six times higher than the surrounding environment, thereby promoting calcification (Venn et al. 2011; Holcomb et al. 2014; Cai et al. 2016; Raybaud et al. 2017). Biological mechanisms enable corals to strongly control the formation, structural, and geochemical composition of their skeleton (Meibom et al. 2007; Brahmi et al. 2012). Numerous studies have demonstrated the species- or genus-specific regulation of calcifying fluid chemistry through biological processes (Comeau et al. 2018; Howells et al. 2018; Martinez et al. 2019; Canesi et al. 2023) sometimes coupled to host and symbiont responses to environmental change (Eagle et al. 2022). However, the underlying physiological mechanisms remain inadequately understood and warrant comprehensive investigation to enhance our understanding of their regulatory mechanisms.

The biomineralization process has been studied mainly in branching corals (because of their rapid growth), but we can assume that the fundamental process is the same in massive corals. Critical processes such as carbonate, bicarbonate, and protons ions transport, and the functioning of proton/Ca^{2+} pumps within coral tissues play pivotal roles in this intricate process. A calcium channel that allows the entry of calcium ions into cells by diffusion has been localized by immunohistochemistry in the calicodermis (Zoccola et al. 1999).

A plasma membrane calcium ATPase responsible for calcium exit from the cells against its electrochemical gradient has been cloned and sequenced and is expressed in the calicodermis (Zoccola et al. 2004). The calcium ATPase likely functions as a proton exchanger and thus could help in regulating pH in the extracellular calcifying medium (Cohen and McConnaughey 2003; Zoccola et al. 2004).

Recent studies have provided evidence of significant variations in $[Ca^{2+}]_{cf}$ compared to seawater (DeCarlo 2018; Sevilgen et al. 2019). It is of utmost importance that this factor is properly identified and addressed, especially as many geochemical studies rely on the assumption of a constant saturation state of the calcifying fluid based on the $[Ca^{2+}]_{cf}$ assumption. To achieve this objective for massive corals, it is essential to develop systematic measurements or evaluation of $[Ca^{2+}]_{cf}$ in both biological studies (e.g., by direct in vivo measurements) and geochemical studies (e.g., isotopic labeling of Ca^{2+} incorporated into the skeleton), as was done for coral ^{86}Sr (Brahmi et al. 2012). This would bring new insights into how Ca^{2+} is incorporated into the calcifying fluid and how its presence affects coral biomineralization and calcification models.

The incorporation of dissolved inorganic carbon (DIC) into the tissues is much more complex as it exists in different forms of carbonate in seawater (i.e., CO_2, HCO_3^-, CO_3^{2-}, depending on pH_{sw}; Gattuso et al. 1999), and as its forms evolve through a series of processes, such as photosynthesis, respiration, and calcification (Fig. 3.2). Understanding the incorporation of these different dissolved carbonate ions and their transfer to form the calcium carbonate is key to understanding the composition and structure of coral skeletons as well as calcification and growth processes and rates. Studies on biological mechanisms are mainly conducted on cultured corals or in situ samples, but rarely on massive species due to their difficult cultivation. However, the examination of coral skeletons obtained from coral cores provides only inert material, limiting direct investigation of the biological processes. Studying the coral skeleton provides additional insights into the long-term composition, structure, and growth of colonies (Lough 2008, 2010; Thompson 2021). This is made possible through various observation techniques (e.g., X-ray, Scanning Electron Microscope, 3D tomography; Rahman et al. 2013; Bautista et al. 2021), geochemical analysis (trace elements and isotopes; Cohen and McConnaughey 2003), and organic analysis (e.g., lipids, organic matrix; Farre et al. 2010).

To achieve a comprehensive understanding of coral calcification, a collaborative and transdisciplinary approach is paramount, bridging the expertise of geochemists, physiologists, and biologists. Geochemist investigations play a crucial role in identifying trace elements and isotopes involved in calcification, while physiologists explore the mechanisms regulating these elements within coral. Integrating physiological data (e.g., nutrient uptake rates, photosynthesis/symbiosis activity, respiration, growth parameters) into geochemical models will provide relevant findings into calcification processes within living corals, enhancing our grasp of these mechanisms. There is a need to prioritize integrated studies of multiple parameters within a single genus/species or colony, over isolated studies, which requires close collaborations, sharing of data and methodologies, and the development of new tools (e.g., the development of the use of in situ microsensors) and techniques capable of harmonizing diverse disciplines, specially designed for massive corals (see Fig. 3.2). This holistic approach will allow us to gain a more sophisticated understanding of the complex mechanisms that drive the integration of geochemical elements into the coral skeleton. It will also improve our ability to predict the responses of these critical ecosystems to global environmental change.

3.6 Species and Genetics Implications

Current studies on the biological processes (i.e., physiological mechanisms) involved in calcification have been conducted mainly on branching corals such as *Pocillopora*, *Acropora*, and *Stylophora* or massive *Porites* corals, as they are very numerous, easily accessible, and exhibit rapid growth (1 cm per month for branching corals and 1 cm per year for *Porites*). However, these physiological mechanisms are known to vary between genera and, thus, cannot always be generalized to other genus/species or require further validation (Gates and Edmunds 1999). Indeed, species-specific differences in biological function and skeletal geochemistry are typically observed (Gibbin et al. 2015; Okazaki et al. 2017; Bahr et al. 2018; Bove et al. 2020) where species composition may influence how reef communities will respond to future climate change. To understand the processes of coral calcification and their driving forces, the influence of genus within scleractinian corals must be assessed. Previous research has focused primarily on the long-lived genus *Porites* in the field of scleractinians, with less studied species such as *Siderastrea* and *Diploastrea* being somewhat neglected due to their limited geographic distribution (Caribbean vs Indo Pacific, respectively). It is, therefore, necessary to extend our investigations to different massive coral genera to investigate the physiological and genetic mechanisms underlying calcification and to better delimit constrain climatic and environmental changes over time independently of the archive studied. This approach will also provide deep insights into coral responses to environmental variability, potential adaptations, and evolutionary responses to global climate change. Long-term studies in controlled

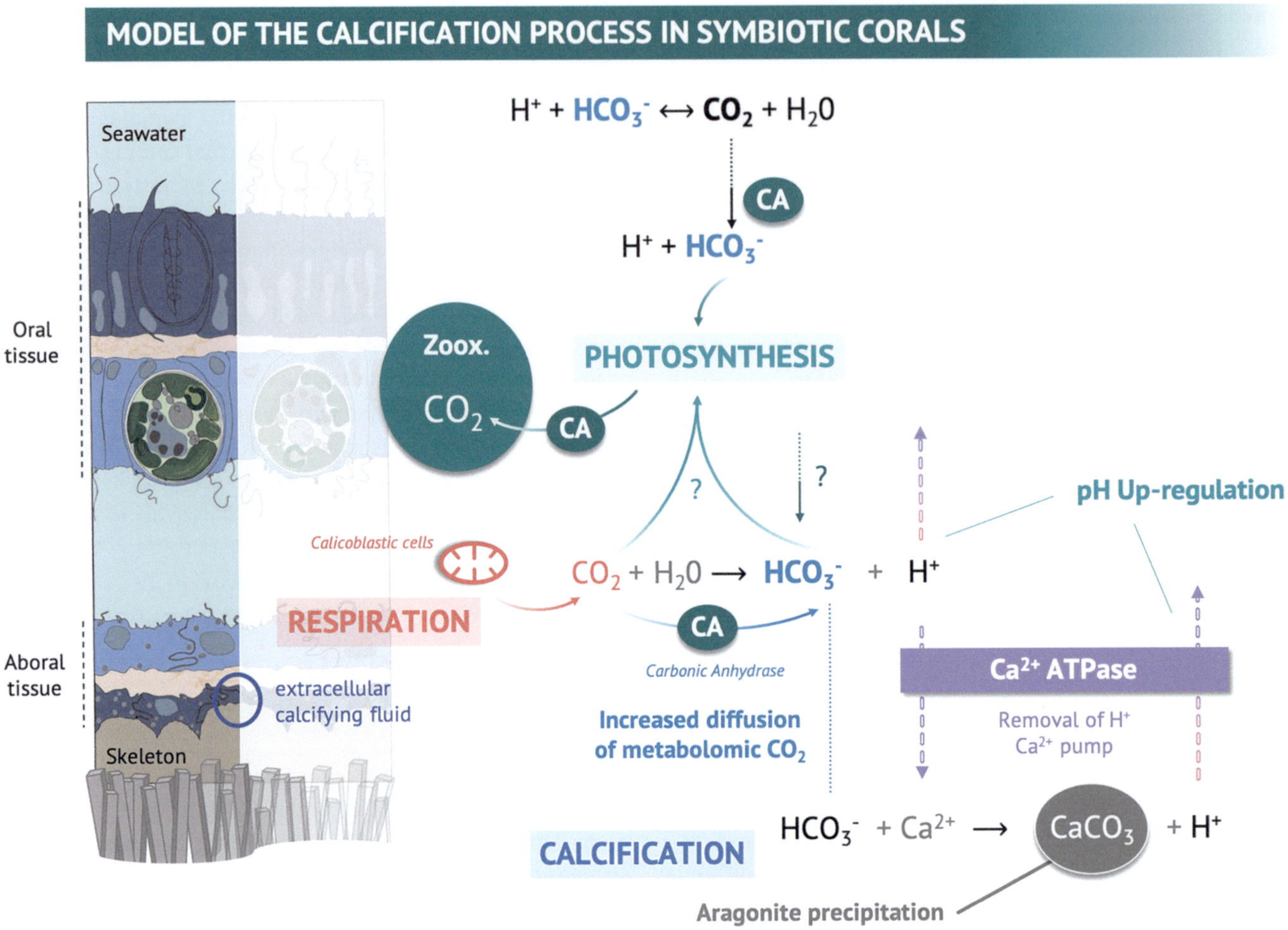

Fig. 3.2 Schematic model of the calcification process in corals occurring within the extracellular calcifying fluid located between the sub-calicoblastic cells and the growing skeleton. (Modified from Furla et al. 2000; Tambutté et al. 2011)

aquarium environments provide a valuable opportunity to investigate calcification in slow-growing corals, such as *Diploastrea*, and to determine whether their responses match with those of faster-growing species under different environmental parameters.

A recent review focused on elucidating the natural processes that have the potential to enhance the adaptive capacity of coral holobionts, with the primary aim of sustaining ecosystem functioning in the face of anticipated oceanic changes (Voolstra et al. 2021). The hypothesis of the natural adaptation capacities of corals deserves to be explored in detail as it holds significant potential for assessing the resistance and adaptive capacity of corals in the face of future global changes (Burt et al. 2020). However, the challenges in understanding the genetic basis of coral calcification and adaptation are complex. The geochemical response of *Porites* to environmental conditions such as a strong seasonality, low pH, and high temperatures remains uncertain, and it is unclear whether corals that have experienced such conditions will respond similarly to those that have not.

The extent to which corals exposed to extreme conditions develop resilience due to genetic heritage remains unclear. A potential memory effect could influence their capacity to adapt to future disturbances, but its impact on geochemical signatures requires further exploration. For instance, a study in Palau showed that *Porites* corals transplanted from an acidified site to a reference site maintained similar calcification rates, whereas those moved from the reference to the acidified site exhibited a decline (Kurihara et al. 2021). Understanding whether corals retain an environmental imprint is crucial for predicting future responses and refining proxy-based reconstructions. If selecting corals for geochemical analysis, would cores from tolerant or sensitive phenotypes yield the most accurate environmental records? Addressing this question is key to improving the reliability of coral-based climate reconstructions.

Comparative genetic studies could help to understand the potential genetic benefits of coral calcification and adaptation to global change. Nevertheless, these studies require a substantial amount of data and are complicated by the fact that genetic traits interact with biological and environmental factors (Putnam 2021), making it challenging to predict how genetic variability will influence coral response to future environmental changes.

3.7 Other Biological Processes Influencing Coral Calcification

Coral holobionts are complex entities in which several essential biological processes of both coral organisms and their symbiotic microalgae occur (such as respiration, photosynthesis, nutrition, and reproduction). Coral-microbiome interactions are closely interwoven to ensure their growth (Fig. 3.3). The process of skeletal formation in corals is characterized by high energy expenditure and is likely related to other energy-dependent processes, including the abovementioned activities. Each of these processes requires energy from the coral, and it is known that their occurrence varies depending on the ecological, biological, and chemical conditions. Furthermore, these biological processes all occur within the coral tissues, which are extremely thin and communicate through the layer walls (Allemand et al. 2011). Given the interconnected circulation of oxygen and carbon within coral tissues, disruptions in one process can impact the others, necessitating varying energy expenditures. A numerical model of energy-dependent processes within a coral polyp has been developed based on the following premise: photosynthesis increases dissolved oxygen levels in the coelenteron (i.e., coral gastrovascular cavity), which, when coupled with light, stimulates respiration; this increased respiration, in turn, fuels calcification through Ca^{2+}-ATPase activity, illustrating the interplay of these processes (Nakamura et al. 2013). The energy produced by corals for their growth is partly attributed to the beneficial symbiosis with photosynthetic dinoflagellates from the Symbiodiniaceae family (Gattuso et al. 1999; Fransolet et al. 2012; Grégoire et al. 2017). Increased photosynthetic activity may result in higher energy and nutrient demands, affecting the availability of these resources for other processes such as calcification. While the impact of photosynthetic activity has been investigated in studies at short temporal resolution, there is a lack of quantification regarding its long-term effects on environmental variations. Hence, the consideration of this parameter in calcification studies and paleoceanography promises new insights.

Moreover, the presence of food sources like *Artemia salina* and copepods also plays a crucial role in coral calcification (Tremblay et al. 2014; Ezzat et al. 2016), and this can vary depending on their feeding frequency, which

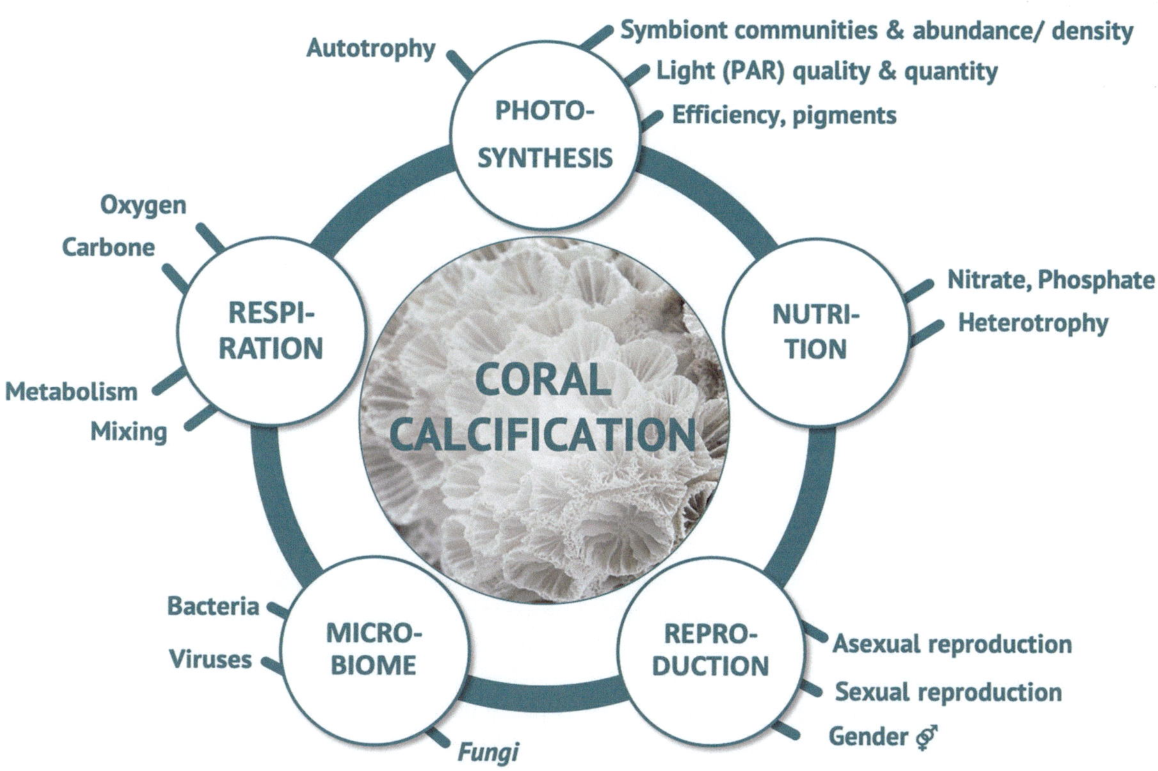

Fig. 3.3 Interactions in the coral meta-organism (holobiont): factors influencing coral calcification and growth

may range from once a week to once a day. Nutrient levels, specifically nitrates and phosphates, also contribute to the coral growth (Ferrier-Pagès et al. 2000). To investigate the impact of these variables, experimental studies can be conducted by systematically modifying individual parameters or by exploring combinations such as light/temperature, light/nutrients, and temperature/nutrients as already suggested in Canesi (2022). By culturing long-life massive corals (e.g., *Diploastrea* and *Porites* colonies) in controlled conditions over an extended period, typically spanning several months, it becomes possible to assess the net influence of these variables, both individually and in combination, on the coral calcification and growth. Analyzing growth parameters and geochemical composition of the coral skeleton at the end of the culture period provides valuable insights into the overall effects of these environmental factors on the skeletal composition.

Furthermore, it has been shown that male *Porites* colonies exhibit larger corallite areas and higher extension and calcification rates than females, indicating a potential gender bias in calcification and growth characteristics which highlights the differential energy requirements for egg and sperm production (Cabral-Tena et al. 2013; Pedraza-Pohlenz et al. 2023). Sequential hermaphroditism, characterized by a transition in sex during growth, is common in scleractinian corals (Harrison 2011), and may potentially introduce natural variations in calcification over time. Another study showed that female corals compensate for the higher energy requirements of gamete production by producing less tissue, leading to variations in the timing of density band formation (Carricart-Ganivet et al. 2013). Physiological changes related to gender or sex change significantly impact coral calcification, yet their integration into climate reconstructions remains overlooked. Investigating isotopic signatures in gonochoric species, such as *Porites*, by comparing males and females within the same environment would provide a first step toward understanding gender-related geochemical variability. This highlights the necessity of sex determination and replication across colonies to refine interpretations, particularly in fossil coral studies.

Definitively, experimental studies conducted in controlled laboratory conditions or seminatural environments, such as mesocosms, offer valuable tools for manipulating and selectively varying individual processes to observe and quantify their impact on coral calcification. By manipulating nutritional inputs (e.g., nitrate, phosphate) or perturbing the coral microbiome, researchers can assess coral responses and establish cause-and-effect relationships with calcification and growth (e.g., Standish et al. 2023). Long-term monitoring studies of coral populations are also crucial for assessing variations in growth and calcification associated with the biological processes studied. This can be achieved through techniques like isotopic labeling, regular growth measurements, and analysis of coral chemical composition in both tissues and skeleton. In addition, advanced molecular techniques like transcriptomic and metabolomic analysis provide insights into the molecular and metabolic changes associated with any biological process. To illustrate, a recent transcriptomic study showed how glyphosate disrupts coral calcification and symbiont nutrient export (Zhou et al. 2022b), while another study revealed altered species interactions between a bioeroding sponge and *Porites* under ocean acidification (DeBiasse et al. 2022). Through a multifaceted approach that integrates different cross-methodologies, it is possible to gain a comprehensive understanding of how biological processes affect coral calcification and growth. This overall perspective allows a thorough exploration of the important biological mechanisms in corals and the identification of key factors that significantly influence their calcification and growth. Conducting trans-disciplinary projects and pooling expertise is essential to unravel the complex relationships among these processes and to incorporate all relevant variables into coral calcification studies. However, a comprehensive, transdisciplinary, and international study of coral calcification and growth on this wild scale has not yet been conducted.

3.8 Conclusion

In light of the discussed limitations and challenges in coral calcification research, it is evident that a comprehensive and multidisciplinary approach is essential for a better understanding of this complex process. Characterization of environmental properties, such as seawater physics, chemistry, and biology, is crucial to unravel the intricate relationships between corals and their surroundings. Additionally, accounting for diagenesis and structural issues at different spatial and temporal scales, as well as skeletal heterogeneity within coral samples, will enhance the accuracy and reliability of paleoceanography or calcification studies and potentially lead to the development of new and/or combined proxies that remove the vital effects. Understanding the effects of coral growth and age-related factors is vital for deciphering the temporal dynamics of calcification and its implications for paleo-reconstructions and reef resilience. Investigating the biological mechanisms underlying the internal regulation of calcifying fluid properties will shed light on the intricate cellular/molecular processes involved in coral skeletal formation. Moreover, considering the species-specific, gender, and genetic aspects will provide insights into the inherent variations and adaptations among different coral taxa. Future studies should not overlook the influence of additional factors on coral growth and calcification, including photosynthesis, nutrition, respiration, reproduction, and the coral microbiome (see Fig. 3.3). Considering these factors in conjunction with environmental parameters will lead to a more comprehensive understanding of the complex interactions governing coral calcification. Strategic samplings, incorporating both spatial and temporal dimensions, will enable a more robust assessment of calcification dynamics in different coral habitats. By addressing these challenges and embracing a whole-system perspective, we can uncover the underlying mechanisms of coral calcification and its responses to future global changes.

3.9 Outlook

To address future challenges in coral calcification research, two promising coral collection strategies have emerged from this synthesis and discussion. Firstly, a wide-scale collection of recent skeletal portions and associated tissues, as described in Fig. 3.4, offers an opportunity for various analyses of both organic tissues and mineral skeletons obtained from the same colonies in different marine environments. This approach will provide new insights into the biological mechanisms involved in calcification modulation, linking them to environmental parameters. Secondly, studying corals grown

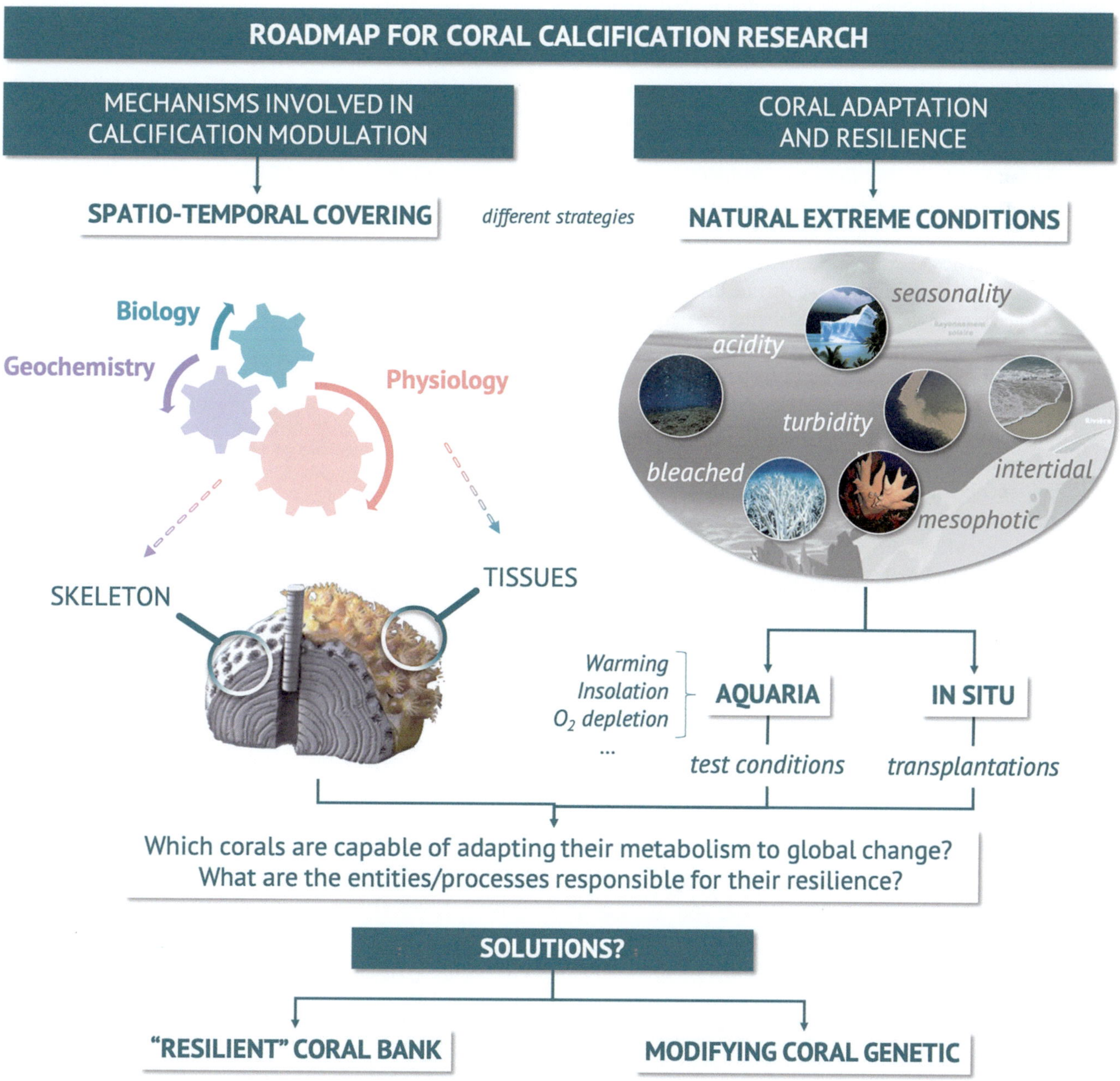

Fig. 3.4 Roadmap for coral calcification research: experimental approaches unveiling biological mechanisms and resilience to global change

under naturally extreme conditions, such as acidified environments, intertidal zones, bleaching events, high turbidity, or high seasonal variations (refer to Fig. 3.4), provides a valuable opportunity to understand the potential resilience and adaptation of corals.

Comparing coral growth in mesophotic and deep-sea environments, while evaluating the impact of light on photosynthesis and translocation, as well as temperature and UV intensity in nontropical settings, offers more than just descriptive insights. This approach can reveal new research avenues into corals' physiological and biological adaptations across diverse environments. By monitoring their growth under controlled conditions in aquariums, where one or more parameters can be manipulated, or through transplantation to different sites in the natural environment, we can achieve higher spatiotemporal resolution and enhance our understanding of how "resilient" corals adapt to changing environments. Furthermore, specific resilience components extracted from these corals could be transplanted into more vulnerable coral tissues to test whether their new resilience capacities are sufficient. This approach has been successfully demonstrated in a recent study, where bacteria were used to increase the climate resilience of corals (Maire and van Oppen 2022). Another promising possibility is to cultivate these "resistant" corals in aquaria, and then transplant them into natural environments (Fig. 3.4).

Numerous initiatives are currently focusing on coral calcification and resilience. For example, the recent program COR-Resilience (2023–2028), funded by the French National Research Agency, fosters global research collaborations and encourages environmentalists, biologists, geochemists, and modelers to work together to understand coral resilience. This collaborative effort, focusing mainly on the massive corals *Porites* and *Dipoastrea,* will unravel the resilience of corals and hopefully enable the development of effective strategies to preserve these vulnerable ecosystems for future generations.

References

Alaguarda D, Brajard J, Coulibaly G et al (2022) 54 years of microboring community history explored by machine learning in a massive coral from Mayotte (Indian Ocean). Front Mar Sci 9. https://doi.org/10.3389/fmars.2022.899398

Allemand D, TambuttE E, Girard JP, Jaubert J (1998) Organic matrix synthesis in the scleractinian coral stylophora pistillata: role in biomineralization and potential target of the organotin tributyltin. J Exp Biol 201(Pt 13):2001–2009. https://doi.org/10.1242/jeb.201.13.2001

Allemand D, Tambutté É, Zoccola D, Tambutté S (2011) Coral calcification, cells to reefs. In: Dubinsky Z, Stambler N (eds) Coral reefs: an ecosystem in transition. Springer, Dordrecht, pp 119–150

Allison N, Finch AA (2009) Reproducibility of minor and trace element determinations in Porites coral skeletons by secondary ion mass spectrometry. Geochem Geophys Geosyst 10. https://doi.org/10.1029/2008GC002239

Allison N, Finch AA, Webster JM, Clague DA (2007) Palaeoenvironmental records from fossil corals: the effects of submarine diagenesis on temperature and climate estimates. Geochim Cosmochim Acta 71:4693–4703. https://doi.org/10.1016/j.gca.2007.07.026

Alpert AE, Cohen AL, Oppo DW et al (2017) Twentieth century warming of the tropical Atlantic captured by Sr-U paleothermometry. Paleoceanography 32:146–160. https://doi.org/10.1002/2016PA002976

Alvarado JJ, Grassian B, Cantera-Kintz JR et al (2017) Coral reef bioerosion in the Eastern Tropical Pacific. In: Glynn PW, Manzello DP, Enochs IC (eds) Coral reefs of the Eastern Tropical Pacific: persistence and loss in a dynamic environment. Springer, Dordrecht, pp 369–403

Badran MI (2001) Dissolved oxygen, chlorophyll a and nutrients: seasonal cycles in waters of the Gulf of Aquaba, Red Sea. Aquat Ecosyst Health Manage 4:139–150. https://doi.org/10.1080/14634980127711

Bagnato S, Linsley B, Howe S et al (2004) Evaluating the use of the massive coral Diploastrea heliopora for paleoclimate reconstruction. https://doi.org/10.1029/2003PA000935

Bahr KD, Rodgers KS, Jokiel PL (2018) Ocean warming drives decline in coral metabolism while acidification highlights species-specific responses. Mar Biol Res 14:924–935. https://doi.org/10.1080/17451000.2018.1551616

Barnes DJ, Lough JM (1996) Coral skeletons: storage and recovery of environmental information. Glob Chang Biol 2:569–582. https://doi.org/10.1111/j.1365-2486.1996.tb00068.x

Bautista VIIAT, Limlingan SJM, Bauyon MMT et al (2021) 3D X-ray computed tomography gray value and age model datasets of coral cores Baler 2 and 3 (Philippines). Data Brief 34:106755. https://doi.org/10.1016/j.dib.2021.106755

Benzerara K, Menguy N, Obst M et al (2011) Study of the crystallographic architecture of corals at the nanoscale by scanning transmission X-ray microscopy and transmission electron microscopy. Ultramicroscopy 111:1268–1275. https://doi.org/10.1016/j.ultramic.2011.03.023

Bove CB, Whitehead RF, Szmant AM (2020) Responses of coral gastrovascular cavity pH during light and dark incubations to reduced seawater pH suggest species-specific responses to the effects of ocean acidification on calcification. Coral Reefs 39:1675–1691. https://doi.org/10.1007/s00338-020-01995-7

Brahmi C, Kopp C, Domart-Coulon I et al (2012) Skeletal growth dynamics linked to trace-element composition in the scleractinian coral Pocillopora damicornis. Geochim Cosmochim Acta 99:146–158. https://doi.org/10.1016/j.gca.2012.09.031

Burt JA, Camp EF, Enochs IC et al (2020) Insights from extreme coral reefs in a changing world. Coral Reefs 39:495–507. https://doi.org/10.1007/s00338-020-01966-y

Cabral-Tena RA, Reyes-Bonilla H, Lluch-Cota S et al (2013) Different calcification rates in males and females of the coral Porites panamensis in the Gulf of California. Mar Ecol Prog Ser 476:1–8. https://doi.org/10.3354/meps10269

Cai W-J, Ma Y, Hopkinson BM et al (2016) Microelectrode characterization of coral daytime interior pH and carbonate chemistry. Nat Commun 7:11144. https://doi.org/10.1038/ncomms11144

Canesi M (2022) Impacts des changements globaux sur les coraux massifs Porites et Diploastrea de l'océan Pacifique. These de doctorat, université Paris-Saclay

Canesi M, Douville E, Bordier L et al (2021) Growth parameters affect geochemical proxies in massive corals: impacts on climate reconstructions [Poster presentation]. Goldschmidt Conference

Canesi M, Douville E, Montagna P et al (2023) Differences in carbonate chemistry up-regulation of long-lived reef-building corals. Sci Rep 13:11589. https://doi.org/10.1038/s41598-023-37598-9

Canesi M, Douville É, Bordier L et al (2024a) Porites' coral calcifying fluid chemistry regulation under normal- and low-pH seawater conditions in Palau Archipelago: impacts on growth properties. Sci Total Environ 911:168552. https://doi.org/10.1016/j.scitotenv.2023.168552

Canesi M, Douville E, Montagna P et al (2024b) Sea surface temperature reconstruction in the Pacific Ocean using multi-elemental proxy in *Porites* and *Diploastrea* corals: application to Palau Archipelago. Chem Geol 645:121884. https://doi.org/10.1016/j.chemgeo.2023.121884

Cantin NE, Lough JM (2014) Surviving coral bleaching events: porites growth anomalies on the great barrier reef. PLoS One 9:e88720. https://doi.org/10.1371/journal.pone.0088720

Carricart-Ganivet JP, Vásquez-Bedoya LF, Cabanillas-Terán N, Blanchon P (2013) Gender-related differences in the apparent timing of skeletal density bands in the reef-building coral Siderastrea siderea. Coral Reefs 32:769–777. https://doi.org/10.1007/s00338-013-1028-y

Chalk TB, Standish CD, D'Angelo C et al (2021) Mapping coral calcification strategies from in situ boron isotope and trace element measurements of the tropical coral Siderastrea siderea. Sci Rep 11:472. https://doi.org/10.1038/s41598-020-78778-1

Clampitt M (2023) Indexation de l'état de santé des coraux par une approche basée sur l'intelligence artificielle. These de doctorat, Université Côte d'Azur

Clode PL, Marshall AT (2002) Low temperature X-ray microanalysis of calcium in a scleractinian coral: evidence of active transport mechanisms. J Exp Biol 205:3543–3552. https://doi.org/10.1242/jeb.205.22.3543

Cohen AL, McConnaughey TA (2003) Geochemical perspectives on coral mineralization. Rev Mineral Geochem 54:151–187. https://doi.org/10.2113/0540151

Colombo-Pallotta MF, Rodríguez-Román A, Iglesias-Prieto R (2010) Calcification in bleached and unbleached Montastraea faveolata: evaluating the role of oxygen and glycerol. Coral Reefs 29:899–907. https://doi.org/10.1007/s00338-010-0638-x

Comeau S, Cornwall CE, DeCarlo TM et al (2018) Similar controls on calcification under ocean acidification across unrelated coral reef taxa. Glob Chang Biol 24:4857–4868. https://doi.org/10.1111/gcb.14379

Cornwall CE, Comeau S, Kornder NA et al (2021) Global declines in coral reef calcium carbonate production under ocean acidification and warming. Proc Natl Acad Sci 118:e2015265118. https://doi.org/10.1073/pnas.2015265118

Cornwall CE, Comeau S, Putnam H, Schoepf V (2022) Impacts of ocean warming and acidification on calcifying coral reef taxa: mechanisms responsible and adaptive capacity. Emerg Top Life Sci 6:1–9. https://doi.org/10.1042/ETLS20210226

Corrège T (2006) Sea surface temperature and salinity reconstruction from coral geochemical tracers. Palaeogeogr Palaeoclimatol Palaeoecol 232:408–428. https://doi.org/10.1016/j.palaeo.2005.10.014

Cuif JP, Dauphin Y (2005) The environment recording unit in coral skeletons–a synthesis of structural and chemical evidences for a biochemically driven, stepping-growth process in fibres. Biogeosciences 2:61–73. https://doi.org/10.5194/bg-2-61-2005

Cuif J-P, Dauphin Y, Doucet J et al (2003) XANES mapping of organic sulfate in three scleractinian coral skeletons. Geochim Cosmochim Acta 67:75–83. https://doi.org/10.1016/S0016-7037(02)01041-4

Cuny-Guirriec K, Douville E, Reynaud S et al (2019) Coral Li/Mg thermometry: caveats and constraints. Chem Geol 523:162–178. https://doi.org/10.1016/j.chemgeo.2019.03.038

D'Olivo JP, Sinclair DJ, Rankenburg K, McCulloch MT (2018) A universal multi-trace element calibration for reconstructing sea surface temperatures from long-lived Porites corals: removing 'vital-effects'. Geochim Cosmochim Acta 239:109–135. https://doi.org/10.1016/j.gca.2018.07.035

D'Olivo JP, Ellwood G, DeCarlo TM, McCulloch MT (2019) Deconvolving the long-term impacts of ocean acidification and warming on coral biomineralisation. Earth Planet Sci Lett 526:115785. https://doi.org/10.1016/j.epsl.2019.115785

de Villiers S, Nelson BK, Chivas AR (1995) Biological controls on coral Sr/Ca and δ18O reconstructions of sea surface temperatures. Science 269:1247–1249. https://doi.org/10.1126/science.269.5228.1247

DeBiasse MB, Stubler AD, Kelly MW (2022) Comparative transcriptomics reveals altered species interaction between the bioeroding sponge Cliona varians and the coral Porites furcata under ocean acidification. Mol Ecol 31:3002–3017. https://doi.org/10.1111/mec.16432

DeCarlo TM (2018) Characterizing coral skeleton mineralogy with Raman spectroscopy. Nat Commun 9:5325. https://doi.org/10.1038/s41467-018-07601-3

DeCarlo TM, Gaetani GA, Holcomb M, Cohen AL (2015) Experimental determination of factors controlling U/Ca of aragonite precipitated from seawater: implications for interpreting coral skeleton. Geochim Cosmochim Acta 162:151–165. https://doi.org/10.1016/j.gca.2015.04.016

DeCarlo TM, Gaetani GA, Cohen AL et al (2016) Coral Sr-U thermometry. Paleoceanogr 31:626–638

DeCarlo TM, Comeau S, Cornwall CE, McCulloch MT (2018) Coral resistance to ocean acidification linked to increased calcium at the site of calcification. Proc R Soc B Biol Sci 285:20180564. https://doi.org/10.1098/rspb.2018.0564

DeLong KL, Flannery JA, Maupin CR et al (2011) A coral Sr/Ca calibration and replication study of two massive corals from the Gulf of Mexico. Palaeogeogr Palaeoclimatol Palaeoecol 307:117–128. https://doi.org/10.1016/j.palaeo.2011.05.005

Descombes P, Wisz MS, Leprieur F et al (2015) Forecasted coral reef decline in marine biodiversity hotspots under climate change. Glob Chang Biol 21:2479–2487. https://doi.org/10.1111/gcb.12868

Douville E, Paterne M, Cabioch G et al (2010) Abrupt sea surface pH change at the end of the Younger Dryas in the central subequatorial Pacific inferred from boron isotope abundance in corals (Porites). Biogeosciences 7:2445–2459. https://doi.org/10.5194/bg-7-2445-2010

Dove SG, Brown KT, Van Den Heuvel A et al (2020) Ocean warming and acidification uncouple calcification from calcifier biomass which accelerates coral reef decline. Commun Earth Environ 1:1–9. https://doi.org/10.1038/s43247-020-00054-x

Duprey NN, Foreman AD, Carriquiry JD et al (2024) Decadal oscillations in the ocean's largest oxygen-deficient zone. Science 386:1019–1024. https://doi.org/10.1126/science.adk4965

Eagle RA, Guillermic M, De Corte I et al (2022) Physicochemical control of Caribbean coral calcification linked to host and symbiont responses to varying pCO2 and temperature. J Mar Sci Eng 10:1075. https://doi.org/10.3390/jmse10081075

Eddy TD, Lam VWY, Reygondeau G et al (2021) Global decline in capacity of coral reefs to provide ecosystem services. One Earth 4:1278–1285. https://doi.org/10.1016/j.oneear.2021.08.016

Emiliani C, Hudson JH, Shinn E, George R (1978) Oxygen and carbon isotopic growth record in a reef coral from the Florida keys and a Deep-Sea Coral from Blake Plateau. Science 202:627. https://doi.org/10.1126/science.202.4368.627

Erez J (1978) Vital effect on stable-isotope composition seen in foraminifera and coral skeletons. Nature 273:199–202. https://doi.org/10.1038/273199a0

Ezzat L, Towle E, Irisson J-O et al (2016) The relationship between heterotrophic feeding and inorganic nutrient availability in the

scleractinian coral T. reniformis under a short-term temperature increase. Limnol Oceanogr 61:89–102. https://doi.org/10.1002/lno.10200
Fabricius KE (2011) Factors determining the resilience of coral reefs to eutrophication: a review and conceptual model. In: Dubinsky Z, Stambler N (eds) Coral reefs: an ecosystem in transition. Springer, Dordrecht, pp 493–505
Farre B, Cuif J-P, Dauphin Y (2010) Occurrence and diversity of lipids in modern coral skeletons. Zoology 113:250–257. https://doi.org/10.1016/j.zool.2009.11.004
Felis T, Pätzold J, Loya Y (2003) Mean oxygen-isotope signatures in Porites spp. corals: inter-colony variability and correction for extension-rate effects. Coral Reefs 22:328–336. https://doi.org/10.1007/s00338-003-0324-3
Ferrier-Pagès C, Gattuso J-P, Dallot S, Jaubert J (2000) Effect of nutrient enrichment on growth and photosynthesis of the zooxanthellate coral Stylophora pistillata. Coral Reefs 19:103–113. https://doi.org/10.1007/s003380000078
Flannery JA, Richey JN, Thirumalai K et al (2017) Multi-species coral Sr/Ca-based sea-surface temperature reconstruction using Orbicella faveolata and Siderastrea siderea from the Florida Straits. Palaeogeogr Palaeoclimatol Palaeoecol 466:100–109. https://doi.org/10.1016/j.palaeo.2016.10.022
Flannery JA, Richey JN, Toth LT et al (2018) Quantifying uncertainty in Sr/Ca-based estimates of SST from the coral Orbicella faveolata. Paleoceanogr Paleoclimatol 33:958–973. https://doi.org/10.1029/2018PA003389
Fowell SE, Sandford K, Stewart JA et al (2016) Intrareef variations in Li/Mg and Sr/Ca sea surface temperature proxies in the Caribbean reef-building coral Siderastrea siderea. Paleoceanography 31:1315–1329. https://doi.org/10.1002/2016PA002968
Fransolet D, Roberty S, Plumier J-C (2012) Establishment of endosymbiosis: the case of cnidarians and Symbiodinium. J Exp Mar Biol Ecol 420–421:1–7. https://doi.org/10.1016/j.jembe.2012.03.015
Furla P, Galgani I, Durand I, Allemand D (2000) Sources and mechanisms of inorganic carbon transport for coral calcification and photosynthesis. J Exp Biol 203:3445–3457. https://doi.org/10.1242/jeb.203.22.3445
Gates RD, Edmunds PJ (1999) The physiological mechanisms of acclimatization in tropical reef corals. Am Zool 39:30–43
Gattuso J-P, Frankignoulle M, Bourge I et al (1998) Effect of calcium carbonate saturation of seawater on coral calcification. Glob Planet Chang 18:37–46. https://doi.org/10.1016/S0921-8181(98)00035-6
Gattuso J-P, Allemand D, Frankignoulle M (1999) Photosynthesis and calcification at cellular, organismal and community levels in coral reefs: a review on interactions and control by carbonate chemistry. Am Zool 39:160–183. https://doi.org/10.1093/icb/39.1.160
Gibbin EM, Putnam HM, Gates RD et al (2015) Species-specific differences in thermal tolerance may define susceptibility to intracellular acidosis in reef corals. Mar Biol 162:717–723. https://doi.org/10.1007/s00227-015-2617-9
Glynn PW, Manzello DP (2015) Bioerosion and coral reef growth: a dynamic balance. In: Birkeland C (ed) Coral reefs in the Anthropocene. Springer, Dordrecht, pp 67–97
Goreau TJ (1977) Coral skeletal chemistry: physiological and environmental regulation of stable isotopes and trace metals in Montastrea annularis. Proc R Soc Lond B Biol Sci 196:291–315. https://doi.org/10.1098/rspb.1977.0042
Grégoire V, Schmacka F, Coffroth MA, Karsten U (2017) Photophysiological and thermal tolerance of various genotypes of the coral endosymbiont Symbiodinium sp. (Dinophyceae). J Appl Phycol 29:1893–1905. https://doi.org/10.1007/s10811-017-1127-1
Guo W, Bokade R, Cohen AL et al (2020) Ocean acidification has impacted coral growth on the great barrier reef. Geophys Res Lett 47:e2019GL086761. https://doi.org/10.1029/2019GL086761
Hall NM, Berry KLE, Rintoul L, Hoogenboom MO (2015) Microplastic ingestion by scleractinian corals. Mar Biol 162:725–732. https://doi.org/10.1007/s00227-015-2619-7
Harrison PL (2011) Sexual reproduction of Scleractinian corals. In: Dubinsky Z, Stambler N (eds) Coral reefs: an ecosystem in transition. Springer, Dordrecht, pp 59–85
Hasumoto H, Imazu T, Miura T, Kogure K (2006) Use of an optical oxygen sensor to measure dissolved oxygen in seawater. J Oceanogr 62:99–103. https://doi.org/10.1007/s10872-006-0036-8
Hayashi E, Suzuki A, Nakamura T et al (2013) Growth-rate influences on coral climate proxies tested by a multiple colony culture experiment. Earth Planet Sci Lett 362:198–206. https://doi.org/10.1016/j.epsl.2012.11.046
Hemming NG, Hanson GN (1992) Boron isotopic composition and concentration in modern marine carbonates. Geochim Cosmochim Acta 56:537–543. https://doi.org/10.1016/0016-7037(92)90151-8
Hoegh-Guldberg O (2011) Coral reef ecosystems and anthropogenic climate change. Reg Environ Chang 11:215–227. https://doi.org/10.1007/s10113-010-0189-2
Holcomb M, Venn AA, Tambutté E et al (2014) Coral calcifying fluid pH dictates response to ocean acidification. Sci Rep 4:5207. https://doi.org/10.1038/srep05207
Hönisch B, Hemming NG, Grottoli AG et al (2004) Assessing scleractinian corals as recorders for paleo-pH: empirical calibration and vital effects. Geochim Cosmochim Acta 68:3675–3685. https://doi.org/10.1016/j.gca.2004.03.002
Houlbrèque F, Tambutté E, Ferrier-Pagès C (2003) Effect of zooplankton availability on the rates of photosynthesis, and tissue and skeletal growth in the scleractinian coral Stylophora pistillata. J Exp Mar Biol Ecol 296:145–166. https://doi.org/10.1016/S0022-0981(03)00259-4
Howells EJ, Dunshea G, McParland D et al (2018) Species-specific coral calcification responses to the extreme environment of the southern Persian Gulf. Front Mar Sci 5:56
Inoue M, Suzuki A, Hantoro WS, Kawahata H (2007) Mg/Ca ratio in a coral skeleton as a possible proxy for the skeletal growth rate. AGU Fall Meeting Abstracts 2007:PP31B–P0417
Kleypas JA, McManus JW, Meñez LAB (1999) Environmental limits to coral reef development: where do we draw the line? Am Zool 39:146–159
Kleypas JA, Buddemeier RW, Gattuso J-P (2001) The future of coral reefs in an age of global change. Int J Earth Sci 90:426–437. https://doi.org/10.1007/s005310000125
Koubarakis M, Caumont H, Daniels U et al (2018) Copernicus app lab: a platform for easy data access connecting the scientific earth observation community with mobile developers. In: Companion proceedings of the the web conference 2018. International World Wide Web Conferences Steering Committee, Republic and Canton of Geneva, CHE, pp 435–436
Kurihara H, Watanabe A, Tsugi A et al (2021) Potential local adaptation of corals at acidified and warmed Nikko Bay, Palau. Sci Rep 11:11192. https://doi.org/10.1038/s41598-021-90614-8
Lamb JB, Willis BL, Fiorenza EA et al (2018) Plastic waste associated with disease on coral reefs. Science 359:460–462. https://doi.org/10.1126/science.aar3320
Lauvset SK, Lange N, Tanhua T et al (2021) An updated version of the global interior ocean biogeochemical data product, GLODAPv2.2021. Earth Syst Sci Data 13:5565–5589. https://doi.org/10.5194/essd-13-5565-2021
Lazareth CE, Soares-Pereira C, Douville E et al (2016) Intra-skeletal calcite in a live-collected *Porites* sp.: impact on environmental proxies and potential formation process. Geochim Cosmochim Acta 176:279–294
Lee J-M, Boyle EA, Suci Nurhati I et al (2014) Coral-based history of lead and lead isotopes of the surface Indian Ocean since the

mid-20th century. Earth Planet Sci Lett 398:37–47. https://doi.org/10.1016/j.epsl.2014.04.030

Levy O, Appelbaum L, Leggat W et al (2007) Light-responsive cryptochromes from a simple multicellular animal, the coral Acropora millepora. Science 318:467–470. https://doi.org/10.1126/science.1145432

Lough JM (2004) A strategy to improve the contribution of coral data to high-resolution paleoclimatology. Palaeogeogr Palaeoclimatol Palaeoecol 204:115–143. https://doi.org/10.1016/S0031-0182(03)00727-2

Lough JM (2008) Coral calcification from skeletal records revisited. Mar Ecol Prog Ser 373:257–264. https://doi.org/10.3354/meps07398

Lough JM (2010) Climate records from corals. WIREs Clim Change 1:318–331. https://doi.org/10.1002/wcc.39

Lough JM, Cooper TF (2011) New insights from coral growth band studies in an era of rapid environmental change. Earth Sci Rev 108:170–184. https://doi.org/10.1016/j.earscirev.2011.07.001

Maier C, Felis T, Pätzold J, Bak RPM (2004) Effect of skeletal growth and lack of species effects in the skeletal oxygen isotope climate signal within the coral genus Porites. Mar Geol 207:193–208. https://doi.org/10.1016/j.margeo.2004.03.008

Maire J, van Oppen MJH (2022) A role for bacterial experimental evolution in coral bleaching mitigation? Trends Microbiol 30:217–228. https://doi.org/10.1016/j.tim.2021.07.006

Martinez A, Crook ED, Barshis DJ et al (2019) Species-specific calcification response of Caribbean corals after 2-year transplantation to a low aragonite saturation submarine spring. Proc R Soc B Biol Sci 286:20190572. https://doi.org/10.1098/rspb.2019.0572

McConnaughey T (1989) Biomineralization mechanisms. In: Crick RE (ed) Origin, evolution, and modern aspects of biomineralization in plants and animals. Springer, Boston, pp 57–73

McCulloch MT, Trotter J, Montagna P et al (2012) Resilience of cold-water scleractinian corals to ocean acidification: boron isotopic systematics of pH and saturation state up-regulationregulation. Geochim Cosmochim Acta 87:21–34. https://doi.org/10.1016/j.gca.2012.03.027

Meibom A, Mostefaoui S, Cuif J-P et al (2007) Biological forcing controls the chemistry of reef-building coral skeleton. Geophys Res Lett 34. https://doi.org/10.1029/2006GL028657

Mishima T, Nakamura H (2009) Pangea: an eager database replication middleware guaranteeing snapshot isolation without modification of database servers. Proc VLDB Endow 2:1066–1077. https://doi.org/10.14778/1687627.1687747

Mollica NR, Guo W, Cohen AL et al (2018) Ocean acidification affects coral growth by reducing skeletal density. PNAS 115:1754–1759. https://doi.org/10.1073/pnas.1712806115

Montagna P, Douville E (2020) Geochemical proxies in marine biogenic carbonates: new developments and applications to global change (preface). Spec Issue Chem Geol 533:119411. https://doi.org/10.1016/j.chemgeo.2019.119411

Montagna P, Mcculloch M, Douville E et al (2014) Li/Mg systematics in scleractinian corals: calibration of the thermometer. Geochim Cosmochim Acta 132:288–310. https://doi.org/10.1016/j.gca.2014.02.005

Muscatine L, Tambutte E, Allemand D (1997) Morphology of coral desmocytes, cells that anchor the calicoblastic epithelium to the skeleton. Coral Reefs 16:205–213. https://doi.org/10.1007/s003380050075

Mw M (1995) Growth of a temperate coral: effects of temperature, light, depth, and heterotrophy. Mar Ecol Prog Ser 122:217–225. https://doi.org/10.3354/meps122217

Nakamura T, Nadaoka K, Watanabe A (2013) A coral polyp model of photosynthesis, respiration and calcification incorporating a transcellular ion transport mechanism. Coral Reefs 32:779–794. https://doi.org/10.1007/s00338-013-1032-2

Nothdurft LD, Webb GE (2009) Earliest diagenesis in scleractinian coral skeletons: implications for palaeoclimate-sensitive geochemical archives. Facies 55:161–201. https://doi.org/10.1007/s10347-008-0167-z

Nothdurft LD, Webb GE, Bostrom T, Rintoul L (2007) Calcite-filled borings in the most recently deposited skeleton in live-collected Porites (Scleractinia): implications for trace element archives. Geochim Cosmochim Acta 71:5423–5438. https://doi.org/10.1016/j.gca.2007.09.025

Okazaki RR, Towle EK, van Hooidonk R et al (2017) Species-specific responses to climate change and community composition determine future calcification rates of Florida keys reefs. Glob Chang Biol 23:1023–1035. https://doi.org/10.1111/gcb.13481

Pakhnevich AV (2021) Bimineral skeleton of Acoscinopleura crassa (Bryozoa, Cheilostomata) from the Maastrichtian of Germany revealed using X-ray microtomography. Paleontol J 55:529–537. https://doi.org/10.1134/S0031030121050117

Paterne M, Druffel ERM, Guilderson TP et al (2023) Pulses of South Atlantic water into the tropical North Atlantic since 1825 from coral isotopes. Sci Adv 9:eadi1687. https://doi.org/10.1126/sciadv.adi1687

Pedraza-Pohlenz RM, Balart EF, Tortolero-Langarica JJA et al (2023) Sexual dimorphism in corallite size and modularity of a broadcast spawning coral, Porites lobata. Front Mar Sci 9:1068391

Porter J, Tougas J (2001) Reef ecosystems: threats to their biodiversity. Enc Biodiv 5:73–95. https://doi.org/10.1016/B0-12-226865-2/00229-7

Putnam HM (2021) Avenues of reef-building coral acclimatization in response to rapid environmental change. J Exp Biol 224:jeb239319. https://doi.org/10.1242/jeb.239319

Puverel S, Tambutté E, Pereira-Mouriès L et al (2005) Soluble organic matrix of two Scleractinian corals: partial and comparative analysis. Comp Biochem Physiol B: Biochem Mol Biol 141:480–487. https://doi.org/10.1016/j.cbpc.2005.05.013

Quinn TM, Sampson DE (2002) A multiproxy approach to reconstructing sea surface conditions using coral skeleton geochemistry. Paleoceanography 17:14–1–14–11. https://doi.org/10.1029/2000PA000528

Quinn TM, Taylor FW (2006) SST artifacts in coral proxy records produced by early marine diagenesis in a modern coral from Rabaul, Papua New Guinea. Geophys Res Lett 33. https://doi.org/10.1029/2005GL024972

Rahman MA, Halfar J, Shinjo R (2013) X-ray diffraction is a promising tool to characterize coral skeletons. Adv Mater Phys Chem 3:120–125. https://doi.org/10.4236/ampc.2013.31A015

Raybaud V, Tambutté S, Ferrier-Pagès C et al (2017) Computing the carbonate chemistry of the coral calcifying medium and its response to ocean acidification. J Theor Biol 424:26–36. https://doi.org/10.1016/j.jtbi.2017.04.028

Reed EV, Thompson DM, Cole JE et al (2021) Impacts of coral growth on geochemistry: lessons from the Galápagos Islands. Paleoceanogr Paleoclimatol 36:e2020PA004051. https://doi.org/10.1029/2020PA004051

Reynaud S, Ferrier-Pagès C, Sambrotto R et al (2002) Effect of feeding on the carbon and oxygen isotopic composition in the tissues and skeleton of the zooxanthellate coral Stylophora pistillata. Mar Ecol Prog Ser 238:81–89. https://doi.org/10.3354/meps238081

Reynaud S, Hemming NG, Juillet-Leclerc A, Gattuso J-P (2004) Effect of pCO2 and temperature on the boron isotopic composition of the zooxanthellate coral Acropora sp. Coral Reefs 23:539–546. https://doi.org/10.1007/s00338-004-0399-5

Reynaud-Vaganay S, Juillet-Leclerc A, Jaubert J, Gattuso J-P (2001) Effect of light on skeletal δ13C and δ18O, and interaction with photosynthesis, respiration and calcification in two zooxanthellate scleractinian corals. Palaeogeogr Palaeoclimatol Palaeoecol 175:393–404. https://doi.org/10.1016/S0031-0182(01)00382-0

Reznikov N, Buss DJ, Provencher B et al (2020) Deep learning for 3D imaging and image analysis in biomineralization research. J Struct Biol 212:107598. https://doi.org/10.1016/j.jsb.2020.107598

Rodolfo-Metalpa R, Peirano A, Houlbrèque F et al (2008) Effects of temperature, light and heterotrophy on the growth rate and budding of the temperate coral Cladocora caespitosa. Coral Reefs 27:17–25. https://doi.org/10.1007/s00338-007-0283-1

Rodrigues LJ, Grottoli AG (2006) Calcification rate and the stable carbon, oxygen, and nitrogen isotopes in the skeleton, host tissue, and zooxanthellae of bleached and recovering Hawaiian corals. Geochim Cosmochim Acta 70:2781–2789. https://doi.org/10.1016/j.gca.2006.02.014

Rollion-Bard C, Blamart D (2015) Possible controls on Li, Na, and Mg incorporation into aragonite coral skeletons. Chem Geol 396:98–111. https://doi.org/10.1016/j.chemgeo.2014.12.011

Rollion-Bard C, Blamart D, Cuif J-P, Juillet-Leclerc A (2003) Microanalysis of C and O isotopes of azooxanthellate and zooxanthellate corals by ion microprobe. Coral Reefs 22:405–415. https://doi.org/10.1007/s00338-003-0347-9

Rollion-Bard C, Chaussidon M, France-Lanord C (2011) Biological control of internal pH in scleractinian corals: implications on paleo-pH and paleo-temperature reconstructions. Compt Rendus Geosci 343:397–405. https://doi.org/10.1016/j.crte.2011.05.003

Ross CL, DeCarlo TM, McCulloch MT (2019) Calibration of Sr/Ca, Li/Mg and Sr-U Paleothermometry in branching and foliose corals. Paleoceanogr Paleoclimatol 34:1271–1291. https://doi.org/10.1029/2018PA003426

Ross CL, Warnes A, Comeau S et al (2022) Coral calcification mechanisms in a warming ocean and the interactive effects of temperature and light. Commun Earth Environ 3:1–11. https://doi.org/10.1038/s43247-022-00396-8

Saha N, Webb GE, Zhao J-X (2016) Coral skeletal geochemistry as a monitor of inshore water quality. Sci Total Environ 566–567:652–684. https://doi.org/10.1016/j.scitotenv.2016.05.066

Sarmento O, Freitas T, Monteiro N et al (2020) Effect of diagenesis on the microstructure of fossil coral skeletons from Atauro Island. Timor-Leste 3:84–92

Schoepf V, McCulloch MT, Warner ME et al (2014) Short-term coral bleaching is not recorded by skeletal boron isotopes. PLoS One 9:e112011. https://doi.org/10.1371/journal.pone.0112011

Schönberg CHL, Fang JKH, Carreiro-Silva M et al (2017) Bioerosion: the other ocean acidification problem. ICES J Mar Sci 74:895–925. https://doi.org/10.1093/icesjms/fsw254

Sevilgen DS, Venn AA, Hu MY et al (2019) Full in vivo characterization of carbonate chemistry at the site of calcification in corals. Sci Adv 5:eaau7447. https://doi.org/10.1126/sciadv.aau7447

Spalding MD (2001) World atlas of coral reefs. UNEP-WCMC

Standish CD, Chalk TB, Saeed M et al (2023) Geochemical responses of scleractinian corals to nutrient stress. Geochim Cosmochim Acta 351:108–124. https://doi.org/10.1016/j.gca.2023.04.011

Stephans CL, Quinn TM, Taylor FW, Corrège T (2004) Assessing the reproducibility of coral-based climate records. Geophys Res Lett 31:L18210. https://doi.org/10.1029/2004GL020343

Stolarski J (2003) Three-dimensional micro- and nanostructural characteristics of the scleractinian coral skeleton: a biocalcification proxy. Acta Palaeontol Pol 48:497–530

Stolarski J, Mazur M (2005) Nanostructure of biogenic versus abiogenic calcium carbonate crystals. Acta Palaeontol Pol 50:847–865

Sugiura M, Yasumoto K, Iijima M et al (2021) Morphological study of fibrous aragonite in the skeletal framework of a stony coral. CrystEngComm 23:3693–3700. https://doi.org/10.1039/D1CE00357G

Suzuki A, Gagan MK, Fabricius K et al (2003) Skeletal isotope microprofiles of growth perturbations in Porites corals during the 1997–1998 mass bleaching event. Coral Reefs 22:357–369. https://doi.org/10.1007/s00338-003-0323-4

Takahashi T, Olafsson J, Goddard JG et al (1993) Seasonal variation of CO2 and nutrients in the high-latitude surface oceans: a comparative study. Glob Biogeochem Cycles 7:843–878. https://doi.org/10.1029/93GB02263

Tambutté E, Allemand D, Zoccola D et al (2007) Observations of the tissue-skeleton interface in the scleractinian coral Stylophora pistillata. Coral Reefs 26:517–529. https://doi.org/10.1007/s00338-007-0263-5

Tambutté S, Holcomb M, Ferrier-Pagès C et al (2011) Coral biomineralization: from the gene to the environment. J Exp Mar Biol Ecol 408:58–78. https://doi.org/10.1016/j.jembe.2011.07.026

Thompson DM (2021) Environmental records from coral skeletons: a decade of novel insights and innovation. WIREs. Climate Change:e745. https://doi.org/10.1002/wcc.745

Tierney JE, Abram NJ, Anchukaitis KJ et al (2015) Tropical Sea surface temperatures for the past four centuries reconstructed from coral archives. Paleoceanography 30:226–252. https://doi.org/10.1002/2014PA002717

Torrano-Silva BN, Ferreira SG, Oliveira MC (2015) Unveiling privacy: advances in microtomography of coralline algae. Micron 72:34–38. https://doi.org/10.1016/j.micron.2015.02.004

Tremblay P, Grover R, Maguer JF et al (2014) Carbon translocation from symbiont to host depends on irradiance and food availability in the tropical coral Stylophora pistillata. Coral Reefs 33:1–13. https://doi.org/10.1007/s00338-013-1100-7

Tribollet A, Golubic S (2011) Reef bioerosion: agents and processes. In: Dubinsky Z, Stambler N (eds) Coral reefs: an ecosystem in transition. Springer, Dordrecht, pp 435–449

Urey HC, Lowenstam HA, Epstein S, McKinney CR (1951) Measurement of paleotemperatures and temperatures of the upper cretaceous of England, Denmark, and the southeastern United States. Geol Soc Am Bull 62:399. https://doi.org/10.1130/0016-7606(1951)62[399:MOPATO]2.0.CO;2

Venn A, Tambutté E, Holcomb M et al (2011) Live tissue imaging shows reef corals elevate pH under their calcifying tissue relative to seawater. PLoS One 6:e20013. https://doi.org/10.1371/journal.pone.0020013

Venn AA, Tambutté E, Holcomb M et al (2013) Impact of seawater acidification on pH at the tissue–skeleton interface and calcification in reef corals. PNAS 110:1634–1639. https://doi.org/10.1073/pnas.1216153110

Voolstra CR, Suggett DJ, Peixoto RS et al (2021) Extending the natural adaptive capacity of coral holobionts. Nat Rev Earth Environ 2:747–762. https://doi.org/10.1038/s43017-021-00214-3

Wang H, Yu K, Tao S et al (2021) New evidence for the periodic bleaching and recovery of Porites corals during the mid-late Holocene in the northern South China Sea. Glob Planet Chang 197:103397. https://doi.org/10.1016/j.gloplacha.2020.103397

Watanabe T, Winter A, Oba T et al (2002) Evaluation of the fidelity of isotope records as an environmental proxy in the coral Montastraea. Coral Reefs 21:169–178. https://doi.org/10.1007/s00338-002-0218-9

Watanabe T, Gagan MK, Corrége T et al (2003) Oxygen isotope systematics in Diploastrea heliopora: new coral archive of tropical paleoclimate. Geochim Cosmochim Acta 67:1349–1358. https://doi.org/10.1016/S0016-7037(02)01221-8

Webb GE, Nothdurft LD, Kamber BS et al (2009) Rare earth element geochemistry of scleractinian coral skeleton during meteoric diagenesis: a sequence through neomorphism of aragonite to calcite. Sedimentology 56:1433–1463. https://doi.org/10.1111/j.1365-3091.2008.01041.x

Wijgerde T, Jurriaans S, Hoofd M et al (2012) Oxygen and heterotrophy affect calcification of the Scleractinian coral Galaxea fascicularis. PLoS One 7:e52702. https://doi.org/10.1371/journal.pone.0052702

Wijgerde T, Silva CIF, Scherders V et al (2014) Coral calcification under daily oxygen saturation and pH dynamics reveals the important role of oxygen. Biol Open 3:489–493. https://doi.org/10.1242/bio.20147922

Wu HC, Dissard D, Douville E et al (2018) Surface Ocean pH variations since 1689 CE and recent ocean acidification in the tropical South Pacific. Nat Commun 9:2543. https://doi.org/10.1038/s41467-018-04922-1

Zhou B, Ku Q, Li C et al (2022a) Single-particle crushing behaviour of carbonate sands studied by X-ray microtomography and a combined finite–discrete element method. Acta Geotech 17:3195–3209. https://doi.org/10.1007/s11440-022-01469-w

Zhou Z, Ni X, Wu Z, Tang J (2022b) Physiological and transcriptomic analyses reveal the threat of herbicides glufosinate and glyphosate to the scleractinian coral Pocillopora damicornis. Ecotoxicol Environ Saf 229:113074. https://doi.org/10.1016/j.ecoenv.2021.113074

Zinke J, d'Olivo J, Gey C et al (2019) Multi-Trace-Element Sea surface temperature coral reconstruction for the southern Mozambique Channel reveals teleconnections with the tropical Atlantic. Biogeosciences 16:695–712. https://doi.org/10.5194/bg-16-695-2019

Zoccola D, Tambutté E, Sénégas-Balas F et al (1999) Cloning of a calcium channel α1 subunit from the reef-building coral, Stylophora pistillata. Gene 227:157–167. https://doi.org/10.1016/S0378-1119(98)00602-7

Zoccola D, Tambutté E, Kulhanek E et al (2004) Molecular cloning and localization of a PMCA P-type calcium ATPase from the coral Stylophora pistillata. Biochim Biophys Acta 1663:117–126. https://doi.org/10.1016/j.bbamem.2004.02.010

Bleaching of Reef Organisms: Current Knowledge, Challenges, and Future Directions into the Anthropocene

4

Caroline E. Dubé, Claudia Pogoreutz, Miriam Reverter, Claire Boitel, Suzanne C. Mills, Aurélie Aqua, Laetitia Hédouin, and Chloé Brahmi

Abstract

Coral reefs worldwide are threatened by climate change and other anthropogenic stressors. A global factor contributing to coral reef decline is the increasing frequency, duration, and intensity of bleaching events leading to extensive mortality of keystone species, which is caused by the loss of their photosymbionts under stressful conditions. The disruption of the symbiosis between scleractinian corals and their dinoflagellate algae has been the primary focus of bleaching studies. However, phototrophic microorganisms live in symbiosis with a diversity of non-scleractinian taxonomic groups, which have received less attention in the past. Here, we provide an overview of the current knowledge on the bleaching responses of a diversity of reef organisms, focusing on four symbiotic metazoan groups: scleractinian corals, sea anemones, giant clams, and sponges. We summarize the methods currently used to detect bleaching events at different spatial and temporal scales, the physiological and molecular mechanisms of bleaching, and the major hypotheses of bleaching. We provide future directions for the implementation of traditional and emerging protection strategies to increase the resilience of coral reefs to increasingly frequent and severe bleaching events. Finally, we discuss the potential for integrating state-of-the-art multi-omics tools to target multiple levels of organismal organization for future research on bleaching in complex coral reef holobionts.

Keywords

Photosymbiosis · Climate change · Adaptation · Holobiont · Monitoring · Mechanisms · Management actions · Multi-omics

4.1 Introduction

Coral reefs are one of the most biologically diverse and productive ecosystems on the planet, providing multiple ecosystem services, that is, supporting, provisioning, regulating, and cultural services (Woodhead et al. 2019; Knowlton et al. 2021). Scleractinian corals are primary ecosystem engineers and foundational species that are responsible for creating most of the three-dimensional structure of coral reefs and thereby provide habitat for more than 25% of all marine species (Fisher et al. 2015). Other organisms, including other cnidarians—octocorals and calcifying hydrocorals—but also bryozoans and crustose coralline algae (CCA), contribute to reef accretion, but to a lesser extent (Montefalcone et al. 2018). Despite their immense ecological importance, coral reefs grow in tropical waters that often offer low levels of nutrients, which require efficient mechanisms for nutrient recycling (Atkinson et al. 2003; de Goeij et al. 2013). The symbiotic association between scleractinian corals and their unicellular dinoflagellates of the family Symbiodiniaceae (LaJeunesse et al. 2018) has shaped the very foundation of coral reef ecosystems through their efficient reciprocal nutrient exchange (Muscatine et al. 1989; Rahav et al. 1989; Tanaka et al. 2015). These Symbiodiniaceae (hereon referred to as "algal symbionts") satisfy up to 95% of their host energetic requirements via the transfer of photosynthetically fixed carbon and the assimilation of nitrogen and phosphorus (Muscatine et al. 1989; Davy et al. 2012; Rädecker et al. 2015), while the host provides a well-protected living space

C. E. Dubé (✉) · C. Brahmi
Univ. Polynésie française, ILM, Ifremer, IRD, UMR SECOPOL, Tahiti, French Polynesia

Laboratoire d'Excellence "CORAIL", Perpignan, France
e-mail: caroline.dube@upf.pf

C. Pogoreutz · C. Boitel · S. C. Mills · A. Aqua · L. Hédouin
Laboratoire d'Excellence "CORAIL", Perpignan, France

UAR 3278 CRIOBE, PSL Research University: EPHE-UPVD-CNRS, Université de Perpignan, Perpignan, France

M. Reverter
School of Biological and Marine Sciences, University of Plymouth, Plymouth, UK

S. Planes (ed.), *The Future of Coral Reefs*, Coral Reefs of the World 21, https://doi.org/10.1007/978-3-031-98584-3_4

and light-rich environment to their endosymbionts, as well as a supply of nutrients and other metabolic by-products vital for the symbiont's photosynthesis (Hughes et al. 2010; Boulotte et al. 2016). This symbiosis is one of the oldest, most widespread, and ecologically successful known mutualisms found in nature (Kirk and Weis 2016; Pogoreutz et al. 2020), but its breakdown can rapidly lead to host starvation and death (Rädecker et al. 2021).

Coral reefs are sentinel ecosystems that are sensitive to both local and global stressors (e.g., increasing seawater temperatures, ocean acidification, pollution, coastal development, and overfishing) (Hughes et al. 2017; Donovan et al. 2021; Knowlton et al. 2021), and as such, are undergoing an unprecedented decline averaging between 30% and 50% loss in scleractinian coral cover globally (Hughes et al. 2018). Trajectories under present CO_2 emission scenarios place coral reefs at high risk, with 60% of all remaining coral reefs critically threatened and more than 70% expected to disappear (Hoegh-Guldberg et al. 2018a; Kleypas et al. 2021) or experience major functional restructuring by 2050 (Williams and Graham 2019). Climate change manifested by the accumulation of CO_2 in the atmosphere and in the oceans is radically altering coral reef ecosystems, mainly through severe increases in sea surface temperature, decreases in ocean pH (Hoegh-Guldberg et al. 2017), and in ocean carbonate saturation state (Kleypas and Langdon 2006).

Since marine invertebrate organisms (including photosymbiotic ones) are ectothermic, they are particularly vulnerable to temperature anomalies. Their increased exposure to conditions above their upper thermal limits threatens coral reefs globally (Smale et al. 2019). Notably, climate-driven marine heat waves—events of extreme seawater temperature lasting days to months—are causing cellular oxidative stress and/or destabilizing nutrient exchanges between partners in marine photosymbioses that can lead to widespread bleaching of a wide range of benthic organisms (e.g., corals, anemones, sponges, giant clams, bryozoans, foraminifera, crustose coralline algae) (Morris et al. 2019; Fig. 4.1). "Bleaching" is caused by the loss of endosymbiotic photoautotrophs under stressful conditions, often resulting in discoloration due to the decrease in the symbionts' pigments (Glynn 1993). Three global coral reef bleaching events triggered by El Niño-driven marine heat waves have already caused devastating coral loss on tropical reefs worldwide (1998, 2010, 2015–2017; Claar et al. 2018). The bleaching event of 1998 alone caused 8% of coral loss worldwide and subsequent events (2009 to 2018) an additional 14% (Souter et al. 2021). Bleaching-induced mortality, and especially mortality of keystone species such as scleractinian corals, can lead to the transition and collapse of some reef ecosystems. Nitrogen fixation rates from epilithic communities on dead corals have been documented to be as much as 30 times higher compared to healthy reefs, favoring an anoxic environment and algal growth (Davey et al. 2008). Newly vacant space, together with nutrient enrichment due to anthropogenic activities facilitate proliferation of opportunistic species (e.g., algae, ascidians, sponges, soft corals, cyanobacteria), sometimes leading to alternative coral reef communities (Tebbett et al. 2019; Reverter et al. 2022). To date, the disruption of the coral-algal symbiosis has been the primary focus of coral reef bleaching studies. Yet, phototrophic microorganisms (Symbiodiniaceae and cyanobacteria) live in symbiosis with a diversity of non-scleractinian taxonomic groups, including sea anemones, zoanthids, octocorals, jellyfish, hydrozoans, sponges, mollusks, and foraminifera, but fewer studies on their bleaching responses are available (Lesser et al. 1990; Richardson 2006; Hill et al. 2016). Since the first reports on bleaching events in the 1980s (Glynn 1983), scientific interest has grown substantially over the last two decades, with nearly half of all articles on bleaching being published within the last 7 years, corresponding to the 2015–2016 global bleaching event (McLachlan et al. 2020) (Fig. 4.2).

In this chapter, we provide an overview of the current knowledge on different photosymbiotic reef organisms, focusing on four taxonomic groups: scleractinian corals, sea anemones, giant clams, and sponges. We then summarize the methods currently used (1) to *detect* bleaching events at different spatial and temporal scales, (2) to *understand* the different physiological and molecular mechanisms triggering the breakdown with photoautotrophic symbionts, and (3) to *protect* coral reefs from increasingly frequent and severe bleaching events. Our main objective is to discuss the state of the art, advantages, and limitations of these methods and provide a synopsis of the challenges, open questions, and future directions for bleaching studies in coral reefs, including multi-omics at the holobiont functional level.

4.2 Biology and Ecology of Photosymbiotic Reef Organisms

To better understand the functional biology of photosymbiosis and inform our decisions for future bleaching studies, we first give an overview of the fundamental understanding of the symbiosis between scleractinian corals, the primary ecosystem engineers of coral reefs, and their intracellular algal symbionts, the Symbiodiniaceae. Then, we draw from examples from a range of non-scleractinian symbioses focusing on three types of photosymbiotic relationships, including actiniarian sea anemones and their intracellular Symbiodiniaceae, giant clams and their extracellular Symbiodiniaceae, sponges and their cyanobacterial symbionts or intracellular Symbiodiniaceae (Table 4.1). We also briefly discuss other mostly overlooked cnidarian-algal symbioses, specifically the upside-down jellyfish (Scyphozoa), fire corals (Hydrozoa), and soft corals (Anthozoa:

Fig. 4.1 Bleaching of photosymbiotic reef organisms. (**a**) Massive bleaching event on the fore reef at Moorea (French Polynesia) in April 2019, (**b**) bleached *Pocillopora* coral and (**c**) *Heteractis magnifica* anemone at Moorea, (**d**) partially bleached giant clam *Tridacna maxima*, and (**e**) *Xestospongia muta* afflicted by annual cyclic bleaching. Photos **a** and **b** are courtesy of Alexis Rosenfield, **c** Yann Lacube, **d** Jean-Claude Gaertner and Nabila Gaertner-Mazouni, and **e** Benjamin Mueller

Octocorallia). Here, we give a brief review on the biology and ecology of these reef organisms and their associated photosymbionts.

4.2.1 Scleractinian Corals

Since their first appearance around 425 million years ago, scleractinian corals (Cnidaria, Anthozoa, Scleractinia) have radiated into more than 1500 species (Cairns 1999, 2009), about 800 of which are symbiotic corals typically containing high densities of endosymbiotic dinoflagellate algae in their tissues and about 700 are aposymbiotic species (Kitahara et al. 2016). Scleractinian corals acquire their algal symbionts in different ways: (1) vertical transmission (i.e., transmission from the mother colony to the eggs or larvae) observed in many brooding species, (2) horizontal transmission (i.e., acquisition from the environment) mainly in broadcast-spawning species, and (3) mixed-mode transmission (Quigley et al. 2018). Within the coral host cell, the algal symbionts reside in a specific membrane-bound symbiosome located in the gastrodermis. This symbiosome contains a vacuolar-type

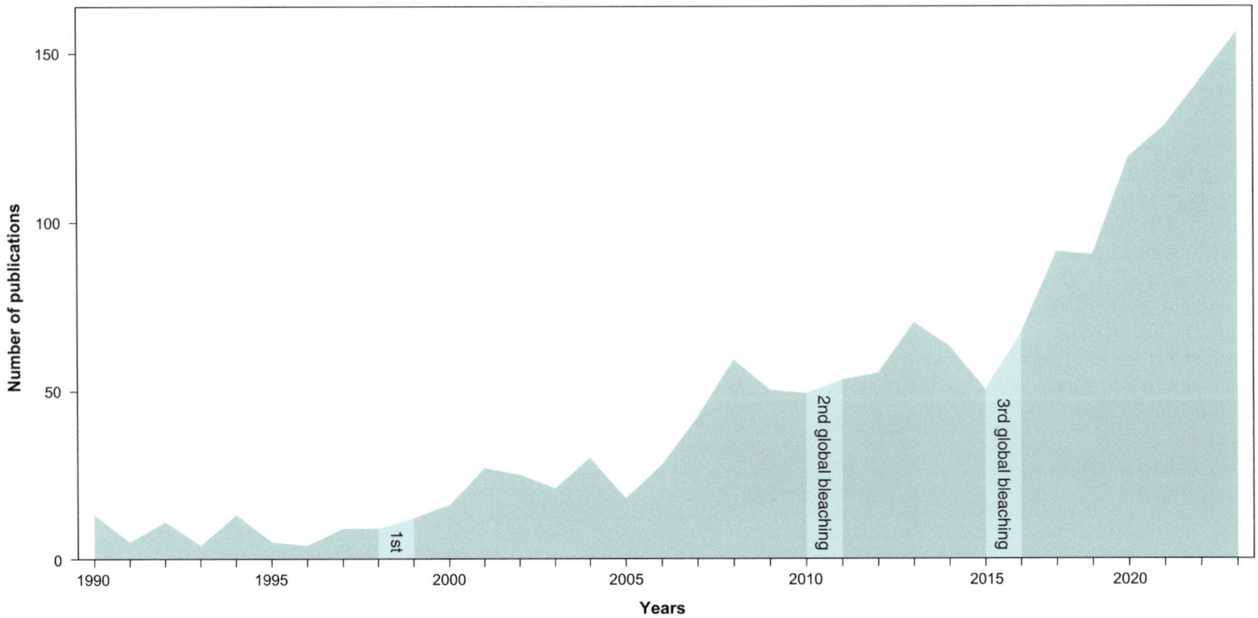

Fig. 4.2 Number of publications on bleaching in coral reefs. Publication records with keywords were recorded from Scopus and plotted according to year of publication (https://www.scopus.com). The keywords used were as follows: (coral and reef) and (bleaching or bleach*) and each publication was checked for an actual observation of a bleaching response. The three global bleaching events are shown

Table 4.1 Comparison of key traits for bleaching studies among four symbiotic metazoan reef groups, scleractinian corals, sea anemones, giant clams, and sponges

	Symbiotic	Intracellular Symbiodiniaceae	Extracellular Symbiodiniaceae	Phototrophic Cyanobacteria	Vacuolar-type H^+-ATPase	Calcifying
 Scleractinian corals	✓	✓	✗	✗	✓	✓
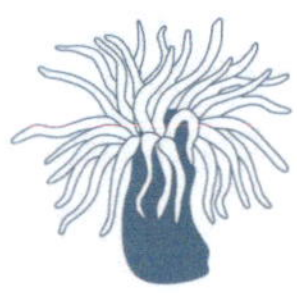 Sea anemones	✓	✓	✗	✗	✓	✗
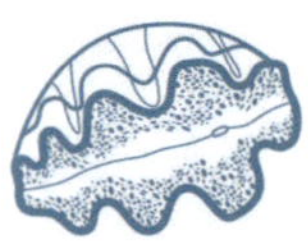 Giant clams	✓	✗	✓	✗	✓	✓
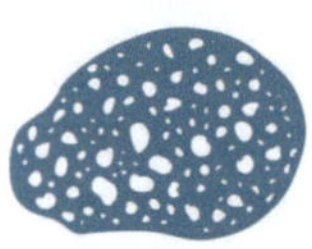 Sponges	✓	✓	✗	✓	NA	✗

H^+-ATPase (VHA), an essential carbon-concentrating mechanism at least in part regulated by the host and which promotes algal photosynthetic performance in symbiotic corals (Barott et al. 2015). This photosynthetic mechanism is highly conserved in Anthozoan photosymbioses (Barott et al. 2022), while a similar mechanism was also reported from distant marine planktonic photosymbioses (Yee et al. 2023). Photosymbiotic corals depend on the photosynthetic activity of their algal symbionts to provide carbon for calcification, a key process to build and maintain the three-dimensional structure of coral reefs. This "light-enhanced calcification" (LEC) is a process in which skeleton formation is stimulated by light, while photosynthesis occurs (Tambutté et al. 2011). One of the main proposed mechanisms involved in LEC is that Symbiodiniaceae may increase pH at the mineralization site (via photosynthetic CO_2 uptake) and influence the equilibrium of dissolved inorganic carbon to promote $CaCO_3$ deposition (Cai et al. 2016; Cohen et al. 2016). Consequently, a decrease in symbiont density and altered photophysiology during bleaching may reduce skeletal growth for several months following the bleaching event (Rodrigues and Grottoli 2006; Colombo-Pallotta et al. 2010; Gold and Palumbi 2018).

4.2.2 Sea Anemones

Sea anemones (order Actiniaria) comprise approximately 1200 species of solitary and skeleton-free polyps and are a diverse member of the anthozoan subclass Hexacorallia. Sea anemones are exclusively marine, occupying benthic marine habitats across all depths and latitudes, and most species engage in symbioses with intracellular Symbiodiniaceae (Barua et al. 2022) that are housed within a symbiosome in the gastrodermis of tentacles exposed to sunlight (Dunn 1981; Barott et al. 2022). Sea anemone cells also contain VAH, which plays a role in the carbon-concentrating mechanism (Barott et al. 2022). Some sea anemones, such as *Entacmaea quadricolor*, acquire Symbiodiniaceae by maternal (vertical) transmission (Scott and Harrison 2007). Symbiont activity in some sea anemones such as *Anthopleura* is susceptible to external food availability, shifting the balance between heterotrophic and autotrophic lifestyles (Bedgood et al. 2020). Sea anemone bleaching has been documented across tropical and subtropical locations throughout their Indo-Pacific distribution (Jones et al. 2008; Hobbs et al. 2013; Beldade et al. 2017; Scott and Hoey 2017). Bleaching diminishes the sea anemones' primary carbon source. Moderate severity can lead to sublethal impacts including oxidative stress and decreased size (Weis 2008; Hobbs et al. 2013), while high severity results in mortality.

4.2.3 Giant Clams

Giant clams (phylum Mollusca) are a taxonomic group composed of two genera, *Hippopus* and *Tridacna*. These organisms play an important ecological role in coral reef ecosystem functioning. For instance, by filtering large volumes of seawater they contribute to environmental nutrient recycling and carbonate budget (Neo et al. 2015). Giant clams are mixotrophic organisms that form a mutualistic symbiosis with Symbiodiniaceae (Mies et al. 2017). Algal symbionts are acquired at the larval stage via horizontal transmission (Morishima et al. 2019) and housed extracellularly in a branched tubular system rising from the stomach and extending to different organs and the mantle, where they are exposed to sunlight (Armstrong et al. 2018). In the lumen of the tubular system, algal symbionts are surrounded by epithelial cells displaying vacuolar H^+-ATPase involved in the carbon concentration mechanism (Armstrong et al. 2018). Through the mutual nutrient exchange between both partners, Symbiodiniaceae translocate photosynthates (e.g., glucose, fatty acids) to their host (Neves et al. 2024), which may satisfy a considerable part of their energy requirements (up to 100%) for metabolism and growth, and the reliability on autotrophy depends on the species, size, and life stage (Klumpp and Griffiths 1994; Yau and Fan 2012). The giant clam mantle displays iridocytes which are cells known to enhance algal symbiont photosynthetic efficiency by providing productive wavelengths for photosynthesis (Holt et al. 2014; Rossbach et al. 2020). Light-enhanced calcification (as described above) and other light-dependent phenomena involving multiple enzymes (such as carbonic anhydrase), transporters, and iridocytes have been documented and underlie the role of light and photosynthesis in giant clam physiology (Ip and Chew 2021). Their calcified shell represents a substratum for epibionts (Vicentuan-Cabaitan et al. 2014), and their fecal pellets, partly containing photosynthetically active algal symbionts, may represent an environmental reservoir of Symbiodiniaceae for other photosymbiotic organisms (Morishima et al. 2019; Umeki et al. 2020). Giant clams also constitute an important food resource, particularly for human populations from remote islands, and economic incomes via fishery and aquaculture activities (Andréfouët et al. 2015; Mies et al. 2017). However, several mass bleaching and mortality events of both wild and cultured giant clams have been reported in the Indo-Pacific region and correlated to abnormally elevated seawater temperatures (Buck et al. 2002; Leggat et al. 2003; Junchompoo et al. 2013; Andréfouët et al. 2013, 2018).

4.2.4 Sponges

Sponges (phylum Porifera) are key ecosystem engineers in a diversity of ecosystems including coral reefs, where they play important roles in nutrient cycling, habitat provisioning, and substrate stabilization and bio-erosion (Wulff 2006; de Goeij et al. 2013). Sponges are traditionally considered to represent the second largest biomass on tropical reefs after corals (Wilkinson 1983), but a recent study suggests that sponge biomass is often underestimated (Kornder et al. 2021). Porifera form close relationships (e.g., symbiosis) with an extraordinary diversity of microorganisms (Thomas et al. 2016), with some of these relationships potentially originating before the Cambrian period over 500 million years ago (Wilkinson and Smith 1997). Although most sponge symbionts remain unculturable, culture-independent approaches (e.g., genomics, transcriptomics) have suggested sponge symbionts benefit their host through metabolic exchange of vitamins, bioactive metabolites, and photosynthetically fixed carbon and nitrogen, contributing to sponge defense and nutrition (Freeman et al. 2013; Pita et al. 2018). Many sponges host photoautotrophic symbionts, with unicellular or filamentous cyanobacteria being the most common (e.g., *Aphanocapsa*, *Synechococcus*, *Prochloron*, *Synechocistis*, *Oscillatoria*) (Rützler 1990; Usher 2008). Cyanosponges (i.e., sponges hosting cyanobacterial symbionts) are reported to normally account for up to 50% of tropical reef sponges, but in some instances, they may constitute up to 80–90% (Wilkinson 1987; Usher 2008). Some cyanosponges have extremely high growth rates (e.g., *Terpios hoshinota* can grow up to 2.5 mm/day), due to the nutritional benefits conferred by the associated cyanobacteria, which can occupy over 50% of the sponge tissue (Rützler and Muzik 1993; Hirose and Murakami 2011). However, nutritional benefits of sponge-associated cyanobacteria seem to vary with symbiont identity and clade (Arillo et al. 1993; Erwin and Thacker 2008). Sponges of the order Hadromerida, and in particular excavating sponges of the genus *Cliona* and of the order Haplosclerida (e.g., some *Haliclona* sponges), host Symbiodiniaceae symbionts, such as those found in scleractinian corals (Garson et al. 1998; van der Windt et al. 2020). Although a vacuolar H^+-ATPase has not yet been described in sponges, the tissue of these organisms contains a cytosolic carbonic anhydrase (CA) (Pomponi 1980), an enzyme known to catalyze CO_2 hydration into H^+ and HCO_3^-, which is consistent with the presence of such active H^+ pumping involved in the carbon-concentrating mechanism for symbiont photosynthesis (Tresguerres 2016). Sponge bleaching was initially described alongside the first coral bleaching reports during the 1980s in the Caribbean (Vicente 1990). Since then, several studies have reported sponge bleaching, and sometimes mass bleaching during global coral reef bleaching events (Fromont and Garson 1999; de Bakker et al. 2016; Hill et al. 2016; Marlow et al. 2018; Mueller et al. 2023). As in corals, elevated temperature has been identified as the main environmental factor driving sponge bleaching (Ramsby et al. 2018; Fortunato et al. 2022). But some studies suggest that other environmental variables (e.g., light, pathogens, or symbiont identity) may also play a role in sponge bleaching susceptibility (McMurray et al. 2011; Curdt et al. 2022; Mueller et al. 2023), although knowledge on the topic remains scarce.

4.2.5 Cassiopea: Apparent Bleaching Resistance and "Invisible Bleaching"

The upside-down jellyfish *(Cassiopea* spp.) is a major laboratory model system for the study of cnidarian-algal symbiosis, including bleaching (Medina et al. 2021). In the wild, scyphozoans (Rhizostomeae family) have a considerable geographic distribution from tropical to subtropical shallow waters, including on coral reefs, and are presumably important for ecosystem functioning due to their role in the food web and biogeochemical cycling (Toullec et al. 2024a). Medusae of *Cassiopea* have a peculiar epibenthic lifestyle, resting their umbrella on the shallow seabed with their oral arms facing up to accommodate the light requirement of their intracellular symbionts. In *Cassiopea*, the majority of the Symbiodiniaceae are housed within specialized host cells, the amoebocytes, which reside in the thick mesoglea (Colley and Trench 1983; Lyndby et al. 2020; Toullec et al. 2024a). Similar to other photosymbiotic cnidarians and giant clams, algal symbionts contribute nutritionally to the metabolism of their host medusa (Klumpp and Griffiths 1994; Lyndby et al. 2020; Toullec et al. 2024a, b). *Cassiopea* have gained a reputation of being tolerant to even high levels of heat stress (33–34 °C), which is mainly due to the fact that few bleaching events of this genus have been reported in the wild (Djeghri et al. 2019; Klein et al. 2019). Recent experimental work, however, showed that despite the absence of a visible bleaching phenotype, symbiotic nutrient cycling between the host and the algae is disrupted (e.g., complete arrest of algal carbon fixation and translocation to the host), and heat-stressed *Cassiopea* exhibit substantial loss of algal symbionts by biomass due to in situ degradation of the algae in the host cells (Toullec et al. 2024b). Further, the medusae exhibited a drop in bell pulsation, and "shrinkage" due to a decrease in wet weight and size. Notably, such shrinkage likely masked the (significant) loss of algal symbionts, resulting in a phenomenon recently described as "invisible bleaching" (Toullec et al. 2024b). In this context, the perceived heat stress tolerance in *Cassiopea* may be relative, and bleaching in wild populations may be more widespread than currently assumed due to this cryptic "invisible" phenotype (Toullec et al. 2024b). Further studies are required for a better understanding on the heat stress response and the "invisible bleaching" phenotype of *Cassiopea*.

4.2.6 Other Cnidarians Reef Organisms

The lack of knowledge on bleaching is also evident in other Anthozoa. For instance, there are few studies investigating the susceptibility and mechanistic of bleaching in zoanthids (Graham and Sanders 2016; Fujimura and Riegl 2017), corallimorpharians (Lin et al. 2019), octocorals (Baker et al. 2015; Steinberg et al. 2022), or in hydrozoan medusozoans such as fire corals (Schwiesow et al. 2021; Hernández-Elizárraga et al. 2019), all of which are cnidarians hosting Symbiodiniaceae within endodermal cells, similar to scleractinian corals.

Fire corals are important calcifying invertebrates of coral reef ecosystems, especially on Caribbean reefs, second in terms of cover to scleractinian corals (Lewis 2006). They are common in shallow and high-light reef environments (Dubé et al. 2019), partially resulting in high susceptibility to bleaching (Loya et al. 2001). Fire corals can rely on behavioral mechanisms of photoacclimation, where algal symbionts can extend throughout the host gastrodermal tissue, creating internal shaded micro-niches (Fitt 2012) that can limit the occurrence of bleaching, as well as physiological mechanisms including heat-shock responses and antioxidant defenses (Hernández-Elizárraga et al. 2019). Fire corals may also rely on heterotrophic feeding as an energy source when suffering from moderate bleaching (Olguín-López et al. 2019). Octocorals are important ecosystem engineers (e.g., forming marine animal forests), providing an attractive habitat for mobile invertebrate and vertebrate species, and food sources for many fish (Epstein and Kingsford 2019). Although octocorals from the Pacific have been described as highly susceptible to increased temperature and bleaching (Maucieri and Baum 2021), Caribbean species are generally more resistant than scleractinian and fire corals (Dias and Gondim 2016). While the consequences of bleaching in octocorals include an increased production of nitric oxide and heat shock proteins by the Symbiodiniaceae (Ross 2014) and changes in lipid, protein, and metabolite content (Farag et al. 2018; Sikorskaya et al. 2020, 2022a), their high bleaching recovery is presumably linked to antioxidant compounds (Sikorskaya et al. 2022b) and bacterial associated lipidomes (Sikorskaya et al. 2022a).

4.3 Methods for Detecting Bleaching: State-of-the-Art, Limitations, and Perspectives

The first step in understanding the functional biology of reef photosymbioses and its breakdown relies on our ability to detect, quantify, and predict the scale of bleaching events, which presents particular challenges, with in situ observations being limited by the spatial and temporal magnitude of such events. To better understand how bleaching is detected and to identify possible research gaps, we provide an overview of the descriptive methodology currently used, including quantitative field-based methods (such as survey and monitoring) and modeling approaches (such as satellite-based remote sensing), and acknowledge the importance of observation consistency and reproducibility, and development of integrative and cost-effective practices.

4.3.1 Field-Based Methods

Sporadic surveys are often used to identify and study the spatial extent and duration of coral bleaching events (Obura et al. 2019; van Woesik et al. 2022). These techniques, if used at an adequate frequency (e.g., surveys are done before, during, and after bleaching events) can also help in quantifying the susceptibility and fate (i.e., recovery/mortality) of different coral reef organisms (Ritson-Williams and Gates 2020). Many coral reef surveys rely on photographic and diver-based documentation, from which quantitative or semiquantitative parameters can be estimated, such as the percentage of bleached corals and/or other health-related visual signs (i.e., symptoms of certain diseases or damage that can be assessed visually) (as in Stuart-Smith et al. 2018).

Observations of coral reef status are often collected and analyzed by humans, which can introduce bias particularly for long-term monitoring considering the lack of factualness and consistent repeatability in data acquisition. Emerging technologies, such as drones, underwater remotely operated vehicles (ROVs), and autonomous underwater vehicles (AUVs), offer the potential to enhance coral reef monitoring by providing cost-effective high-resolution imagery of large-scale reef areas (Teague et al. 2022), but can be limited by water turbidity (Li et al. 2020). Dead, algae-covered, bleached, and healthy corals can be easily identified from these remote platforms because of their distinctive spectral signature (Li et al. 2020), but such approaches are still lacking for other reef organisms, such as giant clams and sea anemones, and should therefore be developed. Likewise, automated image analysis methods, such as CoralNet, have the capacity to generate fully automated benthic cover estimates and differential bleaching status more consistently than human observers (Williams et al. 2019). Unmanned underwater surveys (based on AUVs and ROVs) also offer the possibility to access and provide comprehensive monitoring of habitats that pose risks to divers, such as mesophotic reefs. Recent advances in diving technologies, such as the use of closed-circuit apparatus, also enables the exploration and monitoring of mesophotic reef habitats (Sieber and Pyle 2010). Including depth in bleaching monitoring is of key importance as mesophotic areas are hypothesized to be protected from bleaching (Bongaerts et al. 2010, 2017), while shallower areas are impacted. This could provide managers valuable information on protected coral oases to include when designing marine protected areas or during spatial planning. For example, mesophotic reefs of French Polynesia host diversified benthic communities of high coral cover (Pérez-Rosales et al. 2022) and escaped the massive bleaching event of 2019 (Pérez-Rosales et al. 2021). Nonetheless, this trend cannot be generalized to all mesophotic zones, as contrasted patterns have been previously documented (Frade et al. 2018; Eyal et al. 2022). Additionally, bleaching may impact reef organisms differently in mesophotic areas, as in 2019 the sea anemone *Heteractis magnifica* bleached in mesophotic areas (48–50 m) in Moorea (Haguenauer et al. 2021), while no signs of bleaching were recorded at that time at 60 m for corals (Pérez-Rosales et al. 2021). This highlights that this sea anemone seems more sensitive than most corals to seawater temperature anomalies, but further studies should investigate the differential bleaching susceptibility of reef organisms with depth to provide accurate data to managers for conservation and protection of reef biodiversity in the context of climate change.

At a large spatial scale or in an extensive territory, monitoring bleaching could be time-consuming for scientists and snapshot views of a few locations do not provide a complete representation of its occurrence. For this reason, participative science, which involves volunteer citizens, is an interesting opportunity to enhance global scientific knowledge (Dickinson et al. 2010; Kobori et al. 2016). Participatory science becomes essential for coral reef monitoring as volunteers can record key information during marine heat waves, as shown through a participative science project conducted in 2016 in Moorea (French Polynesia) providing scientists with information about bleaching from 23 islands (https://www.unoeilsurlecorail.org). Currently, published works on aggregated data from scientific and participatory approaches are becoming an essential pairing for large-scale bleaching monitoring (Hédouin et al. 2020).

While short-term reef surveys provide snapshots of coral reef status, establishing coral reef long-term monitoring programs is essential to assess the extent and consequences of bleaching events at relevant timescales (i.e., research has shown that coral reefs may take up to 10 years, or longer, to recover from acute disturbances; McClanahan et al. 2007; Lamy et al. 2016) and to help establish management and conservation practices that aim to preserve reefs facing ongoing climate trajectories and anthropogenic pressures. For example, considerable continuous monitoring efforts have been undertaken locally in French Polynesia with the Polynesia-Mana long-term monitoring program (17 monitored islands, SNO CORAIL, France) and the Long-Term Ecological Research (LTER) network (United States) or the AIMS Long-Term Monitoring programs (Australia) and the Atlantic and Gulf Rapid Reef Assessment (AGGRA) program (Caribbean). Moreover, our ability to aggregate data from local monitoring programs to a global level is equally important to scale global trajectories of coral reefs sustainability. To this end, the Global Coral Reef Monitoring Network (GCRMN) is the first monitoring strategy developed in the 1990s, providing a robust description of reef health worldwide and the last report (GCRMN2020) revealed that marine heat waves have driven a global decline of coral reefs worldwide, with 14% of coral reefs lost since 2009 (Souter et al. 2021). However, other non-scleractinian photosymbiotic organisms are often not recorded in both local and global monitoring. Overlooking non-coral benthic components can lead to biased perceptions of the impact of heat waves at the scale of coral reef ecosystems (Reverter et al. 2020) and could reduce the effectiveness of coral reef protection strategies by not considering relevant biotic interactions. Integrating a greater number of reef organisms in bleaching monitoring is therefore paramount to obtain a more complete picture of coral reef responses to rising seawater temperatures and other stressors.

4.3.2 Satellite-Based Remote Sensing

As corals respond to ocean warming, field surveys have the potential to provide accurate data for coral reef monitoring but have several limitations that restrict its use. They are time-consuming, limited in temporal repeatability, and often cover a limited reef area, providing scattered information on bleaching occurrences in time and space. To address these limitations, researchers have begun to explore the use of remote sensing approaches to complement traditional field surveys and provide a more comprehensive view of coral reef status. Given the physical scale and the distribution of coral reefs, remote sensing techniques have enabled researchers to quantify the constantly changing environment, examine large-scale atmospheric and oceanographic processes (Shchepetkin and McWilliams 2005), monitor bleaching patterns spatially and temporally, and develop models to provide accurate predictions related to thermal stress on coral reefs (Gonzalez-Espinosa and Donner 2021; Johnson et al. 2022).

Long-term and remote monitoring of bleaching through satellite imagery has become an important tool for understanding the global impact of climate change on coral reefs, offering a rapid, cost-effective, and nondestructive approach for multiple reefs over large spatial and temporal scales. Thermal disturbances and anomalously high sea surface temperatures (SSTs) are commonly used as predictors of the likelihood of coral bleaching globally but not all reefs are equally exposed to such severe heat stress events (Hughes et al. 2017). Therefore, the relationship between increased temperatures and bleaching varies over space and time. Recent research has shown that bleaching events are less common on reefs with a high variance in SST, while the probabilities of such events are highest for tropical midlatitude reefs (Sully et al. 2019), which highlight the importance of recording this parameter globally. The NOAA Coral Reef Watch program (CRM) is the most valuable set of sea surface temperature (SST) implemented for coral reef monitoring (https://coralreefwatch.noaa.gov). During the last 20 years, this program has created algorithms allowing the detection and mapping of high positive SST anomalies relative to long-term averages and has provided local coral bleaching risk alerts at a global scale. Since recent bleaching events have demonstrated a clear link between heat accumulation and bleaching (Hughes et al. 2017), Degree Heating Weeks (DHW) has become a standard global predictor of bleaching on coral reefs (Gleeson and Strong 1995), with 1 DHW defined as on week of 1 °C above the long-term climatology for the warmest month at a given locality.

However, inconsistency in the occurrences of bleaching between models based on historical satellite observations of SST metrics and field surveys (e.g., Reefbase) underscores the importance to develop better models, incorporating more metrics such as cloudiness, and analysis of high-resolution satellite imagery to improve accuracy of bleaching predictions (Safaie et al. 2018; Gonzalez-Espinosa and Donner 2021). The development of such models, combined with long-term monitoring networks (e.g., GCRMN) and physiological metrics of bleaching, could help identify coral reef locations that can withstand the effects of increased temperatures, referred to as climate refugia (e.g., localities with upwelling, high historical thermal exposure, high cloud cover, and temporal variability) (Gonzalez-Espinosa and Donner 2021; Dixon et al. 2022), and build capacity in management responses to promote resilience and recovery to future bleaching events (Marshall and Schuttenberg 2006).

4.4 Mechanisms of Bleaching: Disparate Knowledge Among Photosymbiotic Reef Organisms

The second step in understanding the functional biology of reef photosymbioses is to identify the mechanisms triggering its breakdown or underlying its persistence under environmental change. Animals and plants are inhabited by microbial organisms that can influence the fitness of their hosts (Rosenberg et al. 2007), forming a complex entity called "holobiont". On coral reefs, cnidarians and many other reef organisms (e.g., octocorals, upside-down jellyfish, hydrocorals, mollusks) host Symbiodiniaceae algae and a considerable diversity of prokaryotic and eukaryotic microorganisms, which includes bacteria, archaea, algae, fungi (and other microeukaryotes), and viruses (Rohwer et al. 2002; Roik et al. 2022; Bonacolta et al. 2023), which collectively constitute the holobiont (Bourne et al. 2016). Variation in holobiont bleaching susceptibility is driven by physiological trade-offs associated with different host organismal fitness traits, including morphology and growth (Loya et al. 2001; van Woesik et al. 2011), and heterotrophic feeding (Grottoli et al. 2006). Extensive research has shown that bleaching is governed by a complex interplay between host genetics and symbiont assemblages, involving a diversity of mechanisms such as front-loading of genes that respond to stress (Barshis et al. 2013), multilocus adaptation (Bay and Palumbi 2014), association with thermally tolerant strains of Symbiodiniaceae (Claar et al. 2020), and, potentially, properties of bacterial communities (Ziegler et al. 2017; Gardner et al. 2019). However, much of this knowledge on the bleaching response has mainly focused on scleractinian corals, while a detailed understanding of bleaching in other reef photosymbioses is still lacking.

To better understand the different mechanisms of bleaching at the physiological, cellular, and molecular levels, in situ and ex situ experiments have been conducted to study the effects of abiotic factors on biological functions of photosymbiotic organisms, and their stress responses. Aquarium-based experiments and reciprocal transplantations have been mostly used to investigate the role of specific drivers in bleaching responses or to identify specific mechanisms of bleaching in the laboratory (e.g., Ziegler et al. 2017; McLachlan et al. 2020; Rädecker et al. 2021). Such experiments involve the manipulation or rearing of reef organisms, inherently creating conditions to which they are not exposed in their natural environment. Working in the field during natural bleaching events thus provides the possibility to track reef organisms when exposed to multiple environmental stressors, allowing a better understanding of the mechanisms leading to the breakdown of the symbiosis and its recovery. Here we focus on the fundamental physiological and fitness aspects under bleaching, together with the molecular mechanisms, of four metazoan reef groups and their associated photosymbionts: scleractinian corals, sea anemones, giant clams, and sponges.

4.4.1 Physiology and Fitness Traits

4.4.1.1 Scleractinian Corals

Despite the growing concerns of repeated (annual) and increasingly severe mass coral bleaching worldwide by the end of 2050 (Hoegh-Guldberg et al. 2018a; Virgen-Urcelay and Donner 2023), inter- and intraspecific variability in bleaching susceptibility and associated physiological and cellular mechanisms remains poorly understood. Additionally, most studies focus on few model scleractinian species, such as *Stylophora pistillata* and members of the genera *Pocillopora* and *Acropora* (LaJeunesse et al. 2010; Dixon et al. 2020; McLachlan et al. 2020; Rouan et al. 2022; Leinbach et al. 2023). However, new coral laboratory model approaches using the easy-to-maintain scleractinian *Galaxea fascicularis* are currently established (Puntin et al. 2023), which may facilitate functional and systems biology approaches, and may be more representative for coral physiology than the popular anemone model *Exaiptasia diaphana*.

Algal symbiont identity is an important factor contributing to bleaching susceptibility in corals, with a given coral-algal assemblage assumed to be the result of host and algal genotypes and the surrounding environment (e.g., Thornhill et al. 2014; Silverstein et al. 2015). Algal symbionts of the family Symbiodiniaceae are genetically diverse, with four genera commonly associating with corals (*Symbiodinium*, *Breviolum*, *Cladocopium*, and *Durusdinium*; occasionally *Fugacium* and *Gerakladium*) (LaJeunesse et al. 2018). These genera exhibit distinct physiological and ecological traits (LaJeunesse et al. 2010; Parkinson and Baums 2014), and differences in Symbiodiniaceae assemblage can manifest in differential holobiont bleaching susceptibilities (Sampayo et al. 2008; Leinbach et al. 2023). Importantly, bleaching can also induce distinct shifts from one predominant algal genus to another, which can lead to increases in thermotolerant species such as *Durusdinium trenchii* (Cunning et al. 2015; Claar et al. 2020). Because *D. trenchii* is less symbiotic than other coral associated Symbiodiniaceae, such changes typically represent a significant physiological trade-off for the host (Little et al. 2004; Cantin et al. 2009; Grottoli et al. 2014) and often revert back to the original association following recovery (LaJeunesse et al. 2010; Hume et al. 2020). Given the central role of the Symbiodiniaceae in coral nutrient cycling, there are many detailed reviews covering the physiological responses of these symbionts and their effects on cellular homeostasis during bleaching (Warner and Suggett 2016; Szabó et al. 2020). As such, we will only briefly touch upon these topics in this chapter.

Coral physiological variability under bleaching is often monitored by measuring the following parameters: algal symbiont density and (photo)physiology, coral tissue thickness and holobiont energy reserves, calcification, respiration, and heterotrophic plasticity (McLachlan et al. 2020; Grottoli et al. 2021). The photophysiological performance of corals often decreases during bleaching, indicating photoinhibition and/or photodamage of the photosynthetic apparatus in response to thermal stress and high irradiance (Gardner et al. 2017; Hoadley et al. 2019). Further, synergistic effects of increased temperature and light can lead to photooxidative stress, persistent photosystem damage, and longer recovery periods (Lesser 2011; Berg et al. 2021).

Overall, bleaching is typically accompanied by reduced rates of inorganic carbon translocation by the algal symbionts, resulting in reduced organic carbon availability for the host metabolism (Hughes et al. 2010; Rädecker et al. 2021, Toullec et al. 2024b). As such, bleaching diminishes the photosynthetically fixed carbon pool available to the coral. Hence, one strategy of corals (and other cnidarians) for survival and recovery relies on the catabolism of the holobiont's own reserves to obtain organic carbon, specifically proteins and lipid reserves (Rodrigues and Grottoli 2007; Ezzat et al. 2019; Rädecker et al. 2021; Toullec et al. 2024b). Physiological and geochemical studies also showed that decreases in dissolved inorganic carbon upregulation during bleaching causes the disruption of important biomineralization mechanisms, resulting in lower calcification and growth rates (D'Olivo and McCulloch 2017; Schoepf et al. 2021).

In principle, bleaching is not always lethal. Indeed, corals can recover from bleaching by repopulating their tissues with Symbiodiniaceae and/or chlorophyll *a*, although this ability depends on the identity and functional traits of the holobiont, that is, the host and algal symbiont (Rodrigues and Grottoli 2007; Schoepf et al. 2015; Levas et al. 2018; Scharfenstein et al. 2022). Such recovery processes can lead to fitness trade-offs, including reduced energy reserves and lower fecundity in *Acropora* corals (Jones and Berkelmans 2011; Leinbach et al. 2021). However, potential signs of physiological acclimatization were reported in *Orbicella faveolata* across recurrent bleaching events, with increased photosynthetic capacities but reduced coral energy resources available for tissue maintenance, defense, and reproduction (Fisch et al. 2019). For instance, *Montipora* corals were shown to rely on autotrophic carbon for provisioning gametes and to prioritize gamete development over tissue maintenance following bleaching (Rodrigues and Padilla-Gamiño 2022). This body of work underscores the importance of studies in a comparative taxonomic framework during and after natural bleaching events to elucidate interactions between heat tolerance and fitness trade-offs important for the overall reef recovery. Such studies should also include more environmental factors that may affect the bleaching response (Fig. 4.3), such as high irradiance and increased nutrients, and expand to vulnerable early ontogenetic stages (Grottoli et al. 2021).

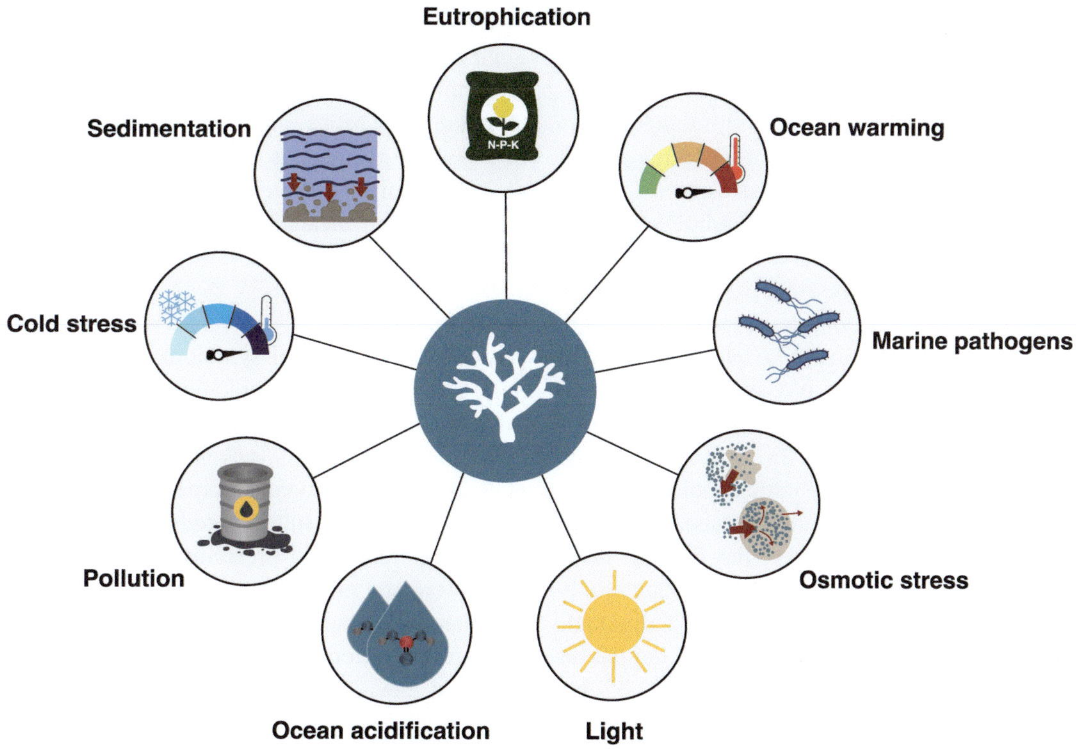

Fig. 4.3 Different factors causing or exacerbating bleaching of scleractinian corals and other photosymbiotic reef organisms

4.4.1.2 Sea Anemones

The main cause of bleaching in anemones is high temperatures, which reduce both symbiont density and chlorophyll *a* and *c* concentrations (Hill and Scott 2012). Anemones can expel large masses of endosymbionts and subsequently lead to bleaching in response to moderate heat stress, while photooxidation and pigment loss of algae are observed under severe heat stress (Hill and Scott 2012). High irradiance further increases thermal stress susceptibility, accelerating the loss of PSII photochemical efficiency and reducing the effective quantum yield of anemone symbionts at high temperatures (Hill and Scott 2012). The influence of ocean acidification on bleaching in sea anemones remains inconclusive. Combined increases in temperature and pCO_2 caused a significant loss of Symbiodiniaceae in sea anemone *Exaiptasia pallida*, but not alone (Hagemeyer 2016). In contrast, *E. quadricolor* bleached at elevated temperature, but not when combined with greater pCO_2, or with greater pCO_2 alone (Pryor et al. 2021).

Bleached sea anemones can recover from bleaching once they have regained their symbiont population, although uptake from free-living symbiotic dinoflagellates is slow (Rodriguez-Lanetty et al. 2003). However, other environmental symbiont sources, such as photosymbiotic invertebrates are present and consumed by the anemones (Rodriguez-Lanetty et al. 2003). Sea anemones could also acquire endosymbionts from viable Symbiodiniaceae present in the feces of anemone-feeding nudibranchs (Seavy and Muller-Parker 2002). Actiniarians are renowned for engaging in symbioses with metazoans including sponges, crustaceans, and fish, in particular anemonefish. All ten sea anemone species that provide habitat for obligate symbiotic anemonefishes are susceptible to bleaching (Dunn 1981). A bleached sea anemone habitat has cascading effects for the resident anemonefishes, including decreased fecundity and reproductive hormones, and increased stress hormones, metabolic demand, and vulnerability to predators due to behavioral changes (Jones et al. 2008; Saenz-Agudelo et al. 2011; Beldade et al. 2017; Norin et al. 2018; Cortese et al. 2021). However, anemonefish facilitate the recovery of their bleached host sea anemones, enhancing resilience to bleaching by excreting waste products absorbed by the sea anemone and their Symbiodiniaceae (Chase et al. 2018; Pryor et al. 2021), which subsequently increase their population density (Cook et al. 1988; Roopin and Chadwick 2009). Finally, bleached sea anemones spent more time with their tentacles expanded when hosting anemonefish, which would enhance light interception and potentially benefit Symbiodiniaceae, further facilitating recovery (Holbrook and Schmitt 2005).

4.4.1.3 Giant Clams

Giant clam bleaching refers to the loss of the extracellular algal symbionts via their expulsion or apoptosis, and a decrease in symbiont pigments (Buck et al. 2002; Leggat et al. 2003). High temperature is one of the main triggers leading to giant clam bleaching (Hernawan 2008; Ramah et al. 2023). Heat stress experiments on naturally and artificially bleached giant clams showed that high temperatures reduce symbiont density and the uptake of inorganic carbon and ammonium, indicating a limited translocation of photosynthates to the host (Leggat et al. 2003), while a twofold increase in symbiont density was observed during the 8-week recovery period (fourfold under nutrient exposure). Brahmi et al. (2021) demonstrated that long-term exposure to high temperatures induced bleaching in *T. maxima*, accompanied by a decrease in symbiont density, oxygen production, and symbiont photosynthetic efficiency (Fv/Fm). These observations are consistent with results from Sayco et al. (2024) on *T. crocea* reporting a decrease in photophysiology performances (i.e., Fv/Fm and gross photosynthesis), but also a reduced reproductive potential, over a 16-day thermal stress exposure inducing bleaching. Interestingly, this latter study showed that phototrophic potential of stressed individuals can be restored after a recovery period (i.e., from day 12 for gross photosynthesis or day 15 for Fv/Fm over a 35-day recovery). The synergistic effect of high light level and temperature resulted in a decrease of symbionts' cell size, density, and chlorophyll concentration in bleached *T. gigas* (Buck et al. 2002). Recently, acute heat stress led to mild bleaching in *T. maxima* but to tissue disintegration in *T. squamosa*, while negative impacts on symbiont photophysiology were found for both species (Ramah et al. 2023). Overall, these experiments highlight differential bleaching responses among giant clam species.

Through a long-term survey encompassing a bleaching event in Thailand, Junchompoo et al. (2013) reported 60% mortality, followed by re-pigmentation of about 40% of the bleached specimens, suggesting that giant clams can regain—at least in part—their algal symbionts within a few months after heat stress. Based on in situ monitoring after a natural bleaching event, Sayco et al. (2023) reported a negative effect on bleached giant clam survivorship and reproductive potential. Additionally, heat stress experiments on early life stages showed a decrease of chlorophyll *a* content in *T. crocea* larvae, resulting in strong bleaching and high mortality (Mies et al. 2018), a decrease in fertilization success in *T. maxima* (Armstrong et al. 2020), and developmental anomalies and a reduced post-settlement survival in *T. gigas* (Enricuso et al. 2019). These latter observations demonstrate a high vulnerability of these organisms to increasing bleaching frequencies and severities, ultimately exacerbating their ongoing worldwide decline. Many other studies have investigated the effects of high temperature, high pCO_2, and hypo- or hypersalinity on giant clam physiology, but no sign of bleaching or discoloration was reported, suggesting the threshold for bleaching was not reached under the tested experimental conditions. Despite the current research effort to understand the effects of thermal and other stresses on giant clams, knowledge of the mechanisms involved in their bleaching responses and recovery remains scarce. Overall, there are similarities in physiological and cellular bleaching responses of giant clams, corals, and sea anemones. However, as giant clams host their symbionts extracellularly, some aspects of their bleaching mechanisms may differ from the other taxa.

4.4.1.4 Sponges

Physiological effects of symbiont loss on sponges are variable. Loss of sponge symbionts led to reductions in sponge mass, area, and growth in the cyanosponges *Lamellodysidea chlorea* and *Lendenfeldia chondrodes* (Thacker 2005; Curdt et al. 2022). Experimental temperature-induced bleaching of *Cliona orientalis* resulted in reduced energy reserves, lower bioerosion rates, and prevailing mortality (Achlatis et al. 2017; Ramsby et al. 2018). Massaro et al. (2012) showed that the Great Barrier Reef sponge *Rhopaloeides odorabile* shifts feeding preferences toward phototrophic bacteria/symbionts at elevated temperatures due to their higher energy content (e.g., *Synechococcus* cyanobacteria are over 20 times as carbon-rich as heterotrophic bacteria), suggesting the sponge may be consuming its cyanobacterial symbionts at elevated temperatures. Multiple stressors (elevated temperature and elevated pCO_2) experiments have shown that ocean acidification (and subsequent elevated inorganic carbon availability) may increase bioeroding sponge resistance to bleaching and decrease the stress on the sponge host temporally (Bennett et al. 2017; Fang et al. 2018). Field reports also showed that bleached *Cliona* aff. *viridis* experienced reductions in size (Marlow et al. 2018). In contrast, studies on the common barrel sponge *Xetospongia muta* have reported that individuals afflicted by annual cyclic bleaching (about 25% of the Caribbean population), recover fully with no known physiological implications on the host, suggesting that cyanobacterial symbionts provide only little benefit (López-Legentil et al. 2008; McMurray et al. 2011). However, about 1% of the population of *X. muta* has been reported to experience fatal bleaching, which induces high hsp70 expression and leads to sponge mortality (López-Legentil et al. 2008). It is worth noting that the traits of giant sponges such as *X. muta* (e.g., growth rates) are significantly different from common encrusting sponges such as the abovementioned (*Cliona*, *Lamellodysidea*, *Lendenfeldia*) (McMurray et al. 2010; Gonzalez-Rivero 2012), and therefore bleaching studies should investigate comparable traits between different sponges to understand the physiological implications of bleaching.

4.4.2 Molecular Responses

4.4.2.1 Scleractinian Corals

Sequencing technologies, including whole genome, transcriptome, and epigenome, ('-omics' tools), are advancing coral research through molecular investigations of underlying mechanisms of bleaching tolerance or susceptibility. Coral thermotolerance is highly polygenic, encoded by many genes across the genome. Genome-wide marker sequencing (RAD-Seq or whole genome sequencing) provided evidence of extensive polymorphism through multilocus selection in corals exposed to recurrent high heat episodes (Bay and Palumbi 2014), and the identification of genetic variants associated with coral bleaching (Fuller et al. 2020). Transcriptomic studies have greatly improved our understanding of the coral physiological responses during bleaching and recovery (Barshis et al. 2013; Bay and Palumbi 2015; Pinzón et al. 2015; Thomas and Palumbi 2017). Oxidative stress during bleaching under heat stress decreased Ca^{2+} homeostasis and the expression of stress-response genes (DeSalvo et al. 2008; Rodriguez-Lanetty et al. 2009). A seemingly universal transcriptional stress response in heat-stressed corals encompasses the downregulation of cell growth gene expression, and upregulation of ribosomal protein biosynthesis, apoptosis, oxidative stress, and immune responses (Dixon et al. 2020). The host transcriptome showed signs of lasting effects on gene expression months after the algal symbiont population has recovered in *O. faveolata* and *A. hyacinthus* (Pinzón et al. 2015; Thomas and Palumbi 2017). Similarly, bleaching-resistant *A. hyacinthus* from naturally extreme thermal environments exhibits a frontloading of stress-response gene expression (heat shock proteins, antioxidant enzymes, and immunity; Barshis et al. 2013), while *Porites astreoides* regulates the gene expression through transcriptomic plasticity (Kenkel and Matz 2017). Recent studies have demonstrated that epigenetic mechanisms (DNA methylation) can dynamically regulate gene expression and play a role in stress-repair mechanisms (Eirin-Lopez and Putnam 2019). The presence of different algal symbionts can also affect the expression of specific host genes (Baumgarten et al. 2018), while symbiont gene expression *in hospite* is more stable than that of their host (Leggat et al. 2011; Barshis et al. 2014).

The breakdown of the coral-algal symbiosis results in marked alterations in the production and translocation of a variety of metabolites, which have major implications on host physiology (e.g., energy budgets, immunity, cell homeostasis) (Hillyer et al. 2018; Williams et al. 2021). Exposure of *A. aspera* to heat stress increased glucose levels, but decreased symbiont de novo biosynthesis pathways of fatty acids and lipids, suggesting a shift to ATP production via alternative pathways, such as the breakdown of energy reserves. This resulted in a reduction of host fatty acids, indicating a decline in the quantity and quality of symbiont-derived metabolites and photosynthates, which could have major implications for the coral energy budget (Hillyer et al. 2018). Using untargeted tandem mass spectrometry, Pei et al. (2022) similarly observed that heat-stressed *Pavona decussata* experienced a decrease in various classes of lipids and amino acids, suggesting activation of gluconeogenesis (through lipid hydrolysis and aminolysis) to meet basic energy demands. Several dipeptides were also found upregulated in bleached corals, a potential strategy to cope with oxidative stress (Williams et al. 2021; Pei et al. 2022).

4.4.2.2 Sea Anemones

While little information on molecular responses of photosymbiotic anemone populations to heat stress and bleaching on the reef are available, recently there has been an increase in molecular studies on the sea anemone model Aiptasia (*Exaiptasia diaphana*; formerly *E. pallida*) due to its amenability for functional studies (Rädecker et al. 2018). Using an ultrastructural approach, Hanes and Kempf (2013) demonstrated that heat stress induces bleaching in Aiptasia through algal symbiont expulsion and active host autophagy. Metabolomic studies demonstrated that prolonged heat stress severely impairs symbiont (re-) colonization by shifting the composition of the algal surface glycome from glycans associated with symbiotic strains to those typically found in free-living strains (Maruyama et al. 2022). An early proteomic study demonstrated the induction of heat shock proteins in heat-stressed coral *O. faveolata* and Aiptasia (Black et al. 1995). Recently, Oakley et al. (2017) additionally showed the global disruption of Aiptasia host proteostasis and stress on the endoplasmic reticulum (ER). As such, ER protein changes may be essential in the molecular heat stress response of symbiotic cnidarians (Oakley et al. 2017). Interestingly, no changes in algal symbiont ROS levels were recently reported for Aiptasia (Dungan et al. 2022); however, different strains of Aiptasia exhibited strong antioxidant gene expression and increases in associated protein abundances in response to heat stress (Cziesielski et al. 2018). Interestingly, symbiotic Aiptasia rapidly drop gene expression of transport proteins for molecules critical for symbiosis support, specifically ammonium, glucose, and sterol (Cleves et al. 2020). Importantly, the downregulation of gene expression of these symbiosis-supporting genes occurs long before bleaching is apparent (Cleves et al. 2020), paralleling the observation of early disruption of symbiotic nutrient cycling in heat-stressed corals (Rädecker et al. 2021). Sydnor et al. (2023) also found that microbiome changes occur during thermally induced bleaching of symbiotic Aiptasia, with a decrease in *Pelobacter*, an anaerobic sulfate reducer potentially involved in nitrogen fixation, suggesting its role in the maintenance of the symbiosis. Finally, a comparison of Aiptasia colonized by native and heterologous symbiont types found that while oxidative stress and apoptosis gene expression are strongly influenced by symbiont type, the overall response to heat stress is dictated by the compatibility of both partners (Cziesielski et al. 2022).

4.4.2.3 Giant Clams

Few studies have investigated the molecular response of the giant clam and its microbial symbionts to temperature-induced bleaching. Alves Monteiro et al. (2019) explored the physiological and transcriptomic responses of *T. maxima* and its algal symbionts to a long-term exposure to thermal stress leading to bleaching. Based on transcriptomic and metabarcoding approaches, and a variety of physiological parameters, they reported enrichment in tryptophan coding genes, as an oxidative stress response under elevated temperature conditions, and impairment of the symbiont's PSII. Additionally, no shift of Symbiodiniaceae communities has been observed in this study under stressful conditions, suggesting *Symbiodinium*-specific fidelity in host clams that may limit the scope for thermal acclimation. Dubousquet et al. (2016) showed that thermal stress induces modification of fatty acid composition, lipid pathways, and an overexpression of genes encoding reactive oxygen species (ROS) scavengers in *T. maxima*. However, they did not observe signs of bleaching but reported important symbiont cell damages and cell death processes. Additionally, acute thermal stress induced symbiont apoptosis and an increase of oxidative stress via a decrease in antioxidant activity (superoxide dismutase), leading to a reduction in symbiont density in *T. crocea* (Zhou et al. 2019).

4.4.2.4 Sponges

Research on the mechanisms triggering the breakdown of the symbiosis between sponges and their photosymbionts remains extremely limited. A transcriptomic study on *R. odorabile* showed that elevated temperature reduces the expression of functions that mediate the partnership between the sponge and their symbionts such as nutritional interdependence and molecular interactions (Fan et al. 2013). To date, no research has investigated the putative mechanisms by which sponges lose their Symbiodiniaceae symbionts. As observed in corals, elevated stress leads to a shift of microbial communities toward scavenging microbes (Webster et al. 2008; Fan et al. 2013; Posadas et al. 2022); however, the role of this shift in the disruption of sponge-photosymbiont relationships remains unknown. It has been hypothesized that sponge holobionts with highly specific microbiomes and low functional redundancy may be more unstable and therefore more susceptible to environmental stressors (Girvan et al. 2005; Posadas et al. 2022). Temperature stress in sponges has also been observed to lead in some cases to downregulation of host-immune related pathways and other key biological functions such as cytoskeleton arrangement, signal transduction, protein synthesis/degradation, oxidative stress, and detoxification (Webster et al. 2013). Posadas et al. (2022) suggested that the suppression of sponge immune pathways could lead to the disruption of the sponge-symbiont interactions due to a lack of efficient symbiont recognition and pathogenic clearance, but experimental evidence of these mechanisms is currently lacking.

4.5 Main Hypotheses of Bleaching

All of the research efforts described previously have provided critical insight into the physiological and molecular mechanisms at play prior to and during bleaching of marine photosymbioses, with much of this work focusing on reef-building corals due to their importance as ecosystem engineers of tropical coral reefs. As a result, several mechanisms have been identified as major drivers of bleaching in corals, anemones, photosymbiotic medusae, giant clams, and sponges. Notably, these mechanisms are likely not exclusive, but complementary, and while some of these mechanisms may take place in all four major focal groups of this chapter (in addition to the cnidarian model *Cassiopea*), some others seem limited to cnidarians (Table 4.2), although future research is needed to confirm this hypothesis. Overall, molecular and ultrastructural evidence strongly suggests that the loss of algal symbiont cells during bleaching is ultimately the result of an innate immune response of the host (Weis 2008) and there is a growing consensus about the different hypotheses of bleaching, which are briefly described below and illustrated in Fig. 4.4.

4.5.1 The Oxidative Theory of Bleaching

For decades, the prevailing theory has been the Oxidative Theory of Bleaching (OTB) (Downs et al. 2002). It suggests that host immune responses are initially catalyzed by the overwhelming of host antioxidant capacity and cellular repair processes in response to the excess release of reactive oxygen species (ROS) by the stressed Symbiodiniaceae photosystem (Downs et al. 2002; Smith et al. 2005) (Fig. 4.4). Production of ROS occurs in the chloroplast by several mechanisms associated with PSI- and PSII-catalyzed electron transfer, most notably of hydrogen peroxide (H_2O_2) by the oxygen-evolving complex (Richter et al. 1990; Tjus et al. 2001). Leakage of algal-generated H_2O_2 into the coral cytoplasm is "neutralized" by (non)enzymatic antioxidant pathways or converted into hydroxyl radicals (Downs et al. 2002). Such "scavenging" by antioxidant defenses can compensate and ameliorate the destructive capacity of ROS. Overwhelming of such defenses due to excess ROS production, however, causes oxidative damage (protein deg-

Table 4.2 Mechanisms of bleaching among five symbiotic reef organisms

Detrimental effects on host physiology	Detrimental effects on photophysiology of symbionts (photoinhibition, damage)	ROS production under bleaching	Host starvation response under bleaching	Bacteria-induced bleaching (dysbiosis)	Shifts in bacterial communities	Shifts or stability in photosymbiont communities

radation, lipid peroxidation, and DNA degeneration, eventually cell death through apoptosis or necrosis; Freeman and Crapo 1982). At this point, corals may eradicate the dominant source of excess ROS by expelling their endosymbiotic algae, resulting in bleaching. This hypothesis has also been proposed to be one of the potential origins of giant clam bleaching (Brahmi et al. 2021), based on the observation of Dubousquet et al. (2016) that heat stress induced an upregulation of antioxidant genes, and of Zhou et al. (2019) who reported increased oxidative stress through reduced superoxide dismutase activity, resulting in the loss of symbionts.

Heat stress increases the release of ROS by algal symbionts. Additionally, the quenching of ROS via the experimental addition of antioxidants can significantly reduce the severity of bleaching in corals (Lesser 1996, 1997). However, more recent findings challenge the proposed importance of photosynthetic ROS production as the main mechanism of coral bleaching (Rädecker et al. 2021). First, bleaching can occur in the dark without photosynthetic ROS production (Tolleter et al. 2013). Second, oxidative stress in heat-stressed coral host cells may occur before any photoinhibition of algal symbionts occurs (Krueger et al. 2015). Third, bleaching was observed in the anemone model system *Exaiptasia diaphana* without concomitant increases in net ROS release of their algal symbionts; at the same time, quenching of ROS with extraneous antioxidants mitigated bleaching while simultaneously increasing ROS release from Symbiodiniaceae (Dungan et al. 2022). Finally, extracellular levels of superoxide do not correlate with symbiont abundance or bleaching status of corals during heat stress (Diaz et al. 2016). In conclusion, the role of excess ROS release in coral bleaching is highly complex and may only be one of several mechanisms (Rädecker et al. 2021).

4.5.2 The Resource Competition Hypothesis (As an Extension of the Arrested Phagosome Hypothesis)

The cellular catalyst of algal symbiont expulsion, and as such, bleaching, remains unknown (Pogoreutz et al. 2020). While ROS levels and oxidative damage in the coral host tissue may be the main cue for coral bleaching (Lesser 1996, 1997; Downs et al. 2002), levels of oxidative damage in the host depend on the interplay of ROS production, antioxidant capacity, and cellular repair processes (Weis 2008). Consequently, reductions in cellular energy availability have direct effects on the host's ability to cope with oxidative stress and thereby determine thermal bleaching thresholds. As such, changes in symbiotic nutrient cycling may constitute a major cue for symbiont expulsion and coral bleaching (Rädecker et al. 2021).

The algal symbionts are typically acquired through phagocytosis by nutritive phagocytes (Fadlallah 1983). In contrast to food particles, however, compatible algal symbionts engulfed by these phagocytes are not digested. Rather, they persist and proliferate within the phagocytes, which at this stage are referred to as "symbiosomes" (Colley and Trench 1983). This process of phagocytosis is a complex part

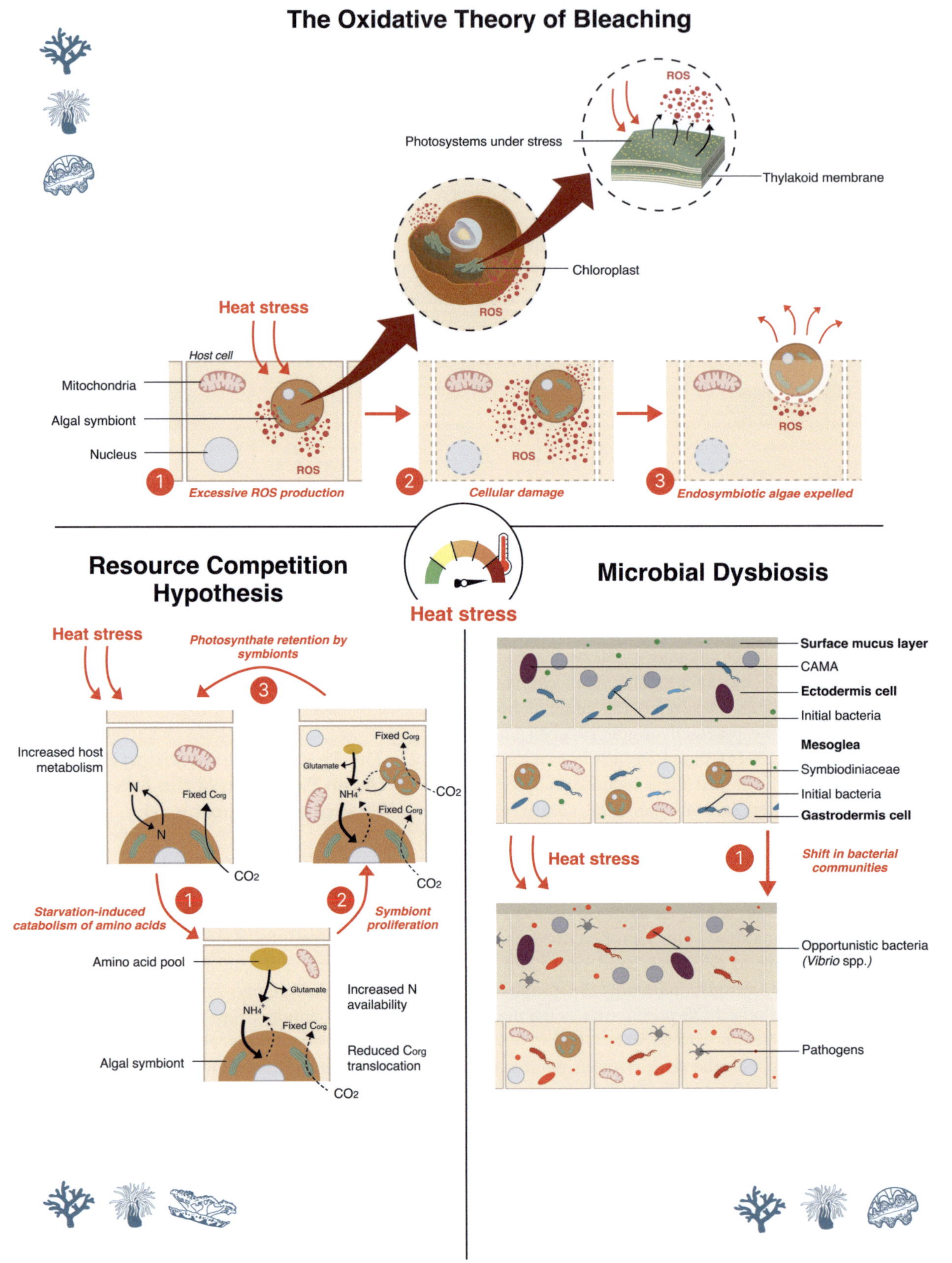

Fig. 4.4 Main hypotheses of bleaching. The oxidative theory of bleaching (top), the resource competition hypothesis (bottom left), and microbial dysbiosis (bottom right) showing the distribution of the Symbiodiniaceae, bacteria, and cell-associated microbial aggregates (CAMAs) in the surface mucus layer, ectodermis, and gastrodermis. All three hypotheses are illustrated for scleractinian corals, while blue icons represent the reef organisms for which each mechanism has been identified as a major driver of bleaching

of innate immunity conserved across the Metazoa (Underhill and Ozinsky 2002). In this context, the Arrested Phagosome Hypothesis (APH) by Hill and Hill (2012) may provide a comprehensive framework to understand a central aspect of bleaching. The APH posits that the continuous release of organic carbon by the Symbiodiniaceae mimics prey digestion, resulting in arrested phagosome maturation. Consequently, reduced carbon release by senescent, stressed, or dying cells with reduced rates of (in)organic carbon fixation, assimilation, and translocation to the host may be insufficient to maintain phagosome arrest, triggering phagosome maturation, resulting in digestion (Hill and Hill 2012; Rädecker et al. 2021).

The APH would therefore suggest that changes to symbiotic nutrient cycling will directly result in symbiont expulsion (Rädecker et al. 2021) (Fig. 4.4). Indeed, recent studies suggest that heat stress undermines symbiotic nutrient cycling in photosymbiotic Cnidaria (corals and *Cassiopea*) by reducing carbon translocation by the algae, ultimately resulting in host starvation (Hughes et al. 2010; Rädecker et al. 2021; Toullec et al. 2024b). Briefly, heat stress results in energetic imbalances (Gillooly et al. 2001; Rodrigues and Grottoli 2007), resulting in the enhanced consumption and depletion of carbon reserves by the host (Davies 1991; Fitt et al. 1993; Rodrigues and Grottoli 2007; Rädecker et al. 2021; Toullec et al. 2024b). Once energy storage compartments (carbohydrates and lipids) are depleted, the energy-deprived host may rapidly enter a progressive, heat-induced metabolic switch from net anabolism (constructive metabolism, i.e., synthesis and assimilation of complex molecules and storage of energy) to catabolism (destructive metabolism, i.e., breakdown of complex molecular and energy release) under heat stress. In photosymbiotic Cnidaria, this is directly reflected in holobiont nitrogen metabolism. Unstressed hosts use carbon backbones for amino acid synthesis through ammonium assimilation, while heat-stressed hosts rapidly switch to a catabolic state. Here, proteins and amino acids may be broken down as a carbon source for energy metabolism, resulting in excess ammonium release by the host (Haunerland 2003), eventually resulting in the breakdown of host protein homeostasis (Rädecker et al. 2021; Toullec et al. 2024b). Importantly, the excess ammonium (cleaved from amino acids during catabolism) may be taken up by the algal symbionts and used for opportunistic growth. This increasingly results in retention of photosynthates for their own metabolism, locking the host in a deadly feedback loop (Rädecker et al. 2021). As a result, starvation of the coral host rather than the loss of algal symbionts may be the direct cause of coral mortality following bleaching (Anthony et al. 2017; Rädecker et al. 2021; Toullec et al. 2024b). Of note, excess nitrogen through stimulated prokaryotic nitrogen fixation may rapidly destabilize the delicate balance in nutrient cycling in corals, rapidly causing coral bleaching even in the absence of heat stress (Pogoreutz et al. 2017). In general, corals with higher energy reserves are more resilient to heat stress (Levas et al. 2018). Importantly, starvation through carbon limitation occurs days before any visible signs of stress (Rädecker et al. 2021).

These mechanistic studies were conducted under controlled laboratory conditions in the absence of other common reef organisms. On the reef, more complex responses may occur due to additional (a)biotic factors and organismal interactions. For instance, a laboratory manipulation found that the presence of ectosymbiotic fish can increase the density of the algal symbionts associated with their anemone host (Roopin and Chadwick 2009). This observation was driven by the transfer of ammonia from clownfish waste metabolism (Godinot and Chadwick 2009). Further, field-based isotopic tracer studies showed that carbon and nitrogen released by ectosymbiotic clownfish are taken up by their anemone hosts and associated intracellular algae (Cleveland et al. 2010), suggesting intricate metabolic interactions that may be of functional importance in host physiology and ecological resilience to environmental stress.

4.5.3 Microbial Dysbiosis As a Driver of Bleaching

The microbial community, or microbiome, is assumed to contribute to the genetic makeup of the holobiont, affecting its health, resilience, and acclimatization/adaptation *sensu latu* (Pogoreutz et al. 2020; Voolstra and Ziegler 2020). Changes to this genetic make-up through shifts in the microbiome may convey functional adjustment to changing environments (Grottoli et al. 2018; Voolstra and Ziegler 2020). Such microbiome changes, however, have largely been observed in stressed, diseased, or bleached corals, sea anemones, and sponges (Bourne et al. 2008; Ziegler et al. 2017; Guibert et al. 2020; Posadas et al. 2022; Sydnor et al. 2023; among many others), likely reflecting an unfavorable "dysbiotic" state (Boilard et al. 2020) (Fig. 4.4). In most cases, it remains unclear whether such microbiome changes are the cause or consequence of stressed, diseased, or bleached phenotypes (Voolstra et al. 2024).

Few studies have explored specific microbial mechanisms implicated in the bleaching response of corals and other photosymbiotic reef organisms. Here, bleaching after the propagation of virulent *V. shiloii* and *V. coralliilyticus* in coral tissues following heat stress has received attention (Kushmaro et al. 1997; Ben-Haim et al. 2003; Rosenberg and Falkovitz 2004), along with *Vibrio* in sea anemones and giant clams (Guibert et al. 2020; Sydnor et al. 2023). Specific mechanisms facilitating *Vibrio*-driven bleaching under elevated temperatures include the production of superoxide dismutase by virulent strains of *V. shiloi,* which facilitate tissue penetration and infection in the Mediterranean coral *Oculina patagonica* (Banin et al. 2003), and of a proline-rich peptide by that inhibits photosynthesis in Symbiodiniaceae (Banin et al. 2001). The latter is also hypothesized to be employed by *V. coralliilyticus*, followed by an increase in bacterial extracellular protease activity effectively resulting in tissue lysis and mortality (Ben-Haim et al. 2003). Another potential virulence factor facilitating infection with (and subsequent bleaching by) *Vibrios* may be flagellar assembly and chemotaxis gene expression (Gao et al. 2021). Mutants of *V. coralliilyticus* lacking flagellae or chemotaxis-related genes are unable to exhibit a chemotactic response toward and to adhere to coral and subsequently cause infection (Meron et al. 2009; Ushijima and Häse 2018). Finally, virulent *V. coralliilyticus* outcompete non-toxigenic *Vibrio* (and potentially other bacteria) through the production of hydrogen peroxide. This metabolite triggers lytic prophage induction in the non-toxigenic *Vibrio*, killing it and resulting in a competitive colonization advantage for the virulent strain (Wang et al. 2022).

It should be noted that, while most research scrutinizing microbial (bacterial) causes of bleaching has focused on *Vibrio* spp., this bacterial genus constitutes only one of the diverse bacteria known to propagate during bleaching. Other bacteria propagating in bleached or stressed corals, include members of the Flavobacteriaceae and Rhodobacteraceae, which remain understudied but deserve further investigation. Finally, bleaching studies should also include other photosymbiotic host-microorganism associations (e.g., interactions, potential nutrient exchange) and symbiont-centric genomic and culture-based approaches to elucidate prokaryotic functional diversity under current and future climate change conditions.

In conclusion, these main hypotheses paint a complex, yet not quite fully resolved picture of bleaching in corals and other photosymbiotic reef holobionts. These hypotheses are likely not exclusive, but complementary. Future work will have to unequivocally identify and mechanistically ground truth the interplay between (a)biotic drivers of bleaching, and might have to take into consideration additional factors currently not yet identified.

4.6 Developing Protection Strategies for Coral Reefs to Severe Future Bleaching Events: A Race Against Time

The descriptive (surveys, monitoring, and modeling) and functional bleaching studies (physiological and molecular mechanisms via field and aquarium experiments) described in this chapter need to be integrated into an intervention framework combining traditional and emerging protection strategies to maintain or even rebuild coral reefs. Here, we address the critical local and global implementation of human interventions to protect coral reefs from the increasing frequency, duration, and intensity of bleaching events, including CO_2-emission reduction policies, manipulation of naturally adapted organisms, and mitigation of bleaching stressors.

4.6.1 Pursuing Global Reef Protection: A Challenging Endeavor

Considering the threats that coral reefs are facing worldwide from ongoing global warming, most coral reef experts expect that passive recovery (i.e., natural recovery without human intervention) will be increasingly insufficient as mass bleaching and mortality events become more frequent and severe (Hughes et al. 2018). Cutting CO_2 emissions has been identified as the primary solution to effectively manage and conserve coral reefs under climate change (Hoegh-Guldberg et al. 2018b; Kleypas et al. 2021). However, the effective implementation of CO_2 emissions reduction policies still remains uncertain. The long-term persistence of climate refugia is thus being questioned as they are not expected to withstand current projected rising temperatures (Dixon et al. 2022). Likewise, adaptation processes of most reef organisms (especially those with long generation times such as corals, 10–100 years) might not keep up with the rapid pace of climate warming (Logan et al. 2014, 2021). Integrating proactive and locally based actions, alongside greenhouse gas emissions reductions, may be the best strategy to help build the resilience of coral reefs to future climate change (Kennedy et al. 2013; Kleypas et al. 2021; Knowlton et al. 2021). In this context, restorative assisted evolution initiatives have been proposed to create more resilient corals to bleaching-inducing conditions (van Oppen et al. 2015; Anthony et al. 2017; Quigley et al. 2019), that could also be applied to other photosymbiotic reef organisms. Furthermore, the establishment of artificial local refugia, acting as gene banks or biodiversity conservatories protected against climate change, seems now more than ever of paramount importance (Zoccola et al. 2020). As such, they could facilitate the gradual adaptation of organisms to rising temperatures, safeguarding existing genetic diversity and ensuring the success of coral reef restoration efforts in the face of climate change. In addition, protecting and aggregating remaining wild individuals would significantly enhance reproductive success (Ramah et al. 2019).

4.6.2 The Need for Locally Based Innovative Solutions

4.6.2.1 Enhancing Thermal Resistance: A Toolkit to Cope with Bleaching

Traditional restoration methods that rely on outplanting preexisting coral genotypes have shown limited effectiveness in enhancing climate resilience, unless sustained efforts are upheld over extensive timeframes and large spatial scales (DeFilippo et al. 2022). In contrast, the use of naturally thermotolerant genotypes has emerged as a more successful approach (Morikawa and Palumbi 2019; DeFilippo et al. 2022) though there is a risk of known and unknown tradeoffs. As such, understanding the thermal history of reefs is crucial, as corals preexposed to higher or more variable temperatures often exhibit better tolerance and resilience to bleaching (McClanahan et al. 2007; Hughes et al. 2019) but not always (Barott et al. 2021). To bolster thermal tolerance in corals, various approaches are being explored, such as selective breeding (Caruso et al. 2021; Drury et al. 2021; Humanes et al. 2021), assisted gene flow (translocation of organisms from warmer reefs or extreme environments; Dixon et al. 2015; Baums et al. 2019; Quigley et al. 2019; Hagedorn et al. 2021), and hybridization with stress-tolerant species (van Oppen et al. 2017; Chan et al. 2018). Additionally, microbiome manipulations through the inoculation of tolerant strains of Symbiodiniaceae (extracted from resistant corals or experimentally heat-evolved) and beneficial bacteria hold great potential to mitigate bleaching and evade mortality (Chakravarti et al. 2017; Chakravarti and van Oppen 2018; Quigley et al. 2021; Doering et al. 2021; Santoro et al. 2021; Peixoto et al. 2022).

These restoration concepts have mostly been tested in scleractinian corals, despite their similar potential for other photosymbiotic reef organisms. Selective breeding programs for giant clams, for instance, have historically prioritized traits desirable for human consumption or the aquarium trade (Zhou et al. 2021). Research efforts are currently underway to identify tolerant symbionts for giant clams (Mies 2019; Guibert et al. 2020), that could be used as beneficial microorganisms to enhance the success of restocking programs (Watson and Neo 2021). Similar efforts are less common for the other organisms considered here. For instance, inoculation of high free radical scavenging bacteria, aimed at metabolizing ROS, has recently been tested on sea anemones but without success as the association did not persist (Dungan et al. 2022). Nonetheless, these innovative strategies, coupled with a comprehensive understanding of thermal history, offer promising interventions for safeguarding coral reefs against bleaching events. Still, whether the study models focus on corals or other photosymbiotic organisms, numerous unanswered questions and challenges must be addressed to ensure their successful implementation in situ and at a large scale (van Oppen et al. 2015; Baums et al. 2019; Peixoto et al. 2021; Doering et al. 2021).

As breeding programs, and subsequent restoration efforts, often face substantial mortality rates and high operational costs (Meesters et al. 2015; Bayraktarov et al. 2019; Hughes et al. 2023), alternative conservation strategies should be considered in conjunction with these approaches to effectively preserve existing wild populations. This can be particularly beneficial for organisms such as giant clams that have low-generation turnover rates, by preventing premature mortality before reaching reproductive maturity (Watson and Neo 2021). Indeed, given adequate time and genetic diversity, restoration approaches that prioritize the conservation of genetic variance may prove more fruitful than those solely reliant on the introduction of heat-tolerant genotypes (DeFilippo et al. 2022).

4.6.2.2 Cooling and Shading: Mitigation of the Main Bleaching-Inducing Stressors

Because bleaching is primarily caused by heat stress and exacerbated under high irradiance (Buck et al. 2002; Hill and Scott 2012), regulating one or both of these stressors has the potential to reduce, or at least buffer the occurrence of bleaching. Inspired by natural upwelling zones where thermal stress can be significantly decreased in duration, severity, and frequency (Chollett et al. 2010; Riegl et al. 2019), artificial upwelling (AU) has shown great potential as a conservation strategy (Feng et al. 2020; Sawall et al. 2020). This involves the use of deep pumped and nutrient-rich waters to cool shallow reefs during peak summer temperatures to avoid reaching bleaching threshold. While natural upwelling zones can only provide effective protection under specific circumstances (upwelling event of sufficient cooling capacity occurring simultaneously with heat stress; Chollett et al. 2010), AU allows for greater control of the cooling process (time, duration, and temperature). AU may therefore offer better potential to create permanent havens for reef organisms compared to climate refugia in upwelling zones (Chollett et al. 2010; Dixon et al. 2022). Yet, most studies rely on modeling approaches (Chollett et al. 2010; Baird et al. 2020; Feng et al. 2020), with no in situ testing so far. Only one experiment has showed that water collected from deeper zones (100 vs. 50 m) has a stronger effect in reducing bleaching but this was in laboratory settings (Sawall et al. 2020).

Naturally occurring shading, such as cloud cover and mangroves, has also been observed to provide some level of protection against bleaching, by reducing its severity or enhancing resilience capacity (Mumby et al. 2001; Yates et al. 2014; Stewart et al. 2021). Likewise, studies comparing shallow and mesophotic reefs have shown a decrease in the prevalence of coral bleaching with increasing depth and subsequent reduced availability of light (Pérez-Rosales et al.

2021; but see Haguenauer et al. 2021). However, recent studies have challenged the notion that depth alone can provide consistent refuge from bleaching, suggesting that many upper mesophotic reefs (<35–40 m) may be as vulnerable to climate change as shallow reefs (Venegas et al. 2019). In light of these findings, both natural (through depth) and artificial shading (through cloth or fogging for instance) have been proposed as a means to reduce light exposure and artificially create refugia for corals (Rau et al. 2012) and giant clams (Andréfouët et al. 2018; Watson and Neo 2021). For example, artificial shading (e.g., through cloth) implemented during and following high-temperature episodes has shown great promise in mitigating bleaching (Coelho et al. 2017; Coles et al. 2018; Berg et al. 2021), and can even outperform other management strategies (Winston et al. 2022).

Therefore, these innovative interventions have the potential to locally limit the primary stressors inducing bleaching. However, the existing research on shading or cooling primarily focuses on its effects on corals, emphasizing the need to investigate the broader impact it can have on the entire ecosystem. Such protection could be deployed over a relatively large area (e.g., 1 to 5 km resolution)—although these interventions are limiting in terms of scalability—and protect all photosymbiotic organisms present, as opposed to the more specific and targeted interventions discussed earlier. As such, these strategies are valuable interventions that warrant serious consideration.

4.6.2.3 Mitigation of Other Local Stressors

On almost all reefs, chronic local stressors, such as coastal development, pollution, eutrophication, overexploitation and destructive fishing practices, act synergistically with climate change to induce bleaching and mortality events that substantially alter coral reef communities with largely unknown consequences for ecosystem resilience (Good and Bahr 2021; Knowlton et al. 2021). These local stressors can intensify coral loss, especially by lowering thermal tolerance thresholds (Carilli et al. 2010). As such, there is an urgent need to identify the local stressors that exacerbate mortality under ocean warming, which in turn could facilitate management strategies to secure coral reefs in the face of climate change. However, a recent systematic review reported that only 10% of the studies on the effects of marine heat waves (20 out 194 articles from 2014–2021) included the influence of local stressors (Baum et al. 2023).

Among these factors, high macroalgae densities, which are promoted by eutrophication and loss of herbivorous animals, can magnify coral loss following bleaching events (Donovan et al. 2021; Smith et al. 2022). Proactive management strategies, including manual removal of macroalgae, introduction of bioagents, and shading, have shown promising results to reduce macroalgal overgrowth (Ceccarelli et al. 2018; Dajka et al. 2021) and increase coral recruitment (Bulleri et al. 2018). Likewise, reducing fisheries of herbivorous species (e.g., fish, urchins), as well as minimizing nutrient pollution to prevent macroalgal proliferation, can enhance reef resilience (Knoester et al. 2023). It is equally important to address the source of pollutants and improve wastewater treatment locally since the reduction of urban runoffs was found to have the greater positive impact on reducing bleaching severity (Winston et al. 2022). More studies should therefore combine a suite of local abiotic (temperature, light, depth, wave exposure, turbidity) and biotic factors (benthic assemblages, recruitment, herbivory) in enhancing reef resilience (as in Donovan et al. 2021).

4.6.2.4 Challenges of Implementing Intervention Strategies on Coral Reefs

Several challenges and considerations arise when implementing interventions to safeguard coral reefs. Prioritization and trade-offs become necessary as it is not possible to save every species (Anthony et al. 2020). However, the selection of one species over another may have strong cascading effects on reef ecosystem functioning. To date, most studies have focused on a limited number of coral species (e.g., McLachlan et al. 2020; Sawall et al. 2020; Tagliafico et al. 2022; Hughes et al. 2023) and have revealed that the success of the strategy varies within and among species. This emphasizes the need for comprehensive studies encompassing a wide range of photosymbiotic organisms (corals included), whose distribution across reefs and hence optimums may differ from the current coral models that are being studied. Today, only a few studies have examined the recovery period following coral bleaching (McLachlan et al. 2020; Tagliafico et al. 2022), and even less for bleaching events of the other photosymbiotic organisms discussed in this chapter (but see Beldade et al. 2017). Therefore, it is necessary to increase the frequency of post-bleaching monitoring to evaluate long-term consequences on population dynamics and to determine the overall resilience of reefs (Grottoli et al. 2021). As coral reefs may simultaneously face multiple environmental stressors, careful consideration is also needed to ensure that these interventions do not inadvertently compromise reef recovery and dynamics (Suggett et al. 2013). Scaling up interventions and addressing practicality pose additional challenges, as they are yet to be fully established for all mentioned strategies (Baird et al. 2020; Tagliafico et al. 2022; Suggett and van Oppen 2022). Addressing these challenges and knowledge gaps requires collaborative research efforts, speeding-up testing to better inform decision-makers through risk assessment, and quicken their deployment in the field pending their approvals. Used collectively, these management strategies, including cutting greenhouse gas emissions, could be able to effectively protect and restore coral reef ecosystems (Kleypas et al. 2021).

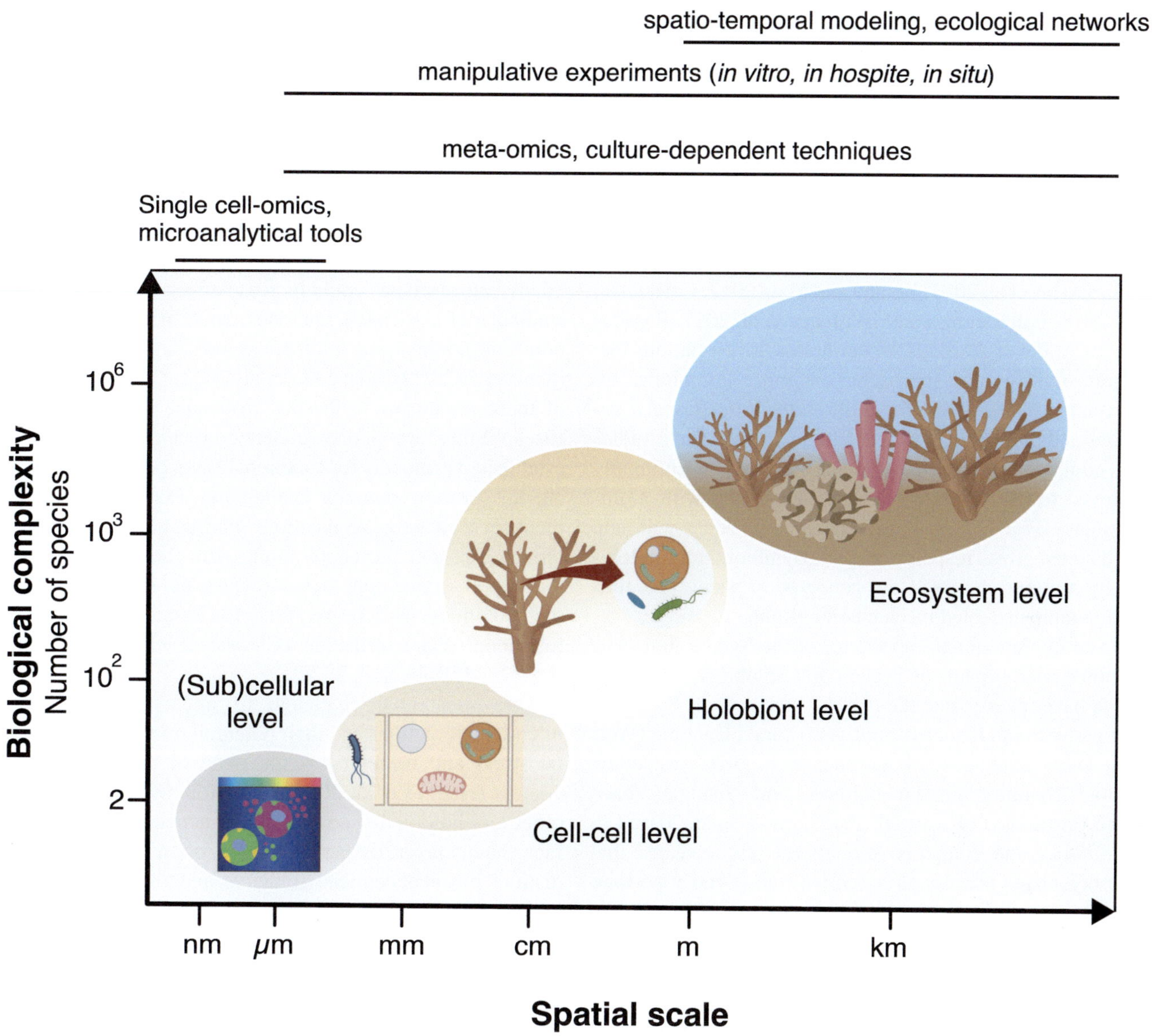

Fig. 4.5 Schematic summary of key research questions and topics to improve our understanding of the mechanisms of bleaching and recovery in photosymbiotic reef organisms and cascading effects on the functioning of coral reef ecosystems

4.7 Roadmap for Future Holobiont Bleaching Studies: Integrating Multi-omics Approaches

Despite the broad examples of bleaching studies explored in this chapter, diverse knowledge gaps remain regarding the functional biology of reef photosymbioses and its breakdown. Here, we provide a conceptual perspective for future holobiont bleaching research by integrating multi-omics tools to multiple levels of organismal organization, ranging from cellular to holobiont, and ultimately impacting the ecological functioning of coral reef ecosystems (Fig. 4.5).

While research focusing on the bleaching response of photosymbiotic reef organisms is steadily increasing, few studies are targeting the entire holobiont as a functional unit (Planes et al. 2019; Belser et al. 2023). Thus, we are potentially missing important molecular interactions among holobiont members that may contribute to (or negatively affect) bleaching tolerance. Indeed, the stress responses of reef organisms are likely not only defined by their genetics, but also their epigenetic responses and the associated microbiome which may confer fitness traits subject to intensified selection by climate change (Voolstra and Ziegler 2020), and as such may be central to bleaching susceptibility. Harnessing the power of multi-omics surveys of major groups of reef organisms along environmental gradients and exposed to various stressors should thus become one of the priorities to understand the molecular and physiological responses of these vulnerable organisms. The application of individual -omics-based approaches, such as (meta) genomics (to elucidate functional potential), (meta) transcriptomics (profiling of expressed functional genes), (meta)proteomics (profiles of protein abundances), epigenomics (patterns of epigenetic modifications), and metabolomics (metabolite profiles) has helped identify molecular signatures associated with the plasticity and adaptation of reef organisms and their associated microbes (Fan et al. 2013; Barshis et al. 2013; Kenkel and Matz 2017; Cziesielski et al. 2018; Fuller et al. 2020; Williams et al. 2021; Posadas et al. 2022; Pei et al. 2022). Nevertheless, investigations that integrate multi-omics approaches of entire holobionts are necessary to facilitate their in-depth study of functional associations and inter-kingdom interactions. Ideally, such studies will combine physiological measurements traditionally employed to characterize bleaching responses with multi-omics data (Fig. 4.5). Such integrative approaches will help elucidate the stress response of photosymbiotic reef holobionts by bridging the gap from genotype to phenotype, thus offering unprecedented mechanistic insight.

Given the functional importance of the Symbiodiniaceae and other microorganisms in the coral holobiont, numerous studies have investigated the cellular mechanisms by which these partners can induce (or mitigate) coral bleaching (Weis 2008; Davy et al. 2012; Pogoreutz et al. 2020). Recently, multiple Symbiodiniaceae cultures and genomes have become available (Xiang et al. 2013; Aranda et al. 2016; Lin et al. 2019), and comparative analyses have revealed that these organisms possess an extensive transporter repertoire for carbon and nitrogen compounds (Aranda et al. 2016). While our understanding of the mechanisms leading to the breakdown of the coral-algal symbiosis builds upon decades of research (Weis 2008; Davy et al. 2012), the role of other microorganisms in bleaching remains poorly understood (Bourne et al. 2008; Ziegler et al. 2017; Boilard et al. 2020). Here, bacterial community compositions and dynamics during bleaching are comparatively well understood (for scleractinian corals; e.g., Ziegler et al. 2017; Ziegler 2019), but our understanding of their functions and interactions within the holobiont remains limited (Pogoreutz et al. 2020, 2022; Voolstra et al. 2024). Future studies will benefit from functional interrogation to identify and test specific interactions (e.g., specific metabolic pathways) by integrating (meta)-omics applications in combination with culture-dependent work (Fig. 4.5), for example, through direct microbiome manipulation of coral reef holobionts and microbial coculturing studies, potentially in conjunction with the characterization of microbial (bio)activities (Li et al. 2022; Pogoreutz et al. 2022; Roik et al. 2022). Here, (un)targeted metabolomics approaches and epigenetics will help advance our knowledge of the molecular responses and physiological implications of bleaching. So far, the use of these approaches in bleaching studies has been limited to scleractinians but offers new avenues to assess microbial roles in the bleaching response of photosymbiotic reef organisms (Barno et al. 2021; Santoro et al. 2021).

While the priority for this proposed work might initially be on entire holobionts along with the main symbionts (e.g., the Symbiodiniaceae and prevalent bacterial symbionts, such as the Endozoicomonadaceae; Pogoreutz et al. 2022; Keller-Costa et al. 2022; Hochart et al. 2023; Pogoreutz and Ziegler 2024), future studies could increase efforts to include understudied reef-associated microbes, such as fungi and other non-Symbiodiniaceae microeukaryotes, as well as viruses (Roik et al. 2022; Bonacolta et al. 2023). The contributions of these organisms to thermal resistance and/or bleaching susceptibility are poorly understood, and the paucity of established protocols for functional (and, partly, taxonomic) characterization remains challenging. As such, improvements in sequencing applications (marker gene development, blocking primer protocols) along with the development of low input -omics approaches (next-generation sequencing, metabolomics) will be required for better taxonomic and functional characterization (Clerissi et al. 2018; Hughes et al. 2022; Roik et al. 2022).

One substantial challenge to understand reef host–microbe interactions and their potential roles in environmental stress and bleaching is the issue of "missing" spatial context (Hughes et al. 2022; Roik et al. 2022). Corals (and other reef holobionts) consist of distinct compartments (surface mucus layer, tissues, skeleton), often characterized by distinct physicochemical gradients and unique microenvironments (Hughes et al. 2022). For sequencing, these compartments are typically homogenized, resulting in the complete loss of spatial information. However, different sampling and analytical platforms increasingly permit the exploration of spatial context in marine host–microbe interactions. One prominent example is the combination of isotopic labeling with correlative electron microscopy and NanoSIMS analysis, that is, "elemental" mapping for subcellular localization of anabolically assimilated compounds (e.g., hot spots of carbon enrichment) within the tissues. Initially developed for geological applications and surface analysis, NanoSIMS has become a powerful tool to assess metabolic interactions in reef host-microbe systems (Yang et al. 2019; Rädecker et al. 2021, 2022; Cárdenas et al. 2022).

Functional interrogation of corals and many other photosymbiotic reef holobionts has been, and still remains, a challenge. One of the main problems is that these non-model marine organisms are often not easy to maintain in captivity: husbandry can be expensive, many organisms are sensitive to minor changes in their environment, and their longevity can be an issue for trans-generational studies (Rädecker et al. 2018). Many corals in particular are not amenable to microbiome manipulation, as they cannot be easily maintained in an "aposymbiotic" (symbiont-free state over meaningful periods of time) (Rädecker et al. 2018), and even the anemone model Aiptasia is difficult to render axenic or gnotobiotic (Hartman et al. 2022). However, the recent establishment of a new coral model system *Galaxea fascicularis* gives hope to facilitate future functional work in corals (Puntin et al. 2023).

Further potential barriers to functional work in marine holobionts are (1) the notoriously low cultivability of marine microbes, including bacteria (Alain and Querellou 2009; Pogoreutz et al. 2022), (2) sequencing approaches to assess microbial functional capacity and activity being challenging due to high proportions of host nucleic acids disproportionately skewing sequencing coverage in "-omics" datasets (Pereira-Marques et al. 2019; Pogoreutz et al. 2022), (3) the lack of appropriate tools and marker genes for targeting specific functional genes and/or for specific microeukaryotic groups, including fungi (Clerissi et al. 2018; Roik et al. 2022; Pogoreutz et al. 2025), and (4) the issue of missing spatial information inherent to sample preparation of high-throughput applications (Hughes et al. 2022). However, given the drive of the international scientific community to increase our understanding of the molecular and physiological mechanisms of bleaching and to develop a decision framework for human inventions that exploit the natural adaptive capacity of these vulnerable reef organisms, there is hope to collectively find a solution for the current coral reef crisis.

References

Achlatis M, Van Der Zande RM, Schönberg CH, Fang JK, Hoegh-Guldberg O, Dove S (2017) Sponge bioerosion on changing reefs: ocean warming poses physiological constraints to the success of a photosymbiotic excavating sponge. Sci Rep 7:10705

Alain K, Querellou J (2009) Cultivating the uncultured: limits, advances and future challenges. Extremophiles 13:583–594

Alves Monteiro HJ, Brahmi C, Mayfield AB, Vidal-Dupiol J, Lapeyre B, Le Luyer J (2019) Molecular mechanisms of acclimation to long-term elevated temperature exposure in marine symbioses. Glob Change Biol 26:1271–1284

Andréfouët S, Van Wynsberge S, Gaertner-Mazouni N, Menkes C, Gilbert A, Remoissenet G (2013) Climate variability and massive mortalities challenge giant clam conservation and management efforts in French Polynesia atolls. Biol Conserv 160:190–199

Andréfouët S, Dutheil C, Menkes CE, Bador M, Lengaigne M (2015) Mass mortality events in atoll lagoons: environmental control and increased future vulnerability. Glob Change Biol 21:95–205

Andréfouët S et al (2018) Adaptative management for the sustainable exploitation of lagoon resources in remote islands: lessons from a massive El Nino-induced giant clam bleaching event in the Tuamotu atolls (French Polynesia). Environ Conservation 45:30–40

Anthony K et al (2017) New interventions are needed to save coral reefs. Nat Ecol Evol 1:1420–1422

Anthony KRN et al (2020) Interventions to help coral reefs under global change – A complex decision challenge. Plos One 15:e0236399

Aranda M et al (2016) Genomes of coral dinoflagellate symbionts highlight evolutionary adaptations conducive to a symbiotic lifestyle. Sci Rep 6:39734

Arillo A, Bavestrello G, Burlando B, Sarà M (1993) Metabolic integration between symbiotic cyanobacteria and sponges: a possible mechanism. Mar Biol 117:159–162

Armstrong EJ, Roa JN, Stillman JH, Tresguerres M (2018) Symbiont photosynthesis in giant clams is promoted by V-type H+-ATPase from host cells. J Exp Biol 221:jeb177220

Armstrong EJ, Dubousquet V, Mills SC, Stillman JH (2020) Elevated temperature, but not acidification, reduces fertilization success in the small giant clam. Tridacna maxima. Mar Biol 167:8

Atkinson M, Falter J, Black K, Shimmield G (2003) Biogeochemistry of marine systems. In: Coral reefs. Blackwell, Oxford, pp 40–64

Baird ME, Green R, Lowe R, Mongin M, Bougeot E (2020) Optimising cool-water injections to reduce thermal stress on coral reefs of the Great Barrier Reef. Plos One 15:e0239978

Baker DM, Freeman CJ, Knowlton N, Thacker RW, Kim K, Fogel M (2015) Productivity links morphology, symbiont specificity and bleaching in the evolution of Caribbean octocoral symbioses. ISME J 9:2620–2629

Banin E, Sanjay KH, Naider F, Rosenberg E (2001) A proline-rich peptide from the coral pathogen *Vibrio shiloithat* inhibits photosynthesis of zooxanthellae. Appl Environ Microbiol 67:1536–1541

Banin E, Vassilakos D, Orr E, Martinez RJ, Rosenberg E (2003) Superoxide dismutase is a virulence factor produced by the coral bleaching pathogen *Vibrio shiloi*. Curr Microbiol 46:0418–0422

Barno AR, Villela HD, Aranda M, Thomas T, Peixoto RS (2021) Host under epigenetic control: A novel perspective on the interaction between microorganisms and corals. BioEssays 43:2100068

Barott KL, Venn AA, Perez SO, Tambutté S, Tresguerres M (2015) Coral host cells acidify symbiotic algal microenvironment to promote photosynthesis. Proc Natl Acad Sci USA 112:607–612

Barott KL et al (2021) Coral bleaching response is unaltered following acclimatization to reefs with distinct environmental conditions. Proc Natl Acad Sci USA 118:e2025435118

Barott KL, Thies AB, Tresguerres M (2022) V-type H+-ATPase in the symbiosome membrane is a conserved mechanism for host control of photosynthesis in anthozoan photosymbioses. R Soc Open Sci 9:211449

Barshis DJ, Ladner JT, Oliver TA, Seneca FO, Traylor-Knowles N, Palumbi SR (2013) Genomic basis for coral resilience to climate change. Proc Natl Acad Sci USA 110:1387–1392

Barshis DJ, Ladner JT, Oliver TA, Palumbi SR (2014) Lineage-specific transcriptional profiles of *Symbiodinium* spp. unaltered by heat stress in a coral host. Mol Biol Evol 31:1343–1352

Barua A, Kashimoto R, Khalturin K, Satoh N, Laudet V (2022) The genetic basis for adaptation in giant sea anemones to their symbiosis with anemonefish and Symbiodiniaceae. bioRxiv 2022-09

Baum JK, Claar DC, Tietjen KL, Magel JM, Maucieri DG, Cobb KM, McDevitt-Irwin JM (2023) Transformation of coral communities subjected to an unprecedented heatwave is modulated by local disturbance. Sci Adv 9:eabq5615

Baumgarten S, Cziesielski MJ, Thomas L, Michell CT, Esherick LY, Pringle JR, Aranda M, Voolstra CR (2018) Evidence for miRNA-mediated modulation of the host transcriptome in cnidarian-dinoflagellate symbiosis. Mol Ecol 27:403–418

Baums IB et al (2019) Considerations for maximizing the adaptive potential of restored coral populations in the western Atlantic. Ecol Appl 29:e01978

Bay RA, Palumbi SR (2014) Multilocus adaptation associated with heat resistance in reef-building corals. Curr Biol 24:2952–2956

Bay RA, Palumbi SR (2015) Rapid acclimation ability mediated by transcriptome changes in reef-building corals. Genome Biol Evol 7:1602–1612

Bayraktarov E et al (2019) Motivations, success, and cost of coral reef restoration. Restor Ecol 27:981–991

Bedgood SA, Mastroni SE, Bracken ME (2020) Flexibility of nutritional strategies within a mutualism: food availability affects algal symbiont productivity in two congeneric sea anemone species. Proc R Soc B 287:20201860

Beldade R, Blandin A, O'Donnell R, Mills SC (2017) Cascading effects of thermally-induced anemone bleaching on associated anemonefish hormonal stress response and reproduction. Nat Commun 8:716

Belser C et al (2023) Integrative omics framework for characterization of coral reef ecosystems from the Tara Pacific expedition. Sci Data 10:326

Ben-Haim Y, Zicherman-Keren M, Rosenberg E (2003) Temperature-regulated bleaching and lysis of the coral *Pocillopora damicornis* by the novel pathogen *Vibrio coralliilyticus*. Appl Environ Microbiol 69:4236–4242

Bennett HM, Altenrath C, Woods L, Davy SK, Webster NS, Bell JJ (2017) Interactive effects of temperature and pCO_2 on sponges: from the cradle to the grave. Glob Change Biol 23:2031–2046

Berg JT, David CM, Gabriel MM, Bentlage B (2021) Fluorescence signatures of persistent photosystem damage in the staghorn coral *Acropora* cf. *pulchra* (Anthozoa: Scleractinia) during bleaching and recovery. Mar Biol Res 16:643–655

Black NA, Voellmy R, Szmant AM (1995) Heat shock protein induction in *Montastraea faveolata* and *Aiptasia pallida* exposed to elevated temperatures. Biol Bull 188:234–240

Boilard A, Dubé CE, Gruet C, Mercière A, Hernandez-Agreda A, Derome N (2020) Defining coral bleaching as a microbial dysbiosis within the coral holobiont. Microorganisms 8:1682

Bonacolta AM, Weiler BA, Porta-Fitó T, Sweet M, Keeling P, del Campo J (2023) Beyond the Symbiodiniaceae: diversity and role of microeukaryotic coral symbionts. Coral Reefs 42:567–577

Bongaerts P, Ridgway T, Sampayo EM, Hoegh-Guldberg O (2010) Assessing the 'deep reef refugia' hypothesis: focus on Caribbean reefs. Coral Reefs 29:309–327

Bongaerts P, Riginos C, Brunner R, Englebert N, Smith SR, Hoegh-Guldberg O (2017) Deep reefs are not universal refuges: reseeding potential varies among coral species. Sci Adv 3:e1602373

Boulotte NM, Dalton SJ, Carroll AG, Harrison PL, Putnam HM, Peplow LM, Van Oppen MJ (2016) Exploring the *Symbiodinium* rare biosphere provides evidence for symbiont switching in reef-building corals. ISME J 10:2693–2701

Bourne DG, Iida Y, Uthicke S, Smith-Keune C (2008) Changes in coral-associated microbial communities during a bleaching event. ISME J 2:350–363

Bourne DG, Morrow KM, Webster NS (2016) Insights into the coral microbiome: Underpinning the health and resilience of reef ecosystems. Annu Rev Microbiol 70:317–340

Brahmi C, Chapron L, Le Moullac G, Soyez C, Beliaeff B, Lazareth CE, Gaertner-Mazouni N, Vidal-Dupiol J (2021) Effects of elevated temperature and pCO_2 on the respiration, biomineralization and photophysiology of the giant clam *Tridacna maxima*. Conserv Physiol 9:coab041

Buck BH, Rosenthal H, Saint-Paul U (2002) Effect of increased irradiance and thermal stress on the symbiosis of *Symbiodinium microadriaticum* and *Tridacna gigas*. Aquat Living Resour 15:107–117

Bulleri F, Thiault L, Mills SC, Nugues MM, Eckert EM, Corno G, Claudet J (2018) Erect macroalgae influence epilithic bacterial assemblages and reduce coral recruitment. Mar Ecol Prog Ser 597:65–77

Cai WJ et al (2016) Microelectrode characterization of coral daytime interior pH and carbonate chemistry. Nat commun 7:11144

Cairns SD (1999) Species richness of recent Scleractinia. Atoll Res Bull 459:1–12

Cairns SD (2009) Phylogenetic list of 722 valid recent azooxanthellate scleractinian species, with their junior synonyms and depth ranges. In: Roberts JM, Wheeler A, Freiwald A, Cairns SD (eds) Cold-water corals: the biology and geology of deep-sea coral habitats. Cambridge University Press, Cambridge. Online appendix

Cantin NE, van Oppen MJH, Willis BL, Mieog JC, Negri AP (2009) Juvenile corals can acquire more carbon from high-performance algal symbionts. Coral Reefs 28:405–414

Cárdenas A, Raina JB, Pogoreutz C, Rädecker N, Bougoure J, Guagliardo P, Pernice M, Voolstra CR (2022) Greater functional diversity and redundancy of coral endolithic microbiomes align with lower coral bleaching susceptibility. ISME J 16:2406–2420

Carilli JE, Norris RD, Black B, Walsh SM, McField M (2010) Century-scale records of coral growth rates indicate that local stressors reduce coral thermal tolerance threshold. Glob Change Biol 16:1247–1257

Caruso C, Hughes K, Drury C (2021) Selecting heat-tolerant corals for proactive reef restoration. Front Mar Sci 8:632027

Ceccarelli DM et al (2018) Rehabilitation of coral reefs through removal of macroalgae: state of knowledge and considerations for management and implementation. Restor Ecol 26:827–838

Chakravarti LJ, van Oppen MJH (2018) Experimental evolution in coral photosymbionts as a tool to increase thermal tolerance. Front Mar Sci 5:227

Chakravarti LJ, Beltran VH, van Oppen MJH (2017) Rapid thermal adaptation in photosymbionts of reef-building corals. Glob Change Biol 23:4675–4688

Chan WY, Peplow LM, Menéndez P, Hoffmann AA, van Oppen MJ (2018) Interspecific hybridization may provide novel opportunities for coral reef restoration. Front Mar Sci 5:160

Chase TJ, Pratchett MS, Frank GE, Hoogenboom MO (2018) Coral-dwelling fish moderate bleaching susceptibility of coral hosts. PloS One 13:e0208545

Chollett I, Mumby PJ, Cortes J (2010) Upwelling areas do not guarantee refuge for coral reefs in a warming ocean. Mar Ecol Prog Ser 416:47–56

Claar DC, Szostek L, McDevitt-Irwin JM, Schanze JJ, Baum JK (2018) Global patterns and impacts of El Niño events on coral reefs: A meta-analysis. PloS One 13:e0190957

Claar DC et al (2020) Dynamic symbioses reveal pathways to coral survival through prolonged heatwaves. Nat Commun 11:6097

Clerissi C, Brunet S, Vidal-Dupiol J, Adjeroud M, Lepage P, Guillou L, Escoubas JM, Toulza E (2018) Protists within corals: the hidden diversity. Front Microbiol 9:2043

Cleves PA, Krediet CJ, Lehnert EM, Onishi M, Pringle JR (2020) Insights into coral bleaching under heat stress from analysis of gene expression in a sea anemone model system. Proc Natl Acad Sci 117:28906–28917

Coelho VR, Fenner D, Caruso C, Bayles BR, Huang Y, Birkeland C (2017) Shading as a mitigation tool for coral bleaching in three common Indo-Pacific species. J Exp Mar Biol Ecol 497:152–163

Cohen I, Dubinsky Z, Erez J (2016) Light enhanced calcification in hermatypic corals: New insights from light spectral responses. Front Mar Sci 2:122

Coles SL, Bahr KD, Ku'ulei SR, May SL, McGowan AE, Tsang A, Bumgarner J, Han JH (2018) Evidence of acclimatization or adaptation in Hawaiian corals to higher ocean temperatures. PeerJ 6:e5347

Colley N, Trench R (1983) Selectivity in phagocytosis and persistence of symbiotic algae by the scyphistoma stage of the jellyfish *Cassiopeia xamachana*. Proc R Soc B 219:61–82

Colombo-Pallotta M, Rodriguez-Roman A, Iglesias-Prieto R (2010) Calcification in bleached and unbleached *Montastraea faveolata*: evaluating the role of oxygen and glycerol. Coral Reefs 29:899–907

Cook CB, D'Elia CF, Muller-Parker G (1988) Host feeding and nutrient sufficiency for zooxanthellae in the sea anemone *Aiptasia pallida*. Mar Biol 98:253–262

Cortese D, Norin T, Beldade R, Crespel A, Killen SS, Mills SC (2021) Physiological and behavioural effects of anemone bleaching on symbiont anemonefish in the wild. Funct Ecol 35:663–674

Cunning R, Silverstein RN, Baker AC (2015) Investigating the causes and consequences of symbiont shuffling in a multi-partner reef coral symbiosis under environmental change. Proc R Soc B 282:20141725

Curdt F, Schupp PJ, Rohde S (2022) Light availability affects the symbiosis of sponge specific cyanobacteria and the common blue aquarium sponge (*Lendenfeldia chondrodes*). Animals (Basel) 12:1283

Cziesielski MJ, Liew YJ, Cui G, Schmidt-Roach S, Campana S, Marondedze C, Aranda M (2018) Multi-omics analysis of thermal stress response in a zooxanthellate cnidarian reveals the importance of associating with thermotolerant symbionts. Proc R Soc B 285:20172654

Cziesielski MJ, Liew YJ, Cui G, Aranda M (2022) Increased incompatibility of heterologous algal symbionts under thermal stress in the cnidarian-dinoflagellate model Aiptasia. Commun Biol 5:760

D'Olivo JP, McCulloch M (2017) Response of coral calcification and calcifying fluid composition to thermally induced bleaching stress. Sci Rep 7:1–5

Dajka J-C, Beasley V, Gendron G, Barlow J, Graham NA (2021) Weakening macroalgal feedbacks through shading on degraded coral reefs. Aquat Conserv: Mar Freshw Ecosyst 31:1660–1669

Davey M, Holmes G, Johnstone R (2008) High rates of nitrogen fixation (acetylene reduction) on coral skeletons following bleaching mortality. Coral Reefs 27:227–236

Davies PS (1991) Effect of daylight variations on the energy budgets of shallow-water corals. Mar Biol 108:137–144

Davy SK, Allemand D, Weis VM (2012) Cell biology of cnidarian-dinoflagellate symbiosis. Microbiol Mol Biol Rev 76:229–261

de Bakker DM, Meesters EH, van Bleijswijk JD, Luttikhuizen PC, Breeuwer HJ, Becking LE (2016) Population genetic structure, abundance, and health status of two dominant benthic species in the Saba Bank National Park, Caribbean Netherlands: *Montastraea cavernosa* and *Xestospongia muta*. PloS One 11:e0155969

de Goeij JM, Van Oevelen D, Vermeij MJ, Osinga R, Middelburg JJ, de Goeij AF, Admiraal W (2013) Surviving in a marine desert: the sponge loop retains resources within coral reefs. Science 342:108–110

DeFilippo LB et al (2022) Assessing the potential for demographic restoration and assisted evolution to build climate resilience in coral reefs. Ecol Appl 32:e2650

DeSalvo MK, Voolstra CR, Sunagawa S, Schwarz JA, Stillman JH, Coffroth MA, Szmant AM, Medina M (2008) Differential gene expression during thermal stress and bleaching in the Caribbean coral *Montastraea faveolata*. Mol Ecol 17:3952–3971

Dias TLP, Gondim AI (2016) Bleaching in scleractinians, hydrocorals, and octocorals during thermal stress in a northeastern Brazilian reef. Mar Biodivers 46:303–307

Diaz JM, Hansel CM, Apprill A, Brighi C, Zhang T, Weber L, McNally S, Xun (2016) Species-specific control of external superoxide levels by the coral holobiont during a natural bleaching event. Nat Commun 7:13801

Dickinson JL, Zuckerberg B, Bonter DN (2010) Citizen science as an ecological research tool: challenges and benefits. Ann Rev Ecol Evol Syst 41:149–172

Dixon GB, Davies SW, Aglyamova GV, Meyer E, Bay LK, Matz MV (2015) Genomic determinants of coral heat tolerance across latitudes. Science 348:1460–1462

Dixon G, Abbott E, Matz M (2020) Meta-analysis of the coral environmental stress response: *Acropora* corals show opposing responses depending on stress intensity. Mol Ecol 29:2855–2870

Dixon AM, Forster PM, Heron SF, Stoner AM, Beger M (2022) Future loss of local-scale thermal refugia in coral reef ecosystems. PloS Climate 1:e0000004

Djeghri N, Pondaven P, Stibor H, Dawson MN (2019) Review of the diversity, traits, and ecology of zooxanthellate jellyfishes. Mar Biol 166:147

Doering T, Wall M, Putchim L, Rattanawongwan T, Schroeder R, Hentschel U, Roik A (2021) Towards enhancing coral heat tolerance: a "microbiome transplantation" treatment using inoculations of homogenized coral tissues. Microbiome 9:102

Donovan MK et al (2021) Local conditions magnify coral loss after marine heatwaves. Science 372:977–980

Downs CA, Fauth JE, Halas JC, Dustan P, Bemiss J, Woodley CM (2002) Oxidative stress and seasonal coral bleaching. Free Radic Biol Med 33:533–543

Drury C, Caruso C, Quigley K (2021) Selective breeding to support the long-term persistence of coral reefs. EcoEvoRxiv

Dubé CE, Bourmaud CAF, Mercière A, Planes S, Boissin E (2019) Ecology, biology and genetics of *Millepora* hydrocorals on coral reefs. In: Ray S (ed) Invertebrates – ecophysiology and management. IntechOpen. https://doi.org/10.5772/intechopen.89103

Dubousquet V, Gros E, Berteaux-Lecellier V, Viguier B, Raharivelomanana P, Bertrand C, Lecellier GJ (2016) Changes in fatty acid composition in the giant clam Tridacna maxima in response to thermal stress. Biol Open 5:1400–1407

Dungan AM, Maire J, Perez-Gonzalez A, Blackall LL, van Oppen MJ (2022) Lack of evidence for the oxidative stress theory of bleaching in the sea anemone, *Exaiptasia diaphana*, under elevated temperature. Coral Reefs 41:1161–1172

Dunn DF (1981) The clownfish sea anemones: Stichodactylidae (Coelenterata: Actiniaria) and other sea anemones symbiotic with pomacentrid fishes. Trans Am Philos Soc 71:3–115

Eirin-Lopez JM, Putnam HM (2019) Marine environmental epigenetics. Ann Rev Mar Sci 11:335–368

Enricuso OB, Conaco C, Sayco SLG, Neo M-L, Cabaitan PC (2019) Elevated seawater temperatures affect embryonic and larval development in the giant clam *Tridacna gigas* (Cardiidae: Tridacninae). J Molluscan Stud 85:66–72

Epstein HE, Kingsford MJ (2019) Are soft coral habitats unfavourable? A closer look at the association between reef fishes and their habitat. Environ Biol Fishes 102:479–497

Erwin PM, Thacker RW (2008) Phototrophic nutrition and symbiont diversity of two Caribbean sponge–cyanobacteria symbioses. Mar Ecol Prog Ser 362:139–147

Eyal G et al (2022) Selective deep water coral bleaching occurs through depth isolation. Sci Total Environ 844:157180

Ezzat L, Maguer JF, Grover R, Rottier C, Tremblay P, Ferrier-Pagès C (2019) Nutrient starvation impairs the trophic plasticity of reef-building corals under ocean warming. Funct Ecol 33:643–653

Fadlallah YH (1983) Coral reefs sexual reproduction, development and larval biology in scleractinian corals: A review. Coral Reefs 2:129–150

Fan L, Liu M, Simister R, Webster NS, Thomas T (2013) Marine microbial symbiosis heats up: the phylogenetic and functional response of a sponge holobiont to thermal stress. ISME J 7:991–1002

Fang JK, Schönberg CH, Mello-Athayde MA, Achlatis M, Hoegh-Guldberg O, Dove S (2018) Bleaching and mortality of a photosymbiotic bioeroding sponge under future carbon dioxide emission scenarios. Oecologia 187:25–35

Farag MA, Meyer A, Ali SE, Salem MA, Giavalisco P, Westphal H, Wessjohann LA (2018) Comparative metabolomics approach detects stress-specific responses during coral bleaching in soft corals. J Proteome Res 17:2060–2071

Feng EY, Sawall Y, Wall M, Lebrato M, Fu Y (2020) Modeling coral bleaching mitigation potential of water vertical translocation–An analogue to geoengineered artificial upwelling. Front Mar Sci 7:556192

Fisch J, Drury C, Towle EK, Winter RN, Miller MW (2019) Physiological and reproductive repercussions of consecutive summer bleaching events of the threatened Caribbean coral *Orbicella faveolata*. Coral Reefs 38:863–876

Fisher R, O'Leary RA, Low-Choy S, Mengersen K, Knowlton N, Brainard RE, Caley MJ (2015) Species richness on coral reefs and the pursuit of convergent global estimates. Curr Biol 25:500–505

Fitt WK (2012) Bleaching of the fire corals *Millepora*. In: Proceedings of 12th international coral reef symposium, ICRS2012_9A_4

Fitt WK, Spero HJ, Halas J, White MW, Porter JW (1993) Recovery of the coral *Montastrea annularis* in the Florida Keys after the 1987 Caribbean "bleaching event". Coral Reefs 12:57–64

Fortunato HFM, Silva AG, Teixeira RPA, Carvalho BC, Fleury BG (2022) Abnormal average increase in sea surface temperature may promote the first documented mortality event of a marine sponge in the Southeastern Brazil. Biota Neotrop 22:e20211298

Frade PR, Bongaerts P, Englebert N, Rogers A, Gonzalez-Rivero M, Hoegh-Guldberg O (2018) Deep reefs of the Great Barrier Reef offer limited thermal refuge during mass coral bleaching. Nat Commun 9:3447

Freeman BA, Crapo JD (1982) Biology of disease: free radicals and tissue injury. Lab Invest 47:412–426

Freeman CJ, Thacker RW, Baker DM, Fogel ML (2013) Quality or quantity: is nutrient transfer driven more by symbiont identity and productivity than by symbiont abundance? ISME J 7:1116–1125

Fromont J, Garson M (1999) Sponge bleaching on the West and East coasts of Australia. Coral Reefs 18:340–340

Fujimura AG, Riegl BM (2017) Effects of water flow on intra-and inter-colonial variability in bleaching of the zoanthid, *Palythoa caribaeorum*. J Exp Mar Biol Ecol 490:29–33

Fuller ZL et al (2020) Population genetics of the coral *Acropora millepora*: toward genomic prediction of bleaching. Science 369:eaba4674

Gao C, Garren M, Penn K, Fernandez VI, Seymour JR, Thompson JR, Raina JB, Stocker R (2021) Coral mucus rapidly induces chemokinesis and genome-wide transcriptional shifts toward early pathogenesis in a bacterial coral pathogen. ISME J 15:3668–3682

Gardner SG, Raina JB, Nitschke MR, Nielsen DA, Stat M, Motti CA, Ralph PJ, Petrou K (2017) A multi-trait systems approach reveals a response cascade to bleaching in corals. BMC Biol 15:117

Gardner SG et al (2019) Coral microbiome diversity reflects mass coral bleaching susceptibility during the 2016 El Niño heat wave. Ecol Evol 9:938–956

Garson MJ, Flowers AE, Webb RI, Charan RD, McCaffrey EJ (1998) A sponge/dinoflagellate association in the haplosclerid sponge *Haliclona* sp.: cellular origin of cytotoxic alkaloids by Percoll density gradient fractionation. Cell Tissue Res 293:365–373

Gillooly JF, Brown JH, West GB, Savage VM, Charnov EL (2001) Effects of size and temperature on metabolic rate. Science 293:2248–2251

Girvan MS, Campbell CD, Killham K, Prosser JI, Glover LA (2005) Bacterial diversity promotes community stability and functional resilience after perturbation. Environ Microbiol 7:301–313

Gleeson MW, Strong AE (1995) Applying MCSST to coral-reef bleaching. Adv Space Res 16:151–154

Glynn PW (1983) Extensive 'bleaching' and death of reef corals on the Pacific coast of Panama. Environ Conserv 10:149–154

Glynn PW (1993) Coral reef bleaching: ecological perspectives. Coral Reefs 12:1–17

Godinot C, Chadwick N (2009) Phosphate excretion by anemonefish and uptake by giant sea anemones: Demand outstrips supply. Bull Mar Sci 85:1–9

Gold Z, Palumbi SR (2018) Long-term growth rates and effects of bleaching in Acropora hyacinthus. Coral Reefs 37:267–277

Gonzalez-Espinosa PC, Donner SD (2021) Cloudiness reduces the bleaching response of coral reefs exposed to heat stress. Glob Change Biol 27:3474–3486

Gonzalez-Rivero M (2012) The Ecology of bioeroding sponges on Caribbean coral reefs. PhD thesis, University of Exeter, Exeter

Good AM, Bahr KD (2021) The coral conservation crisis: interacting local and global stressors reduce reef resiliency and create challenges for conservation solutions. SN Applied Sciences 3:1–14

Graham ER, Sanders RW (2016) Species-specific photosynthetic responses of symbiotic zoanthids to thermal stress and ocean acidification. Mar Ecol 37:442–458

Grottoli AG, Rodrigues LJ, Palardy JE (2006) Heterotrophic plasticity and resilience in bleached corals. Nature 440:1186–1189

Grottoli AG, Warner ME, Levas SJ, Aschaffenburg MD, Schoepf V, McGinley M, Baumann J, Matsui Y (2014) The cumulative impact of annual coral bleaching can turn some coral species winners into losers. Glob Change Biol 20:3823–3833

Grottoli AG et al (2018) Coral physiology and microbiome dynamics under combined warming and ocean acidification. PloS ONE 13:e0191156

Grottoli AG et al (2021) Increasing comparability among coral bleaching experiments. Ecol Appl 31:e02262

Guibert I, Lecellier G, Torda G, Pochon X, Berteaux-Lecellier V (2020) Metabarcoding reveals distinct microbiotypes in the giant clam *Tridacna maxima*. Microbiome. 8:1–4

Hagedorn M et al (2021) Assisted gene flow using cryopreserved sperm in critically endangered coral. Proc Natl Acad Sci 118:e2110559118

Hagemeyer J (2016) Entering the Anthropocene: how ocean acidification and warmer temperatures affect the symbiotic sea anemone *Exaiptasia pallida*. Doctoral dissertation, University of Delaware

Haguenauer A, Zuberer F, Siu G, Cortese D, Beldade R, Mills SC (2021) Deep heat: A comparison of water temperature, anemone bleaching, anemonefish density and reproduction between shallow and mesophotic reefs. Fishes 6:37

Hanes SD, Kempf SC (2013) Host autophagic degradation and associated symbiont loss in response to heat stress in the symbiotic anemone, *Aiptasia pallida*. Invertebr Biol 132:95–107

Hartman LM, Blackall LL, van Oppen MJH (2022) Antibiotics reduce bacterial load in *Exaiptasia diaphana*, but biofilms hinder its development as a gnotobiotic coral model. Access Microbiol 4:000314

Haunerland NH (2003) Invertebrate metabolism. In: eLS. Wiley

Hédouin L et al (2020) Contrasting patterns of mortality in Polynesian coral reefs following the third global coral bleaching event in 2016. Coral Reefs 39:939–952

Hernández-Elizárraga VH et al (2019) Comparative analysis of the soluble proteome and the cytolytic activity of unbleached and bleached *Millepora* ("Fire Coral") from the Mexican Caribbean. Mar Drugs 17:393

Hernawan UE (2008) Symbiosis between the giant clams (Bivalvia: Cardiidae) and zooxanthellae (Dinophyceae). Biodiversitas 9:53–58

Hill M, Hill A (2012) The magnesium inhibition and arrested phagosome hypotheses: New perspectives on the evolution and ecology of *Symbiodinium* symbioses. Biol Rev Cambridge Philos Soc 87:804–821

Hill R, Scott A (2012) The influence of irradiance on the severity of thermal bleaching in sea anemones that host anemonefish. Coral Reefs 31:273–284

Hill M, Walter C, Bartels E (2016) A mass bleaching event involving clionaid sponges. Coral Reefs 35:153–153

Hillyer KE, Dias D, Lutz A, Roessner U, Davy SK (2018) $_{13}$C metabolomics reveals widespread change in carbon fate during coral bleaching. Metabolomics 14:12

Hirose E, Murakami A (2011) Microscopic anatomy and pigment characterization of coral-encrusting black sponge with cyanobacterial symbiont, *Terpios hoshinota*. Zool Sc 28:199–205

Hoadley KD et al (2019) Host–symbiont combinations dictate the photo- physiological response of reef-building corals to thermal stress. Sci Rep 9:1–15

Hobbs JP, Frisch AJ, Ford BM, Thums M, Saenz-Agudelo P, Furby KA, Berumen ML (2013) Taxonomic, spatial and temporal patterns

of bleaching in anemones inhabited by anemonefishes. PloS One 8:e70966
Hochart C et al (2023) Ecology of Endozoicomonadaceae in three coral genera across the Pacific Ocean. Nat Commun 14:3037
Hoegh-Guldberg O, Poloczanska ES, Skirving W, Dove S (2017) Coral reef ecosystems under climate change and ocean acidification. Front Mar Sci 4:158
Hoegh-Guldberg O et al (2018a) Impacts of 1.5°C global warming on natural and human systems. Global warming of 1.5°C. An IPCC Special Report on the impacts of global warming of 1.5°C above pre-industrial levels and related global greenhouse gas emission pathways, in the context of strengthening the global response to the threat of climate change, sustainable development, and efforts to eradicate poverty. Masson-Delmotte V et al (eds). World Meteorological Organization, Geneva
Hoegh-Guldberg O, Kennedy EV, Beyer HL, McClennen C, Possingham HP (2018b) Securing a long-term future for coral reefs. Trends Ecol Evol 33:936–944
Holbrook SJ, Schmitt RJ (2005) Growth, reproduction and survival of a tropical sea anemone (Actiniaria): benefits of hosting anemonefish. Coral Reefs 24:67–73
Holt AL, Vahidinia S, Gagnon YL, Morse DE, Sweeney AM (2014) Photosymbiotic giant clams are transformers of solar flux. J R Soc Interface 11:20140678
Hughes AD, Grottoli AG, Pease TK, Matsui Y (2010) Acquisition and assimilation of carbon in non-bleached and bleached corals. Mar Ecol Prog Ser 420:91–101
Hughes TP et al (2017) Global warming and recurrent mass bleaching of corals. Nature 543:373–377
Hughes TP et al (2018) Spatial and temporal patterns of mass bleaching of corals in the Anthropocene. Science 359:80–83
Hughes TP et al (2019) Ecological memory modifies the cumulative impact of recurrent climate extremes. Nat Clim Change 9:40–43
Hughes DJ, Raina JB, Nielsen DA, Suggett DJ, Kühl M (2022) Disentangling compartment functions in sessile marine invertebrates. Trends Ecol Evol 37:740–748
Hughes TP, Baird AH, Morrison TH, Torda G (2023) Principles for coral reef restoration in the Anthropocene. One Earth
Humanes A et al (2021) An experimental framework for selectively breeding corals for assisted evolution. Front Mar Sci 8:669995
Hume BCC, Mejia-Restrepo A, Voolstra CR, Berumen ML (2020) Fine-scale delineation of Symbiodiniaceae genotypes on a previously bleached central Red Sea reef system demonstrates a prevalence of coral host-specific associations. Coral Reefs 39:583–601
Ip YK, Chew SF (2021) Light-dependent phenomena and related molecular mechanisms in giant clam-dinoflagellate associations: A review. Front Mar Sci 8:627722
Johnson JV, Dick JTA, Pincheira-Donoso D (2022) Local anthropogenic stress does not exacerbate coral bleaching under global climate change. Glob Ecol Biogeogr 31:1228–1236
Jones AM, Berkelmans R (2011) Tradeoffs to thermal acclimation: energetics and reproduction of a reef coral with heat tolerant *Symbiodinium* type-D. J Mar Biol 2011:185890
Jones AM, Gardner S, Sinclair W (2008) Losing 'Nemo': bleaching and collection appear to reduce inshore populations of anemonefishes. J Fish Biol 73:753–761
Junchompoo C, Sinrapasan N, Penpain C, Patsorn P (2013) Changing seawater temperature effects on giant clams bleaching, Mannai Island, Rayong Province, Thailand. In: Proceedings of the design symposium on conservation of ecosystem. The 12th SEASTAR2000 workshop
Keller-Costa T et al (2022) Metagenomics-resolved genomics provides novel insights into chitin turnover, metabolic specialization, and niche partitioning in the octocoral microbiome. Microbiome 10:151
Kenkel CD, Matz MV (2017) Gene expression plasticity as a mechanism of coral adaptation to a variable environment. Nat Ecol Evol 1:0014
Kennedy EV et al (2013) Avoiding coral reef functional collapse requires local and global action. Curr Biol 23:912–918
Kirk NL, Weis VM (2016) Animal–Symbiodinium symbioses: foundations of coral reef ecosystems. In: The mechanistic benefits of microbial symbionts. Springer, Cham, pp 269–294
Kitahara MV, Fukami H, Benzoni F, Huang D (2016) The new systematics of Scleractinia: integrating molecular and morphological evidence. In: The Cnidaria, past, present and future. Springer, Berlin, pp 41–59
Klein SG, Pitt KA, Lucas CH, Hung SH, Schmidt-Roach S, Aranda M, Duarte CM (2019) Night-time temperature reprieves enhance the thermal tolerance of a symbiotic cnidarian. Front Mar Sci 6:453
Kleypas JA, Langdon C (2006) Coral reefs and changing seawater carbonate chemistry. In Coral reefs and coastal change: science and management, Coastal and Estuaries Studies, Phinney JT, Hoegh-Guldberg O, Kleypas JA, Skirving W. Strong A (eds) 61:73–110
Kleypas J et al (2021) Designing a blueprint for coral reef survival. Biol Conserv 257:109107
Klumpp DW, Griffiths CL (1994) Contributions of phototrophic and heterotrophic nutrition to the metabolic and growth requirements of four species of giant clam (Tridacnidae). Mar Ecol Prog Ser 115:103–115
Knoester EG, Plug VE, Murk AJ, Sande SO, Osinga R (2023) Fisheries restrictions and their cascading effects on herbivore abundance and macroalgae removal at Kenyan coral reefs. J Exp Mar Biol Ecol 559:151850
Knowlton N et al (2021) Rebuilding coral reefs: a decadal grand challenge. International Coral Reef Society and Future Earth Coasts. 56 pp
Kobori H et al (2016) Citizen science: a new approach to advance ecology, education, and conservation. Ecol Res 31:1–19
Kornder NA, Cappelletto J, Mueller B, Zalm MJ, Martinez SJ, Vermeij MJ, Huisman J, de Goeij JM (2021) Implications of 2D versus 3D surveys to measure the abundance and composition of benthic coral reef communities. Coral Reefs 40:1137–1153
Kushmaro A, Rosenberg E, Fine M, Loya Y (1997) Bleaching of the coral *Oculina patagonica* by *Vibrio* AK-1. Mar Ecol Prog Ser 147:159–165
LaJeunesse TC et al (2010) Host–symbiont recombination versus natural selection in the response of coral–dinoflagellate symbioses to environmental disturbance. Proc R Soc B 277:2925–2934
LaJeunesse TC, Parkinson JE, Gabrielson PW, Jeong HJ, Reimer JD, Voolstra CR, Santos SR (2018) Systematic revision of Symbiodiniaceae highlights the antiquity and diversity of coral endosymbionts. Curr Biol 28:2570–2580
Lamy T, Galzin R, Kulbicki M, Lison de Loma T, Claudet J (2016) Three decades of recurrent declines and recoveries in corals belie ongoing change in fish assemblages. Coral Reefs 35:293–302
Leggat WP, Buck BH, Grice A, Yellowlees D (2003) The impact of bleaching on the metabolic contribution of dinoflagellate symbionts to their giant clam host. Plant Cell Environ 26:1951–1961
Leggat W, Seneca F, Wasmund K, Ukani L, Yellowlees D, Ainsworth TD (2011) Differential responses of the coral host and their algal symbiont to thermal stress. PloS One 6:e26687
Leinbach SE, Speare KE, Rossin AM, Holstein DM, Strader ME (2021) Energetic and reproductive costs of coral recovery in divergent bleaching responses. Sci Rep 11:23546
Leinbach SE, Speare KE, Strader ME (2023) Reef habitats structure symbiotic microalgal assemblages in corals and contribute to differential heat stress responses. Coral Reefs 42:205–217
Lesser MP (1996) Elevated temperatures and ultraviolet radiation cause oxidative stress and inhibit photosynthesis in symbiotic dinoflagellates. Limnol Oceanogr 41:271–283

Lesser MP (1997) Oxidative stress causes coral bleaching during exposure to elevated temperatures. Coral Reefs 16:187–192

Lesser MP (2011) Coral bleaching: causes and mechanisms. In: Coral reefs: an ecosystem in transition, pp 405–419

Lesser MP, Stochaj WR, Tapley DW, Schick JM (1990) Bleaching in coral reef anthozoans: effects of irradiance, ultraviolet radiation and temperature on the activities of protective enzymes against active oxygen. Coral Reefs 8:225–232

Levas S, Schoepf V, Warner ME, Aschaffenburg M, Baumann J, Grottoli AG (2018) Long-term recovery of Caribbean corals from bleaching. J Exp Mar Biol Ecol 506:124–134

Lewis JB (2006) Biology and ecology of the hydrocoral *Millepora* on coral reefs. Adv Mar Biol 50:1–55

Li J, Fabina NS, Knapp DE, Asner GP (2020) The sensitivity of multispectral satellite sensors to benthic habitat change. Remote Sens 12:532

Li J et al (2022) Cultured bacteria provide insight into the functional potential of the coral-associated microbiome. Msystems 7:e00327–e00322

Lin MF, Takahashi S, Forêt S, Davy SK, Miller DJ (2019) Transcriptomic analyses highlight the likely metabolic consequences of colonization of a cnidarian host by native or non-native *Symbiodinium* species. Biol Open 8:bio038281

Little AF, van Oppen MJH, Willis BL (2004) Flexibility in algal endosymbiosis shapes growth in reef corals. Science 304:1492–1494

Logan CA, Dunne JP, Eakin CM, Donner SD (2014) Incorporating adaptive responses into future projections of coral bleaching. Glob Change Biol 20:125–139

Logan CA, Dunne JP, Ryan JS, Baskett ML, Donner SD (2021) Quantifying global potential for coral evolutionary response to climate change. Nat Clim Change 11:537–542

López-Legentil S, Song B, McMurray SE, Pawlik JR (2008) Bleaching and stress in coral reef ecosystems: hsp70 expression by the giant barrel sponge *Xestospongia muta*. Mol Ecol 17:1840–1849

Loya Y, Sakai K, Yamazato K, Nakano Y, Sambali H, Van Woesik R (2001) Coral bleaching: the winners and the losers. Ecol Lett 4:122–131

Lyndby NH, Rädecker N, Bessette S, Søgaard Jensen LH, Escrig S, Trampe E, Kühl M, Meibom A (2020) Amoebocytes facilitate efficient carbon and nitrogen assimilation in the Cassiopea-Symbiodiniaceae symbiosis. Proc R Soc B 287:20202393

Marlow J, Davy SK, Shaffer M, Haris A, Bell JJ (2018) Bleaching and recovery of a phototrophic bioeroding sponge. Coral Reefs 37:565–570

Marshall P, Schuttenberg H (2006) Adapting coral reef management in the face of climate change. In Coral reefs and climate change: science and management 61:223–241

Maruyama S et al (2022) Heat stress of algal partner hinders colonization success and alters the algal cell surface glycome in a cnidarian-algal symbiosis. Microbiol Spectr 10:e01567–e01522

Massaro AJ, Weisz JB, Hill MS, Webster NS (2012) Behavioral and morphological changes caused by thermal stress in the Great Barrier Reef sponge *Rhopaloeides odorabile*. J Exp Mar Biol Ecol 416:55–60

Maucieri DG, Baum JK (2021) Impacts of heat stress on soft corals, an overlooked and highly vulnerable component of coral reef ecosystems, at a central equatorial Pacific atoll. Biol Conserv 262:109328

McClanahan TR, Ateweberhan M, Ruiz Sebastian C, Graham NA, Wilson SK, Bruggemann JH, Guillaume MM (2007) Predictability of coral bleaching from synoptic satellite and in situ temperature observations. Coral Reefs 26:695–701

McLachlan RH, Price JT, Solomon SL, Grottoli AG (2020) Thirty years of coral heat-stress experiments: a review of methods. Coral Reefs 39:885–902

McMurray SE, Henkel TP, Pawlik JR (2010) Demographics of increasing populations of the giant barrel sponge Xestospongia muta in the Florida Keys. Ecology 91:560–570

McMurray SE, Blum JE, Leichter JJ, Pawlik JR (2011) Bleaching of the giant barrel sponge *Xestospongia muta* in the Florida Keys. Limnol Oceanogr 56:2243–2250

Medina M et al (2021) The upside-down jellyfish *Cassiopea xamachana* as an emerging model system to study cnidarian-algal symbiosis. In: Handbook of marine model organisms in experimental biology. CRC Press, pp 149–171

Meesters HWG, Smith SR, Becking LE (2015) A review of coral reef restoration techniques. Research report. Institute for Marine Resources (IMARES), Wageningen

Meron D, Efrony R, Johnson WR, Schaefer AL, Morris PJ, Rosenberg E, Greenberg EP, Banin E (2009) Role of flagella in virulence of the coral pathogen *Vibrio coralliilyticus*. Applied and Environmental Microbiology 75:5704–5707

Mies M (2019) Evolution, diversity, distribution and the endangered future of the giant clam–Symbiodiniaceae association. Coral Reefs 38:1067–1084

Mies M, van Sluys MA, Metcalfe CJ, Sumida PYG (2017) Molecular evidence of symbiotic activity between *Symbiodinium* and *Tridacna maxima* larvae. Symbiosis 72:13–22

Mies M, Güth AZ, Castro CB, Pires DO, Calderon EN, Pompeu M, Sumida PY (2018) Bleaching in reef invertebrate larvae associated with *Symbiodinium* strains within clades A–F. Mar Biol 165:6

Montefalcone M, Morri C, Bianchi CN (2018) Long-term change in bioconstruction potential of Maldivian coral reefs following extreme climate anomalies. Glob Change Biol 24:5629–5641

Morikawa MK, Palumbi SR (2019) Using naturally occurring climate resilient corals to construct bleaching-resistant nurseries. Proc Natl Acad Sci 116:10586–10591

Morishima SY, Yamashita H, O-hara S, Nakamura Y, Quek VZ, Yamauchi M, Koike K (2019) Study on expelled but viable zooxanthellae from giant clams, with an emphasis on their potential as subsequent symbiont sources. PloS One 14:e0220141

Morris LA, Voolstra CR, Quigley KM, Bourne DG, Bay LK (2019) Nutrient availability and metabolism affect the stability of coral-Symbiodiniaceae symbioses. Trends Microbiol 27:678–689

Mueller JS, Grammel P-J, Bill N, Rohde S, Schupp PJ (2023) Mass mortality event of the giant barrel sponge *Xestospongia* sp.: population dynamics and size distribution in Koh Phangan, Gulf of Thailand. PeerJ 11:e16561

Mumby PJ, Chisholm JR, Edwards AJ, Andréfouët S, Jaubert J (2001) Cloudy weather may have saved Society Island reef corals during the 1998 ENSO event. Mar Ecol Prog Ser 222:209–216

Muscatine L, Falkowski PG, Dubinsky Z, Cook PA, McCloskey LR (1989) The effect of external nutrient resources on the population dynamics of zooxanthellae in a reef coral. Proc R Soc B 236:311–324

Neo ML, Eckman W, Vicentuan K, Teo SL-M, Todd PA (2015) The ecological significance of giant clams in coral reef ecosystems. Biol Conserv 181:111–123

Neves JR, Taniguchi S, Bícego MC, Sumida PWG, Mies M (2024) Symbiodiniaceae-derived fatty acids are stored differentially across giant clam species and organs. Mar Ecol 45:e12845

Norin T, Mills SC, Crespel A, Cortese D, Killen SS, Beldade R (2018) Anemone bleaching increases the metabolic demands of symbiont anemonefish. Proc R Soc B 285:20180282

Oakley CA, Durand E, Wilkinson SP, Peng L, Weis VM, Grossman AR, Davy SK (2017) Thermal shock induces host proteostasis disruption and endoplasmic reticulum stress in the model symbiotic cnidarian Aiptasia. J Proteome Res 16:2121–2134

Obura DO et al (2019) Coral reef monitoring, reef assessment technologies, and ecosystem-based management. Front Mar Sci 6:580

Olguín-López N, Hérnandez-Elizárraga VH, Hernández-Matehuala R, Cruz-Hernández A, Guevara-González R, Caballero-Pérez J, Ibarra-Alvarado C, Rojas-Molina A (2019) Impact of El Niño-Southern Oscillation 2015-2016 on the soluble proteomic profile and cytolytic activity of *Millepora alcicornis* ("fire coral") from the Mexican Caribbean. PeerJ 7:e6593

Parkinson E, Baums IB (2014) The extended phenotypes of marine symbioses: ecological and evolutionary consequences of intraspecific genetic diversity in coral–algal associations. Front Microbiol 5:445

Pei JY, Yu WF, Zhang JJ, Kuo TH, Chung HH, Hu JJ, Hsu CC, Yu KF (2022) Mass spectrometry–based metabolomic signatures of coral bleaching under thermal stress. Anal Bioanal Chem 414:7635–7646

Peixoto RS, Sweet M, Villela HD, Cardoso P, Thomas T, Voolstra CR, Høj L, Bourne DG (2021) Coral probiotics: premise, promise, prospects. Ann Rev Anim Biosci 9:265–288

Peixoto RS et al (2022) Harnessing the microbiome to prevent global biodiversity loss. Nat Microbiol 7:1726–1735

Pereira-Marques J, Hout A, Ferreira RM, Weber M, Pinto-Ribeiro I, Van Doorn LJ, Knetsch CW, Figueiredo C (2019) Impact of host DNA and sequencing depth on the taxonomic resolution of whole metagenome sequencing for microbiome analysis. Front Microbiol 10:1277

Pérez-Rosales G, Rouzé H, Torda G, Bongaerts P, Pichon M, Under The Pole Consortium, Parravicini V, Hédouin L (2021) Mesophotic coral communities escape thermal coral bleaching in French Polynesia. R Soc Open Sci 8:210139

Pérez-Rosales G et al (2022) Mesophotic coral ecosystems of French Polynesia are hotspots of alpha and beta generic diversity for scleractinian assemblages. Divers Distrib 28:391–1403

Pinzón JH, Kamel B, Burge CA, Harvell CD, Medina M, Weil E, Mydlarz LD (2015) Whole transcriptome analysis reveals changes in expression of immune-related genes during and after bleaching in a reef-building coral. R Soc Open Sci 2:140214

Pita L, Rix L, Slaby BM, Franke A, Hentschel U (2018) The sponge holobiont in a changing ocean: from microbes to ecosystems. Microbiome 6:46

Planes S et al (2019) The Tara Pacific expedition – A pan-ecosystemic approach of the "-omics" complexity of coral reef holobionts across the Pacific Ocean. PloS Biol 17:e3000483

Pogoreutz C, Ziegler M (2024) Frenemies on the reef? Resolving the coral-*Endozoicomonas* association. Trends Microbiol 32:422–434

Pogoreutz C, Rädecker N, Cardenas A, Gärdes A, Voolstra CR, Wild C (2017) Sugar enrichment provides evidence for a role of nitrogen fixation in coral bleaching. Glob Change Biol 23:3838–3848

Pogoreutz C, Voolstra CR, Rädecker N, Weis V, Cardenas A (2020) Raina JB (2020) The coral holobiont highlights the dependence of cnidarian animal hosts on their associated microbes. In: Bosch TCG, Hadfield MG (eds) Cellular dialogues in the holobiont. CRC Press, pp 91–118

Pogoreutz C et al (2022) Coral holobiont cues prime Endozoicomonas for a symbiotic lifestyle. ISME J 16:1883–1895

Pogoreutz C, Reverter M, Delgadillo-Ordoñez N, Roik A (2025) Fungi and fungi-like entities. In: Peixoto R, Voolstra CR (eds) Coral reef microbiome. Springer, Cham, pp 65–91

Pomponi SA (1980) Cytological mechanisms of calcium carbonate excavation by boring sponges. Int Rev Cytol 65:301–319

Posadas N, Baquiran JIP, Nada MAL, Kelly M, Conaco C (2022) Microbiome diversity and host immune functions influence survivorship of sponge holobionts under future ocean conditions. ISME J 16:58–67

Pryor SH, Andrews L, Kelaher BP, Tagliafico A, Scott A (2021) Ocean temperature, but not acidification, causes sea anemone bleaching under a near-future climate scenario. Coral Reefs 40:355–364

Puntin G, Craggs J, Hayden R, Engelhardt KE, McIlroy S, Sweet M, Baker DM, Ziegler M (2023) The reef-building coral *Galaxea fascicularis*: a new model system for coral symbiosis research. Coral Reefs 42:239–252

Quigley KM, Warner PA, Bay LK, Willis BL (2018) Unexpected mixed-mode transmission and moderate genetic regulation of *Symbiodinium* communities in a brooding coral. Hered 121:524–536

Quigley KM, Bay LK, van Oppen MJH (2019) The active spread of adaptive variation for reef resilience. Ecol Evol 9:11122–11135

Quigley KM, Alvarez Roa C, Beltran VH, Leggat B, Willis BL (2021) Experimental evolution of the coral algal endosymbiont, *Cladocopium goreaui*: lessons learnt across a decade of stress experiments to enhance coral heat tolerance. Restor Ecol 29:e13342

Rädecker N, Pogoreutz C, Voolstra CR, Wiedenmann J, Wild C (2015) Nitrogen cycling in corals: The key to understanding holobiont functioning? Trends Microbiol 23:490–497

Rädecker N, Raina JB, Pernice M, Perna G, Guagliardo P, Kilburn MR, Aranda M, Voolstra CR (2018) Using Aiptasia as a model to study metabolic interactions in cnidarian-*Symbiodinium* symbioses. Front Physiol 9:214

Rädecker N et al (2021) Heat stress destabilizes symbiotic nutrient cycling in corals. Proc Natl Acad Sci 118:e2022653118

Rädecker N et al (2022) Heat stress reduces the contribution of diazotrophs to coral holobiont nitrogen cycling. ISME J 16:1110–1118

Rahav O, Dubinsky Z, Achituv Y, Falkowski PG (1989) Ammonium metabolism in the zooxanthellate coral, *Stylophora pistillata*. Proc R Soc B 236:325–337

Ramah S, Taleb-Hossenkhan N, Todd PA, Neo ML, Bhagooli R (2019) Drastic decline in giant clams (Bivalvia: Tridacninae) around Mauritius Island, Western Indian Ocean: implications for conservation and management. Mar Biodiv 49:815–823

Ramah S, Kaullysing D, Soondur M, Taleb-Hossenkhan N, Bhagooli R (2023) Differential photo-physiological responses of two giant clam species to elevated temperature stress from Rodrigues Island, Western Indian Ocean. Indo Pac J Ocean Life 7:64–70

Ramsby BD, Hoogenboom MO, Whalan S, Webster NS (2018) Elevated seawater temperature disrupts the microbiome of an ecologically important bioeroding sponge. Mol Ecol 27:2124–2137

Rau GH, McLeod EL, Hoegh-Guldberg O (2012) The need for new ocean conservation strategies in a high-carbon dioxide world. Nat Clim Change 2:720–724

Reverter M, Jackson M, Daraghmeh N, von Mach C, Milton N (2020) 11-yr of coral community dynamics in reefs around Dahab (Gulf of Aqaba, Red Sea): the collapse of urchins and rise of macroalgae and cyanobacterial mats. Coral Reefs 39:1605–1618

Reverter M, Helber SB, Rohde S, de Goeij JM, Schupp PJ (2022) Coral reef benthic community changes in the Anthropocene: Biogeographic heterogeneity, overlooked configurations, and methodology. Glob Change Biol 28:1956–1971

Richardson S (2006) Endosymbiont-bleaching in epiphytic populations of *Sorites dominicensis*. Symbiosis 42:103–117

Richter M, Ruhle W, Wild A (1990) Studies on the mechanism of photosystem II photoinhibition: II. The involvement of toxic oxygen species. Photosynth Res 24:237–243

Riegl B, Glynn PW, Banks S, Keith I, Rivera F, Vera-Zambrano M, D'Angelo C, Wiedenmann J (2019) Heat attenuation and nutrient delivery by localized upwelling avoided coral bleaching mortality in northern Galapagos during 2015/2016 ENSO. Coral Reefs 38:773–785

Ritson-Williams R, Gates RD (2020) Coral community resilience to successive years of bleaching in Kāne'ohe Bay, Hawai'i. Coral Reefs 39:757–769

Rodrigues LJ, Grottoli AG (2006) Calcification rate and the stable carbon, oxygen, and nitrogen isotopes in the skeleton, host tissue, and zooxanthellae of bleached and recovering Hawaiian corals. Geochim Cosmochim Acta 70:2781–2789

Rodrigues LJ, Grottoli AG (2007) Energy reserves and metabolism as indicators of coral recovery from bleaching. Limnol Oceanogr 52:1874–1882

Rodrigues LJ, Padilla-Gamiño JL (2022) Trophic provisioning and parental trade-offs lead to successful reproductive performance in corals after a bleaching event. Sci Rep 12:18702

Rodriguez-Lanetty M, Chang SJ, Song JI (2003) Specificity of two temperate dinoflagellate–anthozoan associations from the north-western Pacific Ocean. Mar Biol 143:1193–1199

Rodriguez-Lanetty M, Harii S, Hoegh-Guldberg O (2009) Early molecular responses of coral larvae to hyperthermal stress. Mol Ecol 18:5101–5114

Rohwer F, Seguritan V, Azam F, Knowlton N (2002) Diversity and distribution of coral-associated bacteria. Mar Ecol Prog Ser 243:1–10

Roik A, Reverter M, Pogoreutz C (2022) A roadmap to understanding diversity and function of coral reef-associated fungi. FEMS Microbiol Rev 46:fuac028

Roopin M, Chadwick NE (2009) Benefits to host sea anemones from ammonia contributions of resident anemonefish. J Exp Mar Biol Ecol 370:27–34

Rosenberg E, Falkovitz L (2004) The *Vibrio shiloi*/*Oculina patagonica* model system of coral bleaching. Ann Rev Microbiol 58:143–159

Rosenberg E, Koren O, Reshef L, Efrony R, Zilber-Rosenberg I (2007) The role of microorganisms in coral health, disease and evolution. Nat Rev Microbiol 5:355–362

Ross C (2014) Nitric oxide and heat shock protein 90 co-regulate temperature- induced bleaching in the soft coral *Eunicea fusca*. Coral Reefs 33:513–522

Rossbach S, Subedi RC, Ng TK, Ooi BS, Durarte CM (2020) Iridocytes mediate photonic cooperation between giant clams (Tridacninae) and their photosynthetic symbionts. Front Mar Sci 7:465

Rouan A et al (2022) Telomere dysfunction is associated with dark-induced bleaching in the reef coral *Stylophora pistillata*. Mol Ecol 31:6087–6099

Rützler K (1990) Associations between Caribbean sponges and photosynthetic organisms. In: Rützler K (ed) New perspectives in sponge biology. Smithsonian Institute Press, Washington, DC, pp 455–466

Rützler K, Muzik KM (1993) *Terpios hoshinota*, a new cyanobacteriosponge threatening Pacific reefs. Sci Mar 57:395–403

Saenz-Agudelo P, Jones GP, Thorrold SR, Planes S (2011) Detrimental effects of host anemone bleaching on anemonefish populations. Coral Reefs 30:497–506

Safaie A et al (2018) High frequency temperature variability reduces the risk of coral bleaching. Nat Commun 9:1671

Sampayo EM, Ridgway T, Bongaerts P, Hoegh-Guldberg O (2008) Bleaching susceptibility and mortality of corals are determined by fine-scale differences in symbiont type. Proc Natl Acad Sci 105:10444–10449

Santoro EP et al (2021) Coral microbiome manipulation elicits metabolic and genetic restructuring to mitigate heat stress and evade mortality. Sci Adv 7:eabg3088

Sawall Y, Harris M, Lebrato M, Wall M, Feng EY (2020) Discrete pulses of cooler deep water can decelerate coral bleaching during thermal stress: implications for artificial upwelling during heat stress events. Front Mar Sci 7:720

Sayco SLG, Cabaitan PC, Kurihara H (2023) Bleaching reduces reproduction in the giant clam Tridacna gigas. Mar Ecol Prog Ser 706:47–56

Sayco SLG, Pomares AA, Cabaitan PC, Kurihara H (2024) Reproductive consequences of thermal stress-induced bleaching in the giant clam *Tridacna crocea*. Mar Environ Res 193:106280

Scharfenstein HJ, Chan WY, Buerger P, Humphrey C, van Oppen MJH (2022) Evidence for de novo acquisition of microalgal symbionts by bleached adult corals. ISME J 16:1676–1679

Schoepf V, Grottoli AG, Levas SJ, Aschaffenburg MD, Baumann JH, Matsui Y, Warner ME (2015) Annual coral bleaching and the long-term recovery capacity of coral. Proc R Soc B 282:20151887

Schoepf V, D'Olivo JP, Rigal C, Jung EM, McCulloch MT (2021) Heat stress differentially impacts key calcification mechanisms in reef-building corals. Coral Reefs 40:459–471

Schwiesow MW, Samayoa AM, Torres J, Leimbach A, Santiago-Rivera G, Tepper CS (2021) *Symbiodinium* distribution patterns in Millepores in the Caribbean: south water cay, Belize and san Salvador, the Bahamas. Mar Sci 9:1–10

Scott A, Harrison PL (2007) Embryonic and larval development of the host sea anemones *Entacmaea quadricolor* and *Heteractis crispa*. Biol Bull 213:110–121

Scott A, Hoey AS (2017) Severe consequences for anemonefishes and their host sea anemones during the 2016 bleaching event at Lizard Island, Great Barrier Reef. Coral Reefs 36:873–873

Seavy BE, Muller-Parker G (2002) Chemosensory and feeding responses of the nudibranch *Aeolidia papillosa* to the symbiotic sea anemone *Anthopleura elegantissima*. Invertebr Biol 121:115–125

Shchepetkin AF, McWilliams JC (2005) The regional oceanic modeling system (ROMS): a split-explicit, free-surface, topography-following-coordinate oceanic model. Ocean Model 9:347–404

Sieber A, Pyle R (2010) A review of the use of closed-circuit rebreathers for scientific diving. Underw Technol 29:73–78

Sikorskaya TV, Ermolenko EV, Imbs AB (2020) Effect of experimental thermal stress on lipidomes of the soft coral *Sinularia* sp. and its symbiotic dinoflagellates. J Exp Mar Biol Ecol 524:151295

Sikorskaya TV, Ermolenko EV, Efimova KV, Dang LT (2022a) Coral holobionts possess distinct lipid profiles that may be shaped by Symbiodiniaceae taxonomy. Mar Drugs 20:485

Sikorskaya TV, Ermolenko EV, Boroda AV (2022b) Recovery of a symbiotic octocoral *Sinularia heterospiculata* after heat stress exposure. Mar Freshw Behav Physiol 55:121–130

Silverstein RN, Cunning R, Baker AC (2015) Change in algal symbiont communities after bleaching, not prior heat exposure, increases heat tolerance of reef corals. Glob Change Biol 21:236–249

Smale DA et al (2019) Marine heatwaves threaten global biodiversity and the provision of ecosystem services. Nat Clim Change 9:306–312

Smith DJ, Suggett DJ, Baker NR (2005) Is photoinhibition of zooxanthellae photosynthesis the primary cause of thermal bleaching in corals? Glob Change Biol 11:1–11

Smith HA, Prenzlau T, Whitman T, Fulton SE, Borghi S, Logan M, Heron SF, Bourne DG (2022) Macroalgal canopies provide corals limited protection from bleaching and impede post-bleaching recovery. J Exp Mar Biol Ecol 553:151762

Souter D, Planes S, Wicquart J, Logan M, Obura D, Staub F (2021) Status of coral reefs of the world: 2020: executive summary. Coral Reef Initiative, Global Coral Reef Monitoring Network, Letöltve, p 11

Steinberg RK, Ainsworth TD, Moriarty T, Bednarek T, Dafforn KA, Johnston EL (2022) Bleaching susceptibility and resistance of octocorals and anemones at the world's southern-most coral reef. Front Physiol 13:804193

Stewart HA, Kline DI, Chapman LJ, Altieri AH (2021) Caribbean mangrove forests act as coral refugia by reducing light stress and increasing coral richness. Ecosphere 12:e03413

Stuart-Smith RD, Brown CJ, Ceccarelli DM, Edgar GJ (2018) Ecosystem restructuring along the Great Barrier Reef following mass coral bleaching. Nature 560:92–96

Suggett DJ, van Oppen MJH (2022) Horizon scan of rapidly advancing coral restoration approaches for 21st century reef management. Emerg Top Life Sci 6:125–136

Suggett DJ, Dong LF, Lawson T, Lawrenz E, Torres L, Smith DJ (2013) Light availability determines susceptibility of reef building corals to ocean acidification. Coral reefs 32:327–337

Sully S, Burkepile DE, Donovan MK, Hodgson G, van Woesik R (2019) A global analysis of coral bleaching over the past two decades. Nat Commun 10:1264

Sydnor JR, Lopez J, Wolfe GV, Ott L, Tran C (2023) Changes in the microbiome of the sea anemone *Exaiptasia diaphana* during bleaching from short-term thermal elevation. Front Mar Sci 10:464

Szabó M, Larkum AW, Vass I (2020) A review: the role of reactive oxygen species in mass coral bleaching. In: Photosynthesis in algae: biochemical and physiological mechanisms, pp 459–488

Tagliafico A, Baker P, Kelaher B, Ellis S, Harrison D (2022) The effects of shade and light on corals in the context of coral bleaching and shading technologies. Front Mar Sci 9:919382

Tambutté S, Holcomb M, Ferrier-Pagès C, Reynaud S, Tambutté É, Zoccola D, Allemand D (2011) Coral biomineralization: From the gene to the environment. J Exp Mar Biol Ecol 408:58–78

Tanaka Y, Grottoli AG, Matsui Y, Suzuki A, Sakai K (2015) Partitioning of nitrogen sources to algal endosymbionts of corals with long-term 15N-labelling and a mixing model. Ecol Model 309:163–169

Teague J, Megson-Smith DA, Allen MJ, Day JC, Scott TB (2022) A review of current and new optical techniques for coral monitoring. Oceans 3:30–45

Tebbett SB, Streit RP, Bellwood DR (2019) Expansion of a colonial ascidian following consecutive mass coral bleaching at Lizard Island, Australia. Mar Environ Res 144:125–129

Thacker RW (2005) Impacts of shading on sponge-cyanobacteria symbioses: A comparison between host-specific and generalist associations. Integr Comp Biol 45:369–376

Thomas L, Palumbi SR (2017) The genomics of recovery from coral bleaching. Proc R Soc B 284:1865

Thomas T et al (2016) Diversity, structure and convergent evolution of the global sponge microbiome. Nat Commun 7:11870

Thornhill DJ, Lewis AM, Wham DC, LaJeunesse TC (2014) Host-specialist lineages dominate the adaptive radiation of reef coral endosymbionts. Evolution 68:352–367

Tjus SE, Scheller HV, Andersson B, Møller BL (2001) Active oxygen species produced during selective excitation of photosystem I is damaging not only to photosystem I, but also to photosystem II. Plant Physiol 125:2007–2015

Tolleter D, Seneca FO, DeNofrio JC, Krediet CJ, Palumbi SR, Pringle JR, Grossman AR (2013) Coral bleaching independent of photosynthetic activity. Curr Biol 23:1782–1786

Toullec G, Lyndby NH, Banc-Prandi G, Pogoreutz C, Martin Olmos C, Meibom A, Rädecker N (2024a) Symbiotic nutrient exchange enhances the long-term survival of cassiosomes, the autonomous stinging-cell structures of Cassiopea. MSphere 9:e00322–e00323

Toullec G et al (2024b) Host starvation and in hospite degradation of algal symbionts shape the heat stress response of the Cassiopea-Symbiodiniaceae symbiosis. Microbiome 12:42

Tresguerres M (2016) Novel and potential physiological roles of vacuolar-type H+-ATPase in marine organisms. J Exp Biol 219:2088–2097

Umeki M, Yamashita H, Suzuki G, Sato T, Ohara S, Koike K (2020) Fecal pellets of giant clams as a route for transporting Symbiodiniaceae to corals. PloS One 15:e0243087

Underhill DM, Ozinsky A (2002) Phagocytosis of microbes: complexity in action. Ann Rev Immunol 20:825–852

Usher KM (2008) The ecology and phylogeny of cyanobacterial symbionts in sponges. Mar Ecol 29:178–192

Ushijima B, Häse CC (2018) Influence of chemotaxis and swimming patterns on the virulence of the coral pathogen *Vibrio coralliilyticus*. J Bacteriol 200:e00791–e00717

van der Windt N, van der Ent E, Ambo-Rappe R, de Voogd NJ (2020) Presence and genetic identity of Symbiodiniaceae in the bioeroding sponge genera *Cliona* and *Spheciospongia* (Clionaidae) in the Spermonde Archipelago (SW Sulawesi). Indonesia. Front Ecol Evol 8:595452

van Oppen MJH, Oliver JK, Putnam HM, Gates RD (2015) Building coral reef resilience through assisted evolution. Proc Natl Acad Sci 112:2307–2313

van Oppen MJH et al (2017) Shifting paradigms in restoration of the world's coral reefs. Glob Change Biol 23:3437–3448

van Woesik R, Sakai K, Ganase A, Loya Y (2011) Revisiting the winners and the losers a decade after coral bleaching. Mar Ecol Prog Ser 434:67–76

van Woesik R et al (2022) Coral-bleaching responses to climate change across biological scales. Glob Change Biol 28:4229–4250

Venegas RM et al (2019) The rarity of depth refugia from coral bleaching heat stress in the western and central Pacific Islands. Sci Rep 9:1–12

Vicente VP (1990) Response of sponges with autotrophic endosymbionts during the coral-bleaching episode in Puerto Rico. Coral Reefs 8:199–202

Vicentuan-Cabaitan K, Neo ML, Eckman W, Teo SL-M, Todd PA (2014) Giant clam shells host a multitude of epibionts. Bull Mar Sci 90:795–796

Virgen-Urcelay A, Donner SD (2023) Increase in the extent of mass coral bleaching over the past half-century, based on an updated global database. PLoS One 18:e0281719

Voolstra CR, Ziegler M (2020) Adapting with microbial help: microbiome flexibility facilitates rapid responses to environmental change. BioEssays 42:2000004

Voolstra CR et al (2024) The coral microbiome in sickness, in health and in a changing world. Nat Rev Microbiol 22:460–475

Wang W et al (2022) The coral pathogen *Vibrio coralliilyticus* kills non-pathogenic holobiont competitors by triggering prophage induction. Nat Ecol Evol 6:1132–1144

Warner ME, Suggett DJ (2016) The photobiology of *Symbiodinium* spp.: linking physiological diversity to the implications of stress and resilience. In: The Cnidaria, past, present and future: the world of Medusa and her sisters, pp 489–509

Watson S-A, Neo ML (2021) Conserving threatened species during rapid environmental change: using biological responses to inform management strategies of giant clams. Conserv Physiol 9:coab082

Webster NS, Cobb RE, Negri AP (2008) Temperature thresholds for bacterial symbiosis with a sponge. ISME J 2:830–842

Webster N, Pantile R, Botte E, Abdo D, Andreakis N, Whalan S (2013) A complex life cycle in a warming planet: gene expression in thermally stressed sponges. Mol Ecol 22:1854–1868

Weis VM (2008) Cellular mechanisms of Cnidarian bleaching: Stress causes the collapse of symbiosis. J Exp Biol 211:3059–3066

Wilkinson CR (1983) Net primary productivity in coral reef sponges. Science 219:410–412

Wilkinson CR (1987) Productivity and abundance of large sponge populations on Flinders Reef flats, Coral Sea. Coral Reefs 5:183–188

Wilkinson CR, Smith DC (1997) Immunological evidence for the Precambrian origin of bacterial symbioses in marine sponges. Proc R Soc Ser B 220:509–518

Williams GJ, Graham NA (2019) Rethinking coral reef functional futures. Funct Ecol 33:942–947

Williams ID, Couch CS, Beijbom O, Oliver TA, Vargas-Angel B, Schumacher BD, Brainard RE (2019) Leveraging automated image analysis tools to transform our capacity to assess status and trends of coral reefs. Front Mar Sci 6:222

Williams A, Chiles EN, Conetta D, Pathmanathan JS, Cleves PA, Putnam HM, Su X, Bhattacharya D (2021) Metabolomic shifts associated with heat stress in coral holobionts. Sci Adv 7:eabd4210

Winston M et al (2022) Coral taxonomy and local stressors drive bleaching prevalence across the Hawaiian Archipelago in 2019. Plos One 17:e0269068

Woodhead AJ, Hicks CC, Norström AV, Williams GJ, Grahams NAJ (2019) Coral reef ecosystem services in the Anthropocene. Funct Ecol 33:1023–1034

Wulff JL (2006) Ecological interactions of marine sponges. Can J Zool 84:146–166

Xiang T, Hambleton EA, DeNofrio JC, Pringle JR, Grossman AR (2013) Isolation of clonal axenic strains of the symbiotic dinoflagellate *Symbiodinium* and their growth and host specificity. J Phycol 49:447–458

Yang SH et al (2019) Metagenomic, phylogenetic, and functional characterization of predominant endolithic green sulfur bacteria in the coral *Isopora palifera*. Microbiome 7:3

Yates KK, Rogers CS, Herlan JJ, Brooks GR, Smiley NA, Larson RA (2014) Diverse coral communities in mangrove habitats suggest a novel refuge from climate change. Biogeosciences 11:4321–4337

Yau AJ-Y, Fan T-Y (2012) Size-dependent photosynthetic performance in the giant clam *Tridacna maxima*, a mixotrophic marine bivalve. Mar Biol 159:65–75

Yee DP et al (2023) The V-type ATPase enhances photosynthesis in marine phytoplankton and further links phagocytosis to symbiogenesis. Curr Biol 33:1–7

Zhou Z, Liu Z, Wang L, Luo J, Li H (2019) Oxidative stress, apoptosis activation and symbiosis disruption in giant clam *Tridacna crocea* under high temperature. Fish Shellfish Immunol 84:451–457

Zhou Z et al (2021) The evaluation of culture performance and mantle coloration of two boring giant clam (*Tridacna crocea*) strains. Aquacult Rep 20:100646

Ziegler M (2019) Coral bacterial community structure responds to environmental change in a host-specific manner. Nat commun 10:3092

Ziegler M, Seneca FO, Yum LK, Palumbi SR, Voolstra CR (2017) Bacterial community dynamics are linked to patterns of coral heat tolerance. Nat commun 8:14213

Zoccola D et al (2020) The World Coral Conservatory (WCC): A Noah's ark for corals to support survival of reef ecosystems. PLoS Biol 18:e3000823

5 Evaluating the Role of Natural and Human-Assisted Coral Reproduction to Maintain and Recover Disturbed Reefs

Alexandre Fellous, Laetitia Hédouin, and Hollie Putnam

Abstract

Global changes highlight a role for phenotypic plasticity as an evolutionary driver of biodiversity under novel environmental conditions. In light of the dire ecological forecast for coral reefs, a demand for understanding acclimatory buffers and the evolutionary implications of climate change has promoted the study of linkages between plasticity and adaptation. In this way, phenotypic plasticity has the potential to direct evolutionary outcomes through modulation of the phenotype, thereby potentially buffering or exacerbating negative ecological performance and also retarding or potentiating adaptive evolution. Work in marine organisms reveal a role for phenotypic plasticity as an evolutionary driver under novel conditions. However, we lack a synthesized understanding of these processes in corals and more critcally across ontogeny. The ultimate issue of the costs or benefits of phenotypic plasticity to adaptive evolution, however, will remain unattainable, as proximate gaps in knowledge of basic dynamics and mechanisms of plasticity remain in the absence of synthesis. It is therefore critical to characterize plasticity as a function of ontogeny within the natural life cycles of coral species. Our goal is to provide longer-term, life-history relevant information to assess the capacity for corals to buffer the negative effects of environmental change through transgenerational or developmental plasticity.

A. Fellous (✉) · L. Hédouin
UAR 3278 CRIOBE, PSL Université – EPHE-UPVD-CNRS, Papetoai, Moorea, French Polynesia
e-mail: Laetitia.hedouin@criobe.pf

H. Putnam
Department of Biological Sciences, University of Rhode Island, Kingston, RI, USA
e-mail: hputnam@uri.edu

Keywords

Corals · Ontogeny · Phenotypic plasticity · Epigenetic · Reproduction · Restoration

5.1 Introduction

Coral reefs provide essential habitat for many marine species and vital resources for a growing proportion of the world's population (Farber et al. 2002; Moberg and Folke 1999). Nevertheless, due to anthropogenic global change, primarily the effects of global warming and marine heat waves (Wecker et al. 2018), corals are threatened by bleaching and mass mortality events (Hughes et al. 2018; Puisay et al. 2018). As a result, corals must always live near their thermal limits, facing substantial challenges to survival, development, and reproduction under all future climate scenarios. Therefore, to preserve and restore reefs, *it is essential that we better understand coral resilience and resistance to climate change* (Duarte et al. 2020). In light of the ecological forecast for coral reefs, a demand for understanding acclimation buffers and the evolutionary implications of climate change has promoted the study of linkages between plasticity and adaptation (Putnam 2021; Torda et al. 2017). In this way, phenotypic plasticity, as a target of selection, has the potential to direct evolutionary outcomes through the modulation of the phenotype, thereby potentially buffering or exacerbating negative ecological performance and also retarding or potentiating adaptive evolution (Ghalambor et al. 2015). Works in urchins, oysters, fish, marine worms, and other taxa reveal a role for phenotypic plasticity as an evolutionary driver under novel conditions. However, we lack a synthesized understanding of these processes in corals and more critically across ontogeny (Putnam 2021). The ultimate issue of the costs or benefits of phenotypic plasticity to adaptive evolution, however, will remain unattainable, as proximate gaps in knowledge of basic dynamics and mechanisms of plasticity remain in the absence of synthesis. It is therefore critical to comprehen-

S. Planes (ed.), *The Future of Coral Reefs*, Coral Reefs of the World 21, https://doi.org/10.1007/978-3-031-98584-3_5

sively characterize plasticity as a function of ontogeny within the natural life cycles of coral species from the lens of reef restoration, as well as basic science.

In order to preserve and/or restore reefs, it is necessary to (1) protect species, (2) limit removal, (3) reduce anthropogenic pressures, and (4) mitigate the effect of climate change. This latter point notably involves processes inherent to the species themselves and to their adaptive capacity in the broad sense (e.g., adaptation/acclimatization) to these changes and to our ability to optimize management measures (Ridlon et al. 2023).

Among the various restoration techniques that currently exist, coral nurseries through fragmentation and asexual multiplication are the most widespread and easy to implement. However, they show limits (loss of genetic diversity, low numbers of individuals, maintenance of structures) and can be ineffective during massive bleaching events (Hugues et al. 2017). In this context, sexual reproduction, specific crossings, and seeding degraded reefs by millions of larvae are poised to be more effective over large areas (Doropoulos et al. 2019a, 2019b; Harrison et al. 2021). Therefore, to preserve and restore reefs, *it is essential that we better understand coral resilience and resistance to climate change through gametogenic and early development plasticity***.** To this end, we provide an overview on the coral reproduction (Sect. 5.2) and embryogenesis (Sect. 5.3), followed by an examination of the role of phenotypic plasticity in restoration strategies (Sect. 5.4), and we discuss the establishment of innovative restoration strategies through enhancing plasticity via reproductive and developmental processes (Sect. 5.5). Lastly, we conclude with a challenge to the field to focus on basic coral biology in the areas of reproduction and development.

5.2 Coral Reproduction

5.2.1 Asexual Reproduction

Figure 5.1 a/asexual mode of reproduction [fragmentation, budding (intra- and extra-tentacular), fission, bailout, asexual planulae), b/table with advantages and disadvantages of asexual reproduction, c/coral tree.

Scleractinian corals can reproduce both sexually and asexually with eleven known modes such as (Fig. 5.1) (Cairn 1988).

- *Budding* allows the coral to grow by the addition of new polyps. Also, budding can be *intra-* or *extra-tentacular*. Intra-tentacular budding occurs when a polyp is divided in two, whereas a new polyp appears between two existing ones in extra-tentacular budding. Also, new polyps can grow over the older one, which then leaves its place to the new buds (*parricid*). It is also possible for newly formed buds to detach from the mother colony together with calcified inclusions to form a new colony, but this is a rarer occurrence (e.g., *Goniopora stokesi*).
- *Fragmentation/fission* takes place when parts of the colony are fragmented or detached. If environmental conditions allow it, fragments can grow via budding and give rise to a new colony. The fragmentation might be *accidental* due to physical actions of the waves for example, but also *non-accidental. Non-accidental fragmentation (transverse division)* is generally genetically programmed at predetermined or random spaces. Also, it seems to be restricted to solitary corals and particularly in Fungiidae. Transverse-division might represent a key innovation that allows specific genotypes to exploit new environments, while being central to adaptative radiation.
- *Polyp bailout* is defined as the abandonment of the mother colony by a polyp in case of a stressful event such as bleaching or water contamination. Then, it can settle and create a new colony (reviewed in Schweinsberg et al. 2021).
- *Asexual planulae* can be produced via *parthenogenesis*. As an example, *Pocillopora damicornis* produces up to 94% of clonal larvae together with sexual planulae (Combosch and Vollmer 2013; Yeoh and Dai 2010). Interestingly, the proportion of asexual/sexual larvae depends on the time of the year and colony size (Combosch and Vollmer 2013). Asexual reproduction occurring more in the biggest ones thus replicating successful genotypes (Combosch and Vollmer 2013).

In asexual reproduction, individuals produced are genetically identical clones and allow the colony to grow (Shikina and Chang 2016; Cairn 1988). While sexual reproduction potentially produces diverse genotypes with the ability to acclimate to a changing environment, asexual reproduction has the advantages to amplify locally well-adapted individuals in a given environment (Cairn 1988; Gorospe and Karl 2013). Asexual production of corals can occur year around in a faster process together with greater chance of survival. Also, it requires less energy and can be accomplished easily in environments where few reproducing colonies remain (Cairn 1988). However, the dispersal capacities might be lower, and range of plasticity of the asexually propagated colonies is unknown.

In the life history of coral, asexual reproduction allows growth, but is also often the predominant mode of propagation, but not the exclusive one, and is genetically determined (Shikina and Chang 2016; Cairn 1988). Therefore, asexual reproduction can result from environmental variations, a genetically controlled process or a coupling environment-genetic. In the context of climate changes and anthropogenic perturbations (IPCC 2018), it appears crucial to consider the

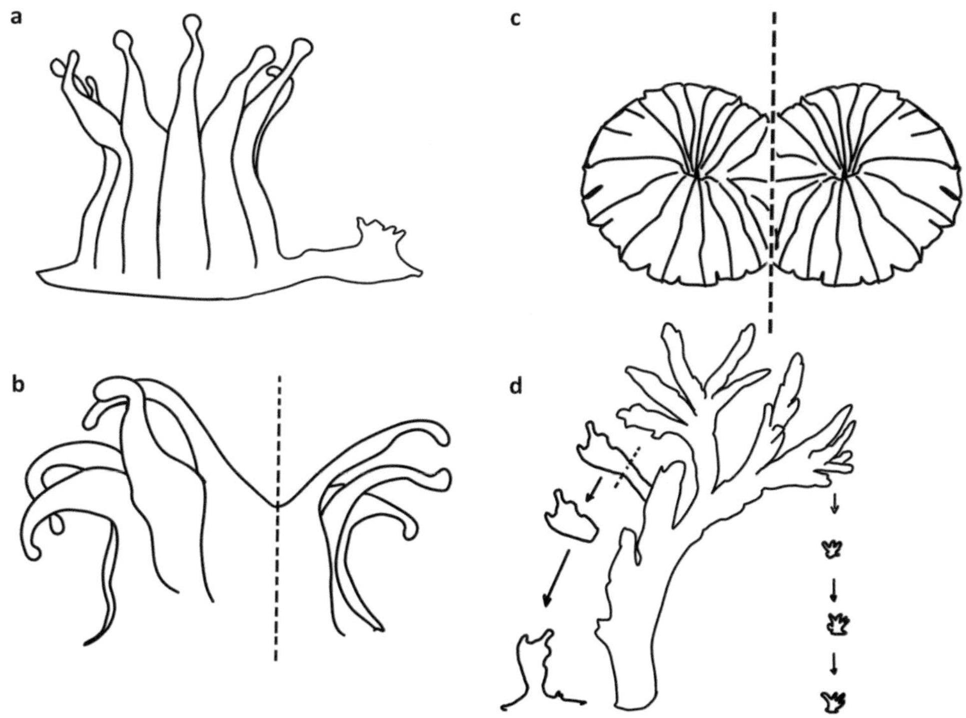

Fig. 5.1 Schematic representation of different asexual modes of reproduction in scleractinian corals with (**a**) extra-tentacular budding, (**b**) intra-tentacular budding, (**c**) fission, (D) fragmentation (on the left) and polyp bailout (on the right)

role of *asexual and sexual reproduction* for the conservation and the restoration of reef ecosystems (Randall et al. 2020).

Asexual reproduction in the form of fragmentation of existing colonies is the most widely used in different restoration efforts through coral gardening, and has been associated with contrasting success (Mcleod et al. 2022). While this technique is easy and considered as "low tech," it has the disadvantages of being time consuming, done on a small scale, in a few species, and typically focuses only on a small number of genotypes (Mcleod et al. 2022). However, new culturing methods are going beyond traditional fragmentation methods to also protect (e.g., via caging) naturally recruiting juvenile corals in order to exponentially increase survival and growth rates, while offering the possibility to create a stock of sexually derived colonies (Linden and Rinkevich 2017). These juveniles then provide the possibility for the nursery to include additional coral genotypes resistant to different environmental changes (Bowden-kerby 2023), and to construct therefore further assist in the novel coral communities (Golomb et al. 2020). Thus, coral gardening in various forms beyond local fragmentation could be used to enhance restoration outcomes via (1) assisted gene flow (more or less 10 km), (2) assisted migration (translocation of individuals in distant population but within the species natural range), (3) species introduction (relocation outside of the species natural range) (Bowden-kerby 2023). However, when conducting translocation, it would be recommended to avoid moving coral from populations with possible disease outbreaks and to directly bring newly collected corals into major gene bank nurseries (Bowden-kerby 2023). Moving corals at a local reef scale (fringing reefs or lagoon to offshore) is generally not very risky in this context, but moving individuals between separated reef systems is riskier and more logistically complicated. Also, in extreme case, long-distance translocation has been proposed locally extinct or very rare species in order to restore sexual reproduction (Bowden-kerby 2023).

5.2.2 Sexual Reproduction

In scleractinian corals, sexual reproduction is not fully understood yet (Chen et al. 2020), and can be done either by internal fertilization (brooding) (with the possible release of sperm into the water column) or external fertilization (broadcasting) (sperm and oocytes are released into the water column) (Harrison 2011). The majority of corals whose mode of reproduction is known is hermaphrodite with external fertilization (69%). These species are followed by gonochoric with external fertilization (21%), hermaphrodite with internal fertilization (7%), and gonochoric with internal fertilization (3%) (Baird et al. 2009; Harrison 2011). Hermaphrodite species can be sequential or synchronous. In synchronic (simultaneously) species, ovaries and testies are actively developing at the same time and form an ovotestis, whereas in sequential corals, individuals can be protogynic (female-to-male) or protandric (male-to-female) (Chen et al. 2020; Harrison 2011). Most sequential species change sexes only once in their life cycle, but others such as Fungiidae exhibit bidirectional sex changes during their life time (Loya and Saka 2008). In gonochoric corals, individuals can be dioecious (female and male), gynodioecious (female and hermaphrodite), and polygamodioecious (male, female, and hermaphroditic polyps) (Chen et al. 2020; Harrison 2011).

The same coral genus does not necessarily have a single mode of reproduction. For the same species, the mode of reproduction can change depending on the geographical location or population (Harrison 2011). However, these differences might also reflect existence of cryptic species (Schmidt-roach et al. 2012). Reproductive cycles are intimately linked to the environment with so-called ultimate factors (exerting selection pressure as seasonal temperatures, winds, tides) and so-called immediate factors (which synchronize the cycles) (Randall et al. 2020). Gametogenesis is generally annual (Fig. 5.2) and depends on environmental signals such as temperature, sunlight, or lunar cycles. Some populations have multiple cycles of gametogenesis and many internally fertilizing corals have overlapping cycles.

In scleractinians, germ cells are located in endodermal mesenteries of polyps as interstitial cells. Oogenesis typically takes place slowly over several months (from 4 to 12 months) (Fig. 5.2). Oogenesis starts with the division of oogonia along the gonadal mesoglea, and is followed by differentiation in oocytes in the meiotic phase (Chiu et al. 2021; Shikina and Chang 2016). Then, oocytes increase in size until maturation by accumulation yolk proteins, lipids, and other materials, and migrate to the mesoglea layer. Spermatogenesis takes place more rapidly (1 to 8 months) and coincides with the final stages of oocyte maturation (Fig. 5.2) (Chiu et al. 2021; Shikina and Chang 2016). At the beginning, active mitotic division of spermatogonia is also located in the gonadal mesogleal layer. As spermatogonia develop, they form many spermatogenic compartments (spermaries). Differentiation into spermiogenesis and spermatocytes take place with spermaries (Chiu et al. 2021; Shikina and Chang 2016). At the molecular level, scleractinian gametogenesis shares many characteristics with other metazoans together with sex- and phase-specific gene expression profiles (Fig. 5.2), but also unique characteristic such as a novel yolk protein (egg protein; Chiu et al. 2021; Shikina and Chang 2016).

Nevertheless, and despite increasing public databases (Baird 2021), the cycles of gametogenesis remain very poorly understood and focused on a few species. A complete understanding of the gametogenesis mechanisms is necessary on a large range of species and this will open coral reef restoration and aquaculture to new perspectives and opportunities. For example, with high resolution and broad species knowledge of gametogenesis, a more efficient propagation of targeted species can be considered through hormonal induction of gametogenesis and spawning in artificial rearing systems (Chiu et al. 2021). As more ex situ spawning system are generated, our understanding of the effects of captivity on gametogenesis (Chiu et al. 2021; Craggs et al. 2017, 2019) and gamete structures (Lin et al. 2018; Lin et al. 2019) can be compared. Also, characterization of stage- and sex-specific germ cell molecular markers would enable the scientific community to evaluate and monitor the gametogenic status of corals cultured in captivity, while providing insight into the evolution of sexual reproduction in metazoans (Chiu et al. 2021).

5.2.3 Impact of Aging on Reproduction

Scleractinian corals are potentially considered as immortal, when taking into account asexual lineage, but it is assumed they undergo an intrinsic aging process (Bythell et al. 2018). Thus, some large massive coral species such as *Porites* spp. (Smith et al. 2021) or *Acropora* spp. (Irwin et al. 2017) attain an age of several hundred years old, up to 1000 years, while many species are short-lived species where older individuals exhibit a high mortality compared to juveniles under environmental stress (Bythell et al. 2018).

What constitutes an individual is a key question in scleractinian corals (Bythell et al. 2018). The size of a colony might indeed be not reflective of its age, and larger colonies are potentially more subject to damage and fragmentation, but are less likely to be killed by it (Bythell et al. 2018). Also, except in solitary species such as some Fungiidae, coral colonies are growing by budding of a primary polyp. Within a colony and in most of the species, polyps remain physiologically connected to the others via a continuous gastrovascular cavity and associated tissues (Bythell et al. 2018). Furthermore, at least three modes of tissue proliferation are recognized in corals, with a growth (1) that occurs equally across all polyps, (2) concentrated in the youngest polyps, or (3) in the oldest polyps (Elahi and Edmunds 2014). Thus, combined with species and location-specific growth rates, aging in corals is often difficult to estimate (Bythell et al. 2018).

Also, fecundity is thought to increase with age, because of an increased number of polyps and/or an increase in polyp fecundity (Harrison and Wallace 1990), but so far, only a few studies have investigated these parameters. Thus, old corals seem to remain sexually active together with a relatively high polyp fecundity, the current data available do not show a decline in reproductive fecundity as the colony age for several of the species (Bythell et al. 2018). However, reproductive decline and dysfunctional gametogenesis are observed in some corals (*Stylophora pistillata*, *Pocillopora damicornis,* and *Acropora tenuis*) at the scale of the polyp, colony, and/or part of the colony (Bythell et al. 2018; Irikawa et al. 2011).

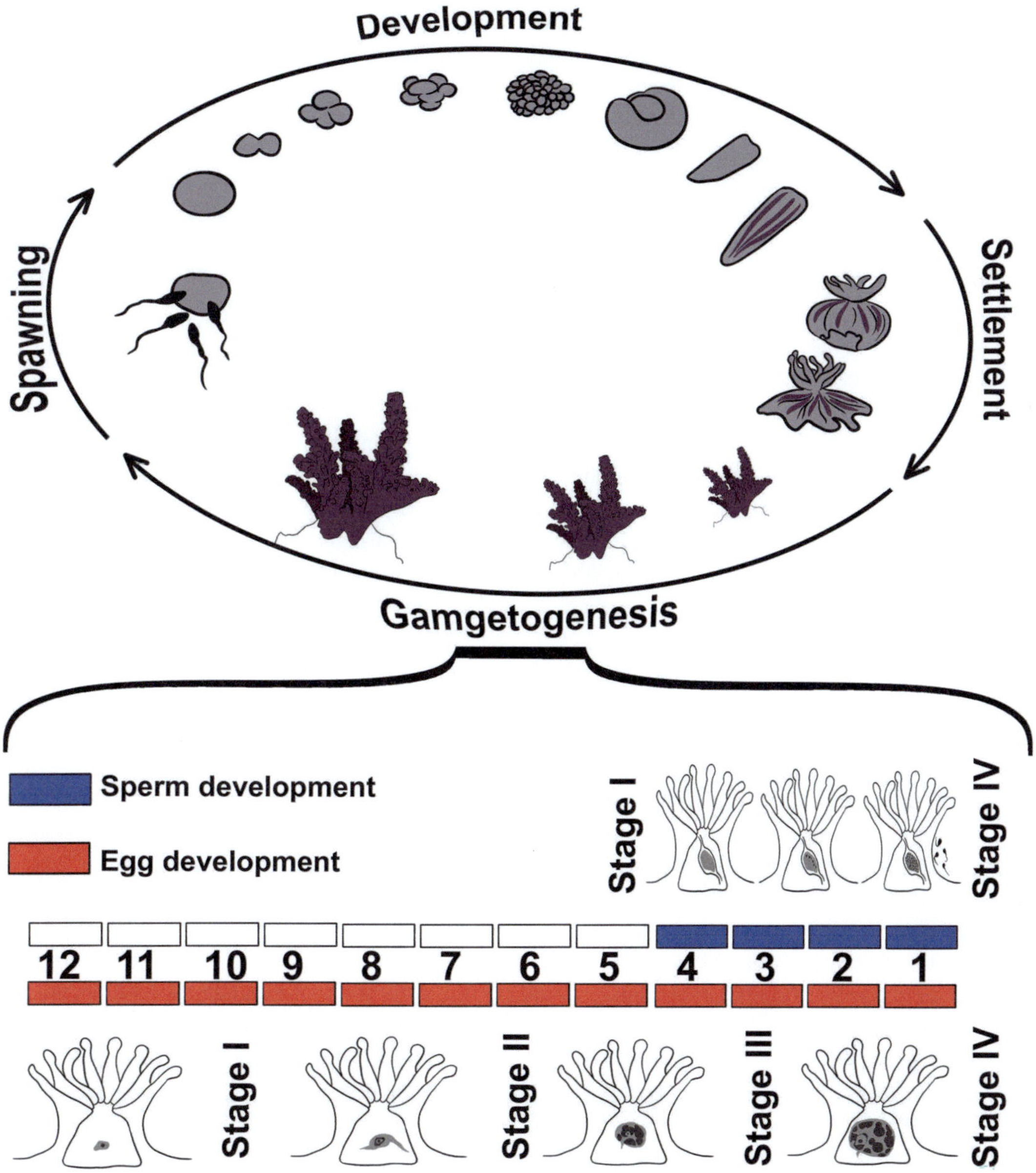

Fig. 5.2 General reproductive cycle and gametogenic timing

In light of the gaps in our knowledge for the intersection of age, growth, and reproduction for restoration via fragmentation, future research should focus on a large spectrum of ages and in multiple species exhibiting different lifespans. While first reproduction seems to be size-dependent, it is necessary to elucidate ontogenic processes such as puberty and aging in order to better characterize the health status of corals and/or to have efficient restoration programs by properly choosing coral colonies for fragmentation, for example. Indeed, fragmented corals seem to "remember" their age (Elahi and Edmunds 2014; Rapuano et al. 2023), while remaining reproductive independently of the fragmentation size (Rapuano et al. 2023). This last point has potentially critical consequences on coral restoration, as coral age might influence reproductive and growth declines together with increased susceptibility in case of environmental stress (Bythell et al. 2018). Further, studies on the rates of somatic mutation and their potential for inheritance remain limited (Olsen et al. 2019; review in Vasquez-Kuntz et al. 2022).

In order to efficiently estimate coral age, the availability of biomarkers such as telomeres is essential (Bythell et al. 2018). Telomere length is indeed associated with senescence and aging in mammals. Also, preliminary studies in corals described differences in telomere length decreasing between sperm, planulae, and adult polyps in *Acropora digitifera* (Tsuta et al. 2014), but not in *Galaxea fascicularis* (Nakamichi et al. 2012). Differences are also observed between *Pocillopora* spp. (short-lived species and environmentally sensitive) and *Porites* spp. (long-lived species and more stress-resistant) (Rouan et al. 2023). These results can be explained in lifespan and/or telomerase differences (Bythell et al. 2018; Nakamichi et al. 2012; Tsuta et al. 2014), and telomere length should be explored and understood in order to properly select corals to be used in restoration strategy.

5.2.4 Deep-Sea: An Almost Unknown Coral World

While recently there has been a growing focus on mesophotic coral ecosystems (from ~30 to 150 m) reproduction (see Sect. 5.2.2), reports of reproduction in the deep sea (defined as coral reefs between 4000 and 6000 m) are rare. In the last 20 years, research on the reproduction of deep-sea coral has increased, but only 4% of all known species have elements of their reproduction reported (review in Waller et al. 2023). While deep-sea coral ecosystems create hot spots of biodiversity together with important services to both the deep ocean and the shallow sea, very few are protected and they are now listed as vulnerable ecosystems. They offer structure and habitat for a variety of obligate and transient species, and are important spawning and nursery site. Main disturbances on these ecosystems are deep fishing and seafloor mining, but also ocean warming. However, our understanding of deep-sea coral systems recovery is even more in its infancy than shallow water reefs. Knowledge about aging, growth rates, and reproduction are lacking, while they represent critical information (review in Waller et al. 2023).

In abyssal coral ecosystems, dominant sexual system among examined species appears to be the gonochorism, and many deep-sea corals might take years, and even decades, before reaching sexual maturity. Even if several species have been characterized as brooder, most species are probably broadcaster corals. Proportions of the asexual mode of reproduction are under-described because of observation difficulties. Deep-sea corals seem to produce larvae with high dispersal capacities explaining why some species are broadly distributed (review in Waller et al. 2023).

Deep-sea coral research suffers from a low degree of spatial and temporal coverage and makes observations difficult to generalize to entire species. Finally, most of the tropical Pacific and Indian oceans are not described for deep-sea coral reproduction and development, while interest for seafloor exploitation is increasing (review in Waller et al. 2023).

5.2.5 Spawning and Fertilization

The coral broadcast spawning phenomenon is often massive and annual (Harrison et al. 1983), it can, however, take place asynchronously and over several months (Foster et al. 2018; Randall et al. 2020). Also, depending on species, spawning can occur during the day or the night and moonrise timing represent a key factor in this biological process (Foster et al. 2018; Lin et al. 2021; Randall et al. 2020). In most cases, spawning is highly synchronized within a species (potentially through the action of neurotransmitters; Taira et al. 2018). Synchronized spawning of a breadth of species has been described in multiple locations like Australia (Simpson 1991), the Gulf of Mexico (Gittings et al. 2018), Okinawa (Hayashibara et al. 1993), Taiwan (Chen et al. 2021), Singapore (Guest et al. 2002), the Solomon islands (Baird et al. 2001), and Palau (Penland et al. 2004), and are called "mass-spawning event" (Harrison 2011; Harrison et al. 1983). However, through a combination of a disconnection between seasonal (such as water temperature) and lunar cues associated with a variation in the initiation and time of oogenesis among colonies, split spawning (or spawning across 2–3 monthly cycles) can occur within a species (Foster et al. 2018). Split spawning seems to occur frequently and is predictable in some locations, and might allow the realignment of spawning dates with favorable conditions (Foster et al. 2018). Also, split spawning seems to increase the source of larval supply, while it might help to counteract the effects of low and variable connections to some populations (Hock et al. 2019). Interestingly, biannual spawning might be a common feature of *Acropora* spp. populations, and might be genetically determined at least for *Acropora tenuis* in Western Australia (Gilmour et al. 2016). While biannual spawning within a species might contribute to reproductive isolation, and then speciation, this isolation seems to be sometimes punctuated by interpopulation breeding when favorable environmental conditions occur (Gilmour et al. 2016). Finally, it has been shown that spawning might also be aperiodic with little linkage to moon cycles in *Montipora abellata* populations in Hawaii (Henley et al. 2022).

After spawning, it is not possible to observe germ line cells in the gonads, which indicate annual gametogenic cycle do not overlap most of the time (Shikina and Chang 2016). Also, it is sometimes possible to observe immature oocytes after spawning as oogenesis requires more than 1 year in a few species (Shikina and Chang 2016).

Synchronized spawning maximizes the chances of fertilization and larvae survival rates, but has also the potential to reduce fertilization success due to sperm limitation when the event is multispecific (Kitanobo et al. 2016; Shikina and Chang 2016) However, hybridization between closely related species is possible, particularly when sperm of conspecific is limited (Kitanobo et al. 2016; Kitanobo et al. 2022). While hybridization might represent a pathway to diversification and speciation (Quattrini et al. 2019; Van Oppen et al. 2002), it seems to be regulated by either egg/sperm recognition together with sperm mobility/inhibition (Morita et al. 2006), or asynchronous spawning times (Gilmour et al. 2016); or prezygotic mechanisms like gamete compatibility (Hatta et al. 1999; Isomura et al. 2013) and/or other molecular mechanisms involved in development following fertilization (Laporte et al. 2019).

Hybridization can be unidirectional or bidirectional with a variable percentage of success depending on species pairs and the direction of hybrid cross (Chan et al. 2019; Hatta et al. 1999; Isomura et al. 2013). However, a low percentage of success in fertilization does not necessarily influence the subsequent development and survival of the larval cohort (Chan et al. 2019), while larval survival, growth, and settlement depend on crossing-family (Chan et al. 2019; Isomura et al. 2013). Hybrids were thought to be sterile with propagation enhanced by asexual reproduction (Vollmer and Palumbi 2002). Recent studies demonstrate that hybrids generally exhibit a low fecundity and a potential difficulty to mate with other hybrids (Kitanobo et al. 2016). However, studies in *Acropora florida* suggest that eggs of this species may have retained a capacity to be fertilized by heterospecific sperm because resulting hybrid offspring have a high fertility (Kitanobo et al. 2016). Finally, genetic data also suggest that some hybrids can backcross with parent species (Kitanobo et al. 2022).

Adult coral hybrid species generally have an intermediate morphology that reflects a paternal or a maternal influence potentially observable at the transcriptomic level (Chan et al. 2020). While adult fitness will depend on crossing, no microbiome difference has been observed between hybrids and purebreds (Chan et al. 2019). Thus in this case, coral host genetic and nongenetic factors are probably responsible for the observed phenotype (Chan et al. 2020). Also, if interspecies breeding is used for a restoration program (Van De Water et al. 2022), trait inheritance and persistence should be studied. Depending on species crossing, inheritance might be maternal, paternal, or a mix of both (Chan et al. 2020). However, while inheritance of parental somatic mutations has been observed in coral, trait selection, understanding of the mechanisms, and persistence across generations are not known. Research in the area of selective breeding is still in their relative infancy for application to large-scale coral restoration programs (Vasquez-Kuntz et al. 2022).

While deployment of a high number of planulae onto degraded reefs has been shown to be potentially efficient in new reef restoration programs (Ridlon et al. 2023), the success of such approaches might be improved by the use of hybrid vigor exhibiting augmented capacities for climate resilience (Chan et al. 2019). However, many gaps remain to make this strategy efficient. Gamete compatibility, parental influences, and molecular mechanisms involved in fertilization and development need to be better characterized (Chan et al. 2019, 2020). While important for gamete recognition and fertilization, little is known about chemoattractant and gamete recognition proteins in corals (Chan et al. 2019). Parental effects have been well-documented for some plants and animals; they remain elusive in corals, but are known to affect morphology (Willis et al. 2006), survival (Chan et al. 2019; Isomura et al. 2013) and thermotolerance (Dixon et al. 2016). There is also some promise in the area of transgenerational acclimatization through the influence of parental conditioning (Putnam 2021; Putnam and Gates 2015). The use of multi-omic approaches should be employed to decipher the parental contribution in different hybrid combinations, but also to understand the genetic and nongenetic (epigenetic) factors related to fitness differences (Chan et al. 2020). While environmentally sensitive genes are important targets to be characterized in order to properly estimate age, health, or resistance to stress in corals to be selected for breeding, most of them remain unknown and constitute a coral molecular dark genes that need to be understood (Cleves et al. 2020).

5.3 Coral Development

5.3.1 Embryonic Development

Embryonic development is the most sensitive life cycle phase of corals (Putnam 2021; Randall et al. 2020). Characterizing these stages and understanding their plasticity and sensitivity to environmental factors is a priority (Randall et al. 2020). In broadcaster corals, the first divisions usually take place at night with lipid-rich embryos floating to the surface and moving passively. However, gametes and embryonic buoyancy will depend on symbiont heritability.

Embryos do not have outer protective membranes, and first cleavage embryos easily fragment into cell clusters in the event of strong winds or large swells. Thus, only the animal hemisphere of the fragmenting embryos can resume their cell division and continue to develop into larval clones proportionally smaller than the larva (Okubo et al. 2016). The subsequent larval survival might depend on blastomere size (Okubo et al. 2013, 2016). Nevertheless, the latter are potentially disadvantaged because the growth time to reach a predation threshold, for example, will potentially be longer (Heyward and Negri 2012). Finally, molecular basis of embryonic regeneration due to fragmentation is still unknown and should be characterized (Okubo et al. 2016).

The first division shares common points with the hydra (holoblastic and progressive development of furrows leading to equal blastomeres) (Okubo et al. 2013, 2016; Okubo and Motokawa 2007). However, these similarities fade after the 8-cell stage when cell division no longer seems particularly orderly. Blastulation and gastrulation are then different depending on coral species (Fig. 5.3) (Okubo et al. 2013, 2016; Okubo and Motokawa 2007). Originally, scleractinian

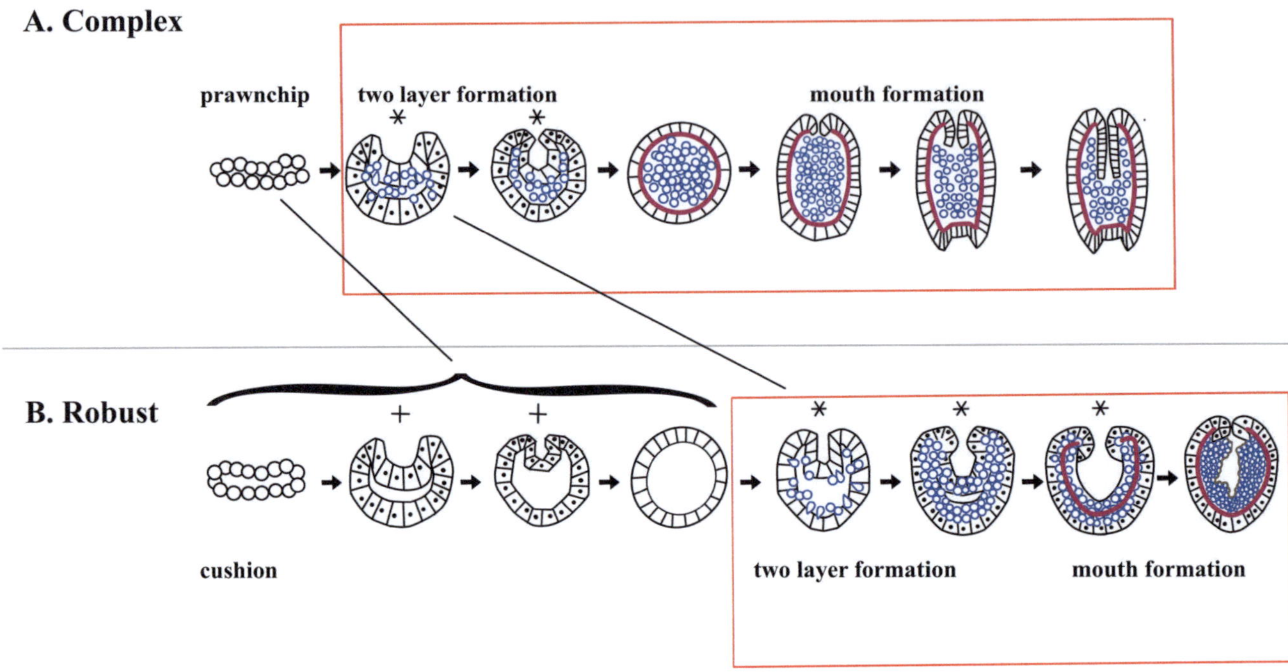

Fig. 5.3 Schematic representation of the blastulation in "complex" corals and "robust" corals. (Adapted from Okubo et al. 2013, 2016; Okubo and Motokawa 2007)

corals were classified based on their morphology only, but recent advances in molecular and developmental biology classified them into "complex" and "robust" species (Ying et al. 2018). In complex corals such as *Acroporidae*, gastrulation occurs through a "prawn shrimp" stage that is characterized by the lack of blastocoel composed of a bilayer of undifferentiated cells (Fig. 5.3). In robust corals such as *Pocilloporidae*, gastrulation is done by invagination of a "cushion" stage assimilated to a spherical blastula (Fig. 5.3) (Ball et al. 2002; Okubo et al. 2013, 2016; Okubo and Motokawa 2007).

Then, the embryo rounds out and cilia develop marking the end of the embryonic phase. The beginning of the larval phase is characterized by rotating swimming larvae with an appearance that varies according to species (between 12 h and 3 days after fertilization). During the last larval stages, the planula tends to elongate and many cell types are differentiated. Planula larvae have a directional swimming activity (within hours to days after rotating swimming stage) and will attach to metamorphose based on chemical signals that emanate from coralline algae and/or associated bacteria (Ball et al. 2002; Harrison 2011; Randall et al. 2020). Brooder corals directly emit planula larvae able to settle and calcify very rapidly (Mass et al. 2016).

Most scleractinian corals acquire zooxanthellae from the environment (horizontal transmission) during the early stages of life, but sometimes they are also inherited vertically through oocytes (Fig. 5.4) (Baird 2021; Hirose et al. 2000; Marlow and Martindale 2007). Also, some species with vertically inherited symbionts exhibit a faster development together with fewer lipidic reserves and a negative buoyancy, like *Pocillopora* spp. (Hirose et al. 2000) but not *Montipora* spp. (Marlow and Martindale 2007) (Fig. 5.4) as a possible mechanism against deleterious UV light (Baird 2021). Symbiont localization in the eggs and embryos is species-dependent, and heritability might have evolved independently in different genera (Marlow and Martindale 2007).

5.3.2 Larval Phase and Dispersal

The planktonic larval phase of scleractinian corals represents a crucial, but also a very critic steps in their life cycle. This phase allows dispersal, while ensuring the genetic diversity among populations (Zhang et al. 2019). In the context of reef degradation due to climate changes and anthropogenic activities, understanding the dynamics that influence larval dispersal appears crucial together with the relatively unknown contribution of genetic and parental effects (Zhang et al. 2019).

Coral populations are indeed highly dependent on the balance between larval dispersal and population distances. While often more complicated on land, dispersal is passively facilitated in the marine environment (Burgess et al. 2015). Long dispersal generally allows spreading across the species range adaptively beneficial alleles, while short ranges maintain locally well genotypes (Palumbi et al. 2023; Strathmann et al. 2002). Scleractinian corals show dispersal range of

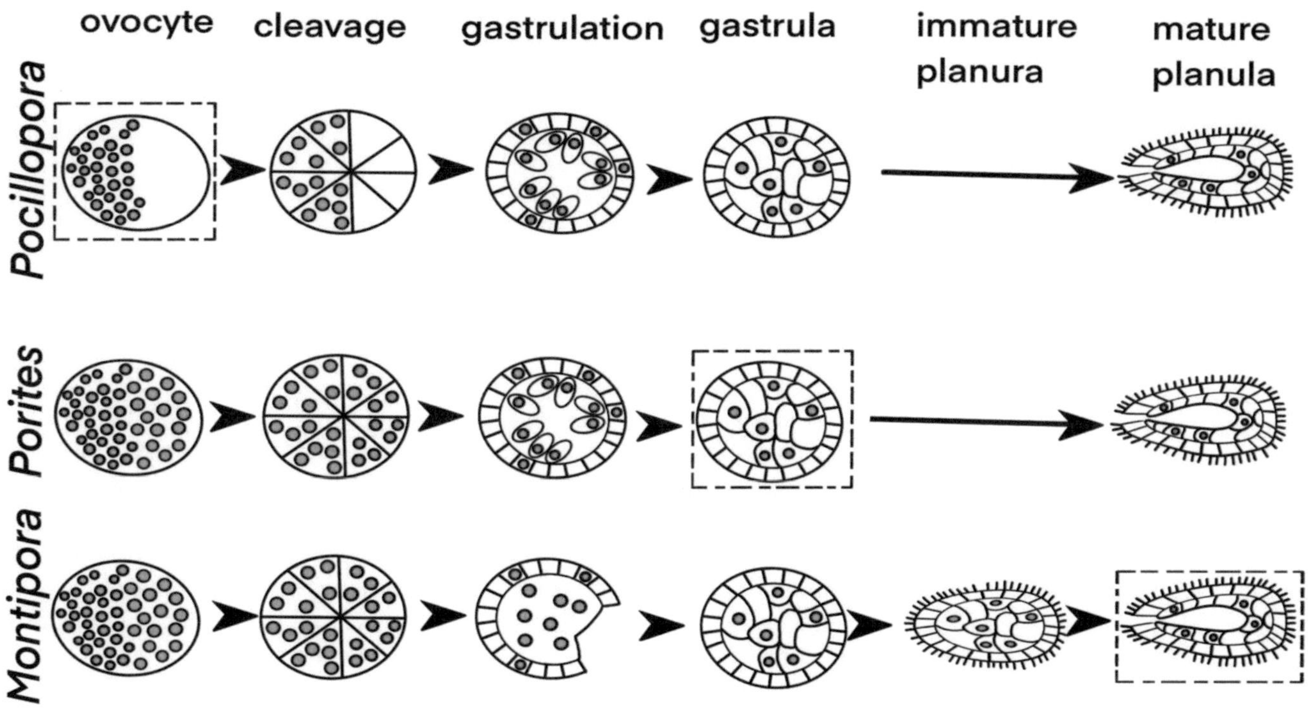

Fig. 5.4 Schematic illustration of the zooxanthellae distribution during the embryonic development of *Pocillopora* sp. ("robust" corals), and *Porites* sp. and *Montipora* sp. ("complex" corals). The stages of development are indicated on the top. The boxes indicate the first stage in which the zooxanthellae are localized in the endoderm in the oocyte or embryo. The timing of zooxanthellae localization and the type of gastrulation differ between the three genera. (Adapted from Hirose and Hidaka 2006)

multiple hundreds of kilometers, but this range is highly species-dependent (Palumbi et al. 2023). Thus, many brooding corals or vertically transmitting symbiont species exhibit short dispersal, while broadcast spawning corals and horizontally transmitting symbiont (acquired during the planula stage) species have longer dispersal (Palumbi et al. 2023). These parameters being of crucial interest to establish efficient restoration and protection programs as they may help to increase the accuracy of assisted migration and the models to predict future coral adaptations (Palumbi et al. 2023). Dispersal of larvae and settlement rates are difficult to track, and not always representative of the final coral communities (Doropoulos and Roff 2022; Sarribouette et al. 2022). Thus, a good evaluation of population connectivity and dispersal is necessary in order to have efficient coral restoration and conservation strategies (Lukoschiek et al. 2016). Coloring larvae with neutral red and Nile blue appears to allow tracking of local dispersal and settlement without apparently impacting neither survival nor recruitment (Doropoulos and Roff 2022). This technique may be promising at a local scale, because the proposed method is simple and rapid, larvae are easy to detect on the field in a high number, the dye is nontoxic in the marine environment, and easily available and cheap (Doropoulos and Roff 2022).

Fluctuations, intensities, and frequencies of environmental conditions can influence larval dispersal, and then patterns of coral biodiversity and population dynamics (Gurdek-Bas et al. 2022). Larval dispersal and duration are indeed associated with oceanographic conditions. While dispersion is generally thought as a horizontal process, it can be made also vertically from the mesophotic environment (Frade et al. 2016; Prasetia et al. 2022). Thus, it has been shown that recolonization of shallow water from the depth might occur through a multigenerational recruitment over time with a proportion of success depending on the species and the family (Prasetia et al. 2022; Prata et al. 2022). Light intensity mainly but also reef complexities (communities) are thought to be the limiting factors explaining this slow recolonization from the depths by mesophotic larvae (Prasetia et al. 2022). While the concept of refuge may be a source of hope and natural recolonization by putatively close populations, the potential of reseeding largely varies among coral species (Bongaerts et al. 2017). Furthermore, mesophotic corals tend to have a shorter reproductive season and smaller planulae compared to their shallow water counterparts (Liberman et al. 2021; Prasetia et al. 2022). Thus, while mesophotic zone could present potential for reef sources for restoration purposes, further studies should be done locally integrating the areas and the species to decipher the potential of mesophotic reefs as protected or resistant oases. Nonetheless, future predictions of climate changes revealed that the exacerbate anthropogenic stressors combined with the increasing frequency and intensity of bleaching events may not protect over the long term the upper mesophotic areas (20 to 40 m depth), which will limit the potential of mesophotic zone opportunity for restorations (Prasetia et al. 2022).

Larval dispersal capacities in shallow and mesophotic habitats not only depend on environmental conditions or the species (Prasetia et al. 2022; Zhang et al. 2019) but also on fluorescence phenotype, which thus complexify our understanding of larval dispersal. Indeed, recent works revealed that dispersion can be influenced by the fluorescence phenotype of the larvae (Puisay et al. 2021). Red fluorescent in *Acropora millepora* has been associated with a greater dispersal explained by a diapauses-like state together with the corresponding transcriptome (Strader et al. 2016). Fluorescence involves a protein family known as fluorescent proteins (FPs) and in corals, FPs have been involved in immunity during development (Palmer et al. 2012), symbiont acquisition, settlement, or thermotolerance (Puisay et al. 2021). Bleaching susceptibility (but not only) of the parents seems to induce larval fluorescence differences (Puisay et al. 2021). Larval fluorescence might indicate greater or lower dispersal capacities (Puisay et al. 2021; Strader et al. 2016). While representing promising noninvasive biomarkers, more studies are needed on this nondestructive proxy to determine its usefulness in restoration programs allowing for selecting high or low larva dispersal capacities or high thermotolerant corals.

In new restoration programs, the use of coral larvae is thought to be an effective method to restore degraded reef systems at a large scale (Doropoulos et al. 2019a, 2019b; Pollock et al. 2017; Ridlon et al. 2023). In existing programs, two large-scale restoration approaches are currently tested with the transplantation of gravid coral directly on the targeted restoration site to provide a high number of larvae (Doropoulos et al. 2019a, 2019b); however, consequences of the translocation of gametes and then on embryos are unknown, or by harvesting and transporting wild coral spawn slick to the targeted reef (Doropoulos et al. 2019a, b). This second option might be done more easily to achieve large-scale restoration at a lower cost with the potential to transport billions of larvae on a vast geographic scale, while having a putatively limited impact on wild populations and retaining the natural species and genetic diversity (Doropoulos et al. 2019a, b). Furthermore, large coral-spawn slicks might be easy to detect using satellites in order to facilitate collection of embryos, and reducing the transport cost (Yamano et al. 2020).

Collection and relocation of coral slick might be done at industrial scale, while reducing larval mortality due to predation (Harrison et al. 2021; Ridlon et al. 2023). However, the technique is more challenging where large larval slicks are absent (most of the Caribbean) (Ridlon et al. 2023). In that case, coral embryos can be collected from restored breeding colonies and cultivated in large facilities. However, while aquaculture already has a known impact on the structure and lipidic reserves of coral eggs (Lin et al. 2019), the impact of closed system on larvae and their survival after introduction into the natural environment is unknown (Ridlon et al. 2023). Also, experiences on teleost fishes have shown at strong and fast impact of aquaculture on larvae biology, but also on survival differences between cultivated and wild fishes after reintroduction of the cultivated once in the environment (Le Luyer et al. 2017). Finally, before any attempt of large-scale restoration programs using coral larvae, targeted sites of restoration should be determined using guidelines provided by the International Coral Reef Initiative (Valderrábano 2021) and integrated with the connectivity of the different surrounding reefs. Some localities are indeed always subject to bleaching events and outbreaks of crown-of-thorns (COTS), while other acts as larval sources to repopulate degraded areas (Hock et al. 2017). Therefore, the history of the restoration site has to be known in order to assess the chance of restoration, and how to restore to a previous coral landscape existing before disturbances.

5.3.3 Metamorphosis, Recruitment, and Growth

The transition from the planula to the adult stage is a very critical phase in scleractinian coral life cycle. This phase occurs in a few hours for internally fertilized corals that directly emit a more or less evolved planula larva that will be able to metamorphose and calcify very rapidly (Mass et al. 2016), but in some case, it could require several days. The planula will attach to the aboral end and contract along the bucco-aboral axis and form a flattened disk. During this process, which is intended to be permanent, the larva metamorphoses into a polyp and calcification begins (Ball et al. 2002; Harrison 2011; Randall et al. 2020).

However, settlement is governed by multiple environmental conditions such as sedimentation (Ricardo et al. 2017), light (Ricardo et al. 2021), water quality (Negri et al. 2005; Negri and Heyward 2000), and algal cover (Hoey et al. 2011; Webster et al. 2015). Thus, metamorphosing larvae can revert back to swimming planulae until favorable conditions (Ishii et al. 2022). Metamorphosis and settlement are regulated at all spatial (topography, from centimeters to multiple kilometers) and temporal scales (season to decades) creating a potentially high spatiotemporal variability also depending on biological inducers (crustose coralline algae—CCA and associated microbiome) or competitors (other recruits, macroalgae, or peyssonnelid algal crusts—PACs; Jouval et al. 2019; Stockton and Edmunds 2021; Wilson et al. 2020). While settlement tends to be inhibited by PACs (Stockton and Edmunds 2021; Wilson et al. 2020), some CCA species and associated microbial biofilms are drivers of coral settlement (Gómez-lemos et al. 2018). However, the exact mechanisms and interactions behind coral settlement must be determined (Gómez-lemos et al. 2018) together with the persistence of CCA in a changing world (Krieger et al. 2023).

Newly settled recruits remain vulnerable to multiple environmental factors such as temperature (Da-anoy et al. 2020) or corallivory by fishes or invertebrates (Ladd et al. 2021), but this vulnerability can be reduced by a natural phenomenon called chimerism (Fig. 5.5) (Puill-Stephan et al. 2012; Vidal-Dupiol et al. 2022). Chimerism results from the fusion of multiple newly settled recruits (Fig. 5.5) (Puill-Stephan et al. 2012; Vidal-Dupiol et al. 2022). While mostly described during the settlement, chimerism might also happen in embryos and pre-metamorphic larvae (Jiang et al. 2015) or even 1–2 years post-settlement (Vidal-Dupiol et al. 2022). Description of chimerism is still in its infancy and is rarely taken into account in most of the studies (Rinkevich 2019), but an increase in size and growth combined with a higher genetic and nongenetic diversity makes chimeras potentially more resilient to environmental changes (Fig. 5.5) (Rinkevich 2019; Vidal-Dupiol et al. 2022). Also, chimerism might be used as an evolutionary rescue in restoration strategies, as it can potentially accelerate adaptative responses to climate changes and anthropogenic perturbations (Rinkevich 2019; Vidal-Dupiol et al. 2022). Chimerism should be considered to precisely understand coral reef population dynamics and as a toolbox in ecological engineering to be developed in active reef restoration (Fig. 5.5) (Rinkevich 2019; Vidal-Dupiol et al. 2022). Chimera can be used through fragmentation to reskin dead skeletons or with pre-settlement larvae as close relatives are potentially able to fuse easily (Mcleod et al. 2022). Each sibling may have distinct traits and allelic combination resulting in a potentially more stress resistant chimera (Mcleod et al. 2022).

Metamorphosis, settlement, and growth/survival of new settle larvae are widely influenced by the macroalgal cover, that is becoming prevalent in many places due to local anthropogenic stressors combined with more frequent bleaching events (Fig. 5.6) (Monteil et al. 2020). While macroalgae generally have a negative impact on coral reproduc-

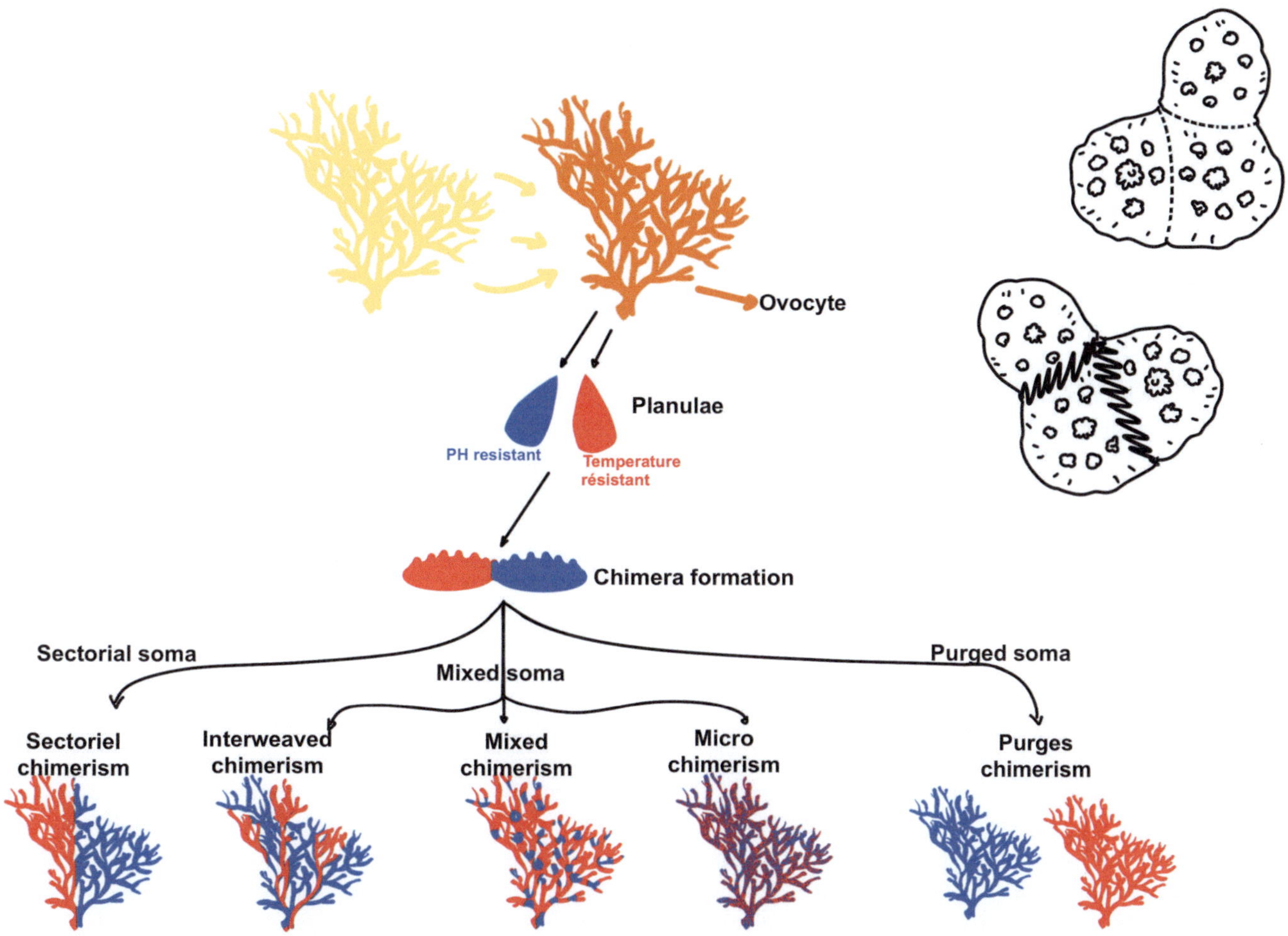

Fig. 5.5 Potential outcomes from chimeric settlements in corals

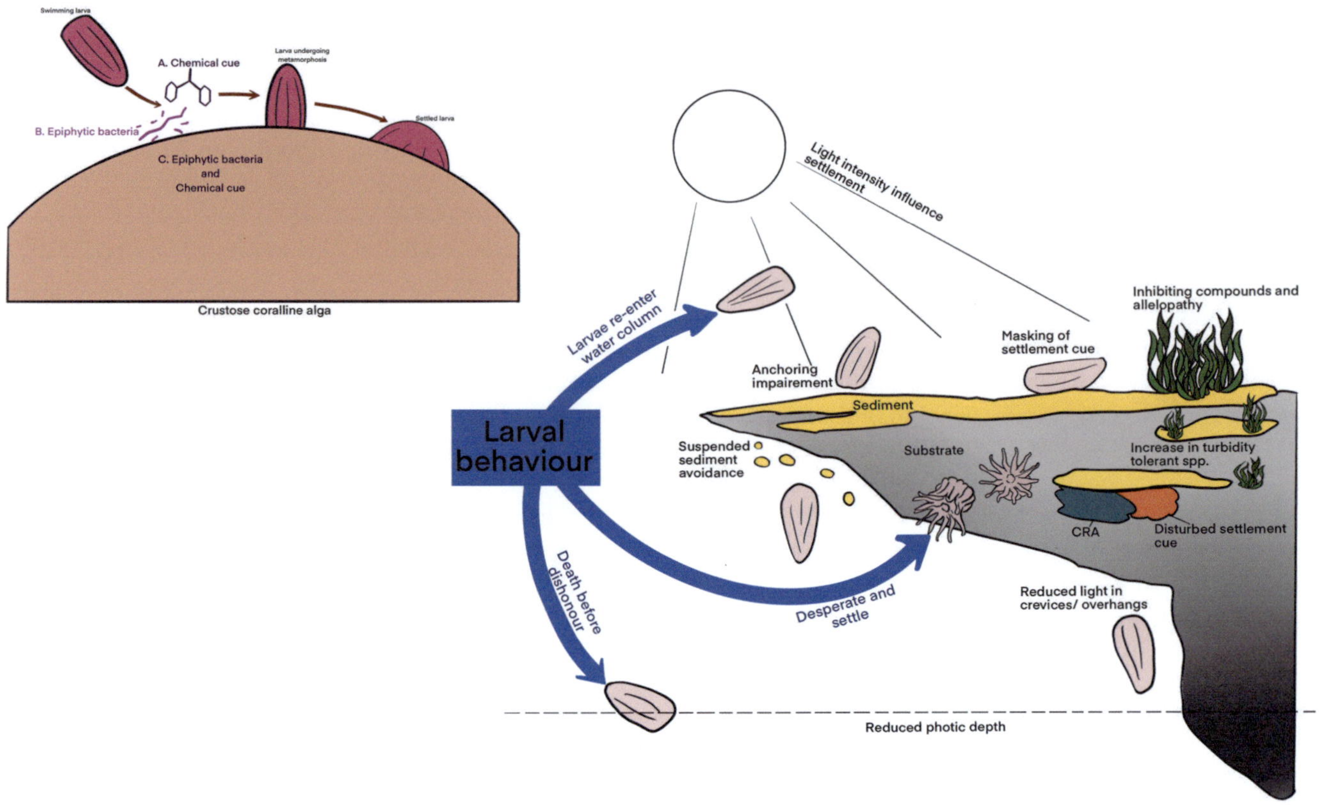

Fig. 5.6 (**a**) Proposed mechanisms of coral settlement facilitated by crustose coralline algae (CCA) with settlement facilitated (**a**) by chemical cues from CCA, (**b**) by the biofilm of the epiphytic bacteria, and (**c**) by the combined action of chemical cues from CCA and of biofilms of the epiphytic bacteria. (Adapted from Gómez-lemos et al. 2018). (**b**) Illustration of possible coral settlement behaviors together with changes when facing environmental perturbations such as sedimentation, algae, or light regimes. (Adapted from Ricardo et al. 2021)

tion (fecundity and egg size; Monteil et al. 2020), impact on early benthic survivorship and growth might be algal- and coral-specific (Bythell et al. 2018; Spiers and Frazer 2023; White and O'Donnell 2010). The macroalgae removal is an activity easily accessible to local communities and has been tested in restoration programs to help facilitate coral recovery (Mcleod et al. 2022). However, removal should be intensive and linked with the addition of herbivorous organisms that generally also declined (Mcleod et al. 2022). To date, the best example is the change in community structure following mass mortality of sea urchins *Diadema anrillarum* in Jamaican coral reefs (Hughes et al. 1987). Thus, many restoration programs successfully use algal removal, grazing activity of sea urchins, and coral reimplantation (Cano et al. 2021). However, in order to maximize coral reef restoration strategies, it is mandatory to have a good understanding of the whole reef ecosystem including the coral species, the density of macroalgae, the herbivorous species, and the potential coral and macroalgal predators. Firstly, historical knowledge of natural macroalgal levels and fluctuations is critical to establish a natural variation of macroalgal cover for a healthy reef and a disturbed reef, in order to reestablish macroalgal levels to previous disturbance or to healthy reef levels (Bruno et al. 2014). Overfishing and anthropogenic activities are often considered as the main factors of increase in macroalgal cover (Knoester et al. 2023), but studies revealed that the existence of a natural algal cycle following a coral mass mortality event in "pristine" (or isolated from human local activities) reefs (Fukunaga et al. 2022) associated with an increase in fish recruitment (Fong et al. 2018). Thus, these natural cycles as well as the influence of algae on coral recruitment must be fundamentally studied together with the source of algal proliferation on chosen sites before starting a large project of algal removal.

While attention is always focused on coral reproduction and survivorship, there is a lack of knowledge on studying the binding organisms [such as some crustose coralline algae (CCA) species] that can support recruitment and regrowth of coral cover (Fig. 5.6) (Fukunaga et al. 2022). However, exact relationships between coral larval settlement and CCA remain to be understood (Fig. 5.6) (Gómez-lemos et al. 2018). In consequence, future restoration projects should include the monitoring of reef binders [exact number of involved CCA and their associated species of corals (together with functional tests of settlement) and herbivorous organisms playing important ecological roles in sediment binding, reef erosion, and algal control, while continuing monitoring basic coral health status (coral growth and algal cover, Fukunaga et al. 2022). Restoring a reef area is a complex approach that will require a combination of approach to be really ecologically and functionally successful. Today, coral transplantation should be combining in a large number of locations with alga biocontrol to ensure chance of success of survival of transplanted coral fragments. To do that, the most efficient native herbivorous species or a combination of herbivorous species (Toh et al. 2013) should be used to enhance the algal biocontrol. In most cases, both sea urchins (Cano et al. 2021) and crabs (Francis et al. 2019) have been identified as the most efficient. While doing herbivorous species reintroduction, their life cycle and persistence in the modified environment should be studied, and their density during the reintroduction phase should be taken into account (Cano et al. 2021), as it could have a positive or negative on coral recruits (Yanovski and Abelson 2019). Predators of corals and of used herbivorous species on restoration sites should be known (Cano et al. 2021) together with susceptibility to outbreaks of the crown-of-thorns (COTS) (Hock et al. 2017) as they can easily annihilate the undertaken effort.

Substrate matters and topography has an incidence of recruitment at different scales. Thus, the use of artificial substrates in restoration programs starts to be highly studied (Dubininkas 2017; Leonard et al. 2021). These works have received exponential interest over the last decade, showing that benthic composition colonizing the support, durability and structural micro-complexity of artificial materials, and the orientation of the surfaces, together with the presence/absence of anti-fouling (AF) coatings, strongly influence coral recruitment depending on the scleractinian family (Harper et al. 2021; Leonard et al. 2021; Levenstein et al. 2022; Roepke et al. 2022). However, restoration programs using artificial substrates have difficulties in scaling solutions to larger areas and in replicating the 3D complexity of the natural coral reef (Levy et al. 2022). While reef restorations often lack data from global reefs to be restored, the use of 3D technologies (such as photogrammetry and 3D printers) might customize efficient artificial structures (Yanovski et al. 2017; Yanovski and Abelson 2019) together with computational fluid dynamic modeling (Levenstein et al. 2022). Thus, eDNA (inventory of the local biodiversity—coral, herbivorous species, predators) and 3D imaging (estimation of the recruitment capacity) might be used to evaluate the species to be reintroduced and the success of this effort (Levenstein et al. 2022). However, while the artificial enhancement of degraded areas might represent a tool for increasing coral recruitment, the removal of anthropogenic stressors (pollution, fishing, etc.) and protection measures should be applied before any interventions in order to minimize the risk of inhibition of the restoration efforts (Levenstein et al. 2022).

5.3.4 Microbiome Dynamics Across Ontogeny

Specific bacteria and bacterial coral communities are hypothesized as crucially important to the reef ecosystem and coral itself. Recently, coral microbiome research has rapidly expanded identifying microbiome variability within the individuals (Epstein et al. 2019), among the colonies (Epstein et al. 2019) and across reefs (Hernandez-agreda et al. 2017). Coral microbiome is also not static but changes dynamically in relation to the seasons (Yu et al. 2021) and the life cycle (Van Oppen and Blackall 2019).

Recent discoveries suggest that microbiome might be acquired vertically by gametes (Ricardo et al. 2017) or horizontally through seawater in broadcast spawning corals (Ceh et al. 2013), while brooder corals would transmit it vertically (Fig. 5.7) (Damjanovic et al. 2019; Van Oppen and Blackall 2019). However, at least for broadcast spawning coral species, dramatic changes in bacterial communities across the different embryonic stages until adulthood indicate that various taxa of bacteria are acquired from the environment (Ceh et al. 2013). Thus, these species may have a mixed transmission strategy, or a strictly horizontal transmission, but this remains to be elucidated (Damjanovic et al. 2019; Van Oppen and Blackall 2019). In coral larvae and newly recruits, the bacterial microbiome is generally more diverse than in later stages (Ceh et al. 2013). This fact indicates a winnowing process necessary to coral maturation in order to potentially fill the need for a particular host in a given environment (Van Oppen and Blackall 2019). However, the factors governing this winnowing process are unknown in corals, even if the host immune system might play a role early in development (Van Oppen and Blackall 2019) and will benefit from further investigation.

Future strategies of coral restoration should consider the modification of the holobiont and the use of probiotic as possible efficient solutions to help coral reefs in the era of global changes and anthropogenic perturbations. However, a trade-off is often found between beneficial bacteria and/or symbionts between, for example, temperature tolerance and reproductive capacities. Also, the research is still in its infancy despite promising results (Barno et al. 2021; Maire and Van Oppen 2021; Rinkevich 2021).

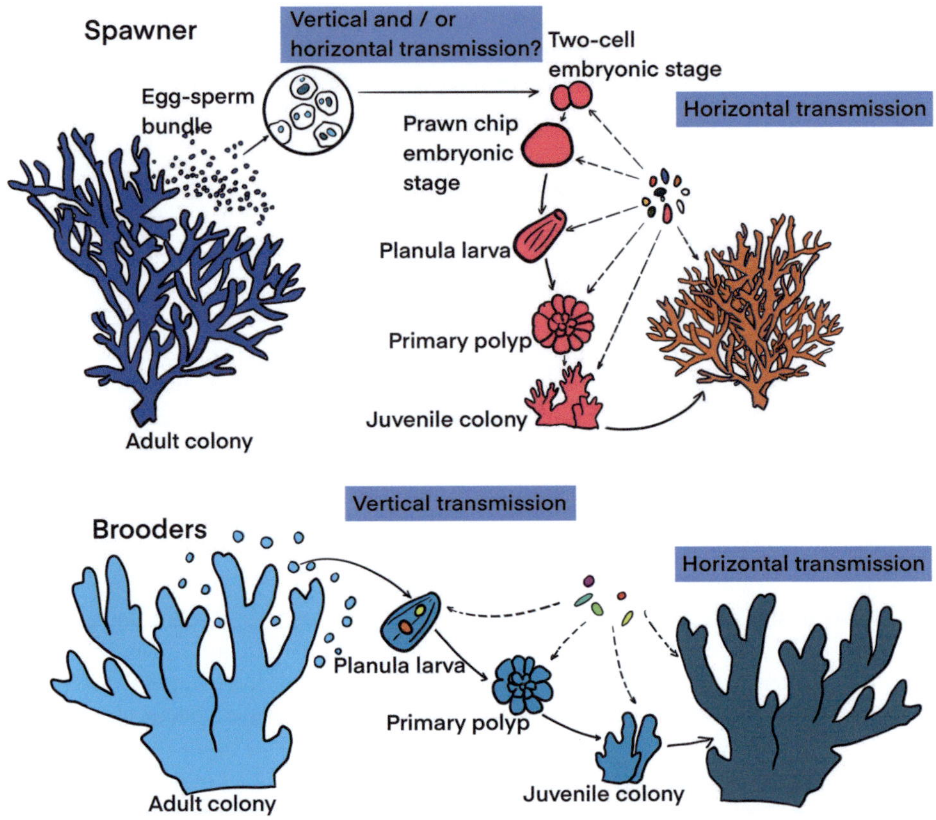

Fig. 5.7 Bacterial acquisition and inheritance throughout the coral life cycle. (Adapted from Van Oppen and Blackall 2019)

5.3.5 Molecular Mechanisms During Coral Development and Their Utility in the Restoration Program

Complex and robust corals not only exhibit distinct embryonic development during the critical steps of blastulation and gastrulation, but also differences in environmental sensitivities with robust corals often presented as "winner" in the context of climate changes and anthropogenic activities (Ball et al. 2002; Okubo et al. 2013, 2016; Okubo and Motokawa 2007). At the molecular level, complex and robust corals possess different metabolic pathways (Ying et al. 2018), and lineage-specific gene families seem to be specific and/or having different expansion histories depending on the examined group and species (Yoshioka et al. 2022). While the scientific community is still looking for good biomarkers of aging (Bythell et al. 2018) and thermal resistance (Louis et al. 2017) in corals, a recent study shows that telomere length is determined by seasonal temperature variation in *Pocillopora* spp. (robust group) and by past thermal anomalies in *Porites* spp. (Rouan et al. 2023). Altogether, developmental and molecular differences might explain that some species appear as "winner" in the context of climate changes and anthropogenic activities, while others are "loser" (Ball et al. 2002; Chan et al. 2020; Okubo et al. 2013, 2016; Okubo and Motokawa 2007).

Recent studies on aquatic and terrestrial organisms underline the importance of studying more globally embryonic development for all taxa in the context of global changes (Burggren 2020, 2021; Eirin-lopez and Putnam 2019; Fellous et al. 2022). The molecular mechanisms regulating the development, the zygotic transition, and gastrulation remain poorly described in scleractinian corals (Chille et al. 2021; Cleves et al. 2020; Cowen and Putnam 2022; Putnam 2021). However, recent studies demonstrate that development in corals is precisely regulated at the molecular level depending on complex and robust groups (Yoshioka et al. 2022). While vertebrate development is based on maternal provisioning and one wave of zygotic transition (Andersen et al. 2013; Vastenhouw et al. 2019), scleractinian corals exhibit two waves of transcriptional zygotic activation (ZGA) (Chille et al. 2021). The first one is occurring between the first cleavages [4 h post-fecundation (hpf)] and the early gastrula (14 hpf), and the second one is happening between the mid-gastrula (22 hpf) and the planula (4 days) (Chille et al. 2021). However, while vertebrates exhibit different ZGA depending on species (Vastenhouw et al. 2019), ZGA diversity, functionality, and sensitivity to environmental changes remain unexplored in corals (Putnam 2021; Chille et al. 2021). Molecular characterization of coral immune system indicates that early stages are able to resist infection, while metabolic studies indicate a shift in embryo metabolism during ZGA and settlement (Huffmyer et al. 2023). During the settlement, the reversible and irreversible phases of metamorphosis are transcriptionally regulated in a distinct way (Ishii et al. 2022) and endosymbiont genes might play a key role (Walker et al. 2019).

Recently, the availability of genomic, transcriptomic, proteomic, or even epigenomic data has increased considerably for different species of scleractinian corals (Cleves et al. 2018, 2020; Mcrae et al. 2021; Trigg et al. 2021) and their associated microbiome (Planes et al. 2019). In addition, a large part of this data is publicly available and directly usable. Functional approaches by CRISPR/Cas9 are also beginning to be applied (Cleves et al. 2018, 2020). These data surprisingly reveal (given the evolutionary distance separating these different taxa) that scleractinian corals and mammals (Bhattacharya et al. 2016), as well as their endosymbiotic algae with mammals and plants (Shoguchi et al. 2013), share many crucial gene networks absent in other taxa such as mollusks or insects. By taking advantage of the knowledge acquired in model organisms, comparative genomics approaches will make it possible to better understand the biology of corals and by taking into account their associated microbiome in the context of climate change. However, these approaches often suffer from incomplete annotation and a large number of genes listed as unknown in corals (Cleves et al. 2020). Moreover, both in corals and their symbionts, and even though they are differentially expressed in the face of stress, a large number of genes have no known functions. For example, 11.7% of the predicted genes in the *Montipora capita* genome have no orthologs and 33% of the dinoflagellate genes have no annotation (1.4% of the proteins encoded by these same genes have a known domain) (Cleves et al. 2020).

Most of the "-omics" studies in scleractinian corals focus on the identification of the main genes or networks involved in the stress response, and on understanding how the observed differences are related to stress tolerance levels among individuals, populations, and species. However, these approaches suffer from the lack of standardized protocols and functional approaches combined to a very few number of species studies to really reflect reef corals diversity (Cleves et al. 2020). Finally, compared to transcriptomic approaches, a very few number of studies has been dedicated to proteomic or epigenomic investigations, and this needs to be considered in the future as these works will be crucial for a complete understanding of the adaptive mechanisms of corals from the transcriptomic approaches (Cleves et al. 2020; Hackerott et al. 2021; Mcrae et al. 2021; Putnam 2021).

5.4 Phenotypic Plasticity and Its Role in the Establishment of Innovative Restoration Strategies

5.4.1 Phenotypic Plasticity

Coral gametogenesis and embryogenesis are very sensitive to environmental modifications and perturbations, such as pollution, artificial light, temperature, or excess nutrients that compromise the formation of gametes, affect the rate of fertilization and cause embryonic abnormalities (Ayalon et al. 2021; Keshavmurthy et al. 2014; Liberman et al. 2021; Randall et al. 2020). While warmer waters accelerate development, these can reduce fertilization rates and compromise embryo and larval survival (Randall et al. 2020). The majority of embryonic abnormalities and losses are occurring at the time of gastrulation (Randall et al. 2020). However, consequences of perturbations on later stages and adult colonies are not understood yet (depending on the environmental factor and/or the exposed life/gametogenic stage) (Putnam 2021), whereas corals exhibit a high degree of heritability for certain traits (Bairos-Nowak et al. 2021) and an "environmental memory" (Hackerott et al. 2021). While reef resilience depends on stress, species, geographical location (Brener-raffalli et al. 2022; Perez-Rosales et al. 2021), or gametogenesis and/or embryonic stages (Hackerott et al. 2021; Putnam 2021), acclimatization mechanisms play a major role in coral resilience (Hackerott et al. 2021; Putnam 2021). Acclimatization occurs when an individual organism experiences a change in its phenotype in response to an environmental stressor (Dixon et al. 2018). This change is reversible and occurs very rapidly, sometimes in a few minutes. In opposition, adaptation occurs when a population experiences a change in its genotype in response to an environmental stressor. Adaptation occurs much more slowly, usually across many generations, and is a result of natural selection (but see Dixon et al. 2018). Among the molecular mechanisms playing a role in phenotypic plasticity and possibly coral resilience and persistence, epigenetic mechanisms appear to be informative (Eirin-lopez and Putnam 2019).

5.4.2 Epigenetic Mechanisms

The development and reproduction of corals result from the implementation of processes of proliferation, migration, cell differentiation, and apoptosis requiring very precise spatiotemporal patterns of gene expression involving the transient establishment of specific transcriptomes. The fine regulation

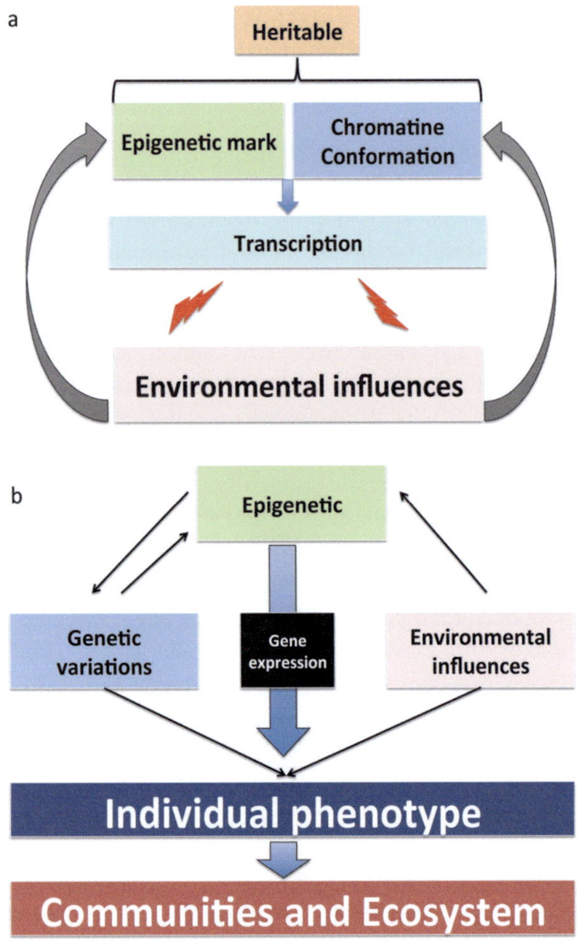

Fig. 5.8 Nongenetic heritability and outcomes for individuals, communities, and ecosystems

of these transcriptomes relies on the implementation of various mechanisms including many epigenetic mechanisms (Fig. 5.8). Extremely important and diversified from yeasts to mammals, epigenetic mechanisms remain poorly understood in marine invertebrates and more particularly in scleractinian corals (Eirin-lopez and Putnam 2019; Putnam 2021).

To date, an increasing number of studies on marine species suggest the possibility of generating environment-specific phenotypes through the control of epigenetic marks for a wide range of physiological traits including sex ratio, growth, or immunity (Eirin-lopez and Putnam 2019; Gavery and Roberts 2017). They can be generally summarized by mechanisms facilitating the phenotypic plasticity of individuals through genotype-epigenotype-environment interactions (Fig. 5.8) (Eirin-lopez and Putnam 2019). Thus, epigenetic mechanisms are defined as changes that can be transmitted without altering the DNA sequence, such as modifications of DNA (methylation/hydroxymethylation), histones (acetylation, methylation, phosphorylation, ubiquitination, etc.), and RNAs (Epitranscriptomics) (Eirin-lopez and Putnam 2019). While data on DNA methylation are becoming available for juvenile and adult aquatic organisms, epigenetic regulation during gametogenesis and embryogenesis and inheritance of these marks are often lacking data (Burggren 2018; 2020, 2021; Fellous et al. 2018, 2022). In mammals, global reprogramming, which occurs twice (once in the germ line and once in the early embryos (Wang et al. 2019), represents a strong barrier to epigenetic inheritance. In aquatic organisms, the precise epigenetic pattern during gametogenesis remains largely unknown (Eirin-lopez and Putnam 2019; Fellous et al. 2022). Nevertheless, in many species such as mollusks (Li et al. 2019; Riviere et al. 2013) or teleosts (Fellous et al. 2018, 2022; Jiang et al. 2013; Ortega-Recalde and Hore 2019; Potok et al. 2013; Wang and Bhandari 2019; Wang and Bhandari 2020), existence of a DNA reprogramming remains controversial. Absence of reprogramming [as possibly in non-bilaterian invertebrates (Xu et al. 2019)] or incomplete reprogramming [as in insects (Yagound et al. 2020)] might signify that environmentally induced epigenetic marks could be heritable more frequently across generations (Ortega-Recalde and Hore 2019) than in mammals for example. These early epigenetic regulations during gamete formation and embryonic development could represent a critical window contributing to the adaptation and phenotypic plasticity of individuals to a fluctuating environment (Fellous et al. 2018, 2022). Consequently, it is essential to describe and understand early epigenetic dynamics in order to better understand their heritability and their link to the environment perturbations (Fig. 5.8) (Lymbery et al. 2021).

Currently, the study of epigenetic mechanisms in corals is an emerging field of research (Cleves et al. 2020; Hackerott et al. 2021; Putnam 2021). To date, DNA methylation is the main mark studied (Eirin-lopez and Putnam 2019; Hackerott et al. 2021; Putnam 2021). While others such as histone and RNA modifications, and their interactions with DNA methylation remain largely unexplored, despite crucial roles in model organisms. Corals, and invertebrates in general, exhibit a mosaic DNA methylation, primarily in intragenic regions, and lower levels of methylation than vertebrates (Feng et al. 2010; Gavery and Roberts 2017; Pillai et al. 2021; Riviere et al. 2017; Rodriguez-Casariego et al. 2021; Rodríguez-casariego et al. 2020; Roquis et al. 2019). However, they have higher methylation levels than *Aiptasia* sp. (model cnidarian) but also exhibit differences among species (particularly between so-called "robust" and "complex" corals, which also have distinct resistance to bleaching) (Rodriguez-Casariego et al. 2021). Additionally, some species show significant levels of methylation at intergenic regions and transposable elements (Li et al. 2018; Rodriguez-Casariego et al. 2021) but the role of DNA methylation in transcription or genome stability remains poorly known in general in cnidarians (Dixon and Matz 2022; Ying et al. 2022; Zhang and Jacobs 2022). These results underline the importance of studying the links between DNA methylation, genome architecture and transcription in corals. Induction of DNA methylation and its association with the phenotype has been demonstrated in cnidarians in response to ocean acidification (Putnam et al. 2016), thermal environment (Dixon et al. 2018), symbiosis (Li et al. 2018; Nawaz et al. 2022; Rodriguez-Casariego et al. 2021; Strader and Quigley 2022), nutrients (Rodríguez-casariego et al. 2018), and seasonal environmental changes (Rodriguez-Casariego et al. 2021). To date, only one study has demonstrated inheritance of DNA methylation from adults to larvae through spermatozoa in the coral *Platygyra daedalea* (Liew et al. 2020). More specifically, genes strongly correlated with thermo-tolerance show a specific methylation profile in adults and offspring (Liew et al. 2020). This inheritance observed in a broadcast corals (*P. daedalea*; Liew et al. 2020) and the transgenerational plasticity (TGP) observed in a brooder coral (*Stylophora pistillata*; Bellworthy et al. 2019; Putnam and Gates 2015) potentially suggest that coral resistance can be obtained via environmental (and controlled) induction of specific epigenetic marks. Nevertheless, the literature also highlights that knowledge of epigenetic regulations during gametogenesis and embryonic development is critically lacking in corals (Chille et al. 2021; Hackerott et al. 2021; Putnam 2021; Vidal-dupiol et al. 2020).

Thus, coral epigenetic studies should focus on the characterization of epigenetic mechanisms regulation during development and gametogenesis to identify if a reprogramming is occurring in corals, and to understand the underlying molecular mechanisms of inheritance. Putting these results together will help to understand the transfer of information from the somatic to the germ cells, and their maintenance in the embryos and adult individuals. However, as epigenetic mechanisms are working through a complex network of interactions between DNA, histone and RNA marks necessary to phenotypic plasticity (Eirin-lopez and Putnam 2019; Fellous et al. 2022), it is important to decipher the exact contributions of each modification to the overall inheritance together with the epigenetic network plasticity under fluctuating environmental conditions at all scales (developmental stages, cell specificity, individual variability, and populations) (Burggren 2021). Finally, functionality of these marks should be studied to understand their role on the phenotype (Cleves et al. 2018, 2020).

5.4.3 Inheritance

While restoration programs through the use of "super corals" are increasing across the world (Bowden-kerby 2023), a major gap remains in coral biology about inheritance and persistence of relevant biological traits (Vasquez-Kuntz et al. 2022). Phenotype inheritance, and particularly inheritance of epigenetic marks, are important gaps that need to be resolved in order to better predict the future of coral reefs and to have efficient restoration programs (Putnam 2021; Vasquez-Kuntz et al. 2022).

In general, germline development and reproductive modes (sexual vs asexual, oviparity vs viviparity, etc.) are useful indicators of epigenetic mark persistence. An other key factor in determining their inheritance together with effects on phenotypes and fitness is environmental predictability (Anastasiadi et al. 2021). Thus, it is necessary to quantify: (a) persistence of the marks across generations, (b) their contributions to fitness and phenotypes, and (c) to compare what is observed across taxa in order to build a general comprehensive model (Anastasiadi et al. 2021).

In animals like mammals, germline segregation and differentiation occur early in development. Then, it is observed two rounds of extensive reprogramming during gametogenesis and embryogenesis, and resulting in the resetting of most epigenetic modifications (Ortega-Recalde and Hore 2019). In other animals such as teleosts, erasure might be absent, incomplete, or understudied (Ortega-Recalde and Hore 2019). Thus, nongenetic inheritance is thought to be more frequent through germline-to-soma transmission (Ortega-Recalde and Hore 2019).

Furthermore, late segregation of the germline is commonly observed in plants, and also occurs in sea urchins, sponges, and cnidarians. This result in a long period in which environmental variations might induce transmissible epigenetic changes (Anastasiadi et al. 2021). In case of late germline segregation, mutations (or epigenetic modifications) could generate adaptive alleles in somatic cells that can be transmitted to the germline, and to the zygote (Vasquez-Kuntz et al. 2022).

Thus, in corals, which have late germline segregation, parental somatic mutations were previously observed. In *Acropora palmata*, it was shown that uniparental offspring (of meiotic origin and not parthenogenesis) inherit parent's somatic mutations (Vasquez-Kuntz et al. 2022). Thus, this type on inheritance might be an important source of adaptation and evolution in corals. However, precise windows of priming must be studied (Hackerott et al. 2021) together with the mechanisms behind this transmission (Vasquez-Kuntz et al. 2022). Also, mutation transmission from somatic cells to germline might have been direct or indirect in case of mutation occurring in somatic cells that dedifferentiate in stem cells before to redifferentiate into germ cells (Vasquez-Kuntz et al. 2022). Importantly, while they have important regenerative properties, identification of stem cells remains as major gap in corals (Vasquez-Kuntz et al. 2022).

5.5 Perspectives for the Establishment of Innovative Restoration Strategies

5.5.1 Overview

Due to increases in anthropogenic perturbations and oceanic warming that drive many reef ecosystems to decline, efforts to conserve and restore coral reefs are facing a myriad of challenges. In order to be efficient in mitigating and/or reversing coral reef decline, employed approaches must be combined as much as possible through a multidisciplinary field (Fig. 5.9) (van Oppen and Lastra 2022). Such approaches are becoming more and more feasible, particularly by using the new advances in informatics and –omic technologies (Lawson et al. 2022; van Oppen and Lastra 2022; Richards et al. 2022) together with the recent discoveries made in epigenetic (Aranda Lastra 2022) and functional biology on model and non-model organisms (Bouwmeester et al. 2022; Cleves 2022). Combining these approaches will allow to go further in the current knowledge about coral holobiont biology (Mcleod et al. 2022), but also to understand reef connectivity and complex multispecies interactions (van Oppen and Lastra 2022). However, newly acquired knowledge in modeling, ecology, and genomic should be translated in formats that can be used by the largest audience possible from nonspecialized scientists to public partners (van Oppen and Lastra 2022). Also, a crucial point (and maybe the main challenge) is the integration by scientists, funders, and public that coral restoration (if feasible and necessary) is a long and complex process that needs time, fundamental researches, and truly open collaboration at a world scale.

Coral reef restoration strategies mainly emerged at the end of the 1980s and involved techniques from coral gardening through fragmentation and floating nursery pools from wild spawn larvae to trials relying on holobiont modifications using assisted evolution and more recently use of probiotics (Mcleod et al. 2022). But, as demonstrated in this chapter, coral biology is complex, and reproduction, development, and inheritance should be taken into account together with view at the ecosystem levels to have efficient restoration and conservation programs (Bowden-kerby 2023; Jacquemont et al. 2022; Mcleod et al. 2022). Thus, while being crucial to species persistence, most of the studies are not including phenotypic plasticity and connectivity (Jacquemont et al. 2022), and their link to each other is still unknown.

5.5.2 Integration of Coral Reef Connectivity, Larval Dispersal, and Cryptic Species

A first point in coral reef restoration is to properly identify the sites to restore and/or protect. However, as we described before, dispersal [at the surface (Palumbi et al. 2023) and from the mesophotic to the shallow environment (Prasetia et al. 2022; Zhang et al. 2019)] together with population connectivity (and sensitivity to disturbances) (Hock et al. 2017) are crucial and complex biological processes that rely on species, reproductive modes and development, and habitat (Palumbi et al. 2023). When studying dispersal and reef connectivity, symbiont dispersal should also be taken into account and not only corals (Riginos and Beger 2022). Interestingly, in marine ecosystems, most of the knowledge comes from fish and invertebrates, but is still in its infancy for coral species (Riginos and Beger 2022). In consequence, coral geneticists and ecosystem modelers should work together. Thus, coral geneticists must improve knowledge on parentages and isolations at a greater spatial resolution together with asymmetric dispersal (Riginos and Beger 2022). This understanding will need the characterization of cryptic species, but also by combined studies of reproductive and dispersal traits to properly identify the dispersal and the distribution of adaptive variations (Riginos and Beger 2022). In parallel, modelers should incorporate in their stimulations asymmetric dispersal of the species of interest (and multiples others) together with oceanographic parameters to properly implement conservation strategies that link population adaptation and connectivity (Riginos and Beger 2022). Then, hybrids (and their viability) and cryptic taxa (and habitat) should be identified as they are important for coral evolution (Riginos and Beger 2022). Taxa misidentifications are indeed parameters that mislead conclusion about species adaptation (Riginos and Beger 2022).

5.5.3 Ecosystem Characterization and Reef Health Through the Use of Environmental Genomic

Ecosystem characterization and species inventory are critical steps, when restoring and/or protecting specific habitats. As we saw previously, reef sensibilities to bleaching and COST invasions probabilities are depending on reef localities (Hock et al. 2017), and presence of predators of herbivorous species and corals might annihilate an efficient restoration project (Cano et al. 2021). One possible and noninvasive way to detect and quantify multispecies would be the use of envi-

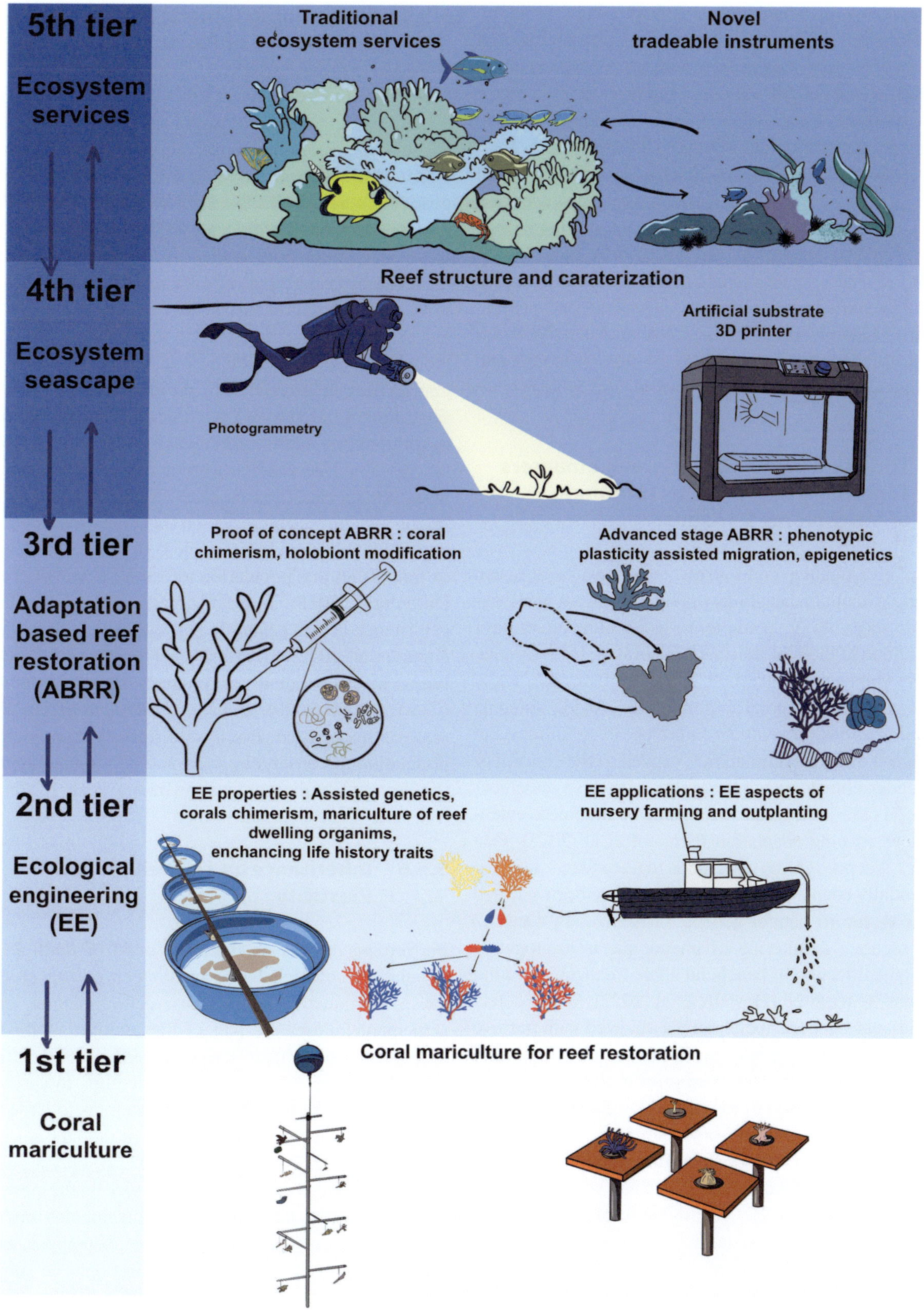

Fig. 5.9 Integrative views of the different stages necessary for an effective restoration of coral reefs. (Adapted from Rinkevich 2021)

ronmental DNA metabarcoding (eDNA) (Richards et al. 2022). However, research on ecosystem characterization with eDNA is only beginning, and no single standard approaches exists, but this weakness can be compensated for through multiple assays and robust sampling designs. Furthermore, availability of genomic databases also heavily influences the quality and the quantity of taxa detected, and researchers must ensure they have strong reference data (molecular and physical specimens) before starting eDNA projects (Richards et al. 2022). In this context, coral phylogeography diversity is important to take into account. Finally, while eDNA starts to be used at the individual and population levels, a combination with eRNA and metabolomics might be efficient to investigate coral stress response and health (Lawson et al. 2022; Richards et al. 2022).

5.5.4 The Need of Validated Coral Biomarkers Together with the Understanding of Microbiome Dynamism

However, investigating coral stress responses, health, and identifying potential adaptations in populations are challenging at this stage. As an example, no good biomarkers have been identified to thermal resistance in corals yet (Louis et al. 2017; Van Oppen and Oakeshott 2020). Furthermore, while marginal and extreme coral communities provide valuable natural laboratories for studying future ocean conditions and the resilience of coral communities, as well as the concept of refugia under climate change stressors, a precise classification of coral reefs as "optimal," marginal, or extreme is essential for investigating adaptation (Schoepf et al. 2023). One major lack that needs to be resolved is to understand if identified potentially adapted coral populations have been selected or if this is due to natural genetic variations. As a consequence, accurate monitoring of phenotypic traits, together with the quantification of beneficial allele frequencies within populations, is required (Howells et al. 2022). While detection of variants and genes potentially associated with positive fitness traits is becoming more easily done with advances in genomic, they need to be validated prior to their application in restoration/conservation programs (Howells et al. 2022). However, corals are not simple organisms. Also, microbiome dynamic and symbiont biology should also be associated with these species traits (Maher et al. 2022; Shah et al. 2022; van Oppen and Nitschke 2022). Once this information has been validated, further actions require translating this complex environmental, genomic, and modeled data into accessible information usable by a larger audience (Howells et al. 2022). Finally, while investigations mainly focus on thermal tolerance traits, corals are affected by multiple environmental variables and stressors that need to be taken into account through large multidisciplinary projects (Howells et al. 2022).

5.5.5 Maintaining Genetic Diversity and the Use of Assisted Gene Flow

Maintaining reef ecosystem functions and preventing genetic diversity loss in corals are two crucial parameters to take into account in restoration/conservation projects (Baums et al. 2022). However, substantial gaps remain to properly manage genetic diversity and limit erosion. Thus, existing data suggest that targeted rearing leads to genetic erosion and potentially important bottlenecks. Action mitigations can be achieved by combining selected breeding and assisted gene flow, but these approaches necessitate good simulation models to guide how this mix of tools should be employed (Baums et al. 2022). Thus, one way possible to prevent genetic lost in a restoration project would be the assisted migration of individuals from specific populations. However, movements of adult colonies across large geographic distances should be avoided as they can bring diseases and/or invasives species (Baums et al. 2022). A possible solution lies on gamete cryopreservation (but long-term consequences on individuals need to be studied) to move identified alleles or traits from one population to another (Baums et al. 2022; Dixon et al. 2018). To do this, gaps on trait inheritance and persistence across generations should be resolved (Vasquez-Kuntz et al. 2022), and functional analysis of these specific biomarkers (or allele of interest) should validate them (Cleves et al. 2020). Finally, question remains about the scale of this assisted gene flow to shift allele frequency in the population to conserve/restore, and how it can be integrated in selective breeding programs (Baums et al. 2022).

5.5.6 Inheritance and Developmental Plasticity

Performing crosses on the same night of mutliple species, population, colonies, while avoiding cryptic species that can cause reproductive mismatches, developmental, and post-settlement failures are part of the multiple methodological bottlenecks occurring in selective breeding (Fig. 5.10) (Drury et al. 2022a, b). Also, this technique aims to spread within populations to restore/protect positive traits in large numbers by implanting a large number of larvae onto the selected sites (Condie et al. 2021; Drury et al. 2022b). Playing on phenotypic plasticity and acclimation is part of the new ways to perform selective breeding. Thus, wild gametes might be collected without any regard for parental phenotypes, and larvae produced from this mass fertilization might be subjected to and primed during particular embryonic windows by exposing them to a stress (Fig. 5.10) (Drury et al. 2022a, b; Fellous et al. 2022). The resulting selection might produce larvae or juveniles putatively more resistant (or tolerant) to the target stress. Priming might be done during embryonic development

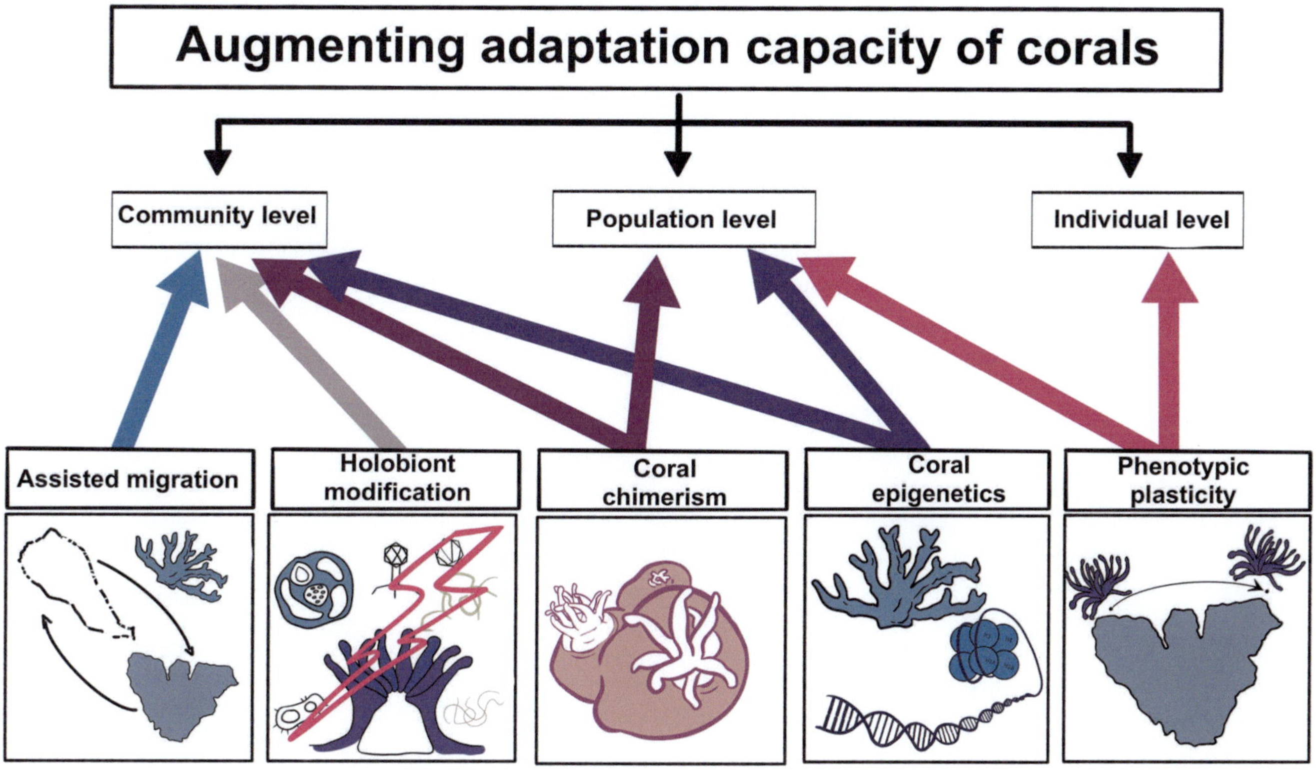

Fig. 5.10 Representation of the different strategies envisaged to increase the adaptive capacities of corals, and the integration of each strategy at the individual, population, and community scales. (Adapted from Rinkevich 2021)

with the advantages of the use of a high number of individuals, but the disadvantages of an unknown optimal window (Drury et al. 2022a, b; Fellous et al. 2022; Trigg et al. 2021), while selection on post-settled juveniles might be more representative of the reef selective pressure (Drury et al. 2022b). This approach is called post hoc selective breeding and can also be realized through captive breeding programs using artificial moon cycle manipulations (Craggs et al. 2017; Craggs et al. 2019). Genetic and environmental drivers of selection can be easily measured, while keeping the technic keep the population-wide genetic diversity (Drury et al. 2022b). However, this approach also necessitates a precise characterization of the nongenetic factors such as epigenetic mechanisms (Fig. 5.10). They represent a huge gap in coral, and mainly taxa, biology, and the main challenges to better predict the future of reef ecosystems (Eirin-lopez and Putnam 2019; Fellous et al. 2022; Putnam 2021). Also, breeding approaches and studies must be accomplished with as many species as possible to maintain ecosystem services, and the required time and monetary investment in testing represent a major challenge (Drury et al. 2022b). Low post-settlement survivorship and adapted substrates (natural or artificial) represent key bottlenecks that need to be solved (Drury et al. 2022b). Studying reef connectivity and larval dispersal on and around selected sites to protect/restore might help to create "larval dispersal hubs," where colonies used in the breeding program will be concentrated in one location before they naturally spawn, develop, disperse, and settle (Amar and Rinkevich 2007; Drury et al. 2022b). While being very promising, this combination of methods needs to be expanded to multiple species, but trade-off on adult individuals (and future generations) and genetic architecture of the selected traits should be evaluated and characterized.

5.6 Conclusions

Human interventions to enhance coral performance under climate changes are broadly called assisted evolution (Van Oppen et al. 2015). These include the use of holobiont natural plasticity (Hackerott et al. 2021; Putnam 2021), but also newly developed molecular techniques such as CRISPRCas9 (Cleves et al. 2018, 2020), cryopreservation (Howells et al. 2021; Kihika et al. 2022). They mainly involve managed or selected breeding (Humanes et al. 2021), preconditioning (reviewed in Hackerott et al. 2021; Putnam 2021), even if it is unclear how trait acquisition will be beneficial, identified, maintained across generations, and if there are potential reductions in other fitness traits; microbiome engineering (highly promising results that are still in its infancy) (Buerger et al. 2020; Chakravarti et al. 2017); 3D printing (Wangpraseurt et al. n.d.); and nanobiotechnology (Roger et al. 2023).

In the future, research in the field of coral restoration should (1) demonstrate that active restoration is meaningful in improving reef health and diversity, when compared to less interventionist strategies; (2) implement the efficient strategies at large scales; and (3) understand and find possibilities to mitigate any potential risks from active restoration. Also, to be efficient, reproductive and developmental characteristics of the targeted reef to restore/protect should be fully integrated. They represent indeed the generational bridge that might help coral reef to acclimate and then to adapt, while being also the most sensitive steps to environmental perturbation in the coral life cycle. Also, a proper understanding of these processes will allow the integration of crucial parameters such as trait inheritance and persistence, but also genetic and nongenetic ways of coral acclimation/adaptation to the Anthropocene. Finally, most, if all, of these initiatives might be useless if our society is not changing and if no actions are seriously taken for the urgent need in the reductions of our greenhouse gas emissions.

References

Amar KO, Rinkevich B (2007) A floating mid-water coral nursery as larval dispersion hub: testing an idea. Mar Biol 151:713–718. https://doi.org/10.1007/s00227-006-0512-0

Anastasiadi D, Venney CJ, Bernatchez L, Wellenreuther M (2021) Epigenetic inheritance and reproductive mode in plants and animals. Trends Ecol Evol 36:1124–1140. https://doi.org/10.1016/j.tree.2021.08.006

Andersen IS, Lindeman LC, Reiner AH, Østrup O, Aanes H, Aleström P, Collas P (2013) Epigenetic marking of the zebrafish developmental program. Curr Top Dev Biol:85–112. https://doi.org/10.1016/B978-0-12-416027-9.00003-6

Aranda Lastra M (2022) Epigenetics and acquired tolerance to environmental stress BT – coral reef conservation and restoration in the omics age. In: van Oppen MJH, Aranda Lastra M (eds) Coral reef conservation and restoration in the omics age. Coral reefs of the world. Springer International Publishing, Cham, pp 135–150. https://doi.org/10.1007/978-3-031-07055-6_9

Ayalon I, Rosenberg Y, Kyba CCM, Cabaitan PC, Levy O, Ayalon I, Rosenberg Y, Benichou JIC, Campos CLD, Sayco SLG, Kyba CCM, Cabaitan PC, Levy O (2021) Coral gametogenesis collapse under artificial light pollution. Curr Biol 31:1–7. https://doi.org/10.1016/j.cub.2020.10.039

Baird AH (2021) An Indo-Pacific coral spawning database. Sci data 1–9. https://doi.org/10.1038/s41597-020-00793-8

Baird A, Sadler C, Pitt M (2001) Synchronous spawning of Acropora in the Solomon Islands. Coral Reefs 19:286. https://doi.org/10.1007/s003380000116

Baird AH, Guest JR, Willis BL (2009) Systematic and biogeographical patterns in the reproductive biology of Scleractinian corals. Annu Rev Ecol Evol Syst 40:551–571. https://doi.org/10.1146/annurev.ecolsys.110308.120220

Bairos-Nowak KR, Hoogenboom M, Van Oppen MJH, Connolly SR (2021) Coral adaptation to climate change: meta-analysis reveals high heritability across multiple traits. Glob Chang Biol 1–17. https://doi.org/10.1111/gcb.15829

Ball EE, Hayward DC, Reece-hoyes JS, Hislop NR, Samuel G, Saint R, Harrison PL, Miller DJ (2002) Coral development: from classical embryology to molecular control. Int J Dev Biol 678:671–678

Barno AR, Villela HDM, Peixoto RS, Aranda M, Thomas T (2021) Host under epigenetic control: a novel perspective on the interaction between microorganisms and corals. BioEssays 43:1–9. https://doi.org/10.1002/bies.202100068

Baums IB, Chamberland VF, Locatelli NS, Conn T (2022) Maximizing genetic diversity in coral restoration projects BT – coral reef conservation and restoration in the omics age. In: van Oppen MJH, Aranda Lastra M (eds) Coral reef conservation and restoration in the omics age. Coral reefs of the world. Springer International Publishing, Cham, pp 35–53. https://doi.org/10.1007/978-3-031-07055-6_3

Bellworthy J, Spangenberg JE, Fine M (2019) Feeding increases the number of offspring but decreases parental investment of Red Sea coral Stylophora pistillata. Ecol Evol 12245–12258. https://doi.org/10.1002/ece3.5712

Bhattacharya D, Agrawal S, Aranda M, Baumgarten S, Belcaid M, Drake JL, Erwin D, Foret S, Gates RD, Gruber DF, Kamel B, Lesser MP, Levy O, Liew YJ, Macmanes M, Mass T (2016) Comparative genomics explains the evolutionary success of reef-forming corals. elife 1–26. https://doi.org/10.7554/eLife.13288

Bongaerts P, Riginos C, Brunner R, Englebert N, Smith SR, Hoegh-guldberg O (2017) Deep reefs are not universal refuges: reseeding potential varies among coral species. Sci Adv 3(2):e1602373

Bouwmeester J, Daly J, Zuchowicz N, Hagedorn M (2022) Cryopreservation to conserve genetic diversity of reef-building corals BT – coral reef conservation and restoration in the omics age. In: van Oppen MJH, Aranda Lastra M (eds) Coral reef conservation and restoration in the omics age. Coral reefs of the world. Springer International Publishing, Cham, pp 225–240. https://doi.org/10.1007/978-3-031-07055-6_14

Bowden-kerby A (2023) Coral-focused climate change adaptation and restoration based on accelerating natural processes: launching the " reefs of Hope " paradigm. Oceans 4:13–26

Brener-raffalli K, Adjeroud M, Brener-raffalli K, Vidal-dupiol J, Rey O, Romans P, Pratlong M, Haguenauer A (2022) Gene expression plasticity and frontloading promote thermotolerance in Pocillopora corals. Peer Community J 2

Bruno JF, Precht WF, Vroom PS, Aronson RB (2014) Coral reef baselines: how much macroalgae is natural? Mar Pollut Bull 80:24–29

Buerger P, Coppin CW, Pearce SL, Chakravarti LJ, Oakeshott JG (2020) Heat-evolved microalgal symbionts increase coral bleaching tolerance. Sci Adv 6:eaba2498

Burgess SC, Baskett ML, Grosberg RK, Morgan SG, Strathmann RR (2015) When is dispersal for dispersal? Unifying marine and terrestrial perspectives. Biol Rev Camb Philos Soc 91:867. https://doi.org/10.1111/brv.12198

Burggren W (2018) Developmental phenotypic plasticity helps bridge stochastic weather events associated with climate change. J Exp Biol 221. https://doi.org/10.1242/jeb.161984

Burggren WW (2020) Phenotypic switching resulting from developmental plasticity: fixed or reversible? Front Physiol Physiol 10:1–13. https://doi.org/10.3389/fphys.2019.01634

Burggren W (2021) Developmental physiology: Grand challenges. Front Physiol 12:1–5. https://doi.org/10.3389/fphys.2021.706061

Bythell JC, Brown BE, Kirkwood TBL (2018) Do reef corals age? Biol Rev 93:1192–1202. https://doi.org/10.1111/brv.12391

Cairn SD (1988) Asexual reproduction in solitary scleractinia. Proceeding 6th IntCoral Reef Symp 2:641–646

Cano I, Sellares-blasco RI, Lefcheck JS, Villalpando MF, Croquer A (2021) Effects of herbivory by the urchin Diadema antillarum on early restoration success of the coral Acropora cervicornis in the Central Caribbean. J Exp Mar Bio Ecol 539:151541. https://doi.org/10.1016/j.jembe.2021.151541

Ceh J, Van Keulen M, Bourne DG (2013) Intergenerational transfer of specific bacteria in corals and possible implications for offspring fitness. Microb Ecol. 65(1):227–231. https://doi.org/10.1007/s00248-012-0105-z

Chakravarti LJ, Beltran VH, Van Oppen MJH (2017) Rapid thermal adaptation in photosymbionts of reef-building corals. Glob Chang Biol. 23(11):4675–4688. https://doi.org/10.1111/gcb.13702

Chan WY, Peplow LM, Van Oppen MJH (2019) Interspecific gamete compatibility and hybrid larval fitness in reef- building corals: implications for coral reef restoration. Sci Rep 1–13. https://doi.org/10.1038/s41598-019-41190-5

Chan WY, Chung J, Peplow LM, Hoffmann AA, Van Oppen MJH (2020) Maternal effects in gene expression of interspecific coral hybrids. Mol Ecol 1–11. https://doi.org/10.1111/mec.15727

Chen C, Chen W, Shikina S, Denis V, Chang C (2020) The plasticity of gonad development of sexual reproduction in a scleractinian coral, Porites lichen. Gen Comp Endocrinol 285:113270. https://doi.org/10.1016/j.ygcen.2019.113270

Chen C, Chen W, Chang C (2021) Multispecies spawning of scleractinian corals in nonreefal coral communities of northern Taiwan in the northwestern Pacific Ocean. Bull Mar Sci 97:351–371

Chille E, Strand E, Neder M, Schmidt V, Sherman M, Mass T, Putnam H (2021) Developmental series of gene expression clarifies maternal mRNA provisioning and maternal-to-zygotic transition in a reef-building coral. BMC Genomics 22:1–17

Chiu YL, Chang CF, Shikina S (2021) Development of an in vitro tissue culture system for hammer coral (Fimbriaphyllia ancora) ovaries. Sci Rep 1–10. https://doi.org/10.1038/s41598-021-03810-x

Cleves PA (2022) A need for reverse genetics to study coral biology and inform conservation efforts BT – coral reef conservation and restoration in the omics age. In: van Oppen MJH, Aranda Lastra M (eds) Coral reef conservation and restoration in the omics age. Coral reefs of the world. Springer International Publishing, Cham, pp 167–178. https://doi.org/10.1007/978-3-031-07055-6_11

Cleves PA, Strader ME, Bay LK, Pringle JR, Matz MV (2018) CRISPR / Cas9-mediated genome editing in a reef-building coral. PNAS 115:5235–5240. https://doi.org/10.1073/pnas.1722151115

Cleves PA, Shumaker A, Lee J, Putnam HM (2020) Unknown to known: advancing knowledge of coral gene function. Trends Genet 36:93–104

Combosch DJ, Vollmer SV (2013) Mixed asexual and sexual reproduction in the indo-Pacific reef coral Pocillopora damicornis. Ecol Evol 3379–3387. https://doi.org/10.1002/ece3.721

Condie SA, Anthony KRN, Babcock RC, Baird ME, Beeden R, Fletcher CS, Gorton R, Harrison D, Hobday AJ, Plagányi ÉE, Westcott DA (2021) Large-scale interventions may delay decline of the Great Barrier Reef. R Soc Open Sci 8

Cowen LJ, Putnam HM (2022) Bioinformatics of corals: investigating heterogeneous omics data from coral holobionts for insight into reef health and resillience. En révision 5:205

Craggs J, Guest JR, Davis M, Simmons J, Dashti E, Sweet M (2017) Inducing broadcast coral spawning ex situ: closed system mesocosm design and husbandry protocol. Ecol Evol 11066–11078. https://doi.org/10.1002/ece3.3538

Craggs J, Guest J, Bulling M, Sweet M (2019) Ex situ co culturing of the sea urchin, Mespilia globulus and the coral Acropora millepora enhances early post-settlement survivorship. Sci Rep 1–12. https://doi.org/10.1038/s41598-019-49447-9

Da-anoy JP, Cabaitan PC, Conaco C (2020) Warm temperature alters the chemical cue preference of Acropora tenuis and Heliopora coerulea larvae. Mar Pollut Bull 161:111755. https://doi.org/10.1016/j.marpolbul.2020.111755

Damjanovic K, Menendez P, Blackall LL, Van Oppen MJH (2019) Early life stages of a common broadcast spawning coral associate with specific bacterial communities despite lack of internalized bacteria. Invertebr Microbiol 79:706

Dixon G, Matz M (2022) Changes in gene body methylation do not correlate with changes in gene expression in Anthozoa or Hexapoda. BMC Genomics 1–11. https://doi.org/10.1186/s12864-022-08474-z

Dixon GB, Bay LK, Matz MV (2016) Evolutionary consequences of DNA methylation in a basal metazoan. Mol Biol Evol 33:2285–2293. https://doi.org/10.1093/molbev/msw100

Dixon G, Liao Y, Bay LK, Matz MV (2018) Role of gene body methylation in acclimatization and adaptation in a basal metazoan. PNAS 115:13342–13346. https://doi.org/10.1073/pnas.1813749115

Doropoulos C, Roff G (2022) Coloring coral larvae allows tracking of local dispersal and settlement. PLoS Biol 1–15. https://doi.org/10.1371/journal.pbio.3001907

Doropoulos C, Elzinga J, Hofstede R, Van Koningsveld M, Babcock RC (2019a) Optimizing industrial-scale coral reef restoration : comparing harvesting wild coral spawn slicks and transplanting gravid adult colonies. Restor Ecol 27:758–767. https://doi.org/10.1111/rec.12918

Doropoulos C, Vons F, Elzinga J, Hofstede R (2019b) Testing industrial-scale coral restoration techniques: harvesting and culturing wild coral-spawn slicks. Front Mar Sci 6:1–14. https://doi.org/10.3389/fmars.2019.00658

Drury C, Bean NK, Harris CI, Hancock JR, Huckeba J, Christian Martin H, Roach TNF, Quinn RA, Gates RD (2022a) Intrapopulation adaptive variance supports tolerance in a reef-building coral. Commun Biol 5:1–10. https://doi.org/10.1038/s42003-022-03428-3

Drury C, Caruso C, Quigley K (2022b) Selective breeding to enhance the adaptive potential of corals BT – coral reef conservation and restoration in the omics age. In: van Oppen MJH, Aranda Lastra M (eds) Coral reef conservation and restoration in the omics age. Coral reefs of the world. Springer International Publishing, Cham, pp 71–84. https://doi.org/10.1007/978-3-031-07055-6_5

Duarte CM, Agusti S, Barbier E, Britten GL, Castilla JC, Gattuso J-P, Fulweiler RW, Hughes TP, Knowlton N, Lovelock CE, Lotze HK, Predragovic M, Poloczanska E, Roberts C, Worm B (2020) Rebuilding marine life. Nature 580:39–51. https://doi.org/10.1038/s41586-020-2146-7

Dubininkas V (2017) Effects of substratum on the growth and survivorship of Montipora capitata and Porites lobata transplants. J Exp Mar Bio Ecol 486:134–139

Eirin-lopez JM, Putnam HM (2019) Marine environmental epigenetics. Annu Rev Mar Sci 11:1–34

Elahi R, Edmunds PJ (2014) Tissue age affects calcification coral Madracis in the Scleractinian mirabilis. Biol Bull 212:20–28

Epstein HE, Torda G, Munday PL, Van Oppen MJH (2019) Parental and early life stage environments drive establishment of bacterial and dinoflagellate communities in a common coral. ISME J 1635–1638. https://doi.org/10.1038/s41396-019-0358-3

Farber SC, Costanza R, Wilson MA (2002) Special issue: the dynamics and value of ecosystem services: integrating economic. Ecol Econ 41:375–392

Fellous A, Labed-Veydert T, Locrel M, Voisin A-S, Earley RL, Silvestre F (2018) DNA methylation in adults and during development of the self-fertilizing mangrove rivulus, Kryptolebias marmoratus. Ecol Evol 8:6016. https://doi.org/10.1002/ece3.4141

Fellous A, Wegner KM, John U, Mark FC, Shama LNS (2022) Windows of opportunity: ocean warming shapes sensitive epigenetic reprogramming and gene expression across gametogenesis and embryogenesis in marine stickleback. Glob Chang Biol 28:54–71. https://doi.org/10.1111/gcb.15942

Feng S, Cokus SJ, Zhang X, Chen P-Y, Bostick M, Goll MG, Hetzel J, Jain J, Strauss SH, Halpern ME, Ukomadu C, Sadler KC, Pradhan S, Pellegrini M, Jacobsen SE (2010) Conservation and divergence of methylation patterning in plants and animals. Proc Natl Acad Sci USA 107:8689–8694. https://doi.org/10.1073/pnas.1002720107

Fong CR, Chancellor KS, Renzi JJ, Robinson DMR, Barber PH, Habtes SY, Fong P (2018) Epibionts on Turbinaria ornata, a secondary foundational macroalga on coral reefs, provide diverse trophic support to fishes. Mar Environ Res 141:39–43. https://doi.org/10.1016/j.marenvres.2018.08.001

Foster T, Heyward AJ, Gilmour JP (2018) Split spawning realigs coral reproduction with optimal environmental windows. Nat Commun 1–8. https://doi.org/10.1038/s41467-018-03175-2

Frade PR, Bongaerts P, Englebert N, Rogers A, Gonzalez-rivero M, Hoegh-guldberg O (2016) Deep reefs of the Great Barrier Reef offer limited thermal refuge during mass coral bleaching. Nat Commun 1–8. https://doi.org/10.1038/s41467-018-05741-0

Francis FT, Filbee-Dexter K, Yan HF, Coté IM (2019) Invertebrate herbivores: overlooked allies in the recovery of degraded coral reefs? Glob Ecol Conserv 17:1–11. https://doi.org/10.1016/j.gecco.2019.e00593

Fukunaga A, Burns JHR, Pascoe KH, Kosaki RK (2022) A remote coral reef shows macroalgal succession following a mass bleaching event. Ecol Indic 142:109175. https://doi.org/10.1016/j.ecolind.2022.109175

Gavery MR, Roberts SB (2017) Epigenetic considerations in aquaculture. PeerJ 5:e4147. https://doi.org/10.7717/peerj.4147

Ghalambor CK, Hoke KL, Ruell EW, Fischer EK, Reznick DN, Hughes KA (2015) Non-adaptive plasticity potentiates rapid adaptive evolution of gene expression in nature. Letter 525:372. https://doi.org/10.1038/nature15256

Gilmour JP, Underwood JN, Howells EJ, Gates E (2016) Biannual spawning and temporal reproductive isolation in Acropora corals. PLoS One 1–14. https://doi.org/10.1371/journal.pone.0150916

Gittings SR, Boland GS, Deslarzes KJP, Combs L, Holland BS, Bright TJ (2018) Mass spawning and reproductive viability of reef corals at the East Flower Garden Bank, Northwest Gulf of Mexico R. Bull Mar Sci 51(3):420–428

Golomb D, Shashar N, Rinkevich B (2020) Coral carpets- a novel ecological engineering tool aimed at constructing coral communities on soft sand bottoms. Ecol Eng 145:105743. https://doi.org/10.1016/j.ecoleng.2020.105743

Gómez-lemos LA, Dor C, Bayraktarov E, Diaz-pulido G (2018) Coralline algal metabolites induce settlement and mediate the inductive effect of epiphytic microbes on coral larvae. Sci Rep 1–11. https://doi.org/10.1038/s41598-018-35206-9

Gorospe KD, Karl SA (2013) Genetic relatedness does not retain spatial pattern across multiple spatial scales: dispersal and colonization in the coral, Pocillopora damicornis. Mol Ecol 3721–3736. https://doi.org/10.1111/mec.12335

Guest J, Baird A, Goh B, Chou L (2002) Multispecific, synchronous coral spawning in Singapore. Coral Reefs 21:422–423. https://doi.org/10.1007/s00338-002-0263-4

Gurdek-Bas R, Benthuysen JA, Harrison HB, Zenger KR, Van Herwerden L (2022) The El Niño Southern oscillation drives multidirectional inter - reef larval connectivity in the Great Barrier Reef. Sci Rep 1–12. https://doi.org/10.1038/s41598-022-25629-w

Hackerott, S., Martell, H.A., Eirin-lopez, J.M., 2021. Coral environmental memory : causes, mechanisms, and consequences for future reefs. Trends Ecol Evol 1–13. https://doi.org/10.1016/j.tree.2021.06.014

Harper LM, Huebner LK, Cain EDO, Ruzicka R, Gleason DF, Fogarty ND (2021) Methodological recommendations for assessing scleractinian and octocoral recruitment to settlement tiles. PeerJ 1–16. https://doi.org/10.7717/peerj.12549

Harrison PL (2011) Sexual reproduction of Scleractinian corals, in: coral Ree: an ecosystem in transition. https://doi.org/10.1007/978-94-007-0114-4

Harrison PI, Wallace CC (1990) Reproduction, dispersal and recruitment of scleractinian corals, in: Coral Reefs

Harrison PL, Barbock RC, Bull GD, Oliver JK, Wallace CC, Willis BL (1983) Mass spawning in tropical reef corals. Science (80-.) 223(4641):1186–1189

Harrison PL, Cruz DW, Cameron KA, Cabaitan PC, Harrison PL (2021) Increased coral larval supply enhances recruitment for coral and fish habitat restoration. Front Mar Sci 8:1–22. https://doi.org/10.3389/fmars.2021.750210

Hatta M, Fukami H, Wang W, Omori M, Shimoike K, Hayashibara T, Ina Y, Sugiyama T (1999) Reproductive and genetic evidence for a reticulate evolutionary history of Mass-spawning corals. Mol Biol Evol 16:1607–1613

Hayashibara T, Shimoike K, Kimura T, Hosaka S, Heyward A, Harrison P, Kudo K, Omori M (1993) Patterns of coral spawning at Akajima Island, Okinawa. Japan Mar Ecol Prog Ser 101:253–262

Henley EM, Quinn M, Bouwmeester J, Daly J, Lager C, Zuchowicz N, Bailey DW, Hagedorn M (2022) Contrasting reproductive strategies of two Hawaiian Montipora corals. Sci Rep 1–15. https://doi.org/10.1038/s41598-022-16032-6

Hernandez-agreda A, Gates RD, Ainsworth TD (2017) Defining the Core microbiome in corals ' microbial soup. Trends Microbiol 25:125–140. https://doi.org/10.1016/j.tim.2016.11.003

Heyward AJ, Negri AP (2012) Turbulence, cleavage, and the naked embryo: a case for coral clones. Science (80-.) 335:1064

Hirose M, Hidaka M (2006) Early development of zooxanthella-containing eggs of the corals Porites cylindrica and Montipora digitata: the endodermal localization of zooxanthellae. Zool Sci 23:873–881. https://doi.org/10.2108/zsj.23.873

Hirose M, Kinzie RA III, Hidaka M (2000) Early development of zooxanthella-containing eggs of the corals Pocillopora verrucosa and P. eydouxi with special reference to the distribution of zooxanthellae. Biol Bull 68–75

Hock K, Wolff NH, Ortiz JC, Condie SA, Kenneth R, Anthony N, Blackwell PG, Mumby PJ (2017) Connectivity and systemic resilience of the Great Barrier Reef. PLoS Biol 1–23. https://doi.org/10.5281/zenodo.1017925

Hock K, Doropoulos C, Gorton R, Condie SA, Mumby PJ (2019) Split spawning increases robustness of coral larval supply and inter-reef connectivity. Nat Commun 1–10. https://doi.org/10.1038/s41467-019-11367-7

Hoey AS, Pratchett MS, Cvitanovic C (2011) High macroalgal cover and low coral recruitment undermines the potential resilience of the world's Southernmost Coral Reef Assemblages. PLoS One 6:1–9. https://doi.org/10.1371/journal.pone.0025824

Howells EJ, Abrego D, Liew YJ, Burt JA, Meyer E, Aranda M (2021) Enhancing the heat tolerance of reef-building corals to future warming. Sci Adv 7(34):eabg6070

Howells EJ, Bay LK, Bay RA (2022) Identifying, monitoring, and managing adaptive genetic variation in reef-building corals under rapid climate warming BT – coral reef conservation and restoration in the omics age. In: van Oppen MJH, Aranda Lastra M (eds) Coral reef conservation and restoration in the omics age. Coral Reefs of the World. Springer International Publishing, Cham, pp 55–70. https://doi.org/10.1007/978-3-031-07055-6_4

Huffmyer AS, Wong KH, Becker DM, Strand E, Mass T, Putnam HM (2023) Multi-omic analysis of coral development elucidates the metabolic transitions from parental provisioning to endosymbiotic nutritional exchange. bioRxiv 1–21

Hughes TP, Reed DC, Boyle M (1987) Herbivory on coral reefs: community structure following mass mortalities of sea urchins. J Exp Mar Bio Ecol 113:39–59

Hughes TP, Anderson KD, Connolly SR, Heron SF, Kerry JT, Lough JM, Baird AH, Baum JK, Berumen ML, Bridge TC, Claar DC, Eakin CM, Gilmour JP, Graham NAJ, Harrison H, Hobbs JPA, Hoey AS, Hoogenboom M, Lowe RJ, McCulloch MT, Pandolfi JM, Pratchett M, Schoepf V, Torda G, Wilson SK (2018) Spatial and temporal patterns of mass bleaching of corals in the Anthropocene. Science (80-.) 359:80–83. https://doi.org/10.1126/science.aan8048

Hugues TP, Kerry JT, Álvarez-noriega M, Álvarez-romero JG, Anderson KD, Baird AH, Babcock RC, Beger M, Bellwood DR, Berkelmans R, Bridge TC, Butler IR, Byrne M, Cantin NE, Comeau S, Connolly SR, Cumming GS, Dalton SJ, Kuo C, Lough JM, Hoey AS, Hobbs JA, Hoogenboom MO, Emma V, Pears RJ, Pratchett MS, Schoepf V, Simpson T, Skirving WJ, Sommer B (2017) Global warming and recurrent mass bleaching of corals. Nature 543:373–377. https://doi.org/10.1038/nature21707

Humanes A, Beauchamp EA, Bythell JC, Carl MK, Craggs JR, Edwards AJ, Golbuu Y, Lachs L, Martinez HM, Palmowski P, Paysinger F, Randle JL (2021) An experimental framework for selectively breeding corals for assisted evolution. Front Mar Sci 1–18. https://doi.org/10.3389/fmars.2021.669995

IPCC (2018) Global warming of 1.5°C. An IPCC special report on the impacts of global warming of 1.5°C above pre-industrial levels

Irikawa A, Casareto BE, Suzuki Y, Agostini S, Hidaka M, Van Woesik R (2011) Growth anomalies on Acropora cytherea corals. Mar Pollut Bull 62:1702–1707. https://doi.org/10.1016/j.marpolbul.2011.05.033

Irwin A, Greer L, Humston R, Cabe P, Lescinsky H, Wirth K, Curran HA, Baums IB, Greer L (2017) Age and intraspecific diversity of resilient Acropora communities in Belize. Coral Reefs 36:1111–1120. https://doi.org/10.1007/s00338-017-1602-9

Ishii Y, Hatta M, Deguchi R, Kawata M, Maruyama S (2022) Gene expression alterations from reversible to irreversible stages during coral metamorphosis. Zool Lett 8:1–11

Isomura N, Iwao K, Fukami H (2013) Possible natural hybridization of two morphologically distinct species of acropora (Cnidaria, Scleractinia) in the Pacific: fertilization and larval survival rates. PLoS One 8:1–8. https://doi.org/10.1371/journal.pone.0056701

Jacquemont J, Blasiak R, Le Cam C, Gouellec L, Claudet J (2022) Ocean conservation boosts climate change mitigation and adaptation. One Earth 5:1126. https://doi.org/10.1016/j.oneear.2022.09.002

Jiang L, Zhang J, Wang J-J, Wang L, Zhang L, Li G, Yang X, Ma X, Sun X, Cai J, Zhang J, Huang X, Yu M, Wang X, Liu F, Wu C-I, He C, Zhang B, Ci W, Liu J (2013) Sperm, but not oocyte, DNA methylome is inherited by zebrafish early embryos. Cell 153:773–784. https://doi.org/10.1016/j.cell.2013.04.041

Jiang L, Lei X, Liu S, Huang H (2015) Fused embryos and pre-metamorphic conjoined larvae in a broadcast spawning reef coral. F1000Res 1–10. https://doi.org/10.12688/f1000research.6136.1

Jouval F, Latreille AC, Bureau S, Adjeroud M, Penin L (2019) Multiscale variability in coral recruitment in the Mascarene Islands: from centimetric to geographical scale. PLoS One 14:1–17

Keshavmurthy S, Fontana S, Mezaki T, Del Cano Gonzalez L, Allen Chen C (2014) Doors are closing on early development in corals facing climate change. Sci Rep 4:18–23. https://doi.org/10.1038/srep05633

Kihika JK, Wood SA, Rhodes L, Smith KF, Miller MR, Pochon X, Thompson L, Butler J, Schattschneider J, Oakley C, Ryan KG (2022) Cryopreservation of six Symbiodiniaceae genera and assessment of fatty acid profiles in response to increased salinity treatments. Sci Rep 1–15. https://doi.org/10.1038/s41598-022-16735-w

Kitanobo S, Isomura N, Fukami H, Iwao K (2016) The reef-building coral Acropora conditionally hybridize under sperm limitation. Biol Lett 12:6–9

Kitanobo S, Iwao K, Fukami H, Isomura N, Morita M (2022) First evidence for backcrossing of – F 1 hybrids in Acropora corals under sperm competition. Sci Rep 1–8. https://doi.org/10.1038/s41598-022-08989-1

Knoester EG, Plug VE, Murk AJ, Sande SO, Osinga R (2023) Fisheries restrictions and their cascading effects on herbivore abundance and macroalgae removal at Kenyan coral reefs. J Exp Mar Bio Ecol 559:151850. https://doi.org/10.1016/j.jembe.2022.151850

Krieger EC, Taise A, Nelson WA, Grand J, Le Ru E, Davy SK, Cornwall CE (2023) Tolerance of coralline algae to ocean warming and marine heatwaves. PLOS Clim 1–26. https://doi.org/10.1371/journal.pclm.0000092

Ladd MC, Winslow EM, Burkepile DE, Lenihan HS (2021) Corallivory varies with water depth to influence the growth of Acropora hyacinthus, a reef-forming coral. Ecosphere 12. https://doi.org/10.1002/ecs2.3623

Laporte M, Le Luyer J, Rougeux C, Krick M, Bernatchez L (2019) DNA methylation reprogramming, TE derepression, and postzygotic isolation of nascent animal species. Sci Adv 5:1–9

Lawson CA, Camp E, Davy SK, Ferrier-Pagès C, Matthews J, Suggett DJ (2022) Informing coral reef conservation through Metabolomic approaches BT – coral reef conservation and restoration in the omics age. In: van Oppen MJH, Aranda Lastra M (eds) Coral reef conservation and restoration in the omics age. Coral reefs of the world. Springer International Publishing, Cham, pp 179–202. https://doi.org/10.1007/978-3-031-07055-6_12

Le Luyer J, Laporte M, Beacham TD, Kaukinen KH, Withler RE, Leong JS (2017) Parallel epigenetic modifications induced by hatchery rearing in a Pacific salmon. PNAS 1–6. https://doi.org/10.1073/pnas.1711229114

Leonard C, Hédouin L, Lacorne MC, Dalle J, Lapinski M, Blanc P, Nugues MM (2021) Performance of innovative materials as recruitment substrates for coral restoration. Restor Ecol 30:1–14. https://doi.org/10.1111/rec.13625

Levenstein MA, Gysbers DJ, Marhaver KL, Kattom S, Id LT, Quinlan Z, Tholen HM, Kelly LW, Mark J, Vermeij A, Johnson AJW, Juarez G (2022) Millimeter-scale topography facilitates coral larval settlement in wave-driven oscillatory flow. PLoS One 1–19. https://doi.org/10.1371/journal.pone.0274088

Levy N, Berman O, Yuval M, Loya Y, Treibitz T, Tarazi E, Levy O (2022) Emerging 3D technologies for future reformation of coral reefs: enhancing biodiversity using biomimetic structures based on designs by nature. Sci Total Environ 830:154749. https://doi.org/10.1016/j.scitotenv.2022.154749

Li Y, Liew YJ, Cui G, Cziesielski MJ, Zahran N, Michell CT, Voolstra CR, Aranda M (2018) DNA methylation regulates transcriptional homeostasis of algal endosymbiosis in the coral model Aiptasia. Sci Adv 4:eaat2142

Li Y, Zhang L, Li Y, Li W, Guo Z, Li R, Hu X, Bao Z (2019) Dynamics of DNA methylation and DNMT expression during gametogenesis and early development of scallop Patinopecten yessoensis, vol 21. Mar Biotechnol, p 196

Liberman R, Fine M, Benayahu Y (2021) Simulated climate change scenarios impact the reproduction and early life stages of a soft coral. Mar Environ Res 163:105215. https://doi.org/10.1016/j.marenvres.2020.105215

Liew YJ, Howells EJ, Wang X, Michell CT, Burt JA, Idaghdour Y, Aranda M (2020) Intergenerational epigenetic inheritance in reef-building corals. Nat Clim Chang 10:254–259. https://doi.org/10.1038/s41558-019-0687-2

Lin C, Zhuo J, Chong G, Wang L, Meng P (2018) The effects of aquarium culture on coral oocyte ultrastructure. Sci Rep 1–13. https://doi.org/10.1038/s41598-018-33341-x

Lin C, Wang L, Thongpoo P, Juri C (2019) Cryopreservation of a Thermotolerant lineage of the coral. Biopreserv Biobank:1–10. https://doi.org/10.1089/bio.2019.0019

Lin C, Takahashi S, Mulla AJ, Nozawa Y (2021) Moonrise timing is key for synchronized spawning in coral Dipsastraea speciosa. PNAS 118. https://doi.org/10.1073/pnas.2101985118

Linden B, Rinkevich B (2017) Elorating an eco-engineering approach for stock enhanced sexually derived coral colonies. J Exp Mar Bio Ecol 486:314–321. https://doi.org/10.1016/j.jembe.2016.10.014

Louis YD, Bhagooli R, Kenkel CD, Baker AC, Dyall SD (2017) Gene expression biomarkers of heat stress in scleractinian corals: promises and limitations. Comp Biochem Physiol Part C 191:63–77. https://doi.org/10.1016/j.cbpc.2016.08.007

Loya Y, Sakai K (2008). Bidirectional sex change in mushroom stony corals. Proceedings of the Royal Society B: Biological Sciences 275(1649):2335–2343. https://doi.org/10.1098/rspb.2008.0675

Lukoschiek V, Riginos C, Van Oppen MJH (2016) Congruent patterns of connectivity can inform management for broadcast spawning corals on the Great Barrier Reef. Mol Ecol 3065–3080. https://doi.org/10.1111/mec.13649

Lymbery RA, Kennington WJ, Evans JP (2021) The thermal environment of sperm affects offspring success: a test of the anticipatory paternal effects hypothesis in the blue mussel. Biol Lett 17:20210213

Maher RL, Epstein HE, Vega Thurber R (2022) Dynamics of bacterial communities on coral reefs: implications for conservation BT – coral reef conservation and restoration in the omics age. In: van Oppen MJH, Aranda Lastra M (eds) Coral reef conservation and restoration in the omics age. Coral reefs of the world. Springer International Publishing, Cham, pp 97–116. https://doi.org/10.1007/978-3-031-07055-6_7

Maire J, Van Oppen MJH (2021) A role for bacterial experimental evolution in coral bleaching mitigation? Trends Microbiol 1–12. https://doi.org/10.1016/j.tim.2021.07.006

Marlow HQ, Martindale MQ (2007) Embryonic development in two species of scleractinian coral embryos: Symbiodinium localization and mode of gastrulation. Evol Dev 9:355–367

Mass T, Putnam HM, Drake JL, Zelzion E, Gates RD, Bhattacharya D, Falkowski PG (2016) Temporal and spatial expression patterns of biomineralization proteins during early development in the stony coral Pocillopora damicornis. Proc Biol Sci 283:20160322

Mcleod IM, Hein MY, Babcock R, Bay L, Bourne G, Id NC, Doropoulos C, Gibbs M, Harrison P, Lockie S, Van Oppen MJH, Mattocks N, Page CA, Randall J, Smith A, Smith HA, Suggett DJ, Taylor B, Vella KJ, Wachenfeld D, Bostro L (2022) Coral restoration and adaptation in Australia: the first five years. PLoS One 1–22. https://doi.org/10.1371/journal.pone.0273325

Mcrae CJ, Mayfield AB, Huang W, Côté IM, Fan T, Ross C (2021) Contrasting proteomic responses of adult and larval coral to high temperatures. Front Mar Sci 8:1–7. https://doi.org/10.3389/fmars.2021.716124

Moberg F, Folke CS (1999) Ecological goods and services of coral reef ecosystems. Ecol Econ 29:215–233. https://doi.org/10.1016/S0921-8009(99)00009-9

Monteil Y, Teo A, Fong J, Bauman AG, Todd PA, La Rochelle UD, Sciences P, De Biologie L, Marins E, Crépeau AM, Rochelle L (2020) Effects of macroalgae on coral fecundity in a degraded coral reef system. Mar Pollut Bull 151:110890. https://doi.org/10.1016/j.marpolbul.2020.110890

Morita M, Nishikawa A, Nakajima A, Iguchi A, Sakai K, Takemura A, Okuno M (2006) Eggs regulate sperm flagellar motility initiation, chemotaxis and inhibition in the coral Acropora digitifera, A. gemmifera and A. tenuis. J Exp Biol 209:4574–4579. https://doi.org/10.1242/jeb.02500

Nakamichi H, Ojimi MC, Isomura N, Hidaka M (2012) Somatic tissues of the coral Galaxea fascicularis possess telomerase activity. J Coral Reef Stud 53–59. https://doi.org/10.3755/galaxea.14.1

Nawaz K, Cziesielski MJ, Mariappan KG, Cui G, Aranda M (2022) Histone modifications and DNA methylation act cooperatively in regulating symbiosis genes in the sea anemone Aiptasia. BMC Biol 1–17. https://doi.org/10.1186/s12915-022-01469-y

Negri AP, Heyward AJ (2000) Inhibition of Fertilization and Larval Metamorphosis of the Coral Acropora millepora (Ehrenberg, 1834) by Petroleum Products. Mar Pollut Bull. 41:420–427

Negri A, Vollhardt C, Humphrey C, Heyward A, Jones R, Eaglesham G, Fabricius K (2005) Effects of the herbicide diuron on the early life history stages of coral. Mar Pollut Bulleti 51:370–383. https://doi.org/10.1016/j.marpolbul.2004.10.053

Okubo N, Motokawa T (2007) Embryogenesis in the reef-building coral Acropora spp. Zool Sci 24:1169–1177. https://doi.org/10.2108/zsj.24.1169

Okubo N, Mezaki T, Nozawa Y, Nakano Y, Lien Y, Fukami H, Hayward DC, Ball EE (2013) Comparative embryology of eleven species of stony corals (Scleractinia). PLoS One 8:1–22. https://doi.org/10.1371/journal.pone.0084115

Okubo N, Hayward DC, Forêt S, Ball EE (2016) A comparative view of early development in the corals Favia lizardensis, Ctenactis echinata, and Acropora millepora – morphology, transcriptome, and developmental gene expression. BMC Evol Biol 16:1–12. https://doi.org/10.1186/s12862-016-0615-2

Olsen KC, Moscoso JA, Levitan DONR (2019) Somatic mutation is a function of clone size and depth in orbicella reef-building corals. Biol Bull. https://doi.org/10.1086/700261

Ortega-Recalde O, Hore TA (2019) DNA methylation in the vertebrate germline: balancing memory and erasure. Essays Biochem 63:1–13

Palmer CV, Graham E, Baird AH (2012) Immunity through early development of coral larvae. Dev Comp Immunol 38:395–399. https://doi.org/10.1016/j.dci.2012.07.008

Palumbi SR, Walker NS, Hanson E, Armstrong K, Lippert M, Cornwell B, Nestor V, Golbuu Y (2023) Small-scale genetic structure of coral populations in Palau based on whole mitochondrial genomes: implications for future coral resilience. Evol Appl 518–529. https://doi.org/10.1111/eva.13509

Penland L, Kloulechad J, Idip D, van Woesik R (2004) Coral spawning in the western Pacific Ocean is related to solar insolation: evidence of multiple spawning events in Palau. Coral Reefs 23:133–140. https://doi.org/10.1007/s00338-003-0362-x

Perez-Rosales G, Brandl SJ, Chancerelle Y, Siu G, Martinez E, Parravicini V, Hédouin L (2021) Documenting decadal disturbance dynamics reveals archipelago-specific recovery and compositional change on Polynesian reefs. Mar Pollut Bull 170:112659. https://doi.org/10.1016/j.marpolbul.2021.112659

Pillai A, Gungi A, Reddy PC, Galande S, Hanna CW (2021) Epigenetic regulation in hydra: conserved and divergent roles. Front Cell Dev Biol 9:1–23. https://doi.org/10.3389/fcell.2021.663208

Planes S, Allemand D, Agostini S, Banaigs B, Id EB, Id EB, Bourdin G, Id CB, Id ED, Id JMF, Forcioli D, Id PF, Id EG (2019) The Tara Pacific expedition — a pan-ecosystemic approach of the "-omics " complexity of coral reef holobionts across the Pacific Ocean. PLoS Biol 17:1–14

Pollock FJ, Katz SM, Van De Water JAJM, Davies SW, Hein M, Torda G, Matz MV, Beltran VH, Buerger P, Puill-stephan E, Abrego D, Bourne DG, Willis BL (2017) Coral larvae for restoration and research: a large-scale method for rearing Acropora millepora larvae, inducing settlement, and establishing symbiosis. PeerJ 1–21. https://doi.org/10.7717/peerj.3732

Potok M, Nix D, Parnell T, Cairns B (2013) Reprogramming the maternal zebrafish genome after fertilization to match the paternal methylation pattern. Cell 153:759–772

Prasetia R, Sinniger F, Nakamura T, Harii S (2022) Limited acclimation of early life stages of the coral Seriatopora hystrix from mesophotic depth to shallow reefs. Sci Rep 1–12. https://doi.org/10.1038/s41598-022-16024-6

Prata KE, Riginos C, Gutenkunst RN, Latijnhouwers KRW, Sánchez JA, Englebert N, Hay KB, Bongaerts P (2022) Deep connections: divergence histories with gene flow in mesophotic Agaricia corals. Mol Ecol 2511–2527. https://doi.org/10.1111/mec.16391

Puill-Stephan E, Van Oppen MJH, Pichavant-Rafini K, Willis BL (2012) High potential for formation and persistence of chimeras following aggregated larval settlement in the broadcast spawning coral, Acropora millepora. Proc R Soc B 279:699–708. https://doi.org/10.1098/rspb.2011.1035

Puisay A, Pilon R, Goiran C, Hédouin L (2018) Thermal resistances and acclimation potential during coral larval ontogeny in Acropora pulchra. Mar Environ Res 135:1–10. https://doi.org/10.1016/j.marenvres.2018.01.005

Puisay A, Elleaume N, Fouqueau L, Lacube Y, Goiran C, Sidobre C, Metian M, Laetitia H (2021) Parental bleaching susceptibility leads to differences in larval fluorescence and dispersal potential in Pocillopora acuta corals. Mar Environ Res 163:105200. https://doi.org/10.1016/j.marenvres.2020.105200

Putnam HM (2021) Avenues of reef-building coral acclimatization in response to rapid environmental change. Co Biol 224. https://doi.org/10.1242/jeb.239319

Putnam HM, Gates RD (2015) Preconditioning in the reef-building coral Pocillopora damicornis and the potential for trans-generational acclimatization in coral larvae under future climate change conditions. J Exp Biol 218:2365–2372. https://doi.org/10.1242/jeb.123018

Putnam HM, Davidson JM, Gates RD (2016) Ocean acidification influences host DNA methylation and phenotypic plasticity in environmentally susceptible corals. Evol Appl 9:1165. https://doi.org/10.1111/eva.12408

Quattrini AM, Wu T, Soong K, Jeng M, Benayahu Y, Mcfadden CS (2019) A next generation approach to species delimitation reveals the role of hybridization in a cryptic species complex of corals. BMC Evol Biol 19:116

Randall CJ, Negri AP, Quigley KM, Foster T, Ricardo GF, Webster NS, Bay LK, Harrison PL, Babcock RC, Heyward AJ (2020) Sexual production of corals for reef restoration in the Anthropocene. Mar Ecol Prog Ser 635:203–232

Rapuano H, Shlesinger T, Roth L, Bronstein O, Loya Y (2023) Coming of age: annual onset of coral reproduction is determined by age rather than size. IScience 106533. https://doi.org/10.1016/j.isci.2023.106533

Ricardo GF, Jones RJ, Nordborg M, Negri AP (2017) Settlement patterns of the coral Acropora millepora on sediment-laden surfaces. Sci Total Environ 609:277–288. https://doi.org/10.1016/j.scitotenv.2017.07.153

Ricardo GF, Harper CE, Negri AP, Luter HM, Azmi M, Wahab A, Jones RJ (2021) Impacts of water quality on Acropora coral settlement: the relative importance of substrate quality and light. Sci Total Environ 777:146079. https://doi.org/10.1016/j.scitotenv.2021.146079

Richards ZT, Stat M, Heydenrych M, DiBattista JD (2022) Environmental DNA for biodiversity monitoring of coral reefs BT – coral reef conservation and restoration in the omics age. In: van Oppen MJH, Aranda Lastra M (eds) Coral reef conservation and restoration in the omics age. Coral reefs of the world. Springer International Publishing, Cham, pp 203–224. https://doi.org/10.1007/978-3-031-07055-6_13

Ridlon AD, Grosholz ED, Hancock B, Miller MW, Bickel A, Froehlich HE, Lirman D, Pollock FJ, Putnam HM, Tlusty MF, Waters TJ, Wasson K, Jackson G, Smyth A, Ridlon AD, Mw M, Bickel A, He F (2023) Culturing for conservation: the need for timely investments in reef aquaculture. Front Mar Sci 1–20. https://doi.org/10.3389/fmars.2023.1069494

Riginos C, Beger M (2022) Incorporating genetic measures of connectivity and adaptation in marine spatial planning for corals BT – coral reef conservation and restoration in the omics age. In: van Oppen MJH, Aranda Lastra M (eds) Coral reef conservation and restoration in the omics age. Coral reefs of the world. Springer International Publishing, Cham, pp 7–33. https://doi.org/10.1007/978-3-031-07055-6_2

Rinkevich B (2019) Coral chimerism as an evolutionary rescue mechanism to mitigate global climate change impacts. Glob Chang Biol 1198–1206. https://doi.org/10.1111/gcb.14576

Rinkevich B (2021) Augmenting coral adaptation to climate change via coral gardening (the nursery phase). J Environ Manag 291:112727. https://doi.org/10.1016/j.jenvman.2021.112727

Riviere G, Wu G-C, Fellous A, Goux D, Sourdaine P, Favrel P (2013) DNA methylation is crucial for the early development in the Oyster C. Gigas. Mar Biotechnol 15:739. https://doi.org/10.1007/s10126-013-9523-2

Riviere G, He Y, Tecchio S, Crowell E, Sourdaine P, Guo X, Favrel P (2017) Dynamics of DNA methylomes underlie oyster development. PLoS Genet 13:1–16

Rodríguez-casariego JA, Ladd MC, Shantz AA, Manjinder L, Bohyun SC, Steven K, James BR, Juan F, Deron A, Jose EB, Lopez ME (2018) Coral epigenetic responses to nutrient stress: histone H2A.X phosphorylation dynamics and DNA methylation in the staghorn coral Acropora cervicornis. Ecol Evol 12193–12207. https://doi.org/10.1002/ece3.4678

Rodríguez-casariego JA, Mercado-molina AE, Garcia-souto D (2020) Genome-wide DNA methylation analysis reveals a conserved epigenetic response to seasonal environmental variation in the staghorn coral Acropora cervicornis. Front Mar Sci 7:1–17. https://doi.org/10.3389/fmars.2020.560424

Rodriguez-Casariego JA, Cunning R, Baker AC, Eirin- JM (2021) Symbiont shuffling induces differential DNA methylation responses to thermal stress in the coral Montastraea cavernosa. Mol Ecol 1–15. https://doi.org/10.1111/mec.16246

Roepke LK, Brefeld D, Soltmann U, Randall CJ, Negri AP, Kunzmann A (2022) Antifouling coatings can reduce algal growth while preserving coral settlement. Sci Rep 1–14. https://doi.org/10.1038/s41598-022-19997-6

Roger LM, Lewinski NA, Putnam HM, Roxbury D, Tresguerres M (2023) Ll perspective Nanobiotech engineering for future coral reefs. One Earth. 6:778. https://doi.org/10.1016/j.oneear.2023.05.008

Roquis D, Picart A, Brener K, Romans P, Masanet P (2019) The tropical coral Pocillopora acuta has a mosaic DNA methylome, an unusual chromatin structure and shows histone H3 clipping. bioRxiv

Rouan A, Pousse M, Djerbi N, Porro B, Bourdin G, Carradec Q, Hume BC, Poulain J, Armstrong E, Agostini S, Salazar G, Ruschemweyh HJ, Aury J-M, Paz-Garcia DA, McMinds R, Giraud-Panis M-J, Deshuraud R, Ottaviani A, Morini LD, Leone C, Wurze L, Tran J, Zoccola D, Pey A, Moulin C, Boissin E, Iwankow G, Romac S, Vargas C d, Banaigs B, Boss E, Bowler C, Douville E, Flores M, Reynaud S, Thomas OP, Troublé R, Thuber RV, Planes S, Allemand D, Pesant S, Galand PE, Wincker P, Sunagawa S, Rottinger E, Furla P, Voolstra CR, Forcioli D, Lombard F, Gilson E (2023) Telomere DNA length regulation is influenced by seasonal temperature differences in short-lived but not in long- lived reef-building corals. Nat Commun 14:3038. https://doi.org/10.1038/s41467-023-38499-1

Sarribouette L, Pedersen NE, Edwards CB, Sandin SA (2022) Post - settlement demographics of reef building corals suggest prolonged recruitment bottlenecks. Oecologia 199:387–396. https://doi.org/10.1007/s00442-022-05196-7

Schmidt-roach S, Miller KJ, Woolsey E, Gerlach G, Baird AH (2012) Broadcast spawning by Pocillopora species on the great barrier reef. PLoS One 7:1–7. https://doi.org/10.1371/journal.pone.0050847

Schoepf V, Baumann JH, Barshis DJ, Browne NK, Camp EF, Comeau S, Cornwall CE, Guzmán HM, Riegl B, Rodolfo-metalpa R, Sommer B (2023) Corals at the edge of environmental limits: a new conceptual framework to re-de fi ne marginal and extreme coral communities. Sci Total Environ 884:163688. https://doi.org/10.1016/j.scitotenv.2023.163688

Schweinsberg M, Gösser F, Tollrian R (2021) The history, biological relevance, and potential applications for polyp bailout in corals. Ecol Evol 8424–8440. https://doi.org/10.1002/ece3.7740

Shah S, Dougan KE, Bhattacharya D, Chan CX (2022) Coral conservation from the genomic perspective on Symbiodiniaceae diversity and function in the Holobiont BT – coral reef conservation and restoration in the omics age. In: van Oppen MJH, Aranda Lastra M (eds) Coral reef conservation and restoration in the omics age. Coral reefs of the world. Springer International Publishing, Cham, pp 85–96. https://doi.org/10.1007/978-3-031-07055-6_6

Shikina S, Chang C (2016) Sexual reproduction in stony corals and insight into the evolution of oogenesis in Cnidaria. In: The Cnidaria, past, present and future. https://doi.org/10.1007/978-3-319-31305-4

Shoguchi E, Shinzato C, Kawashima T, Gyoja F, Mungpakdee S, Koyanagi R, Takeuchi T, Hisata K, Tanaka M, Fujiwara M, Hamada M, Seidi A, Fujie M, Usami T, Goto H, Yamasaki S, Arakaki N, Suzuki Y, Sugano S, Toyoda A, Kuroki Y, Coffroth MA, Fujiyama A, Bhattacharya D, Satoh N (2013) Draft assembly of the Symbiodinium minutum nuclear genome reveals dinoflagellate gene structure. Curr Biol 1399–1408. https://doi.org/10.1016/j.cub.2013.05.062

Simpson CJ (1991) Mass spawning of corals on Western Australian reefs and comparisons with the Great Barrier Reef. J R Soc West Aust:85–91

Smith A, Cook N, Cook K, Brown R, Woodgett R, Veron J, Saylor V (2021) Field measurements of a massive Porites coral at Goolboodi (Orpheus Island), Great Barrier Reef. Sci Rep 1–6. https://doi.org/10.1038/s41598-021-94818-w

Spiers LJ, Frazer TK (2023) How herbivores reshape a macroalgal community on a little Cayman coral reef: the role of herbivore type and density. J Exp Mar Bio Ecol 562:151884. https://doi.org/10.1016/j.jembe.2023.151884

Stockton L, Edmunds PJ (2021) Spatially aggressive peyssonnelid algal crusts (PAC) constrain coral recruitment to Diadema grazing halos on a shallow Caribbean reef. J Exp Mar Bio Ecol. 541:151569

Strader ME, Quigley KM (2022) The role of gene expression and symbiosis in reef-building coral acquired heat tolerance. Nat Commun 13:1–8. https://doi.org/10.1038/s41467-022-32217-z

Strader ME, Aglyamova GV, Matz MV (2016) Red fluorescence in coral larvae is associated with a diapause-like state. Mol Ecol 559–569. https://doi.org/10.1111/mec.13488

Strathmann RR, Hughes TP, Kuris AM, Lindeman KC, Morgan SG, Pandolfi JM, Warner RR (2002) Evolution of local recruitment and its consequences for marine populations. Bull Mar Sci 70:377–396

Taira J, Higa I, Tsuchida E, Isomura N, Iguchi A (2018) Biochemical and Biophysical Research Communications Neurotransmitters in hermatypic coral, Acropora spp., and its contribution to synchronous spawning during reproductive event. Biochem Biophys Res Commun 501:80–84. https://doi.org/10.1016/j.bbrc.2018.04.170

Toh TC, Soon C, Ng L, Guest J, Chou LM (2013) Grazers improve health of coral juveniles in ex situ mariculture. Aquaculture 414–415:288–293. https://doi.org/10.1016/j.aquaculture.2013.08.025

Torda G, Donelson JM, Aranda M, Barshis DJ, Bay L, Berumen ML, Bourne DG, Cantin N, Foret S, Matz M, Miller DJ, Moya A, Putnam HM, Ravasi T, Van Oppen MJH, Thurber RV, Vidal-Dupiol J, Voolstra CR, Watson SA, Whitelaw E, Willis BL, Munday PL (2017) Rapid adaptive responses to climate change in corals. Nat Clim Chang 7:627–636. https://doi.org/10.1038/nclimate3374

Trigg SA, Venkataraman YR, Gavery MR, Roberts SB, Bhattacharya D, Eirin- JM, Johnson LKM, Lotterhos KE, Puritz JB, Putnam HM (2021) Invertebrate methylomes provide insight into mechanisms of environmental tolerance and reveal methodological biases. Mol Ecol Resour 1–15. https://doi.org/10.1111/1755-0998.13542

Tsuta H, Shinzato C, Satoh N, Hidaka M, Tsuta H, Shinzato C, Satoh N, Hidaka M (2014) Telomere shortening in the colonial coral Acropora digitifera during development. Zool Sci 31:129–134. https://doi.org/10.2108/zsj.31.129

Valderrábano M (2021) Turning the tide for coral reefs. Int coral reef Initiat

Van De Water JAJM, Tignat-perrier R, Allemand D, Ferrier-page C (2022) Coral holobionts and biotechnology: from blue economy to coral reef conservation. Curr Opin Biotechnol 110–121. https://doi.org/10.1016/j.copbio.2021.10.013

Van Oppen MJH, Blackall LL (2019) Coral microbiome dynamics, functions and design in a changing world. Nat Rev Microbiol 17:557. https://doi.org/10.1038/s41579-019-0223-4

van Oppen, M.J.H., Lastra MA (2022) Introduction to coral reef conservation and restoration in the omics age

Van Oppen MJH, Oakeshott JG (2020) A breakthrough in understanding the molecular basis of coral heat tolerance. PNAS 117:28546–28548. https://doi.org/10.1073/pnas.2020201117

Van Oppen MJH, Willis BL, Van Rheede T, Miller DJ (2002) Spawning times, reproductive compatibilities and genetic structuring in the Acropora aspera group: evidence for natural hybridization and semi-permeable species. Mol Ecol 11:1363–1376

Van Oppen MJH, Oliver JK, Putnam HM, Gates RD (2015) Building coral reef resilience through assisted evolution. Proc Natl Acad Sci USA 112:2307–2313. https://doi.org/10.1073/pnas.1422301112

Vasquez-Kuntz KL, Kitchen SA, Conn TL, Vohsen SA, Chan AN, Vermeij MJA, Page C, Marhaver KL, Baums IB (2022) Inheritance of somatic mutations by animal offspring. Sci Adv 0707:1–11

Vastenhouw NL, Cao WX, Lipshitz HD (2019) The maternal-to-zygotic transition revisited. Development 146. https://doi.org/10.1242/dev.161471

Vidal-dupiol J, Chapparo C, Pratlong M, Pontarotti P, Grunau C, Mitta G (2020) Sequencing, de novo assembly and annotation of the genome of the scleractinian coral, Pocillopora acuta. bioRxiv:1–28

Vidal-Dupiol J, Harscouet E, Negri AP, Shefy D, Toulza E, Rey O, Allienne JF, Mitta G, Rinkevich B (2022) Frontloading of stress response genes enhances robustness to environmental change in chimeric corals. BMC Biol 1–18. https://doi.org/10.1186/s12915-022-01371-7

Vollmer SV, Palumbi SR (2002) Hybridization and the evolution of reef coral diversity. Science (80-.). https://doi.org/10.1126/science.1069524

Walker NS, Giribet G, Fernández R, Sneed JM, Paul VJ, Combosch DJ (2019) Differential gene expression during substrate probing in larvae of the Caribbean coral Porites astreoides. Mol Ecol 4899–4913. https://doi.org/10.1111/mec.15265

Waller RG, Goode S, Tracey D, Johnstone J, Mercier A (2023) A review of current knowledge on reproductive and larval processes of deep - sea corals. Mar Biol 170. https://doi.org/10.1007/s00227-023-04182-8

Wang X, Bhandari RK (2019) DNA methylation dynamics during epigenetic reprogramming of medaka embryo. Epigenetics:1–12. https://doi.org/10.1080/15592294.2019.1605816

Wang X, Bhandari RK (2020) The dynamics of DNA methylation during epigenetic reprogramming of primordial germ cells in medaka (Oryzias latipes). Epigenetics 15:483–498. https://doi.org/10.1080/15592294.2019.1695341

Wang Y, Liu Q, Tang F, Yan L, Qiao J (2019) Epigenetic regulation and risk factors during the development of human gametes and early embryos. Annu Rev Genomics Hum Genet 20:1–20

Wangpraseurt D, You S, Azam F, Jacucci G, Gaidarenko O, Hildebrand M, Kühl M, Smith AG, Davey MP, Smith A, Deheyn DD, Chen S, Vignolini S (n.d.) Bionic 3D printed corals. Nat Commun 11:1748. https://doi.org/10.1038/s41467-020-15486-4

Webster FJ, Babcock RC, Van Keulen M, Loneragan NR (2015) Macroalgae inhibits larval settlement and increases recruit mortality at Ningaloo Reef, Western Australia. PLoS One 1–14. https://doi.org/10.1371/journal.pone.0124162

Wecker P, Lecellier G, Guibert I, Zhou Y, Bonnard I, Berteaux-Lecellier V (2018) Exposure to the environmentally-persistent insecticide chlordecone induces detoxification genes and causes polyp bail-out in the coral P. damicornis. Chemosphere 195:190–200. https://doi.org/10.1016/j.chemosphere.2017.12.048

White J-SS, O'Donnell JL (2010) Indirect effects of a key ecosystem engineer alter survival and growth of foundation coral species. Ecology 91:3538–3548

Willis BL, Van Oppen MJH, Miller DJ, Vollmer SV, Ayre DJ (2006) The role of hybridization in the evolution of reef corals. Annu Rev Ecol Evol Syst 37:489. https://doi.org/10.1146/annurev.ecolsys.37.091305.110136

Wilson B, Fan CM, Edmunds PJ (2020) An unusual microbiome characterises a spatially - aggressive crustose alga rapidly overgrowing shallow Caribbean reefs. Sci Rep 1–13. https://doi.org/10.1038/s41598-020-76204-0

Xu X, Li G, Li C, Zhang J, Wang Q, Simmons DK, Chen X, Wijesena N, Zhu W, Wang Z, Wang Z, Ju B, Ci W, Lu X, Yu D, Wang Q-f, Aluru N, Oliveri P, Zhang YE (2019) Evolutionary transition between invertebrates and vertebrates via methylation reprogramming in embryogenesis. Natl Sci Rev 6:993–1003. https://doi.org/10.1093/nsr/nwz064

Yagound B, Remnant EJ, Buchmann G, Oldroyd BP (2020) Intergenerational transfer of DNA methylation marks in the honey bee. PNAS 1–9. https://doi.org/10.1073/pnas.2017094117

Yamano H, Sakuma A, Harii S (2020) Remote sensing of environment coral-spawn slicks: reflectance spectra and detection using optical satellite data. Remote Sens Environ 251:112058. https://doi.org/10.1016/j.rse.2020.112058

Yanovski R, Abelson A (2019) Structural complexity enhancement as a potential coral-reef restoration tool. Ecol Eng 132:87

Yanovski R, Nelson PA, Abelson A (2017) Structural complexity in coral reefs: examination of a novel evaluation tool on different spatial scales. Front Ecol Evol 5:1–9. https://doi.org/10.3389/fevo.2017.00027

Yeoh S, Dai C (2010) The production of sexual and asexual larvae within single broods of the The production of sexual and asexual larvae within single broods of the scleractinian coral, Pocillopora damicornis. Mar Biol 157:351–359. https://doi.org/10.1007/s00227-009-1322-y

Ying H, Cooke I, Sprungala S, Wang W, Hayward DC, Tang Y, Huttley G, Ball EE, Forêt S, Miller DJ (2018) Comparative genomics reveals the distinct evolutionary trajectories of the robust and complex coral lineages. Genome Biol 19:1–24

Ying H, Hayward DC, Klimovich A, Bosch TCG, Baldassarre L, Neeman T, For S, Huttley G, Reitzel AM, Fraune S, Ball EE (2022) The role of DNA methylation in genome defense in Cnidaria and other invertebrates. Mol Biol Evol 39:1–14. https://doi.org/10.1093/molbev/msac018

Yoshioka Y, Suzuki G, Zayasu Y, Yamashita H, Shinzato C (2022) Comparative genomics highlight the importance of lineage - specific gene families in evolutionary divergence of the coral genus, Montipora. BMC Ecol Evol 1–16. https://doi.org/10.1186/s12862-022-02023-8

Yu X, Yu K, Liao Z, Chen B, Deng C, Yu J, Yao Q (2021) Seasonal fluctuations in symbiotic bacteria and their role in environmental adaptation of the scleractinian coral Acropora pruinosa in high-latitude coral reef area of the South China Sea. Sci Total Environ 792:148438. https://doi.org/10.1016/j.scitotenv.2021.148438

Zhang X, Jacobs D (2022) A broad survey of gene body and repeat methylation in Cnidaria reveals a complex evolutionary history. Genome Biol Evol 14:1–14

Zhang Y, Chiu Y, Chen C, Ho Y, Shinzato C (2019) Discovery of a receptor guanylate cyclase expressed in the sperm flagella of stony corals. Sci Rep 1–9. https://doi.org/10.1038/s41598-019-51224-7

Vulnerability of Coral Reefs to Contaminants of Emerging Concern

6

T. Gardon, C. R. Dromard, J. C. Gaertner, and N. Gaertner-Mazouni

Abstract

Coral reefs are highly vulnerable to contaminants of emerging concern (CECs), posing substantial threats to their sustainability within the context of global climate change. This chapter focuses on three major classes of CECs: micro-nanoplastics (MNPs), persistent organic pollutants (POPs), and trace metallic elements (TMEs). It provides an overview of their occurrence in coral reefs across the Pacific, Atlantic, and Indian Oceans, along with their potential impacts on key organisms such as corals and fish. Drawing upon insights from both laboratory and field investigations, we assess existing research gaps and explore new prospects for effective risk management strategies. The contamination of coral reefs by these pollutants, driven by a combination of local and global anthropogenic pressures, can impair marine species at multiple biological levels. Moreover, their simultaneous presence may amplify ecological disruptions, affecting primary production and reef ecosystem functioning. Given the complexity of these interactions, additional research is required to better understand the fate, transport, and synergistic effects of these contaminants in reef environments. Such efforts are crucial to improve exposure assessments, refine conservation strategies, and develop targeted mitigation measures. Addressing these challenges requires interdisciplinary collaboration, standardized monitoring protocols, and cooperation among research institutions to ensure the long-term resilience of coral reef ecosystems worldwide.

Keywords

Anthropogenic stressors · Ecotoxicology · Pollutant interactions · Ecological disruptions · Reef conservation strategies

6.1 Introduction

In the present Anthropocene epoch, multiple stressors are driving ecological changes with significant global consequences, particularly with respect to the widespread degradation of coral reef habitats (Hughes et al. 2017). While the primary stressors linked to climate change, such as warming and seawater acidification, have been extensively documented (Hughes et al. 2017), other anthropogenic stressors, such as hazardous contaminants, remain poorly studied and potentially underestimated. Yet the interplay between global and local stressors yields an exposome paradigm, where multiple stressors interact and may affect coral reef ecosystems simultaneously and synergistically (Brown et al. 2013). For instance, some coral species display heightened sensitivity to seawater warming when already physiologically stressed by nitrogen pollution (Donovan et al. 2020). Moreover, environmental changes not only push some species to their physiological limits, but also reshape interaction networks within ecosystems, triggering cascading effects

T. Gardon (✉)
USR 3278 CRIOBE, PSL University of Paris, EPHE-UPVD-CNRS, Papetoai, Moorea, French Polynesia

Laboratoire d'Excellence "CORAIL", Paris, France
e-mail: tony.gardon1@upf.pf

C. R. Dromard
Laboratoire d'Excellence "CORAIL", Paris, France

UMR BOREA, CNRS 7208, MNHN, UPMC, UCBN, IRD 207, University of Antilles, Campus de Fouillole, Pointe-à-Pitre, Guadeloupe, France
e-mail: charlotte.dromard@univ-antilles.fr

J. C. Gaertner
Laboratoire d'Excellence "CORAIL", Paris, France

IRD, University of French Polynesia-IFREMER-ILM, SECOPOL UMR 241, Tahiti, French Polynesia
e-mail: jean-claude.gaertner@ird.fr

N. Gaertner-Mazouni
Laboratoire d'Excellence "CORAIL", Paris, France

University of French Polynesia, IFREMER-ILM-IRD, SECOPOL UMR 241, Tahiti, French Polynesia
e-mail: nabila.gaertner-mazouni@upf.pf

S. Planes (ed.), *The Future of Coral Reefs*, Coral Reefs of the World 21, https://doi.org/10.1007/978-3-031-98584-3_6

that drive large community changes (Gilman et al. 2010). Consequently, anthropogenic disturbances are reshaping coral reefs worldwide, altering ecosystem functions and processes at various ecological levels and over different temporal and spatial scales. Considering the dynamic interplay among various stressors, it is crucial to account for both existing and emerging contaminants due to their widespread presence and potential detrimental effects, which could exacerbate the vulnerability of coral reefs (Kroon et al. 2020).

Contaminants of emerging concern (CECs) are defined as "naturally occurring, manufactured or manmade chemicals or materials which have now been discovered or are suspected present in various environmental compartments and whose toxicity or persistence are likely to significantly alter the metabolism of a living being" (Sauvé and Desrosiers 2014). CECs primarily originate from chemical pollution resulting from inadequate land management practices, including inefficient wastewater management and urban runoff, as well as activities in agriculture and industry. They can be broadly categorized into several chemical groups, including persistent organic pollutants (POPs), pharmaceuticals and personal care products (PPCPs), various trace elements including trace metallic elements (TMEs), plasticizers, and more recently, micro- and nanoplastics (MNPs) (Sauvé and Desrosiers 2014; Mofijur et al. 2021).

Due to plastic's widespread use, mismanagement, and enduring nature, an astonishing 6900 million tons (Mt) of plastic waste were generated from 1950 to 2018. By 2019, it was estimated that around 140 Mt of plastic had infiltrated rivers and oceans, resulting in environmental pollution (Geyer 2020; OECD 2022). Notably, plastic waste constitutes the largest proportion of marine litter, accounting for up to 60%–80% of all marine debris, and sources of plastic input continue to expand from terrestrial and marine origins (Jambeck et al. 2015). Plastic pollution can now be regarded as an emerging threat concerning associated diseases in coral reefs. It can lead to physical damage to coral tissue and can carry pathogens across reefs, promoting their spread and elevating the risk of infection (Lamb et al. 2018). In the broader context, studies suggest that the surface of the sea holds an estimated 93,000–236,000 tons of floating plastic, amounting to 5000–50,000 billion fragments, 92% of which are microparticles of plastic (Eriksen et al. 2014; van Sebille et al. 2015). Microplastics (MPs, <5 mm) and even smaller nanoplastics (NPs, <0.1–1 μm) have drawn attention due to their considerable potential for biological and chemical interactions. Their small size enables ingestion by numerous species, raising concerns about possible bioaccumulation and biomagnification (Miller et al. 2023). Additionally, their structural and surface attributes make them prone to sorbing and to releasing environmental chemicals, including other CECs from the surrounding water, as well as leaching additives (e.g., phthalates and flame retardants) initially incorporated during plastic manufacturing (Hermabessiere et al. 2017). Consequently, beyond their physical impact upon ingestion (e.g., gut obstruction), MNPs can serve as vectors for a cocktail of contaminants, including POPs and TMEs, triggering chemical effects (e.g., endocrine disruption) in marine organisms (Zhang et al. 2022).

Primarily stemming from anthropogenic activities, POPs and TMEs are pervasive contaminants that pose a global concern. Increasing degrees of agricultural, industrial, and technological pursuits have given rise to substantial emissions of these pollutants across the world (Tufail et al. 2022). A recent study focusing on 25 POPs indicated that up to 2020, a cumulative total of 31.3 Mt had been produced and distributed globally throughout history, leading to cumulative releases of 20.3 Mt into the global environment (Li et al. 2023). Meanwhile, naturally occurring TMEs present in the environment are also classified as inorganic pollutants, with their anthropogenic inputs exacerbating environmental concentrations. Due to their resistance to degradation, both POPs and TMEs persist as contaminants, potentially altering or even compromising the quality of the Earth's systems (Tufail et al. 2022). These substances exhibit both bioaccumulative and toxic properties, endangering all organisms within the food chain and giving rise to diverse health risks and detrimental impacts on ecosystems worldwide (Brodie et al. 2019).

CECs are thought to induce detrimental effects in aquatic organisms even at environmentally relevant concentrations. Understanding their origins, transport, fate, exposure pathways, and effects is therefore crucial. Over the past decade, the occurrence and effects of emerging pollutants have been studied extensively in aquatic environments (Yadav et al. 2021). However, no comprehensive synthesis has yet examined the occurrence and fate of MNPs, POPs, and TMEs, and their interactions within coral reefs. This chapter provides the first overview of these key CECs, focusing on (i) their sources, abundance, distribution, and characteristics across coral reefs globally, and (ii) their interactions and potential impacts on living organisms. To conclude, this chapter addresses knowledge gaps based on its findings, shedding light on the challenges and implications for coral reef conservation, and paving the way for forthcoming research strategies.

6.2 Micro-nanoplastics and Chemicals As Complex Threats to Coral Reefs

According to the U.S. Environmental Protection Agency's Emerging Contaminants Workgroup (2008), sources of CECs are as varied as the contaminants themselves. The primary contributors of CECs in aquatic environments are inefficient wastewater treatment processes in wastewater treatment plants (WWTPs) (Fig. 6.1). Additionally, terrestrial runoff from rivers and streams, polluted by agricultural, industrial, or urban activities, serves as significant pathways for CECs to enter marine waters (Yadav et al. 2021) (Fig. 6.1). However, the factors that determine the impact of CECs on coral reefs may be the result of onetime inputs or may stem from extended diffusion processes, including marine resurgences, atmospheric transport, and the movement of wildlife, which, in general, remain poorly understood (Fig. 6.1).

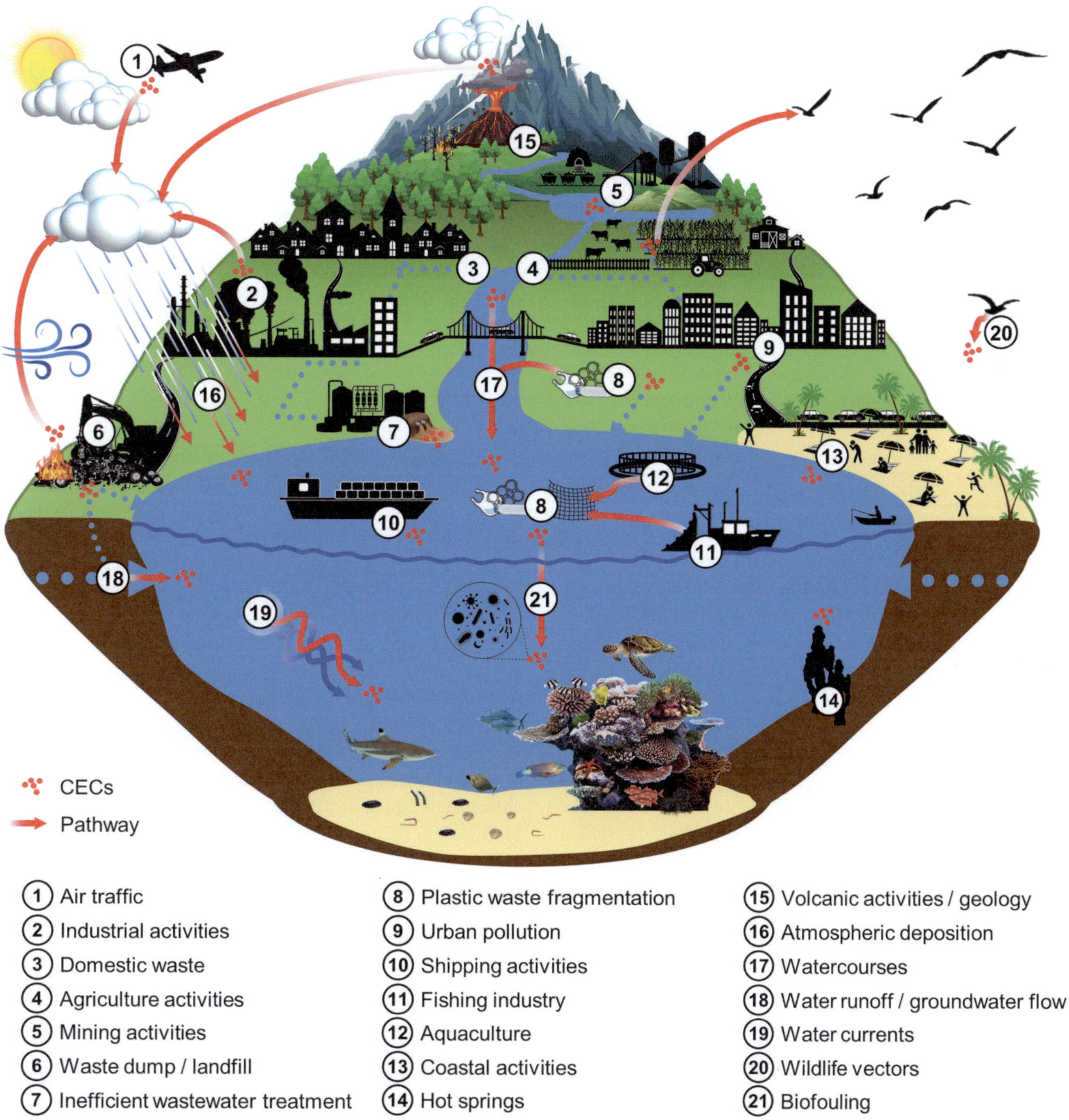

Fig. 6.1 Synthetic scheme of major sources and pathways of CECs (MNPs, POPs, and TMEs) in coral reef ecosystems. Several sources and pathways of CECs can be specific to, or common among, MNPs, POPs, and TMEs. The complexity and challenge of determining the exact origins of these CECs come from a combination of factors. These factors include the wide range of anthropogenic sources (1–13) and natural sources (related to some TMEs; 14 and 15), the co-occurrence and intrinsic interactions of these CECs, and the diverse pathways (16–21) that play a key role in their transport and distribution in natural environments before ultimately reaching coral reef habitats

Consequently, as CECs in reef environments may have originated from distant sources and thus would have traveled over extended distances, managing such pollution at a local scale presents a number of challenges.

6.2.1 Characteristics and Emission Sources

6.2.1.1 Micro-nanoplastics

Micro-nanoplastics can be categorized as primary or secondary depending on their origins. The majority of MPs (and likely NPs) found in the ocean result from the gradual fragmentation of larger items, transforming into secondary MNPs (Fig. 6.1). This process is influenced by a combination of environmental factors, including mechanical, chemical, biological, and photooxidative degradation (Andrady 2011) (Fig. 6.2). Conversely, primary MPs are directly released into the environment in their initial form, often as microbeads, pellets, and microfibers (Fu and Wang 2019). MNPs can further be characterized by their polymer type (e.g., polyethylene (PE), polypropylene (PP), polystyrene (PS)), shape and structure (e.g., sphere, fragment, fiber, film), chemical composition, surface properties, density, rigidity, and color (Fig. 6.2), influencing their interactions with marine biota (Hartmann et al. 2019).

While numerous techniques have been developed for the separation, identification, and quantification of MPs in different environmental matrices, many of these methods are ineffective for detecting NPs due to their nanoscale dimensions. To date, assessing the sources and fate of NPs in the environment remains inherently challenging due to methodological limitations in detecting and quantifying particles smaller than 10 μm (Schwaferts et al. 2019). Nevertheless, investigating the sources of MPs and larger debris can provide a preliminary basis for understanding the overarching patterns of MNPs. It is estimated that approximately 70%–80% of marine plastic debris originates from land-based sources, with ocean-based sources contributing the remaining 20%–30% (Li et al. 2016b; Lebreton et al. 2018). Within terrestrial sources, a significant proportion of MPs come from the weathering and fragmentation of protective plastic sheeting in agricultural fields, plastic litter, and discarded plastic items in landfills (Fig. 6.1) (Rochman 2018). Additionally, fibers from synthetic textiles, microbeads from products like facial exfoliating soaps and toothpaste, and preproduction pellets used in plastic manufacturing constitute substantial sources of primary MPs. These materials are introduced into the aquatic environment through household wastewater systems and other drainage pathways (Fu and Wang 2019) (Fig. 6.1). Notably, microfibers originating from domestic and industrial washing of synthetic textiles account for up to 35% of primary MPs in marine environments (Liu et al. 2021). For instance, Napper and Thompson (2016) revealed that a single 6 kg domestic wash could release over 700,000 fibers into sewage. Despite the importance of wastewater treatment for achieving sustainable water management, current systems remain inefficient in preventing CECs from entering the effluent of WWTPs (Yadav et al. 2021) (Fig. 6.1). Moreover, MNPs can also be transported by wind, washed away from soil surfaces by rainfall or stormwater runoff, and subsequently enter aquatic environments (Lambert et al. 2014; Jambeck et al. 2015) (Fig. 6.1). Rivers serve as major routes for transferring plastic waste from land to sea (Fig. 6.1), contributing between 1.15 and 2.41 Mt of plastic waste to the world's oceans annually (Lebreton et al. 2017). While ocean-based sources contribute less to MNPs pollution compared to land-based sources, their impact remains noteworthy. Among maritime activities, commercial fishing is the leading contributor, accounting for 10%–18% of plastic waste at sea (Lebreton et al. 2018) (Fig. 6.1). Richardson et al. (2022) estimated that 2% of all fishing gear is lost to the ocean each year. Additionally, waste from other activities including maritime trade, aquaculture, coastal tourism, and offshore industries adds to this burden (Lebreton et al. 2018) (Fig. 6.1).

6.2.1.2 Persistent Organic Pollutants

Persistent organic pollutants are classified by the Stockholm Convention (2001) into three categories based on their origin. The convention's objective is to regulate the production, usage, and release of these chemicals (Matei et al. 2023). The first category, the most prominent, encompasses synthetic chemicals primarily designed for agricultural purposes (Fig. 6.1). This group includes organochlorine pesticides (OCPs), such as aldrin, chlordecone, dieldrin, dichlorodiphenyltetrachloroethane (DDT), or even α-, β-, and γ-hexachlorocyclohexane (HCH). The second category involves chemicals produced for diverse industrial applications or released as industrial waste (Fig. 6.1), such as polychlorinated biphenyls (PCBs), hexachlorobenzene (HCB), or brominated flame retardants (BFRs). The third category comprises chemicals formed through incomplete combustion of organic matter (Fig. 6.1), such as polycyclic aromatic hydrocarbons (PAHs) and polychlorinated dibenzodioxins/dibenzofurans (PCDDs/PCDFs) (Matei et al. 2023). Consequently, POPs can arise both intentionally and unintentionally from various anthropogenic activities. These chemicals mainly stem from agricultural and industrial processes, as well as activities involving materials like paints, electronics, rubbers, and plastics. This is particularly relevant for compounds, such as OCPs, PCBs, and PAHs, which are most often residues from pesticides and other chlorinated organic compounds, as well as during plastic production (Matei et al. 2023). In this context, MNPs play a significant role as sources of POPs regarding theses interactions that occur before plastic production, during their usage (via

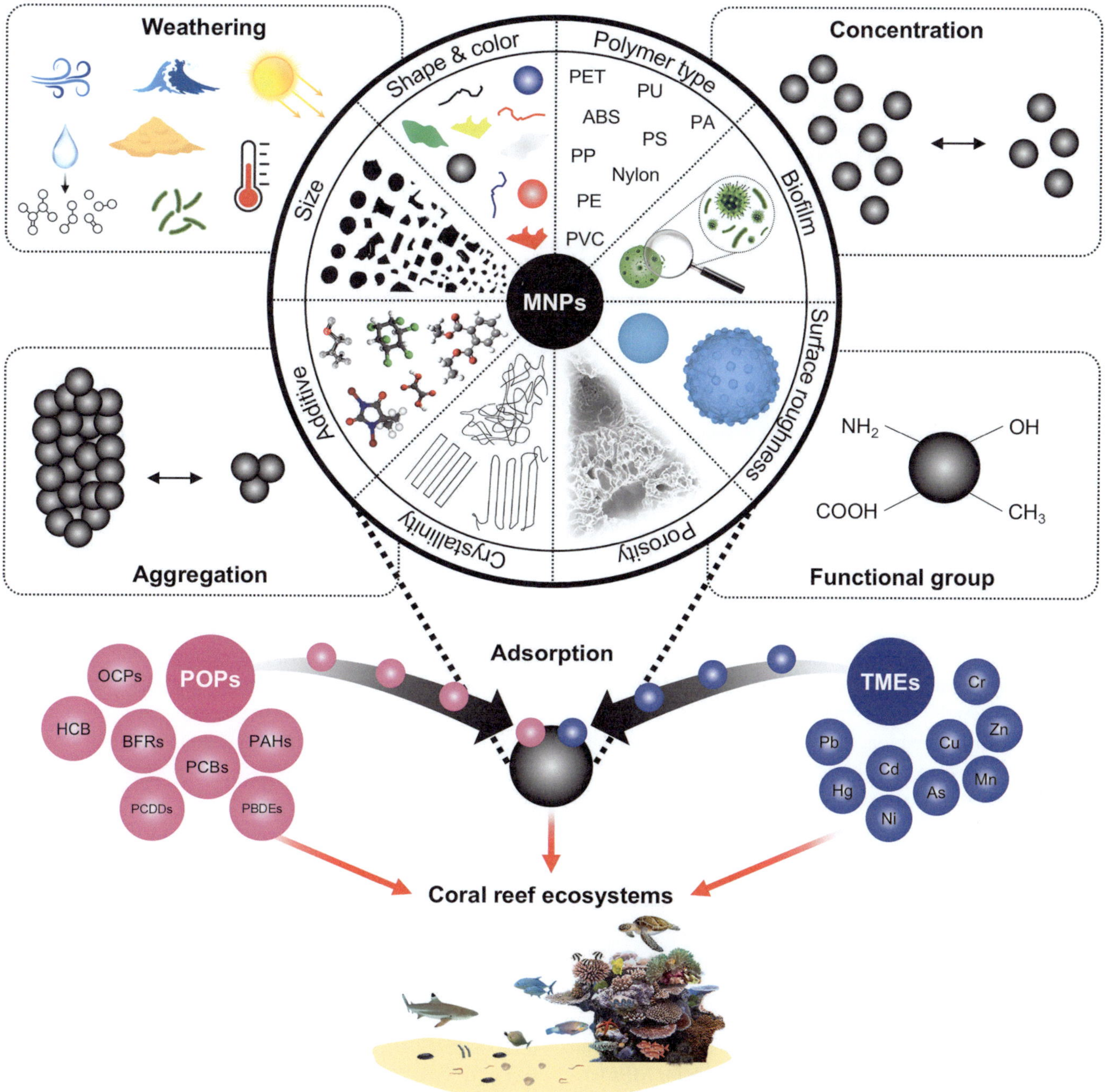

Fig. 6.2 Factors influencing MNP characteristics and their capacity to adsorb and co-transport POPs and TMEs, thereby shaping synergistic, additive, or antagonistic effects via multiple exposure pathways on coral reef ecosystems

wastewater, urban runoff, or landfill leachate), and after their release into the marine environment (Rodrigues et al. 2019) (Fig. 6.2).

6.2.1.3 Trace Metallic Elements

Heavy metals are naturally occurring elements characterized by relatively high densities (≥5 g cm^{-3}), atomic weights (63.5–200.6 g mol^{-1}), and atomic numbers (≥20). Here, we employ the term "trace metallic elements" as the term "heavy metals" is no longer in favor due to its imprecise usage in scientific literature (Pourret et al. 2021). Common TMEs in the environment are chromium (Cr), manganese (Mn), nickel (Ni), copper (Cu), zinc (Zn), cadmium (Cd), lead (Pb), and arsenic (As). Many TMEs naturally exist in the biosphere, primarily in water, soils, and rocks, originating from processes like volcanic activity (Fig. 6.1), lithogenesis, weathering, erosion, and other geological phenomena (Masindi and Muedi 2018). Additionally, human activities significantly contribute to environmental TME contamination. These activities include metal resource extraction, metallurgy, foundries, and other metal-based industries. Leaching of metals from landfills, automobiles, and roadworks further compounds this contamination (Fig. 6.1). Agriculture, utilizing pesticides, insecticides, and fertilizers containing metalloids, is also a prominent source of TME pollution (Masindi and Muedi 2018) (Fig. 6.1). Due to the substantial surge in human activities involving metallic substances within terrestrial and aquatic environments, TMEs are now recognized as inorganic pollutants directly linked to anthropogenic influence. Moreover, the accumulation of TMEs in soil and water is influenced by factors including POPs and MNPs (Fig. 6.2), promoting their presence on a global scale (Menéndez-Pedriza and Jaumot 2020).

6.2.2 Occurrence in Coral Reef Ecosystems

MNPs, POPs, and TMEs have been reported in all major oceans and seas worldwide. They are ubiquitous and found across a wide array of marine habitats from epipelagic, pelagic, and mesopelagic zones down through the hadopelagic zones and deep-sea habitats to benthic ecosystems. They have been found everywhere, from populated coastal environments to the most remote areas (Ajith et al. 2020; Jones 2021). Aware of the extent of this pervasive contamination, the scientific community has progressively begun to investigate the distribution, characteristics, and prevalence of these contaminants in coral reef regions. Coral reef distribution is broadly categorized into five regions spanning the Indo-Pacific and Atlantic areas: Southeast Asia (~32% of total coral reef areas), the Pacific Ocean including Australia (~41%), the Indian Ocean encompassing Middle Eastern seas (~19%), and the Atlantic Ocean, predominantly represented by the Caribbean Sea (~8%) (Spalding et al. 2001). The environmental matrices investigated within coral reefs encompass seawater (Gardon et al. 2021; Miller et al. 2023), sediments (Sandre et al. 2019; Patterson et al. 2020), and associated species including corals and fishes (Ding et al. 2019; Markic et al. 2020). Analytical methods for MPs have primarily involved sampling techniques (e.g., manta trawl and planktonic net, pump), followed by extraction processes (e.g., digestion of organic matter, density separation, filtration), visual characterization using microscope, and polymer identification methods such as Raman and FTIR spectroscopy. Concerning POPs and TMEs, conventional analytical methods largely revolve around the extraction of these substances from environmental samples (e.g., Soxhlet, ASE, and SPE). This extraction can be carried out actively (e.g., ISCO sampler and core tube) or passively (e.g., POCIS). Subsequently, the samples are subjected to chromatography and mass spectrometry analyses (e.g., HPLC, GC-MS/MS, and ICP-MS). These analyses may be complemented by the integration of stable isotope tracers (e.g., carbon $\delta^{13}C$ and nitrogen $\delta^{15}N$) to track their transfer along the food chain.

Southeast Asia, representing the largest coral reef region, stands as a hot spot for MP field investigations. Interestingly, Ding et al. (2019) evaluated MP abundance in seawater, coral, and fish samples from three atolls within the Xisha Islands in the South China Sea. Concentrations ranging from 200 to 12,200 MPs m^{-3} were measured in the water column of outer reef slopes and reef flats, while lagoons exhibited levels of 1000–45,200 MPs m^{-3}. Lagoons, characterized by a semi-enclosed nature, were thought to exacerbate MP accumulation (Ding et al. 2019). The prevalence of MPs in fish and coral samples was notable, with 84% and 100% of samples containing MPs, at concentrations of 0–4.7 and 0.02–1.3 MP g^{-1}, respectively. Interestingly, 74% of the MPs found in corals were located within the skeleton, while the remainder were attached to the skeletal surface (Ding et al. 2019). Laboratory experiments indicated that corals can directly and/or indirectly interact with MNPs through chemoreception, leading to active ingestion and/or passive adhesion to coral reef structures considered as a significant sink for MPs suspended in the water column (Corona et al. 2020). Furthermore, Ding et al. (2019) highlighted that MPs' characteristics (e.g., size, polymer type, shape, color) in seawater were more similar to those in fish gills rather than in fish gastrointestinal tracts or coral samples. This implies that MPs in seawater can be passively captured by gills through filtering, a process more transient and nonselective than MP ingestion (Yin et al. 2022). Nie et al. (2019) found blue microbeads to be the predominant MP type in the surface water of Nanxun reef (Nansha Islands, South China Sea), while blue and transparent fibers dominated the digestive tracts of fish species. The authors suggested that the slender shape of microfibers resembled zooplankton, leading to mistaken ingestion (Nie et al. 2019). In contrast, a different study conducted in remote and uninhabited reef atolls within the Nansha Islands reported significantly lower MP concentrations ranging from 0.01 to 0.15 MPs m^{-3}, with microfibers (<3 mm) constituting over 70% of MPs detected (Tan et al. 2020). Likewise, in the Pacific Islands, surface water of the fringing reef of Moorea in French Polynesia exhibited low concentrations of MPs, measuring 0.74 MPs (0.05–5 mm) m^{-2} (Connors 2017). On the same island, Garnier et al. (2019) identified MPs (0.03–2.4 mm) in the digestive tracts of 21% of four coral reef-associated fish species. In contrast, Markic et al. (2018) found MPs (0.1–5 mm) in nearly half (49.2%) of the commercial marine fish species examined near the remote island of Rapa Nui. This prevalence can be attributed to its location within the South Pacific subtropical gyre, where high plastic concentrations prevail. While remote regions tend to exhibit low MP concentrations due to limited anthropogenic influence, they are not immune to plastic debris transport or localized human pressures. This is particularly true in remote territories of French Polynesia, especially pearl farming atoll lagoons, where the widespread MP contamination in surface water (0.2–8.4 MPs m^{-3}), water column (4.0–716.2 MPs m^{-3}), and cultivated pearl oyster (2.1–125.0 MPs g^{-1} dry weight) was demonstrated, with 20–200 μm fragmented particles dominating (Gardon et al. 2021). The most commonly identified MPs were black fragments of PE and PP, matching the polymer and color of PE ropes and PP spat collectors, two prevalent sources of macroplastic waste in these farming areas, suggesting a pearl farming origin (Gardon et al. 2021; Crusot et al. 2023). Focusing on the Australian reef regions, considerable effort has also been directed toward assessing MP contamination in the Great Barrier Reef (GBR). For example, Jensen et al. (2019) collected surface water samples from both inshore and offshore reef locations within the GBR, measuring concentrations of 0.04–0.48 MPs (0.355–5 mm) m^{-3}. Inshore locations were primarily characterized by microfiber-dominated concentrations, suggesting riverine discharges as significant sources of inshore contamination. Based on its small home range criteria, Lemon damselfish (*Pomacentrus moluccencis*) from both study sites exhibited no significant difference in median MP abundance (0.355–5 mm), measuring 4.5 MPs $fish^{-1}$ at inshore and 4.0 MPs $fish^{-1}$ at offshore reef locations (Jensen et al. 2019). Similar to Nie et al., Jensen et al. observed significant differences between the characteristics of MPs in fish gastrointestinal tracts and those in surface waters. However, while foraging preferences influence MP ingestion (Peters et al. 2017), it is worth noting that both studies evaluated MP contamination in surface waters. Assessing MPs in the water column could offer further insight, as MP characteristics (e.g., density, shape, biofouling) influence their vertical distribution (Karkanorachaki et al. 2021), and subsequently, interactions with marine biota. Recent research has focused on comprehensively evaluating MP bioaccumulation and biomagnification within a simple coral reef food web (Miller et al. 2023). This study sampled taxa representing three trophic levels including zooplankton, benthic crustaceans, and reef fish, from two mid-shelf reefs in the central GBR, encompassing both water column and sediment environments. Results indicated the bioconcentration of MPs, with consistent concentrations, polymer types, sizes, shapes, and colors across trophic levels and environments (Miller et al. 2023). However, despite differences in MP contamination among trophic levels, no significant increase in concentrations occurred within the food web, suggesting limited MP biomagnification (Miller et al. 2023). In the Caribbean Sea, a study explored connectivity among Albuquerque atoll habitats (beach sediment, surface water, and reef) to identify potential marine litter sources, including MPs (Portz et al. 2020). This investigation reported an average concentration of 0.06 MPs (>80 μm) m^{-3}, dominated by fragmented particles with diverse colors, making source identification challenging (Portz et al. 2020). Finally, in Indian Ocean reef regions, significant MP concentrations were documented in surface waters and sediments of the Gulf of Mannar (Southeast India), ranging from 60,000 to 126,600 MP m^{-3} and 50 to 104 MP kg^{-1} with fibers (1–3 mm) and fragments (3–5 mm) predominating, respectively (Patterson et al. 2020). In the Faafu atoll (Maldives), lower MP concentrations were reported, averaging 0.32 MPs m^{-3} in surface water and 22.8 MPs m^{-2} in beach sediments (Saliu et al. 2018). Both studies explored the relationship between MP concentrations and levels of POPs and TMEs in the investigated environmental matrices. Patterson et al. (2020) discovered that some metals (Zn, Cd, Pb, Ni, Fe) associated with MPs in Gulf of Mannar sediments exceeded levels directly measured in sediments, suggesting that MPs could be sources of metal pollution and/or that metals from the sediment preferentially partition to MPs. Indeed, previous research demonstrated that the organic pollutant adsorption capacity of MPs is approximately two to six orders of magnitude higher than that of sediments and seawater, respectively (Fu et al. 2021). Furthermore, although the transfer of adsorbed environmental pollutants by MNPs involves various natural pathways at sea (e.g., detritus, colloids, bacteria, phytoplankton, organic matter), this is typically not the case for plastic additives (Koelmans et al. 2016). Saliu et al. (2018) demonstrated MP occurrence in both seawater and beach sediments in the Faafu atoll (Maldives) and the subsequent presence of plastic-derived phthalate acid esters (PAEs) was measured at the ng g^{-1} scale in plankton and the coral *Acropora muricata*. The highest concentration recorded in *A. muricata* was 24.1 ng g^{-1} for the total sum of six PAEs (Σ_6PAEs). This concentration was found to be seven times higher (Σ_5PAEs = 172.4 ng g^{-1}) when Montano et al. (2020) conducted the same survey at the same study site on three different scleractinian coral species 2 years later.

Various POPs and TMEs have been detected in a range of reef species worldwide. Significant monitoring efforts have been directed toward the GBR and Torres Strait (Australia), where elevated TME levels were observed near urban and industrial areas, likely due to anthropogenic activities (Kroon et al. 2020). Studies have reported the presence of metallic elements in several organisms within these regions, including seagrass (*Zostera capricorni*), crayfish (*Panulirus ornatus*), oysters (*Crassostrea echinata*, *Saccostrea amasa*, *S. commercialis*, *S. echinata* and *S. glomerata*), mud whelk (*Telescopium telescopium*), mud crab (*Scylla serrata*), tuxedo crab (*Australoplax tridentata*), fishes (such as barramundi, *Lates calcarifer*, parrotfish, *Scarus dimidiatus*, and mullet, *Valamugil seheli*), dugong (*Dugong dugon*), and green turtle (*Chelonia mydas*) (as reviewed by Kroon et al. 2020). Research has also been carried out in Pacific reef regions, such as French Polynesia, where the giant clam *Tridacna maxima* was used as a biointegrator (Godéré et al.

2023). Similarly, studies in New Caledonia demonstrated the contamination of TMEs in various components of the coral reef trophic networks in the southwest lagoon (Briand et al. 2018). The authors found that the trophic levels within sedimentary and reef benthic food webs were primarily affected by TMEs originating from mining activities, notably nickel factories, leading to the presence of elements like Ni and associated trace elements (Co, Cr, Fe, and Mn). However, the majority of TMEs decreased with higher trophic levels along food webs, except for Hg, which exhibited marked biomagnification with higher levels measured in predator fish (Briand et al. 2018). Given their high ecological significance as top predators in marine ecosystems, Shipley et al. (2021) conducted a study to monitor the concentrations of TMEs in sharks from six species across the Bahamas reef regions, with a specific focus on the Caribbean reef shark (*Carcharhinus perezi*) and the tiger shark (*Galeocerdo cuvier*). The authors discovered that Caribbean reef sharks exhibited the highest concentrations of four TMEs (Pb, Cr, As, and Hg). Total Hg (THg) concentrations were up to 1.5 times higher in Caribbean Reef sharks (with a mean of 16.5 mg kg^{-1}) than in any other species. Elevated concentrations of Pb, Cr, and Cu were observed as individuals reached sexual maturity. This outcome was likely due to significant correlations between the concentrations of different TME species, influenced by factors such as foraging dynamics, longevity, physiology, and a slower growth rate in older individuals of *Carcharhinus perezi*. In contrast, despite their larger size, tiger sharks exhibited the lowest THg concentrations among all species (with a mean of 4.4 mg kg^{-1}). Additionally, this study highlighted the potential toxicity of shark meat, which could have substantial implications for human consumers (Shipley et al. 2021). Turning attention to POPs, particularly organochlorine pesticides, a study conducted in the French West Indies assessed the contamination of marine organisms by chlordecone, a highly persistent pesticide used in banana fields from 1972 to 1993 (Dromard et al. 2016). The findings revealed that contamination levels in marine organisms decreased with increasing distance from the coast and pollution sources. For example, the average chlordecone concentration within the coral reef food web was 61.1 μg kg^{-1} (Dromard et al. 2018), which is three times lower than in littoral habitats like mangroves (Dromard et al. 2016). This study demonstrated that reef organisms become contaminated through contact with the polluted ambient environment (referred to as "bathing" contamination), but more significantly through trophic contamination (arising from consuming contaminated prey). As a result, the least contaminated organisms (below the authorized limit for marketing and human consumption, i.e., <20 μg kg^{-1}) were herbivorous fish, such as *Scarus taeniopterus* with 8 μg kg^{-1}. Conversely, the most contaminated species were found at the top of the food chain, such as *Lutjanus apodus* and *Sphyraena barracuda*, with concentrations of 290 and 318 μg kg^{-1}, respectively (Dromard et al. 2018). Detritivore crustaceans also exhibited high concentrations (e.g., *Panulirus guttatus*, 148 μg kg^{-1}) due to their diverse diet and taxonomic group (Dromard et al. 2018). However, much of the research on coral reef ecosystems has only focused on a few species or specific types of POPs and TMEs. A noteworthy study investigated the concentrations of POPs [polycyclic aromatic hydrocarbons (PAHs), polychlorinated biphenyls (PCBs), and polychlorinated naphthalenes (PCNs)] in coral tissues, symbiotic algae (zooxanthellae), sediments, and seawater within the coral reefs of Lark and Kharg in Iran (Ranjbar Jafarabadi et al. 2018). This study involved extensive sampling efforts to assess the contamination across eight species from six genera and three families of hard corals, as well as one family of soft coral. The findings revealed that the bioaccumulation of POPs was twice as high in soft corals compared to hard corals, and the total concentrations (Σ_{40}PAHs, Σ_{22}PCBs, and Σ_{20}PCNs) were ranked as follows in decreasing order: zooxanthellae > coral tissue > coral skeleton > reef sediment > seawater. The authors proposed that PAHs found in coral tissues, zooxanthellae, and reef sediments at both sites originated from petroleum sources, such as exploration, exploitation, and crude oil activities. Another study examined the concentrations of pesticides, PCBs, and TMEs in coral reef communities across three Pacific zones (French Polynesia, Wallis, and New Caledonia) (Fey et al. 2019). For each site, the authors explored the relationship between contaminant concentrations and the trophic level of species (δ^{15}N), used as a predictor of biomagnification or bioreduction. The results indicated a decrease in the concentrations of most TMEs (e.g., Cd, Cr, Cu, Fe, Mn, Ni, Pb) with increasing trophic levels (i.e., bioreduction) across all three islands. However, mercury (Hg) and selenium (Se) exhibited biomagnification. Nonetheless, diverse patterns were observed for pesticides and PCBs on different islands, underscoring the significance of considering local characteristics (Fey et al. 2019).

6.2.3 Potential Harmful Effects on Coral Reef Ecosystems

The impacts of CECs on marine organisms have been recognized in various coral reef biota spanning different taxonomic groups and trophic levels (Ouédraogo et al. 2023). MNPs, POPs, TMEs, and their interactions are extremely complex and lead to multidimensional contaminants with a wide range of effects on living organisms. The following presents a summary of potential mechanisms for toxicological effects derived from both laboratory exposures and field observations, with a particular emphasis on key reef organisms.

6.2.3.1 Effects of Micro-nanoplastics

Unlike other CECs, MNPs act as both chemical and physical stressors, which poses challenges in assessing their related toxicity. The clear evidence for the toxicological potential of MNPs is intricate and multifaceted, involving numerous biotic and abiotic factors, as well as particle characteristics (Jeong and Choi 2019) (Fig. 6.2). Organism toxicity in response to MNPs largely depends on their intrinsic properties, with particle shape, size, polymer type, and environmental chemistry acting as the primary factors that cause harm (Bucci and Rochman 2022) (Fig. 6.2). Exposure to MNPs can directly and/or indirectly induce various effects, encompassing physical damage, biological toxicity, and chemical effects in organisms. The most immediate impact occurs through ingestion, while indirect effects stem from adhesion, chemical leaching, and even colonization of MNPs by invasive species and pathogens (Zhang et al. 2022) (Fig. 6.3). These biological processes broadly initiate pathways of toxicity in marine biota, setting off domino effects across different levels of biological organization. Ultimately, these mechanisms may result in unpredictable risks for coral reef ecosystems (Fig. 6.3).

Physical impacts of MNPs that arise from ingestion, either directly or via transfer from prey, include intestinal blockage, internal injury, inflammation, food dilution, or even tissue translocation. Food dilution is probably one of the most significant driving determinants in MNP toxicity, resulting in reduced food intake and/or nutrient acquisition, which in turn leads to various biological dysfunctions (Thornton Hampton et al. 2022) (Fig. 6.3). For instance, Rotjan et al. (2019) reported that the ingestion of biofilmed-microbeads made of PE (170.5–230.8 μm) by the coral *Astrangia poculata* inhibited the consumption of natural brine shrimp eggs (*Artemia salina*) of the same diameter, due to a false perception of satiety. Another study demonstrated that PS-MPs (microbeads of 6 and 10 μm) at concentrations of 0.25, 2.5, and 25 μg L^{-1} altered the assimilation efficiency of microalgae (*Isochrysis galbana* T-Iso and *Chaetoceros gracilis*) in the pearl oyster *Pinctada margaritifera* (Gardon et al. 2018). Such physiological alterations affect the overall energy budget, with chronic depletion potentially leading to changes in behavior, respiration, decreased growth and development, altered reproduction, reduced fecundity, and even decreased survival (Thornton Hampton et al. 2022) (Fig. 6.3). For example, Yin et al. (2018) reported a significant reduction in feeding activity of Korean rockfish (*Sebastes schlegelii*) following exposure to PS-MPs (10^6 microbeads L^{-1}), which in turn impaired their exploration and hunting behavior. Beyond these physiological effects, chronic energy depletion may also lead to functional impairments in key ecosystem services and aquaculture outputs. In a recent 5-month study simulating ecological conditions of pearl farming, *P. margaritifera* was exposed to MNPs (0.4–200 μm) derived from aquaculture gear at environmentally relevant concentrations (0.025 and 1 μg L^{-1}) (Gardon et al. 2024). The authors further found that MNP exposure impaired energy metabolism by reducing microalgal assimilation efficiency, which in turn led to compensatory increases in ingestion at 1 μg L^{-1} and depletion of muscle glycogen reserves at 0.025 μg L^{-1}. These metabolic disturbances also degraded pearl quality—yielding thinner aragonite crystals and a higher incidence of Keshi pearls—underscoring the vulnerability of pearl oysters and the potential risks to the pearl industry in polluted French Polynesian waters (Gardon et al. 2024). Energy depletion resulting from malnutrition may result from digestive processes interference and/or gastrointestinal microbiota disruption (Zhang et al. 2022). A study on juvenile orange-spotted groupers (*Epinephelus coioides*) exposed for 14 days to PS-NPs (~100 nm; 300 and 3000 μg mL^{-1}) demonstrated reduced growth rates, decreased digestive enzyme activities (lipase, trypsin, and lysozyme), and shift in gut microbial communities—most notably increases in *Vibrio* and *Aliivibrio* (Wang et al. 2022). Such adverse effects can inevitably trigger stress response mechanisms, including oxidative stress, which is linked to the imbalance between the production of damaging reactive oxygen species (ROS) and the organism's ability to manage them (Prokić et al. 2019) (Fig. 6.3). The control of oxidative stress by the biochemical cellular antioxidant system is crucial for an organism's normal functioning. When ROS production exceeds the capacity of antioxidant defense mechanisms, there is a subsequent increase in ROS interactions within the internal environment, leading to oxidative damage to important biomolecules such as DNA, proteins, and lipids (Prokić et al. 2019) (Fig. 6.3). For example, a study reported that short-term (48 h) exposure to PE-MPs (50–70 mg L^{-1}) in the soft coral *Coelogorgia palmosa* led to an increase in oxidative stress, characterized by a significant rise in antioxidant enzyme activity (glutathione reductase, catalase, and superoxide dismutase), triggering cellular damage through lipid peroxidation (Montalbetti et al. 2022). Gardon et al. (2020b) also demonstrated that genes involved in the detoxification process were impacted by long-term (2 months) exposure to PS-MPs (0.25, 2.5, and 25 μg L^{-1}) in *P. margaritifera*, resulting in a decrease in antioxidant response functioning. The authors suggested dose-specific responses, mainly due to the energy-limited capacity of *P. margaritifera* to cope with prolonged stress, as exposed individuals were already impaired by a dose-dependent decrease in energy balance (Gardon et al. 2018). Moreover, MNPs can also exert external stress via passive adhesion, as demonstrated in the symbiotic dinoflagellate *Cladocopium goreaui* exposed to PS-MPs (1 μm) over a 7-day period (9.0×10^9 microbeads L^{-1}) (Su et al. 2020). Results revealed oxidative stress enrichments that suppressed the detoxification response of crucial metabolic enzymes (i.e., glutathione

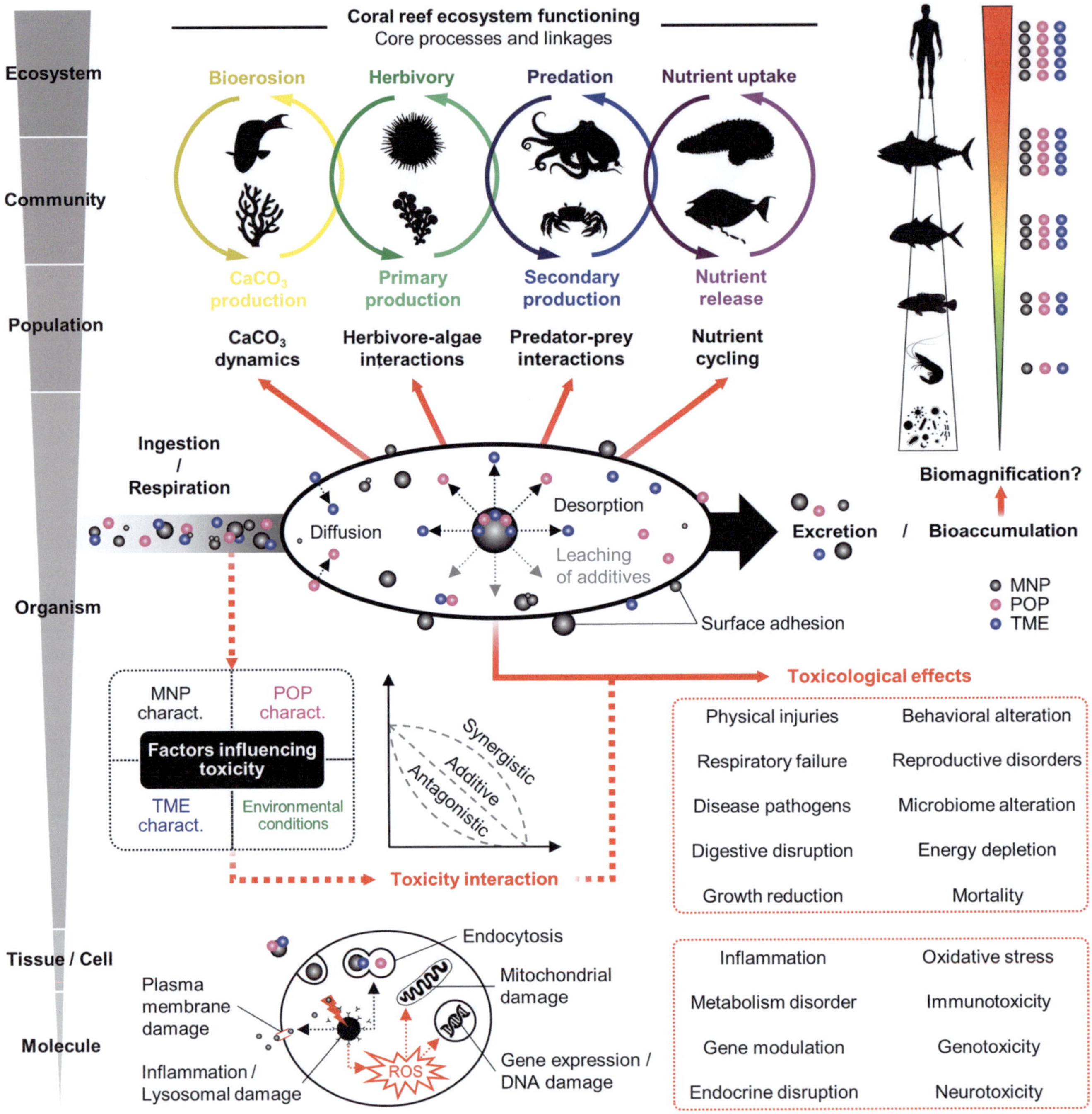

Fig. 6.3 Overview of the toxicological effects induced by MNPs, POPs, TMEs, and their mixtures at the individual, cellular, and molecular scales, and their potential to disrupt core processes and linkages underlying coral reef ecosystem functioning. The representation of coral reef ecosystem functioning was adapted from Brandl et al. (2019)

S-transferase), potentially elevating the apoptosis level of Symbiodiniaceae and leading to a decline in their growth (Su et al. 2020). In this way, Tang et al. (2021) suggested that MNPs are likely to disrupt the in situ host-symbiont relationships between the scleractinian coral *Pocillopora damicornis* and Symbiodiniaceae by inducing apoptosis in the coral. Furthermore, a laboratory experiment was conducted to explore the multiple impacts of MPs on the octocoral *Corallium rubrum*, closely mimicking natural exposure condition. This involved using MP mixtures selected to reflect the size, polymeric composition, and concentrations present in highly contaminated coastal waters over a 2-week period (Corinaldesi et al. 2021). This study highlighted, among other findings, that MPs (~20–1000 μm; 60, 300 and 600 MPs L^{-1}) directly damaged coral tissues, creating tears and incisions down to the skeleton and sclerites. The highest concentration resulted in more than 50% of the tissue surface being damaged (Corinaldesi et al. 2021). Several investigations have revealed that MNPs can cause abrasion, physical damage, and tissue necrosis, thereby promoting mucus overproduction, pathogen invasion, or immune system activation (reviewed in Huang et al. 2021). The biofilm on MNP surfaces may become enriched with foreign microbial communities, such as Vibrionaceae, Rhodobacteraceae, and Flavobacteraceae, which are pathogens known to cause coral tissue damage and bleaching (Feng et al. 2020). Consequently, the presence of MNPs in coral reef ecosystems is likely to exacerbate disease risk during coral recovery following stressful events and, more broadly, to hinder coral reef resilience. Moreover, in addition to affecting the first physical (i.e., tissue) and chemical (i.e., mucus) lines of defense that prevent disease, MNPs may induce adverse impacts on an organism's immune system, such as blood cell immunotoxicity, lysosomal membrane instability, or even a decrease in phagocytic activity (Zhang et al. 2022) (Fig. 6.3). Particle size is a critical determinant for such toxicological outcomes, especially with regard to NPs, as they can directly interact with cells (Fig. 6.3). Research on NPs is a very recent area in environmental and laboratory sciences, but they are expected to be more hazardous than MPs due to their size, which facilitates a wider and deeper distribution in body organs and tissues by passing through cell membranes (Fig. 6.3). Once translocated, NPs may enter the blood circulation and immune system through the gastrointestinal tract, gills, or superficial tissues, and eventually accumulate in various internal organs and tissues (Thornton Hampton et al. 2022).

Chemical impacts of MNPs may also arise from the release of plastic additives and toxic monomers (Fig. 6.3). Depending on the type of polymer, it has been reported that plastic may contain up to 60%, by weight, of additives such as polyvinyl chloride (PVC) (reviewed in Hermabessiere et al. 2017). Evidence of indirect MNP toxicity has been increasing in recent years and has been demonstrated through leaching experiments that primarily focus on early life stages. For example, Li et al. (2016a) used seven plastic bags of various polymer types, including high and low density PE (HDPE and LDPE), PP, PVC, PS, polyethylene terephthalate (PET), and polycarbonate (PC), to evaluate the impact of associated leachate (i.e., 24 h leaching of plastic pieces) on the survival and settlement of barnacle *Amphibalanus amphitrite* larvae. Settlement was inhibited by all plastic leachates (−30% to −42%), with significantly higher mortality (+11% to +22%) at the highest concentrations (0.1 and 0.5 m^2 L^{-1}), and PVC exhibiting the highest toxicity (Li et al. 2016a). Focusing on a specific coral reef species, Hamlin et al. (2015) demonstrated the effects of nonylphenol on the marine fish orchid dottyback (*Pseudochromis fridmani*) after exposure to leachates produced from two different PE plastic bags (PE1 and PE2). No significant mortality was observed following 48 h of exposure to PE1 leachate (11%), while it resulted in a 60% mortality in PE2 leachate. Hence, significantly higher concentrations in nonylphenol were measured in leachates from PE2 than PE1, with a mean of 41.0 and 2.5 μg L^{-1}, respectively (Hamlin et al. 2015). This result highlighted that an identical polymer type may result in drastically different toxicological outcomes. This is likely due to the diverging composition of plastic additives in both PE bags, the exact formulation of which remains frequently unknown and specific to the plastic manufacturer (Hamlin et al. 2015). In this study, the authors focused their chemical analyses on nonylphenol and did not consider the content of other additives by studying the overall composition of leachates. Yet, phthalates acid esters (PAEs) account for 95% of all plasticizers and are widely spread in the marine environment (Hermabessiere et al. 2017). Hazardous effects induced by PAEs have been widely suggested in toxicological research. For example, a study suggested that the difference in toxicity observed on *Pinctada margaritifera* embryo-larval development was due to higher levels of PAEs in leachates from new (6.7–9.1 μg L^{-1}), rather than weathered (0.4–0.5 μg L^{-1}), plastic gear used in pearl farming (Gardon et al. 2020a). Bioassays showed higher mortality at lower concentrations (0.1, 1, and 10 g L^{-1}) in leachates from new (26%–86%), rather than aged (17%–28%) plastics, while similar mortality (90%–100%) occurred at 100 g L^{-1} (Gardon et al. 2020a). In addition to assessing apical end points (i.e., feeding, mortality, growth, developmental, and reproductive toxicity), other research has documented molecular mechanisms triggered by additives [e.g., PAEs, BFRs, OPFRs, and bisphenol A (BPA)] and reported endocrine disrupting effects, immunotoxicity, or even neurotoxicity (Hermabessiere et al. 2017). Overall, these results suggest that over their lifetime, plastic materials may release significant amounts of additives into the marine environment, which may represent a serious threat to marine life.

6.2.3.2 Effects of POPs and TMEs

Although chemical pollutants are likely to enter organisms' bodies through diffusion, adsorption, or ingestion (Fig. 6.3), their occurrence in coral reef species remains unintentional and dependent on the surrounding environment. Several studies have documented that both POPs and TMEs induce adverse effects on reef organisms, ranging from metabolic disorders, interruption of neurological function, changes in gene expression, and carcinogenesis, to endocrine disruption and interference with reproduction and natural behavior (reviewed in Brodie et al. 2019). According to a recent systematic review on experimental studies that assessed the toxicity of multiple chemical pollutants to tropical reef-building corals, pesticides were the most investigated exposure category (Ouédraogo et al. 2023). For example, a laboratory experiment revealed that exposure to chlorpyrifos at an environmentally relevant concentration (i.e., 1 μg L^{-1}) impairs visual lateralization in the larval coral reef fish *Acanthurus triostegus* (Besson et al. 2017). The authors suggested that chlorpyrifos may induce brain dysfunctions in larval fishes during reef colonization (i.e., movement of larvae from planktonic to reef environments), which is a critical life-history transition, and may potentially impact recruitment success (Besson et al. 2017). Another study demonstrated that exposure to chlordecone (30 and 300 μg L^{-1}) induces polyp bailout in *Pocillopora damicornis*, resulting from the overexpression of transcripts involved in apoptosis with subsequent degradation of cellular matrix proteins leading to coenosarc dissociation and loss (Wecker et al. 2018). A number of field investigations have also documented the impacts of herbicides on coral reef species. Among the toxic effects observed in situ, Photosystem II (PSII) inhibitor herbicides (e.g., diuron, atrazine, simazine, and Irgarol 1051) exhibit phytotoxicity by inhibiting electron transport through the photosystem, thereby negatively impacting the photosynthetic capacity of coral symbionts that provide up to 90% of the coral's energy (Brodie et al. 2019). As a result, the phytotoxicity of PSII-inhibiting herbicides on zooxanthellae has repercussions on the coral host, whose energy acquisition may decrease, potentially leading to a long-term reduction in reproductive output (Sokolova 2013). However, such physiological effects are likely to be influenced by a broad range of POPs, given the observed toxic effects on photosynthetic performance and/or the density of zooxanthellae within corals. Ranjbar Jafarabadi et al. (2018) reported in situ levels of POPs and health assessment in scleractinian corals and their zooxanthellae in two distinct coral reefs of the Persian Gulf. Results indicated significantly negative correlations between the total concentrations of POPs, composed of PAHs (Σ_{40}), PCBs (Σ_{22}), and PCNs (Σ_{20}), and both the zooxanthellae density and concentrations of chlorophyll-a/-c2 in corals at both reefs (Ranjbar Jafarabadi et al. 2018). Toxic effects of hydrocarbons such as PAHs have also been widely investigated (Ouédraogo et al. 2023), especially in filter-feeding marine bivalves, considered both bioindicators and model organisms for research on environmental stressors. Subsequent experiments demonstrated that PAHs bioaccumulate in the soft tissues of oysters with sublethal effects, including increases in xenobiotic and detoxification responses, enzyme activities (e.g., cytochrome P450 enzymes and glutathione S-transferase), deterioration of DNA integrity and infection prevalence, as well as reductions in and/or abnormal larval growth (Gan et al. 2021). Many of these responses may be attributed to either oxidative damage or energy reallocation toward health maintenance to the detriment of vital biological processes (i.e., reproduction and calcification). Such metabolic disorder may be exacerbated by alterations in feeding, as reported in the milkfish *Chanos chanos,* when exposed to sublethal concentrations of anthracene (1.0, 2.0, 3.0, 6.0, and 12.0 μg L^{-1}) and benzo[a]pyrene (0.3, 0.7, 1.4, 2.8, and 5.6 μg L^{-1}), leading to a reduction in growth rate after a 28-day PAH exposure (Palanikumar et al. 2013). Among other chemical stressors, TMEs have received a great deal of attention to date, especially Cu, which can become very toxic, causing deleterious effects on coral reef biota when reaching excessive concentrations (Ouédraogo et al. 2023). A laboratory experiment on the coral *Acropora muricata* revealed that Cu exposure (36 h) to 32 and 65 μg L^{-1} not only resulted in bleaching but also altered the composition of both the eukaryote and bacterial communities of the coral's microbiome (Gissi et al. 2019). Several TMEs may also have the potential to affect nutrient cycling, cell growth, and regeneration, as well as reproductive cycles and photosynthetic potential (Fig. 6.3). A field investigation revealed that increasing concentrations of As, Cr, and Mn led to the inhibition of carbonic anhydrase (CA) activity, suggested as a stress response mechanism of coral symbionts (ROS production), thus increasing oxidative damage in the host coral *Millepora alcicornis* collected from Abrolhos Reef Banks in Brazil (Dal Pizzol et al. 2022). The same survey was conducted on the coral *Mussismilia harttii*, and no impairment of CA activity was observed, highlighting species-dependent responses and the potential implication of the associated antioxidant system efficiency in TME stress conditions (Dal Pizzol et al. 2022). Furthermore, some metals such as Hg have the potential to cause delayed effects due to progressive bioaccumulation in the tissues of marine organisms, resulting in an increase in concentration at each successive trophic level along the food chain (Fig. 6.3). For example, Hg can accumulate in various tissues, such as the liver, kidney, gonads, or even the muscle, leading to genetic, biochemical, physiological, morphological, and behavioral changes in marine fish (reviewed in Zheng et al. 2019). Such metallic contamination could also directly impact human health (Fig. 6.3), as fish constitute an essential food resource for small island populations (Gaertner et al. 2023).

Multiple POPs and TMEs can bioaccumulate and trigger similar responses in reef species through alternative pathways, causing sublethal effects and a reduction in organism fitness, especially in key reef organisms like reef-building corals. Coral bleaching, a common endpoint following exposure to both contaminants, has been demonstrated, which is considered a primary consequence of global warming and the coral response to elevated temperatures. These toxic contaminants can interfere with primary production and may exert a bottom-up pressure on coral reef ecosystems. This interference is likely to alter resilience processes to other environmental stressors due to cumulative or even synergistic toxic effects that can occur (Brown et al. 2013). This vulnerability raises questions regarding implications for interconnected coral reef communities and intrinsic drivers ensuring core processes of ecosystem functioning (Fig. 6.3).

6.2.3.3 Effects of MNPs Combined with Sorbed POPs and TMEs

Recent studies have raised concerns about the ability of MNPs to adsorb a wide range of environmental contaminants on their surfaces, posing an additional environmental issue for marine ecosystems (Rodrigues et al. 2019). As a result, MNPs can serve as vectors for lipophilic pollutants, including POPs and TMEs, potentially intensifying the overall toxicity of MNPs through additional indirect effects on contaminated biota (Figs. 6.2 and 6.3). In addition to addressing general exposure pathways for POPs and TMEs, such as seawater and sediments, it is crucial to consider the vector effect of MNPs and describe the combined effects of interactions between POPs, TMEs, and MNPs.

Only a limited number of publications have analyzed the toxicity of TME–MNP interactions on reef species, whereas POPs are the group of chemicals that has been most extensively studied in terms of their interaction toxicity with MNPs (Menéndez-Pedriza and Jaumot 2020). For instance, the immunotoxicity of MPs (PS microbeads) and two POPs (benzo[a]pyrene and 17β-estradiol) was investigated both individually and in combination in the blood clam *Tegillarca granosa* (Tang et al. 2020). The results indicated that POP toxicity was generally exacerbated by smaller MPs (0.5 μm) and alleviated by larger MPs (30 μm), suggesting a size-dependent effect on POP toxicity arising from size-dependent interactions between MPs and POPs (Tang et al. 2020). On the contrary, another experiment revealed that MPs (PE micronized powder) reduced the toxicity induced by chlorpyrifos alone (2–3 mg L^{-1}) on microalgae growth (*Isochrysis galbana* clone T-Iso) because the microplastics enriched with chlorpyrifos were not small enough (mean size of 2–6 μm) to penetrate the microalgal cell wall and cause any damage (Garrido et al. 2019). Similarly, Guven et al. (2018) investigated the individual and interactive effects of MPs (PS-divinylbenzene microbeads of 97 μm) and pyrene (PAH) on the swimming behavior and predatory performance of juvenile barramundi (*L. calcarifer*). The authors found that the combined exposure to pyrene and MPs (0.1 μM pyrene +100 microplastics L^{-1}) did not amplify the acute single impact of PAH (0.1 μM pyrene). Considering the outcomes of toxicological studies assessing interactions between POPs, TMEs, and MNPs, both synergistic and antagonistic effects have been demonstrated on marine biota, suggested to be influenced by several factors related to experimental procedures (e.g., microplastic sizes, targeted species, exposure times, and concentration levels) (Menéndez-Pedriza and Jaumot 2020). The interplay between these three CECs presents a complex and multifaceted challenge, requiring comprehensive research and holistic approaches to fully understand their combined effects on coral reefs and associated species.

6.3 Implications for Coral Reef Conservation: Research Gaps and New Prospects

A comprehensive understanding of the risks and impacts posed by MNPs, POPs, and TMEs on coral reefs is crucial for the development of effective management strategies to protect and conserve these vulnerable ecosystems. This understanding is essential to safeguard their long-term health and resilience in the face of environmental challenges. While extensive research has been conducted on some of these contaminants, several challenges and research gaps persist in specific and overlapping areas, including environmental science, ecology, and toxicology field investigations (Fig. 6.4).

6.3.1 Ecologically Relevant Data to Inform Effective Research Strategies and Decision-Making

Collectively, ecological data provide evidence that the intensification of human activities, including wastewater discharges, agricultural runoff, and other pressures, may be the primary factor influencing the presence of CECs in coral reef regions. However, further research is required to gain a better understanding of contamination sources and pathways, spatial and temporal variabilities, degradation and transformation processes, as well as bioaccumulation and biomagnification processes. Enhancing monitoring techniques and strategies, broadening sampling coverage across different environmental levels, and developing standardized methods for data collection and analysis are key paths to address research gaps (Fig. 6.4). This is particularly crucial for MNP monitoring, which lacks standardization for sampling, extraction and analysis methods, as well as reported units, thereby limiting accurate comparisons between sites and organisms (Lv et al. 2021). The establishment of standardized protocols is an essential step toward enabling meaningful comparisons and reliable assessment of MNP pollution levels. Furthermore, many methods for isolating and identifying MNPs require laboratory-based treatments and analyses. This limitation hinders real-time field assessments and increases the risk of sample contamination, which continues to pose significant challenges (e.g., from airborne particles, synthetic fibers from clothing, or plastic-based equipment). Innovative portable and field-deployable analytical techniques, including new sensors, are expected to significantly improve the precision of on-site identification and quantification of MNPs (Blevins et al. 2021) (Fig. 6.4). Currently, the analysis of the MNP chain is a laborious and time-intensive process, which is susceptible to biases such as the limited capability to analyze particles with sizes ≤10 μm, or

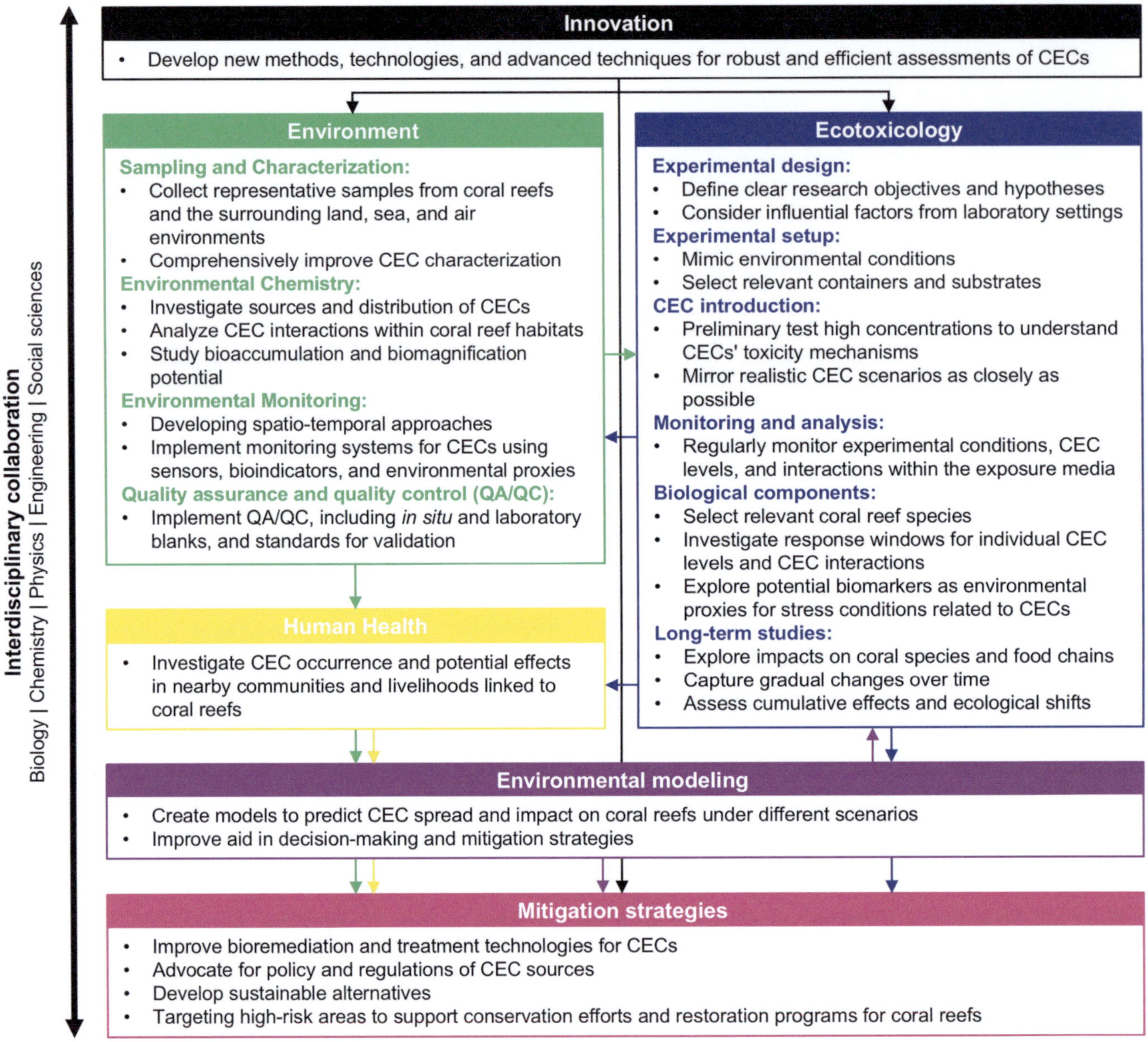

Fig. 6.4 Paradigms to assess the vulnerability of coral reefs to contaminants of emerging concern (CECs)

even the extrapolation of polymer identification based on subsampling, which may not accurately represent the full particle population. Furthermore, due to methodological and technical constraints, subsampling typically focuses on larger particles, leading to a margin of error influenced by the proportion of particles included in the subsample (Paul-Pont et al. 2018; Schwaferts et al. 2019). Regarding POPs, long-term monitoring studies are imperative to assess their spatial and temporal trends, particularly concerning the mechanisms of long-range transport to coral reefs and their interactions with various environmental matrices (e.g., air, water, and sediments) (van der Schyff et al. 2021) (Fig. 6.4). Sampling strategies that consider the potential sources and transport pathways of POPs could be developed. This would involve selecting study sites both proximate to coral reefs and in areas where the transport of POPs from distant sources is likely to occur, such as downstream or downwind locations. Additionally, the atmosphere significantly contributes to the long-range transport of POPs (Kallenborn et al. 2015). Therefore, it would be ecologically relevant to combine seawater and sediment sampling with high-volume air samplers or passive air sampling techniques (e.g., polyurethane foam or diffusive samplers) to collect air samples. Subsequently, source fingerprinting techniques could be used to compare the chemical composition and profiles of POPs in the sampled reefs with those from suspected source regions or pollution hotspots. Other modeling techniques could also be developed and employed to predict the transport and dispersion of POPs over long distances and to estimate their deposition patterns in reef ecosystems, such as the multimedia mass-balance model (Guardo et al. 2018) (Fig. 6.4). Furthermore, there are existing research gaps concerning TMEs, and an urgent need exists to focus specific attention on their intricate transformation mechanisms (e.g., chemical reactions or interactions with other compounds) that influence their toxicity and environmental behavior (Miranda et al. 2021). Research is therefore necessary to develop speciation analysis techniques to identify and quantify different TMEs, as this information is critical for evaluating their bioavailability, transformation, and potential risks. Adopting integrated multimedia monitoring approaches (i.e., encompassing air, water, soil, sediment, and biota) might offer a comprehensive understanding of TME contamination pathways, their bioavailability, and potential impacts on ecosystem functioning and human health (Hao et al. 2022). Bioaccumulation and biomagnification processes raise public health concerns as humans are the ultimate consumers (Fig. 6.3). This concern is particularly significant for populations living in remote territories, such as the numerous small tropical Pacific islands, where the population's diet (specifically protein intake) relies heavily on seafood resources, and where ocean contamination has emerged as a challenge for food security (Gaertner et al. 2023). For instance, substantial levels of Hg (~11.5 $\mu g\ L^{-1}$) were detected in the blood of human populations located in Hao and Makemo (St-Jean et al. 2019), two insular atolls of French Polynesia where local fish consumption (especially pelagic fish-top predators such as tuna and marlin) can reach up to 146 kg per inhabitant per year (Bell et al. 2011). The Hg levels reported by St-Jean et al. among adolescent and adult populations were approximately 20-fold higher than those recently documented in Canada and the United States. Investigating the potential for bioaccumulation and biomagnification within food webs, as well as the interactions among MNPs, POPs, and TMEs, is therefore essential for comprehending the extent of pollution and predicting potential impacts on living organisms under controlled laboratory conditions (Fig. 6.4).

6.3.2 Achieving Realistic Laboratory Conditions for Relevant Toxicological Knowledge

To date, substantial research gaps persist regarding the potential impacts of CECs under laboratory settings. This is particularly evident in toxicological studies involving model taxa that are phylogenetically distant from the organisms monitored in the field. In fact, there is a lack of effective bioindicator organisms that can accurately reflect local contamination and serve as model species in laboratory experiments to establish coherent and appropriate risk assessment strategies across coral reef regions (Zhang et al. 2022) (Fig. 6.4). Numerous studies have highlighted the suitability of filter-feeding bivalves as bioindicators of environmental stressors in marine ecosystems. These organisms possess a high capacity for contaminant bioaccumulation, limited mobility, wide distribution, and close interactions with marine predators and human health (Gan et al. 2021). Marine bivalves can thus serve as effective tools for monitoring spatial and temporal variations in contaminant levels, as well as assessing corresponding ecological and physiological impacts. On the other hand, corals play a crucial role in reef-building and maintain complex symbiotic relationships with zooxanthellae. This unique aspect makes corals essential organisms, forming a pivotal link in the biological equilibrium of both coral reef ecosystems and oceans. Therefore, understanding the complexity of symbiotic partnerships that govern coral health is of significant interest. As such, a scientific consortium could focus on selecting representative sentinel species that are phylogenetically close across various coral reef regions, employing epidemiological approaches along anthropogenic gradients. An emblematic organism like the photosymbiotic bivalve of the genus *Tridacna* sp. offers an additional opportunity for investigating host–symbiont relationships and serves as a bioindicator for environmental surveys within coral reef ecosystems (Gaertner et al. 2023). Additionally, macroalgae also emerge as compelling bioindicators due to their ease of collection, abundance, widespread distribution across coral reef regions, and their propensity for high bioaccumulation. For instance, taxa such as *Dictyota* sp. hold promise as valuable bioindicators (Contarini and Dromard 2021). By combining the deployment of suitable sentinel species in their natural habitat with a robust and continuous monitoring of their surrounding environment, it is possible to explore correlation patterns between recorded stressors (and their levels) and effective biomarker responses. Drawing from laboratory experiments, these ecological biomarkers can complement the toxicological outcomes observed in model species exposed to single and combined contaminants at realistic concentrations (Fig. 6.4). This strategy aims to unravel the specific, synergistic, and/or antagonistic effects induced by multiple stressors. To achieve this, such integrative approaches necessitate the meticulous evaluation of environmental parameters and the implementation of tightly controlled laboratory conditions, facilitating a comprehensive understanding of toxicological mechanisms. However, uncertainties persist regarding the limited field evaluation of MNPs <10 μm (Schwaferts et al. 2019). These missing data limit our knowledge of the actual exposure levels that reef organisms are facing and constrain our ability to replicate environmental scenarios in laboratory conditions as closely as possible. At present, the majority of laboratory experiments centered around MNPs employ unrealistic scenarios, particularly concerning concentrations that are orders of magnitude higher (in terms of particle count per unit volume) than those encountered in situ (Lenz et al. 2016). This gap arises partly from an extrapolation of estimated mass concentrations primarily based on particles >335 μm, corresponding to the standardized mesh size employed in surface water monitoring methods (i.e., manta trawls). In contrast, the MNPs used in previous toxicological investigations were substantially smaller in size. This research gap gives rise to crucial questions about the relevance of the tested concentrations, considering that small particles contribute minimally in terms of volume/weight but significantly in terms of particle count, thus influencing interactions with organisms (Lenz et al. 2016). These unrealistic conditions are further compounded by the characteristics of the MNPs used, which predominantly consist of single polymer types with a spherical shape (i.e., beads considered as model particles) and specific sizes. These characteristics are far short of capturing the real complexity of environmental MNPs, which encompass various polymer types, particle shapes, sizes, surface properties, chemical compositions, biofilm presence, and more (Rochman et al. 2019). Nevertheless, studies that involve controlled parameters with limited MNP characteristics represent a crucial step in understanding the factors influencing toxicological kinetics (Paul-Pont et al. 2018) (Fig. 6.4). Consequently, although some of these laboratory experiments may involve constrained MNP features, they provide valuable data regarding the potential risks that MNPs could pose to coral reefs, anticipating more comprehensive field information. Given that the topic of "microplastics" is an ever-evolving area, this information holds significance for the progression of our future knowledge. Furthermore, in the absence of effective waste management to address the accumulation of thousands of tons of plastic waste in the marine environment over the years, the cumulative quantity of MNPs is projected to increase in the coming years due to ongoing aging and degradation processes (Lebreton et al. 2019). Nonetheless, scientists are now advocating for a proper and unambiguous methodology to study the potential impact of MNPs on marine organisms, aiming to conduct ecotoxicological tests that are as controlled and accurate as possible (Paul-Pont et al. 2018; Gardon et al. 2022) (Fig. 6.4). On the other hand, while the toxic effects of certain POPs and TMEs are well-established, a deeper understanding of their underlying mechanisms of toxicity is needed, including those induced by MNPs. Unraveling the specific molecular and cellular pathways through which these contaminants exert their adverse effects can provide valuable insights into their toxicity mechanisms and guide the development of targeted interventions (Fig. 6.4). Traditional toxicology studies often employ high-dose exposures to establish dose–response relationships in individual organisms. Consequently, future research should prioritize investigating the long-term impacts of low-dose exposures (i.e., chronic exposure) to gain valuable insights into the cumulative and delayed effects of these contaminants on population, community, and ecosystem levels. Furthermore, understanding the combined effects and potential synergistic interactions of MNPs, POPs, and TMEs is vital to fully grasp the spectrum of their toxicological impacts and to assess the risks associated with complex mixtures for vulnerable species. This includes the need to explore differential toxicity and susceptibility among vulnerable populations and to target specific toxicity thresholds (Fig. 6.4). For instance, Ouédraogo et al. (2023) compared the toxicity thresholds of POPs and TMEs, as reviewed in the literature, with the reference values established by environmental agencies for the protection of marine organisms. These reference values proved to be protective for corals with respect to most contaminants, except for three, which included the metal copper and the pesticides diuron and Irgarol 1051 (Ouédraogo et al. 2023). This outcome underscores the importance of understanding specific risks and formulating focused risk assessment approaches. This becomes particularly pertinent when considering emerging POPs resulting from industrial activities, technological advancements, or environmental changes (Fiedler et al. 2019). Thus, maintaining continuous monitoring and conducting ongoing toxicological studies are essential to ensure timely regulatory action and effective risk management.

Overall, these research gaps underscore the necessity for comprehensive monitoring and surveillance programs aimed at enhancing our understanding of the fate and transport mechanisms of MNPs, POPs, and TMEs within coral reef ecosystems, as well as their potential synergistic effects. Furthermore, addressing these challenges requires interdisciplinary collaboration, the establishment of monitoring networks, and organizations dedicated to these contaminants to share data, methodologies, and expertise. By closing these gaps, scientists, stakeholders, and decision-makers can make more effective contributions to risk assessment, management strategies, and regulatory measures aimed at safeguarding both human health and the health of coral reef ecosystems from the adverse impacts of these contaminants (Fig. 6.4).

6.4 Conclusion

This review of existing literature underscores the extensive contamination of coral reef ecosystems by MNPs, POPs, and TMEs, while also highlighting that this contamination is likely to be nonuniform in terms of distribution, concentrations, characteristics, and overlap with regions of high biodiversity. It is imperative that continuous research efforts and collaborative engagement among scientists, policymakers, and stakeholders be pursued to address current gaps in knowledge and to establish a more comprehensive understanding of the vulnerability of coral reefs to these CECs. To predict and mitigate the potential loss of crucial ecosystem services, conservation strategies must encompass an assessment of the nature and extent of anthropogenic disturbances acting as local stressors within the broader context of global climate change. Such an approach will provide insights into the spatiotemporal variability of coral reef structure and function across environmental gradients, both sheltered and exposed to specific anthropogenic disturbances. By identifying local stressors and potential synergisms between MNPs, POPs, and TMEs, effective management strategies can be prioritized to mitigate the most severe interactions. A comprehensive grasp of the abundance, distribution, fate, and interactions of these CECs is urgently required to proactively anticipate and address the associated risks to ecosystem structures, functions, and trophic transfer dynamics within coral reefs.

References

Ajith N, Arumugam S, Parthasarathy S, Manupoori S, Janakiraman S (2020) Global distribution of microplastics and its impact on marine environment—a review. Environ Sci Pollut Res 27:25970–25986. https://doi.org/10.1007/s11356-020-09015-5

Andrady AL (2011) Microplastics in the marine environment. Mar Pollut Bull 62:1596–1605. https://doi.org/10.1016/j.marpolbul.2011.05.030

Bell JD, Johnson JE, Ganachaud PC, Gehrke PC, Hobday AJ, Hoegh-Guldberg O, Le Borgne R, Lehoday P, Lough JM, Pickering T, Pratchett MS, Waycott M (2011) Vulnerability of tropical pacific fisheries and aquaculture to climate change: summary for Pacific Island countries and territories, Johann D. Bell Johanna E. Johnson and Alistair J. Hobday. Secretariat of the Pacific Community, Noumea

Besson M, Gache C, Bertucci F, Brooker RM, Roux N, Jacob H, Berthe C, Sovrano VA, Dixson DL, Lecchini D (2017) Exposure to agricultural pesticide impairs visual lateralization in a larval coral reef fish. Sci Rep 7:9165. https://doi.org/10.1038/s41598-017-09381-0

Blevins MG, Allen HL, Colson BC, Cook A-M, Greenbaum AZ, Hemami SS, Hollmann J, Kim E, LaRocca AA, Markoski KA, Miraglia P, Mott VL, Robberson WM, Santos JA, Sprachman MM, Swierk P, Tate S, Witinski MF, Kratchman LB, Michel APM (2021) Field-portable microplastic sensing in aqueous environments: a perspective on emerging techniques. Sensors 21:3532. https://doi.org/10.3390/s21103532

Brandl SJ, Rasher DB, Côté IM, Casey JM, Darling ES, Lefcheck JS, Duffy JE (2019) Coral reef ecosystem functioning: eight core processes and the role of biodiversity. Front Ecol Environ 17:445–454. https://doi.org/10.1002/fee.2088

Briand MJ, Bustamante P, Bonnet X, Churlaud C, Letourneur Y (2018) Tracking trace elements into complex coral reef trophic networks. Sci Total Environ 612:1091–1104. https://doi.org/10.1016/j.scitotenv.2017.08.257

Brodie J, Johnson J, Waterhouse J, Erdmann S (2019) Wastewater pollution on coral reefs: supporting science. C2O (Coasts Climate Oceans) https://oursharedseas.com/wp-content/uploads/2020/08/UNEP_2019_Wastewater-pollution-and-coral-reefs.pdf

Brown CJ, Saunders MI, Possingham HP, Richardson AJ (2013) Managing for interactions between local and global stressors of ecosystems. PLoS One 8:e65765. https://doi.org/10.1371/journal.pone.0065765

Bucci K, Rochman CM (2022) Microplastics: a multidimensional contaminant requires a multidimensional framework for assessing risk. Microplast Nanoplast 2:7. https://doi.org/10.1186/s43591-022-00028-0

Connors EJ (2017) Distribution and biological implications of plastic pollution on the fringing reef of Mo'orea, French Polynesia. PeerJ 5:e3733. https://doi.org/10.7717/peerj.3733

Contarini P-E, Dromard CR (2021) Biosorption capacity of genus *Dictyota* facing organochlorine pesticide pollutions in coastal areas of the Lesser Antilles. Aquat Bot 169:103346. https://doi.org/10.1016/j.aquabot.2020.103346

Corinaldesi C, Canensi S, Dell'Anno A, Tangherlini M, Di Capua I, Varrella S, Willis TJ, Cerrano C, Danovaro R (2021) Multiple impacts of microplastics can threaten marine habitat-forming species. Commun Biol 4:1–13. https://doi.org/10.1038/s42003-021-01961-1

Corona E, Martin C, Marasco R, Duarte CM (2020) Passive and active removal of marine microplastics by a mushroom coral (*Danafungia scruposa*). Front Mar Sci 7. https://doi.org/10.3389/fmars.2020.00128

Crusot M, Gaertner J-C, Rodriguez T, Lo C, Gaertner-Mazouni N (2023) Assessment of plastic waste generated by the aquaculture industry: the case study of pearl farming in French Polynesia. J Clean Prod 425:138902. https://doi.org/10.1016/j.jclepro.2023.138902

Dal Pizzol JL, Marques JA, da Silva FJ, Costa PG, Bianchini A (2022) Metal accumulation induces oxidative stress and alters carbonic anhydrase activity in corals and symbionts from the largest reef complex in the South Atlantic Ocean. Chemosphere 290:133216. https://doi.org/10.1016/j.chemosphere.2021.133216

Ding J, Jiang F, Li J, Wang Z, Sun C, Wang Z, Fu L, Ding NX, He C (2019) Microplastics in the coral reef systems from Xisha Islands of South China Sea. Environ Sci Technol 53:8036–8046. https://doi.org/10.1021/acs.est.9b01452

Donovan MK, Adam TC, Shantz AA, Speare KE, Munsterman KS, Rice MM, Schmitt RJ, Holbrook SJ, Burkepile DE (2020) Nitrogen pollution interacts with heat stress to increase coral bleaching across the seascape. Proc Natl Acad Sci 117:5351–5357. https://doi.org/10.1073/pnas.1915395117

Dromard CR, Bodiguel X, Lemoine S, Bouchon-Navaro Y, Reynal L, Thouard E, Bouchon C (2016) Assessment of the contamination of marine fauna by chlordecone in Guadeloupe and Martinique (Lesser Antilles). Environ Sci Pollut Res 23:73–80. https://doi.org/10.1007/s11356-015-4732-z

Dromard CR, Bouchon-Navaro Y, Cordonnier S, Guéné M, Harmelin-Vivien M, Bouchon C (2018) Different transfer pathways of an organochlorine pesticide across marine tropical food webs assessed with stable isotope analysis. PLoS One 13:e0191335. https://doi.org/10.1371/journal.pone.0191335

Eriksen M, Lebreton LCM, Carson HS, Thiel M, Moore CJ, Borerro JC, Galgani F, Ryan PG, Reisser J (2014) Plastic pollution in the world's oceans: more than 5 trillion plastic pieces weighing over 250,000 tons afloat at sea. PLoS One 9:e111913. https://doi.org/10.1371/journal.pone.0111913

Feng L, He L, Jiang S, Chen J, Zhou C, Qian Z-J, Hong P, Sun S, Li C (2020) Investigating the composition and distribution of microplastics surface biofilms in coral areas. Chemosphere 252:126565. https://doi.org/10.1016/j.chemosphere.2020.126565

Fey P, Bustamante P, Bosserelle P, Espiau B, Malau A, Mercader M, Wafo E, Letourneur Y (2019) Does trophic level drive organic and metallic contamination in coral reef organisms? Sci Total Environ 667:208–221. https://doi.org/10.1016/j.scitotenv.2019.02.311

Fiedler H, Kallenborn R, de Boer J, Sydnes LK (2019) The Stockholm Convention: a tool for the global regulation of persistent organic pollutants. Chem Int 41:4–11. https://doi.org/10.1515/ci-2019-0202

Fu Z, Wang J (2019) Current practices and future perspectives of microplastic pollution in freshwater ecosystems in China. Sci Total Environ 691:697–712. https://doi.org/10.1016/j.scitotenv.2019.07.167

Fu L, Li J, Wang G, Luan Y, Dai W (2021) Adsorption behavior of organic pollutants on microplastics. Ecotoxicol Environ Saf 217:112207. https://doi.org/10.1016/j.ecoenv.2021.112207

Gaertner J-C, Markic A, Godéré I, Belamy T, Gaertner-Mazouni N (2023) Ocean contamination: a challenge for food security in small and remote tropical islands. In: Brown P, Gaertner-Mazouni N (eds) Small islands, big issues. Peter Lang edition, Oxford, pp 19–36, 435

Gan N, Martin L, Xu W (2021) Impact of polycyclic aromatic hydrocarbon accumulation on oyster health. Front Physiol 12:734463

Gardon T, Reisser C, Soyez C, Quillien V, Le Moullac G (2018) Microplastics affect energy balance and gametogenesis in the pearl oyster *Pinctada margaritifera*. Environ Sci Technol 52:5277–5286. https://doi.org/10.1021/acs.est.8b00168

Gardon T, Huvet A, Paul-Pont I, Cassone A-L, Sham Koua M, Soyez C, Jezequel R, Receveur J, Le Moullac G (2020a) Toxic effects of leachates from plastic pearl-farming gear on embryo-larval development in the pearl oyster *Pinctada margaritifera*. Water Res 179:115890. https://doi.org/10.1016/j.watres.2020.115890

Gardon T, Morvan L, Huvet A, Quillien V, Soyez C, Le Moullac G, Le Luyer J (2020b) Microplastics induce dose-specific transcriptomic disruptions in energy metabolism and immunity of the pearl oyster *Pinctada margaritifera*. Environ Pollut 266:115180. https://doi.org/10.1016/j.envpol.2020.115180

Gardon T, El Rakwe M, Paul-Pont I, Le Luyer J, Thomas L, Prado E, Boukerma K, Cassone A-L, Quillien V, Soyez C, Costes L, Crusot M, Dreanno C, Le Moullac G, Huvet A (2021) Microplastics contamination in pearl-farming lagoons of French Polynesia. J Hazard Mater 419:126396. https://doi.org/10.1016/j.jhazmat.2021.126396

Gardon T, Paul-Pont I, Le Moullac G, Soyez C, Lagarde F, Huvet A (2022) Cryogrinding and sieving techniques as challenges towards producing controlled size range microplastics for relevant ecotoxicological tests. Environ Pollut 315:120383. https://doi.org/10.1016/j.envpol.2022.120383

Gardon T, Le Luyer J, Le Moullac G, Soyez C, Lagarde F, Dehaut A, Paul-Pont I, Huvet A (2024) Pearl farming micro-nanoplastics affect oyster physiology and pearl quality. Environ Sci Technol 58:207–218. https://doi.org/10.1021/acs.est.3c06684

Garnier Y, Jacob H, Guerra AS, Bertucci F, Lecchini D (2019) Evaluation of microplastic ingestion by tropical fish from Moorea Island, French Polynesia. Mar Pollut Bull 140:165–170. https://doi.org/10.1016/j.marpolbul.2019.01.038

Garrido S, Linares M, Campillo JA, Albentosa M (2019) Effect of microplastics on the toxicity of chlorpyrifos to the microalgae *Isochrysis galbana*, clone t-ISO. Ecotoxicol Environ Saf 173:103–109. https://doi.org/10.1016/j.ecoenv.2019.02.020

Geyer R (2020) Chapter 2 – Production, use, and fate of synthetic polymers. In: Letcher TM (ed) Plastic waste and recycling. Academic Press, Amsterdam, pp 13–32

Gilman SE, Urban MC, Tewksbury J, Gilchrist GW, Holt RD (2010) A framework for community interactions under climate change. Trends Ecol Evol 25:325–331. https://doi.org/10.1016/j.tree.2010.03.002

Gissi F, Reichelt-Brushett AJ, Chariton AA, Stauber JL, Greenfield P, Humphrey C, Salmon M, Stephenson SA, Cresswell T, Jolley DF (2019) The effect of dissolved nickel and copper on the adult coral *Acropora muricata* and its microbiome. Environ Pollut 250:792–806. https://doi.org/10.1016/j.envpol.2019.04.030

Godéré I, Gaertner J-C, Dassié EP, Belamy T, Maihota N, Baudrimont M, Gaertner-Mazouni N (2023) Metallic trace element contamination of the giant clam *Tridacna maxima* in French Polynesia. Mar Pollut Bull 196:115639. https://doi.org/10.1016/j.marpolbul.2023.115639

Guardo AD, Gouin T, MacLeod M, Scheringer M (2018) Environmental fate and exposure models: advances and challenges in 21st century chemical risk assessment. Environ Sci Process Impacts 20:58–71. https://doi.org/10.1039/C7EM00568G

Guven O, Bach L, Munk P, Dinh KV, Mariani P, Nielsen TG (2018) Microplastic does not magnify the acute effect of PAH pyrene on predatory performance of a tropical fish (*Lates calcarifer*). Aquat Toxicol 198:287–293. https://doi.org/10.1016/j.aquatox.2018.03.011

Hamlin HJ, Marciano K, Downs CA (2015) Migration of nonylphenol from food-grade plastic is toxic to the coral reef fish species *Pseudochromis fridmani*. Chemosphere 139:223–228. https://doi.org/10.1016/j.chemosphere.2015.06.032

Hao M, Zuo Q, Li J, Shi S, Li B, Zhao X (2022) A comprehensive exploration on distribution, risk assessment, and source quantification of heavy metals in the multi-media environment from Shaying River Basin, China. Ecotoxicol Environ Saf 231:113190. https://doi.org/10.1016/j.ecoenv.2022.113190

Hartmann NB, Hüffer T, Thompson RC, Hassellöv M, Verschoor A, Daugaard AE, Rist S, Karlsson T, Brennholt N, Cole M, Herrling MP, Hess MC, Ivleva NP, Lusher AL, Wagner M (2019) Are we speaking the same language? Recommendations for a definition and categorization framework for plastic debris. Environ Sci Technol 53:1039–1047. https://doi.org/10.1021/acs.est.8b05297

Hermabessiere L, Dehaut A, Paul-Pont I, Lacroix C, Jezequel R, Soudant P, Duflos G (2017) Occurrence and effects of plastic additives on marine environments and organisms: a review. Chemosphere 182:781–793. https://doi.org/10.1016/j.chemosphere.2017.05.096

Huang W, Chen M, Song B, Deng J, Shen M, Chen Q, Zeng G, Liang J (2021) Microplastics in the coral reefs and their potential impacts on corals: a mini-review. Sci Total Environ 762:143112. https://doi.org/10.1016/j.scitotenv.2020.143112

Hughes TP, Barnes ML, Bellwood DR, Cinner JE, Cumming GS, Jackson JBC, Kleypas J, van de Leemput IA, Lough JM, Morrison TH, Palumbi SR, van Nes EH, Scheffer M (2017) Coral reefs in the Anthropocene. Nature 546:82–90. https://doi.org/10.1038/nature22901

Jambeck JR, Geyer R, Wilcox C, Siegler TR, Perryman M, Andrady A, Narayan R, Law KL (2015) Plastic waste inputs from land into the ocean. Science 347:768–771. https://doi.org/10.1126/science.1260352

Jensen LH, Motti CA, Garm AL, Tonin H, Kroon FJ (2019) Sources, distribution and fate of microfibres on the Great Barrier Reef, Australia. Sci Rep 9:9021. https://doi.org/10.1038/s41598-019-45340-7

Jeong J, Choi J (2019) Adverse outcome pathways potentially related to hazard identification of microplastics based on toxicity mechanisms. Chemosphere 231:249–255. https://doi.org/10.1016/j.chemosphere.2019.05.003

Jones KC (2021) Persistent organic pollutants (POPs) and related chemicals in the global environment: some personal reflections. Environ Sci Technol 55:9400–9412. https://doi.org/10.1021/acs.est.0c08093

Kallenborn R, Hung H, Brorström-Lundén E (2015) Chapter 13 – Atmospheric long-range transport of persistent organic pollutants (POPs) into polar regions. In: Zeng EY (ed) Comprehensive analytical chemistry. Elsevier, Amsterdam, pp 411–432

Karkanorachaki K, Syranidou E, Kalogerakis N (2021) Sinking characteristics of microplastics in the marine environment. Sci Total Environ 793:148526. https://doi.org/10.1016/j.scitotenv.2021.148526

Koelmans AA, Bakir A, Burton GA, Janssen CR (2016) Microplastic as a vector for chemicals in the aquatic environment: critical review and model-supported reinterpretation of empirical studies. Environ Sci Technol 50:3315–3326. https://doi.org/10.1021/acs.est.5b06069

Kroon FJ, Berry KLE, Brinkman DL, Kookana R, Leusch FDL, Melvin SD, Neale PA, Negri AP, Puotinen M, Tsang JJ, van de Merwe JP, Williams M (2020) Sources, presence and potential effects of contaminants of emerging concern in the marine environments of the Great Barrier Reef and Torres Strait, Australia. Sci Total Environ 719:135140. https://doi.org/10.1016/j.scitotenv.2019.135140

Lamb JB, Willis BL, Fiorenza EA, Couch CS, Howard R, Rader DN, True JD, Kelly LA, Ahmad A, Jompa J, Harvell CD (2018) Plastic waste associated with disease on coral reefs. Science 359:460–462. https://doi.org/10.1126/science.aar3320

Lambert S, Sinclair C, Boxall A (2014) Occurrence, degradation, and effect of polymer-based materials in the environment. In: Whitacre DM (ed) Reviews of environmental contamination and toxicology, vol 227. Springer International Publishing, Cham, pp 1–53

Lebreton LCM, van der Zwet J, Damsteeg J-W, Slat B, Andrady A, Reisser J (2017) River plastic emissions to the world's oceans. Nat Commun 8:15611. https://doi.org/10.1038/ncomms15611

Lebreton L, Slat B, Ferrari F, Sainte-Rose B, Aitken J, Marthouse R, Hajbane S, Cunsolo S, Schwarz A, Levivier A, Noble K, Debeljak P, Maral H, Schoeneich-Argent R, Brambini R, Reisser J (2018) Evidence that the Great Pacific Garbage Patch is rapidly accumulating plastic. Sci Rep 8:4666. https://doi.org/10.1038/s41598-018-22939-w

Lebreton L, Egger M, Slat B (2019) A global mass budget for positively buoyant macroplastic debris in the ocean. Sci Rep 9:12922. https://doi.org/10.1038/s41598-019-49413-5

Lenz R, Enders K, Nielsen TG (2016) Microplastic exposure studies should be environmentally realistic. Proc Natl Acad Sci 113:E4121–E4122. https://doi.org/10.1073/pnas.1606615113

Li H-X, Getzinger GJ, Ferguson PL, Orihuela B, Zhu M, Rittschof D (2016a) Effects of toxic leachate from commercial plastics on larval survival and settlement of the barnacle *Amphibalanus amphitrite*. Environ Sci Technol 50:924–931. https://doi.org/10.1021/acs.est.5b02781

Li WC, Tse HF, Fok L (2016b) Plastic waste in the marine environment: a review of sources, occurrence and effects. Sci Total Environ 566–567:333–349. https://doi.org/10.1016/j.scitotenv.2016.05.084

Li L, Chen C, Li D, Breivik K, Abbasi G, Li Y-F (2023) What do we know about the production and release of persistent organic pollutants in the global environment? Environ Sci Adv 2:55–68. https://doi.org/10.1039/D2VA00145D

Liu J, Liang J, Ding J, Zhang G, Zeng X, Yang Q, Zhu B, Gao W (2021) Microfiber pollution: an ongoing major environmental issue related to the sustainable development of textile and clothing industry. Environ Dev Sustain 23:11240–11256. https://doi.org/10.1007/s10668-020-01173-3

Lv L, Yan X, Feng L, Jiang S, Lu Z, Xie H, Sun S, Chen J, Li C (2021) Challenge for the detection of microplastics in the environment. Water Environ Res Res Publ Water Environ Fed 93:5–15. https://doi.org/10.1002/wer.1281

Markic A, Niemand C, Bridson JH, Mazouni-Gaertner N, Gaertner J-C, Eriksen M, Bowen M (2018) Double trouble in the South Pacific subtropical gyre: increased plastic ingestion by fish in the oceanic accumulation zone. Mar Pollut Bull 136:547–564. https://doi.org/10.1016/j.marpolbul.2018.09.031

Markic A, Gaertner J-C, Gaertner-Mazouni N, Koelmans AA (2020) Plastic ingestion by marine fish in the wild. Crit Rev Environ Sci Technol 50:657–697. https://doi.org/10.1080/10643389.2019.1631990

Masindi V, Muedi KL (2018) Environmental contamination by heavy metals. In: Heavy metals. IntechOpen, London

Matei M, Zaharia R, Petrescu S-I, Radu-Rusu CG, Simeanu D, Mierliță D, Pop IM (2023) Persistent organic pollutants (POPs): a review focused on occurrence and incidence in animal feed and cow milk. Agriculture 13:873. https://doi.org/10.3390/agriculture13040873

Menéndez-Pedriza A, Jaumot J (2020) Interaction of environmental pollutants with microplastics: a critical review of sorption factors, bioaccumulation and ecotoxicological effects. Toxics 8:40. https://doi.org/10.3390/toxics8020040

Miller ME, Motti CA, Hamann M, Kroon FJ (2023) Assessment of microplastic bioconcentration, bioaccumulation and biomagnification in a simple coral reef food web. Sci Total Environ 858:159615. https://doi.org/10.1016/j.scitotenv.2022.159615

Miranda LS, Wijesiri B, Ayoko GA, Egodawatta P, Goonetilleke A (2021) Water-sediment interactions and mobility of heavy metals in aquatic environments. Water Res 202:117386. https://doi.org/10.1016/j.watres.2021.117386

Mofijur M, Ahmed SF, Rahman SMA, Arafat Siddiki SKY, Islam ABMS, Shahabuddin M, Ong HC, Mahlia TMI, Djavanroodi F, Show PL (2021) Source, distribution and emerging threat of micro- and nanoplastics to marine organism and human health: socio-economic impact and management strategies. Environ Res 195:110857. https://doi.org/10.1016/j.envres.2021.110857

Montalbetti E, Isa V, Vencato S, Louis Y, Montano S, Lavorano S, Maggioni D, Galli P, Seveso D (2022) Short-term microplastic exposure triggers cellular damage through oxidative stress in the soft coral *Coelogorgia palmosa*. Mar Biol Res 18:495–508. https://doi.org/10.1080/17451000.2022.2137199

Montano S, Seveso D, Maggioni D, Galli P, Corsarini S, Saliu F (2020) Spatial variability of phthalates contamination in the reef-building corals *Porites lutea*, *Pocillopora verrucosa* and *Pavona varians*. Mar Pollut Bull 155:111117. https://doi.org/10.1016/j.marpolbul.2020.111117

Napper IE, Thompson RC (2016) Release of synthetic microplastic plastic fibres from domestic washing machines: effects of fabric type and washing conditions. Mar Pollut Bull 112:39–45. https://doi.org/10.1016/j.marpolbul.2016.09.025

Nie H, Wang J, Xu K, Huang Y, Yan M (2019) Microplastic pollution in water and fish samples around Nanxun Reef in Nansha Islands, South China Sea. Sci Total Environ 696:134022. https://doi.org/10.1016/j.scitotenv.2019.134022

OECD (2022) Global plastics outlook: economic drivers, environmental impacts and policy options. Organisation for Economic Co-operation and Development, Paris

Ouédraogo D-Y, Mell H, Perceval O, Burga K, Domart-Coulon I, Hédouin L, Delaunay M, Guillaume MMM, Castelin M, Calvayrac C, Kerkhof O, Sordello R, Reyjol Y, Ferrier-Pagès C (2023) What are the toxicity thresholds of chemical pollutants for tropical reef-building corals? A systematic review. Environ Evid 12:4. https://doi.org/10.1186/s13750-023-00298-y

Palanikumar L, Kumaraguru AK, Ramakritinan CM, Anand M (2013) Toxicity, feeding rate and growth rate response to sub-lethal concentrations of anthracene and benzo [a] pyrene in milkfish *Chanos chanos* (Forsskal). Bull Environ Contam Toxicol 90:60–68. https://doi.org/10.1007/s00128-012-0895-1

Patterson J, Jeyasanta KI, Sathish N, Edward JKP, Booth AM (2020) Microplastic and heavy metal distributions in an Indian coral reef ecosystem. Sci Total Environ 744:140706. https://doi.org/10.1016/j.scitotenv.2020.140706

Paul-Pont I, Tallec K, Gonzalez-Fernandez C, Lambert C, Vincent D, Mazurais D, Zambonino-Infante J-L, Brotons G, Lagarde F, Fabioux C, Soudant P, Huvet A (2018) Constraints and priorities for conducting experimental exposures of marine organisms to microplastics. Front Mar Sci 5:252. https://doi.org/10.3389/fmars.2018.00252

Peters CA, Thomas PA, Rieper KB, Bratton SP (2017) Foraging preferences influence microplastic ingestion by six marine fish species from the Texas Gulf Coast. Mar Pollut Bull 124:82–88. https://doi.org/10.1016/j.marpolbul.2017.06.080

Portz L, Manzolli RP, Herrera GV, Garcia LL, Villate DA, Ivar do Sul JA (2020) Marine litter arrived: distribution and potential sources on an unpopulated atoll in the Seaflower Biosphere Reserve, Caribbean Sea. Mar Pollut Bull 157:111323. https://doi.org/10.1016/j.marpolbul.2020.111323

Pourret O, Bollinger J-C, Hursthouse A (2021) Heavy metal: a misused term? Acta Geochim 40:466–471. https://doi.org/10.1007/s11631-021-00468-0

Prokić MD, Radovanović TB, Gavrić JP, Faggio C (2019) Ecotoxicological effects of microplastics: examination of biomarkers, current state and future perspectives. TrAC Trends Anal Chem 111:37–46. https://doi.org/10.1016/j.trac.2018.12.001

Ranjbar Jafarabadi A, Riyahi Bakhtiari A, Aliabadian M, Laetitia H, Shadmehri Toosi A, Yap CK (2018) First report of bioaccumulation and bioconcentration of aliphatic hydrocarbons (AHs) and persistent organic pollutants (PAHs, PCBs and PCNs) and their effects on alcyonacea and scleractinian corals and their endosymbiotic algae from the Persian Gulf, Iran: inter and intra-species differences. Sci Total Environ 627:141–157. https://doi.org/10.1016/j.scitotenv.2018.01.185

Richardson K, Hardesty BD, Vince J, Wilcox C (2022) Global estimates of fishing gear lost to the ocean each year. Sci Adv 8:eabq0135. https://doi.org/10.1126/sciadv.abq0135

Rochman CM (2018) Microplastics research—from sink to source. Science 360:28–29. https://doi.org/10.1126/science.aar7734

Rochman CM, Brookson C, Bikker J, Djuric N, Earn A, Bucci K, Athey S, Huntington A, McIlwraith H, Munno K, Frond HD, Kolomijeca A, Erdle L, Grbic J, Bayoumi M, Borrelle SB, Wu T, Santoro S, Werbowski LM, Zhu X, Giles RK, Hamilton BM, Thaysen C, Kaura A, Klasios N, Ead L, Kim J, Sherlock C, Ho A, Hung C (2019) Rethinking microplastics as a diverse contaminant suite. Environ Toxicol Chem 38:703–711. https://doi.org/10.1002/etc.4371

Rodrigues JP, Duarte AC, Santos-Echeandía J, Rocha-Santos T (2019) Significance of interactions between microplastics and POPs in the marine environment: a critical overview. TrAC Trends Anal Chem 111:252–260. https://doi.org/10.1016/j.trac.2018.11.038

Rotjan RD, Sharp KH, Gauthier AE, Yelton R, Lopez EMB, Carilli J, Kagan JC, Urban-Rich J (2019) Patterns, dynamics and consequences of microplastic ingestion by the temperate coral, *Astrangia poculata*. Proc R Soc B Biol Sci 286:20190726. https://doi.org/10.1098/rspb.2019.0726

Saliu F, Montano S, Garavaglia MG, Lasagni M, Seveso D, Galli P (2018) Microplastic and charred microplastic in the Faafu Atoll, Maldives. Mar Pollut Bull 136:464–471. https://doi.org/10.1016/j.marpolbul.2018.09.023

Sandre F, Dromard CR, Le Menach K, Bouchon-Navaro Y, Cordonnier S, Tapie N, Budzinski H, Bouchon C (2019) Detection of adsorbed chlordecone on microplastics in marine sediments in Guadeloupe: a preliminary study. Gulf Caribb Res 30:GCFI8–GCFI14. https://doi.org/10.18785/gcr.3001.14

Sauvé S, Desrosiers M (2014) A review of what is an emerging contaminant. Chem Cent J 8:15. https://doi.org/10.1186/1752-153X-8-15

Schwaferts C, Niessner R, Elsner M, Ivleva NP (2019) Methods for the analysis of submicrometer- and nanoplastic particles in the environment. TrAC Trends Anal Chem 112:52–65. https://doi.org/10.1016/j.trac.2018.12.014

Shipley ON, Lee C-S, Fisher NS, Sternlicht JK, Kattan S, Staaterman ER, Hammerschlag N, Gallagher AJ (2021) Metal concentrations in coastal sharks from the Bahamas with a focus on the Caribbean Reef shark. Sci Rep 11:218. https://doi.org/10.1038/s41598-020-79973-w

Sokolova IM (2013) Energy-limited tolerance to stress as a conceptual framework to integrate the effects of multiple stressors. Integr Comp Biol 53:597–608. https://doi.org/10.1093/icb/ict028

Spalding M, Ravilious C, Green EP (2001) World atlas of coral reefs. University of California Press, Berkeley

St-Jean A, Suhas E, De Pina J-J, Cordier S, Lucas M, Ayotte P (2019) Exposure to toxic metals and polychlorinated biphenyls of adolescents and adults from two atolls in the Tuamotu Archipelago (French Polynesia). Sci Total Environ 695:133791. https://doi.org/10.1016/j.scitotenv.2019.133791

Su Y, Zhang K, Zhou Z, Wang J, Yang X, Tang J, Li H, Lin S (2020) Microplastic exposure represses the growth of endosymbiotic dinoflagellate *Cladocopium goreaui* in culture through affecting its apoptosis and metabolism. Chemosphere 244:125485. https://doi.org/10.1016/j.chemosphere.2019.125485

Tan F, Yang H, Xu X, Fang Z, Xu H, Shi Q, Zhang X, Wang G, Lin L, Zhou S, Huang L, Li H (2020) Microplastic pollution around remote uninhabited coral reefs of Nansha Islands, South China Sea. Sci Total Environ 725:138383. https://doi.org/10.1016/j.scitotenv.2020.138383

Tang Y, Rong J, Guan X, Zha S, Shi W, Han Y, Du X, Wu F, Huang W, Liu G (2020) Immunotoxicity of microplastics and two persistent organic pollutants alone or in combination to a bivalve species. Environ Pollut 258:113845. https://doi.org/10.1016/j.envpol.2019.113845

Tang J, Wu Z, Wan L, Cai W, Chen S, Wang X, Luo J, Zhou Z, Zhao J, Lin S (2021) Differential enrichment and physiological impacts of ingested microplastics in scleractinian corals *in situ*. J Hazard Mater 404:124205. https://doi.org/10.1016/j.jhazmat.2020.124205

Thornton Hampton LM, Brander SM, Coffin S, Cole M, Hermabessiere L, Koelmans AA, Rochman CM (2022) Characterizing microplastic hazards: which concentration metrics and particle characteristics are most informative for understanding toxicity in aquatic organisms? Microplast Nanoplast 2:20. https://doi.org/10.1186/s43591-022-00040-4

Tufail MA, Iltaf J, Zaheer T, Tariq L, Amir MB, Fatima R, Asbat A, Kabeer T, Fahad M, Naeem H, Shoukat U, Noor H, Awais M, Umar W, Ayyub M (2022) Recent advances in bioremediation of heavy metals and persistent organic pollutants: a review. Sci Total Environ 850:157961. https://doi.org/10.1016/j.scitotenv.2022.157961

van der Schyff V, du Preez M, Blom K, Kwet Yive NSC, Klánová J, Přibylová P, Audy O, Martiník J, Bouwman H (2021) Chlorinated and brominated persistent compounds in hard coral, soft coral, and parrotfish from remote Mascarene islands. Chemosphere 267:129316. https://doi.org/10.1016/j.chemosphere.2020.129316

van Sebille E, Wilcox C, Lebreton L, Maximenko N, Hardesty BD, van Franeker JA, Eriksen M, Siegel D, Galgani F, Law KL (2015) A global inventory of small floating plastic debris. Environ Res Lett 10:124006. https://doi.org/10.1088/1748-9326/10/12/124006

Wang Q, Huang F, Liang K, Niu W, Duan X, Jia X, Wu X, Xu P, Zhou L (2022) Polystyrene nanoplastics affect digestive function and growth in juvenile groupers. Sci Total Environ 808:152098. https://doi.org/10.1016/j.scitotenv.2021.152098

Wecker P, Lecellier G, Guibert I, Zhou Y, Bonnard I, Berteaux-Lecellier V (2018) Exposure to the environmentally-persistent insecticide chlordecone induces detoxification genes and causes polyp bail-out in the coral *P. damicornis*. Chemosphere 195:190–200. https://doi.org/10.1016/j.chemosphere.2017.12.048

Yadav D, Rangabhashiyam S, Verma P, Singh P, Devi P, Kumar P, Hussain CM, Gaurav GK, Kumar KS (2021) Environmental and health impacts of contaminants of emerging concerns: recent treatment challenges and approaches. Chemosphere 272:129492. https://doi.org/10.1016/j.chemosphere.2020.129492

Yin L, Chen B, Xia B, Shi X, Qu K (2018) Polystyrene microplastics alter the behavior, energy reserve and nutritional composition of marine jacopever (*Sebastes schlegelii*). J Hazard Mater 360:97–105. https://doi.org/10.1016/j.jhazmat.2018.07.110

Yin X, Wu J, Liu Y, Chen X, Xie C, Liang Y, Li J, Jiang Z (2022) Accumulation of microplastics in fish guts and gills from a large natural lake: selective or non-selective? Environ Pollut 309:119785. https://doi.org/10.1016/j.envpol.2022.119785

Zhang Q, He Y, Cheng R, Li Q, Qian Z, Lin X (2022) Recent advances in toxicological research and potential health impact of microplastics and nanoplastics in vivo. Environ Sci Pollut Res 29:40415–40448. https://doi.org/10.1007/s11356-022-19745-3

Zheng N, Wang S, Dong W, Hua X, Li Y, Song X, Chu Q, Hou S, Li Y (2019) The toxicological effects of mercury exposure in marine fish. Bull Environ Contam Toxicol 102:714–720. https://doi.org/10.1007/s00128-019-02593-2

7 The Use of New Biologically Based Technological Development in the Management of Crown-of-Thorns Sea Stars

Lucy M. Gorman, Maria Byrne, Ricardo Beldade, Mehdi Adjeroud, Pascal Dumas, and Suzanne C. Mills

Abstract

Crown-of-thorns sea star (CoTS) outbreaks represent one of the biggest threats to scleractinian coral reefs across the Indo-Pacific, and this threat is further exacerbated by ocean warming, cyclones and ocean acidification. There is currently a large collective effort by stakeholders across Indo-Pacific nations to develop technologies to accurately manage (predict, monitor and reduce) these outbreaks. Emerging technologies include: low-cost and high-throughput eDNA profiling of seawater (biosensors) to accurately count the abundance of CoTS in a local area; developing chemosensory attractants and dispersants to herd CoTS for efficient removal; desynchronising CoTS reproduction by altering the timing of CoTS reproductive cues; producing CoTS with lower fecundity through gene editing/contraceptives; using machine learning technologies and robots to automatically detect and cull CoTS; making CoTS more susceptible to diseases through microbiome manipulation to reduce population sizes; strategically planting corals more resilient to CoTS predation, e.g., more venomous corals; and facilitating the predation of CoTS by increasing the numbers of CoTS predators through marine protected area (MPA) designation and habitat restoration. In this chapter, we discuss the emerging technologies and considerations that should be taken into account to effectively manage CoTS outbreaks for future reef resilience.

Keywords

Crown-of-thorns sea star (CoTS) · Chemosensory cues · Machine learning · Reef management · Gene editing · Reproduction

L. M. Gorman (✉)
UAR 3278 CRIOBE, PSL Université Paris: EPHE-UPVD-CNRS, Papetoai, Moorea, French Polynesia

Laboratoire d'Excellence "CORAIL", Paris Cedex, France
e-mail: lucy.gorman@ephe.psl.eu

M. Byrne
Marine Invertebrate Futures Group, School of Life and Environmental Sciences, Marine Science Institute, The University of Sydney, Sydney, NSW, Australia

R. Beldade
Facultad de Ciencias Biologicas, Pontificia Universidad Católica de Chile, Santiago, Chile

M. Adjeroud
Laboratoire d'Excellence "CORAIL", Paris Cedex, France

ENTROPIE, Institut de Recherche pour le Développement (IRD), Université de la Réunion, CNRS, IFREMER, Université de la Nouvelle-Calédonie, Perpignan, France

P. Dumas
Laboratoire d'Excellence "CORAIL", Paris Cedex, France

Institut des Sciences Exactes et Appliquées, University of New Caledonia (UNC), Nouméa, New Caledonia

S. C. Mills
PSL Université Paris: EPHE-UPVD-CNRS; Centre de Recherche Insulaire et Observatoire de L'environnement (CRIOBE), Papetoai, Moorea, French Polynesia

Laboratoire d'Excellence "CORAIL", Paris Cedex, France

Institut Universitaire de France (IUF), Paris, France

7.1 Introduction

Crown-of-thorns sea stars (CoTS; Fig. 7.1) belong to the *Acanthaster* species complex (Haszprunar et al. 2017) and are some of the most intensively studied coral reef organisms due to their dramatic population density fluctuations and cascading consequences (Uthicke et al. 2009; Pratchett et al. 2017a). The *Acanthaster* species complex occurs from the Red Sea across to the Indo-Pacific and this corallivorous group has four recognised clades: (1) the recently named Red Sea species *Acanthaster benziei*; (2) a Pacific species that has two available species names (*solaris* or *echinatus*), with *Acanthaster* cf. *solaris* in current use; (3) a Southern Indian Ocean species *Acanthaster mauritiensis*; and (4) a Northern Indian Ocean species *Acanthaster planci* sensu stricto (Vogler et al. 2008; Haszprunar et al. 2017; Wörheide

S. Planes (ed.), *The Future of Coral Reefs*, Coral Reefs of the World 21, https://doi.org/10.1007/978-3-031-98584-3_7

Fig. 7.1 Close-up image of the Pacific species *Acanthaster* cf. *solaris* of the crown-of-thorns sea star (CoTS) feeding on an *Acropora* coral. (Image courtesy of Institut de Recherche pour le Développement (IRD), New Caledonia/Pascal Dumas)

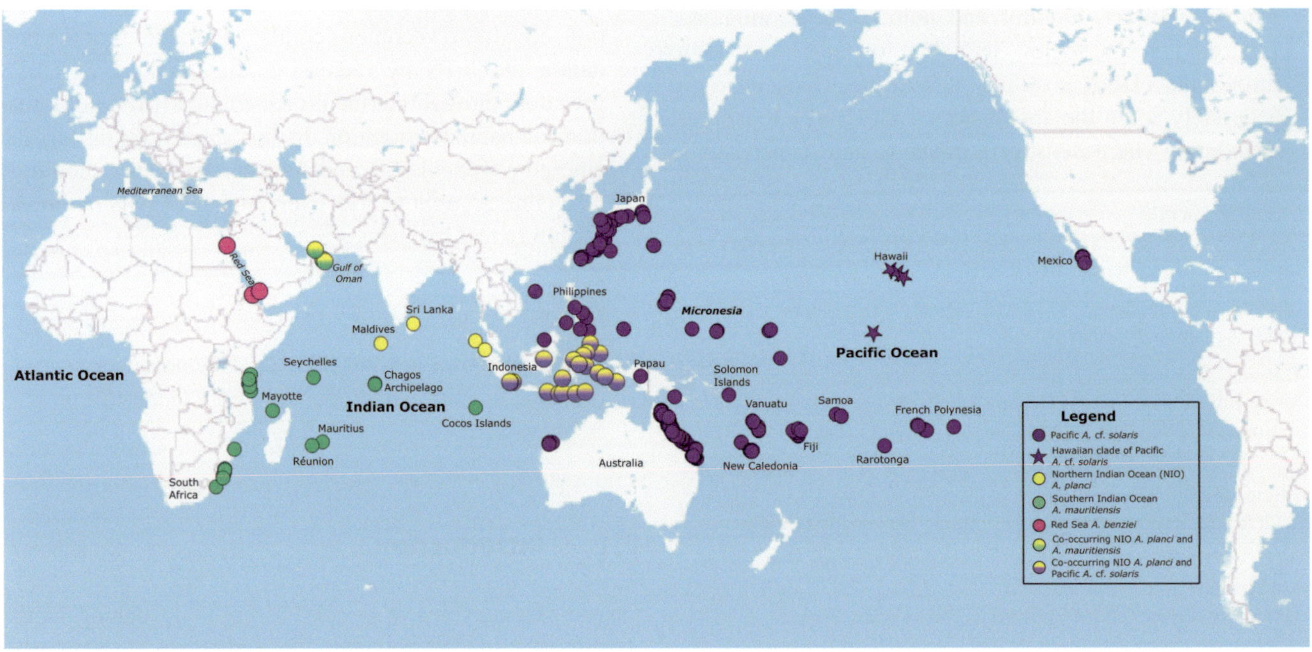

Fig. 7.2 Global distribution of CoTS species outbreaks based on current CoTS species knowledge (Vogler et al. 2008; Vogler et al. 2012; Haszprunar et al. 2017; Yasuda et al. 2022) including two sympatric areas (Haszprunar et al. 2017) updated from Pratchett et al. (2014) to include new outbreaks and newly revised species identification. Pink circles: Red Sea species *Acanthaster benziei*. Purple circles: Pacific species *Acanthaster* cf. *solaris*. Purple stars: Hawaiian clade of Pacific *Acanthaster* cf. *solaris*. Green circles: South Indian Ocean species *Acanthaster mauritiensis*. Yellow circles: Northern Indian Ocean species *Acanthaster planci*. Green and yellow circles: *Acanthaster mauritiensis* and *Acanthaster planci*. Yellow and purple circles: *Acanthaster planci* and *Acanthaster* cf. *solaris*

et al. 2022; Fig. 7.2). The sister taxon to the *Acanthaster* species complex, *Acanthaster brevispinus*, inhabits deeper waters (20–70 m) and has also shown to be corallivorous (Keesing et al. 2023), although not exclusively so, preying on a wide range of prey including molluscs and soft corals (Yuasa et al. 2017; Keesing et al. 2023).

CoTS boom-and-bust cycles (Uthicke et al. 2009) involve rapid population density increases, reaching up to 15,000 sea stars per ha^{-1} (Dumas et al. 2022) during "outbreaks"

Fig. 7.3 CoTS outbreak in New Caledonia (August 2018). (Image courtesy of Institut de Recherche pour le Développement (IRD), New Caledonia/Jean-Michel Boré)

(Fig. 7.3), followed by a "non-outbreak" state where CoTS densities are usually one to ten sea stars per ha^{-1} (Pratchett et al. 2014; Dumas et al. 2016).

The current time between boom-and-bust cycles ranges from 15 to 17 years (Babcock et al. 2020), with 16 years between outbreaks on both the Great Barrier Reef (GBR) (Fabricius et al. 2010) and in French Polynesia (2006–2021, Fig. 7.4; Lamy et al. 2016). This contrasts with previous predictions of 50–80 years on the GBR (Fabricius et al. 2010) and 26 years in French Polynesia (Lamy et al. 2016; 1980–2006, Fig. 7.4), suggesting that outbreaks are becoming more frequent. Each individual CoTS is estimated to consume 5–12 m^2 of coral surface per year (Chesher 1969; Pearson and Endean 1969; Dana and Wolfson 1970), making CoTS one of the most efficient and prolific corallivores. The knock-on effects of a reduction in coral cover by CoTS predation during an outbreak spill over into other trophic levels of the reef community, with a decrease in coral cover and an increase in algal cover causing cascading effects on coral reef benthic invertebrates and fish assemblages (Birkeland 1989a; Lamy et al. 2016). These consequences make CoTS population outbreaks a major threat to coral reefs and the main cause of reduced coral reef resilience (Vercelloni et al. 2017; Haywood et al. 2019). Unsurprisingly, CoTS outbreaks are estimated to account for a 34.5–45.6% and a 42% reduction in scleractinian coral cover in French Polynesia and the GBR, respectively (De'Ath et al. 2012; Lamy et al. 2016), and even >90% reduction in corals on a single reef (Kayal et al. 2012). However, CoTS outbreaks are not the only stress to reefs, with outbreaks occurring in complex environmental contexts where other concomitant threats are also driving coral mortality, e.g., back-to-back bleaching events and cyclones across the Pacific (Hughes et al. 2017; Adjeroud et al. 2018; Byrne et al. 2025; Fig. 7.4).

In contrast to outbreaks, at non-outbreak densities, disturbances from CoTS do not have long-term impacts on hard coral cover, with reefs recovering readily (GBRMPA 2017). At low outbreak intensities (low CoTS numbers), outbreaks may even increase coral diversity in some areas, with CoTS mainly feeding on faster growing/more abundant coral species, e.g., *Acropora* spp., reducing the competition for space and allowing slower growing/rare coral to grow and form colonies (Done and Potts 1992; Bellwood et al. 2024). Furthermore, CoTS may even contribute to reef health by releasing biologically active substances in their mucus that can help protect fish from disease (Miura et al. 2023). Although as most studies of CoTS demography focus on population outbreaks at high intensity (Pratchett et al. 2021), there is a lack of knowledge on the ecological differences between areas with stable non-outbreak populations/low outbreak intensities of CoTS and those undergoing a highly intense population outbreak (Pratchett et al. 2021). Albeit due to the effects of climate change, a recent study has shown that it is unlikely that these outbreaks lead to increased coral

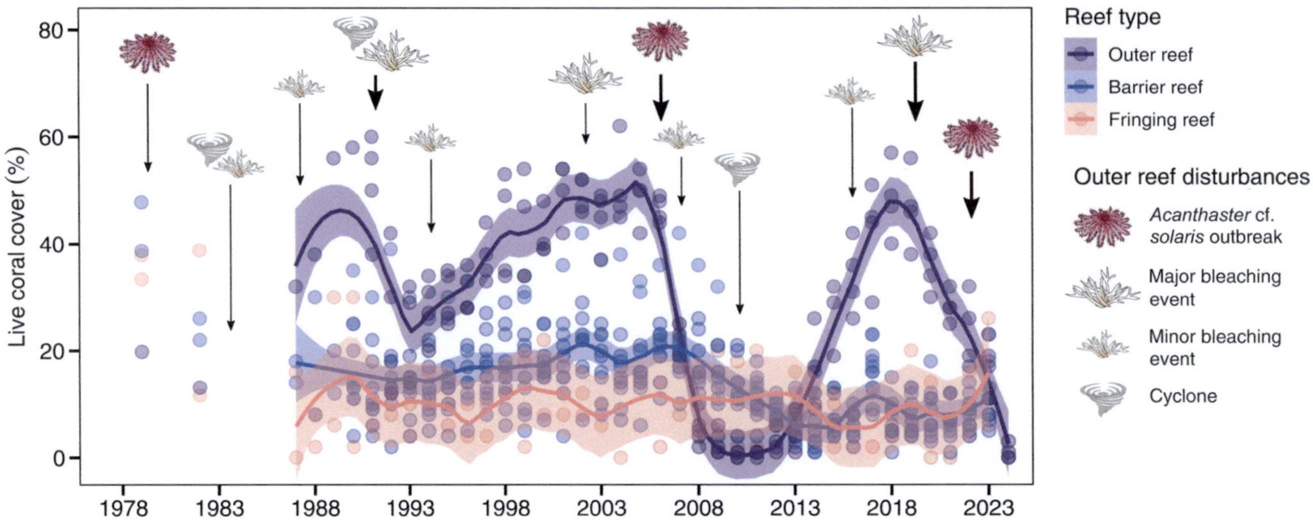

Fig. 7.4 Variation in live coral cover at the outer (purple; 12 m depth), barrier (blue) and fringing (pink) reefs of Moorea Island, French Polynesia over the past 43 years. Best fit locally estimated scatterplot smoothing (LOESS) curve (span = 0.2) ± CI (shading). All major disturbance events are shown with arrows next to their corresponding cause. Arrow thickness corresponds to disturbance intensity on coral cover. Major disturbances that triggered declines in coral cover are highlighted with black arrows. Four of these disturbances were considered major: a CoTS outbreak in 1980 (Berumen and Pratchett 2006); a cyclone and a bleaching event in 1991 (Salvat 1992); one of the most extreme declines in coral reef cover ever recorded from 2006 to 2010, which was first initiated by a CoTS outbreak in 2006 and then followed by cyclone Oli in 2010 (Leray et al. 2012; Adjeroud et al. 2018); and the most recent ongoing decline, caused by a bleaching event in 2019 (Haguenauer et al. 2021) followed by a CoTS outbreak in 2022 (pers. comm. LMG & SCM). (Data from le Service National d'Observation (SNO) CORAIL and figure created by Daphne Cortese)

species richness in the Anthropocene but rather exacerbate the decline in overall hard coral cover (Foo et al. 2024).

CoTS have been recorded at reef surveys as far back as 1924 in the Pacific (Domantay and Roxas 1938), being cited as "common" on reef habitats (Mortensen 1931; Domantay and Roxas 1938; Hayashi 1938; Dana 1970). However, whether these reports were from sites experiencing population outbreaks or non-outbreak densities of CoTS remains unknown (Dana 1970), with the earliest confirmed scientific recording of CoTS population outbreaks dating back to the 1950s and 1960s (Barnes and Endean 1964; Endean 1969; Pearson and Endean 1969; Endean and Stablum 1973). However, it should be noted that indigenous Polynesian fishers encountered large aggregations of CoTS as far back as the 1930s (Vine 1973). CoTS outbreaks have been reported throughout the Pacific Ocean (Fig. 7.2), for example, French Polynesia (Kayal et al. 2012), New Caledonia (Dumas et al. 2020), Vanuatu (Dumas et al. 2016, 2020), Fiji (Dumas et al. 2020), Indonesia (Plass-Johnson et al. 2015), Japan (Yasuda 2018), Xisha Islands (Xiao et al. 2022) Australia (De'Ath et al. 2012) and Hawaii (Timmers et al. 2011). Additionally, although not as systemic, outbreaks/high densities of CoTS have been reported across the Indian Ocean (Fig. 7.2), for example, the Gulf of Oman (Mendonça et al. 2010; Seveso et al. 2024), South Africa (Celliers and Schleyer 2006), Mauritius (Fagoonee 1990), Réunion (Tourrand et al. 2013), Sri Lanka (Dissanayak and Terney Pradeep Kumara 2015), the Maldives (Saponari et al. 2015), the Cocos (Keeling) Islands (Hobbs et al. 2005), the Chagos Archipelago (Roche et al. 2015), the Seychelles (Obura et al. 2017) and the Red Sea (Abou Zaid and Kotb 2000; Fig. 7.2). Contrastingly, CoTS are absent from the Atlantic and Caribbean oceans (Pratchett et al. 2014).

CoTS outbreaks in the Indian Ocean and the Red Sea are not regarded to be as common or as intense as in the Pacific (Vogler et al. 2012; Pratchett et al. 2014). It is unclear, however, why this difference in outbreak frequency and intensity exists. Whether the different CoTS species show different life history traits or whether the difference in environmental factors influencing outbreaks, e.g., the number of CoTS predators present (Ormond et al. 1990), influences the outbreak frequency, and thus, the intensity in these locations warrants future examination (Haszprunar et al. 2017). Additionally, the bias towards CoTS research occurring predominantly in the Pacific (Vogler et al. 2008, 2012; Haszprunar et al. 2017) may play a role by allowing more Pacific *A.* cf. *solaris* outbreaks to be recorded and monitored. Unsurprisingly, *A.* cf. *solaris* is the best-studied CoTS species (Haszprunar et al. 2017). There is also a bias within the Pacific, with more research occurring in highly economically developed countries, particularly Australia, due to the economic, environmental and social significance of the GBR allowing extensive funding into CoTS research projects from government-funded organisations (Milne et al. 2023).

Fig. 7.5 Snorkelers from Vanuatu Fisheries Department injecting CoTS. (Images courtesy of Institut de Recherche pour le Développement (IRD), New Caledonia/Pascal Dumas)

Contrastingly, low-income Pacific Island nations rely heavily on local support and citizen science (Dumas et al. 2015, 2016, 2020; Fig. 7.5). This bias towards CoTS found on the GBR possibly stems from earlier research regarding all CoTS as one single species (Haszprunar et al. 2017). However, it is becoming increasingly evident that this bias is weighting CoTS biology to observations made on *A.* cf. *solaris* on the GBR and these observations cannot be extrapolated to other CoTS species or to *A.* cf. *solaris* in other oceanographic locations (Vogler et al. 2012; Haszprunar et al. 2017; Milne et al. 2023). This is especially true when recent genome analysis shows species-specific selection in CoTS genes (Yuasa et al. 2021).

Nevertheless, the ease of access to resources that allow the close monitoring of CoTS outbreaks on the GBR has facilitated the discovery of valuable insights regarding the triggers and onset of outbreaks as well as their movement. On the GBR, outbreaks start off at mid-shelf reefs located across an area called the "initiation box" as it is suggested to be the location where primary outbreaks originate (Pratchett et al. 2014; Vanhatalo et al. 2017). CoTS outbreaks can be primary or secondary, with primary outbreaks being standalone, exponential population booms in one geographic location (Horoiwa et al. 2022). Areas can also experience secondary outbreaks, where oceanographic currents transport CoTS larvae from the outbreak site (source) to nearby reefs (sinks) (Horoiwa et al. 2022). If these sink reefs receiving the larvae experience high levels of larval recruitment, it can lead to subsequent outbreaks. The hypotheses supporting the "initiation box" as the site of primary CoTS outbreaks across the GBR include its high dispersive potential, promoting connectivity among reefs through the transportation of CoTS larvae during an outbreak to neighbouring reefs, causing secondary outbreaks at these sites (Hock et al. 2014; Matthews et al. 2020a). It is also suggested that the "initiation box" enhances the survival of CoTS larvae due to increased food levels (phytoplankton blooms), which in the past, were hypothesised to be generated from terrestrial nutrient run-off (Wooldridge and Brodie 2015). However, a link between nutrient run-off and CoTS outbreaks has not yet been observed (Pratchett et al. 2021), with a recent study showing that terrestrial nutrient run-off provides a negligible influx to the overall nutrient levels witnessed in the "initiation box" (Kroon et al. 2023). Once established, the CoTS population expands latitudinally (Kenchington 1977) at a rate estimated to be 1–2° (100–200 km) every 3 years but can reach speeds of up to 60 km.year^{-1} (Reichelt et al. 1990; Vanhatalo et al. 2017). On insular island reef systems such as Moorea, CoTS outbreaks start off in the deep, ocean-exposed reef base and then subsequently migrate to shallow waters of the reef lagoon (Kayal et al. 2012).

The triggers of CoTS outbreaks have been extensively debated and include hypotheses ranging from anthropogenic impacts to the life-history traits of CoTS, for example, predator removal hypothesis (Endean 1969; Kroon et al. 2021), nutrient run-off hypothesis (Birkeland 1982; Fabricius et al. 2010; Mendoza-Porras et al. 2023; also see Clements et al. 2022), coral prey availability (Ling et al. 2020; Pratchett et al. 2021), high fecundity (Babcock et al. 2016b), aggregation behaviour (Dana et al. 1972) and juveniles lying in waiting (Deaker et al. 2020). Until recently, the nutrient run-off hypothesis and success of the larval stage have obtained the greatest support with respect to the underlying trigger of CoTS outbreaks, but outbreaks occur in areas very distant from land-based run-off (Roche et al. 2015). As a result, the planktotrophic larval stages of CoTS (Fig. 7.6) have been the focus of research. However, for CoTS outbreaks to occur, all life stages (Fig. 7.6) must be successful. Recent research revisiting the predator removal hypothesis has found evidence of a broad suite of predators of CoTS on the GBR (see section 7.4.1.1.). Overall, the current consensus is that there is not one cause of CoTS outbreaks but rather several causes which can be exploited by CoTS due to their phenotypic plasticity, triggering outbreaks (see Pratchett et al. (2021) and Deaker and Byrne (2022) for in-depth reviews).

Regardless of the causes of CoTS population outbreaks, their frequency and intensity are considered to have increased (Fabricius et al. 2010). This increase, compounded with the

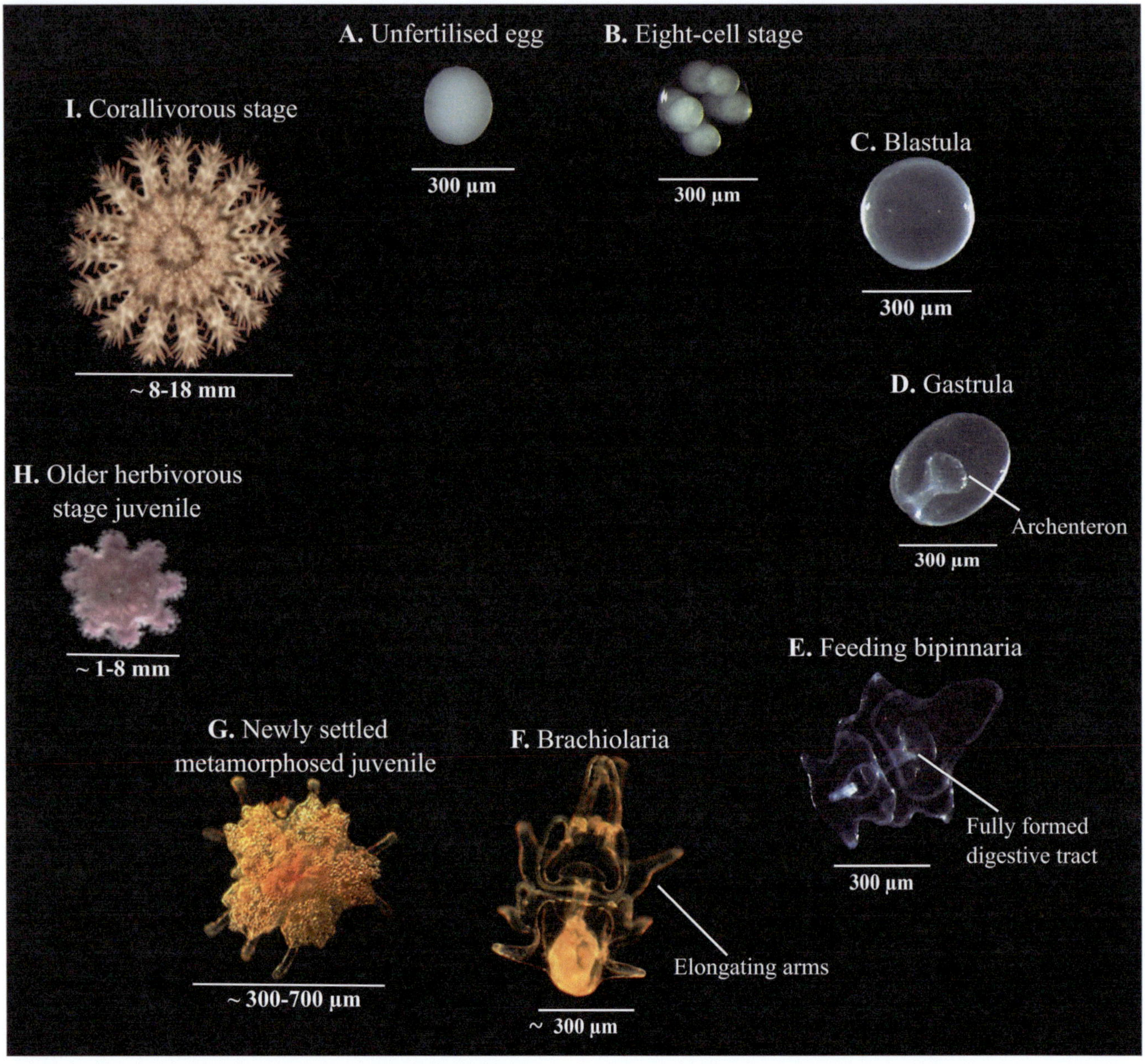

Fig. 7.6 CoTS life cycle. Based on the CoTS life cycle figure from Deaker and Byrne (2022). (**a**) Unfertilised egg. (**b**) The eight-cell stage. (**c**) Blastula. (**d**) Gastrula with the archenteron. (**e**) Feeding bipinnaria with a fully developed digestive tract. (**f**) Brachiolaria with elongating arms. (**a–f**) ~ 2–3 weeks. (**g**) Newly settled metamorphosed juvenile; metamorphosis occurs ~2 days post-settlement. (**h**) Older herbivorous stage juvenile; indeterminate age—see Deaker et al. (2020). (**i**) The corallivorous stage; ~4–6 months old*. Reproductive maturity reached at ~2 years old* (*if growth is continuous without a "lie in wait" phase at stage (**h**), see Deaker et al. (2020)). (Images: (**a–e**) supplied by Dione Deaker; (**f**) and (**g**): supplied by Australian Institute of Marine Science (AIMS, available at: www.eatlas.org.au); (**h**) and (**i**) supplied by Matthew Clements)

impacts from coral bleaching and cyclones, pose a significant threat to the survival of reefs (Hughes et al. 2017). Alarmingly, when the added complexity of climate change is considered, recent evidence suggests that under future climate change scenarios, although biologically stressed, CoTS can acclimatise to these conditions over generations albeit with an increased metabolic rate (Kamya et al. 2018; Hue et al. 2022; Lang et al. 2022). This increased metabolic rate will probably be fuelled by increased consumption of corals, ultimately amplifying the impacts of outbreaks (Lang et al. 2022). Several studies show that CoTS larvae (Uthicke et al. 2013; Lamare et al. 2014; Hue et al. 2022; Lang et al. 2023), juveniles (Kamya et al. 2014; Byrne et al. 2023) and adults (Lang et al. 2022) are more resilient to heatwave conditions than corals. The plasticity of CoTS coupled with their life-history traits allows them to be elusive, opportunistic coral predators, able to lie in wait at high abundances for years at reefs during unfavourable conditions, waiting for the perfect moment to strike (Deaker et al. 2020; Deaker and Byrne 2022). Importantly, the increase in coral rubble generated by post bleaching mortality also increases the extent of habitat for juvenile CoTS (Wolfe and Byrne 2024). Thus, insights into their biology to reduce their elusiveness are paramount to informing effective management strategies that curb these outbreaks (Babcock et al. 2020).

7.2 Current Methods in Studying CoTS Outbreaks

Proactive and effective management strategies are reported to be the best way to reverse scleractinian coral cover loss in the Indo-Pacific (De'Ath et al. 2012; Westcott et al. 2020). Control is the first step and most effective strategy in managing CoTS outbreaks (Westcott et al. 2020).

7.2.1 Monitoring

To detect areas that require control intervention, CoTS monitoring methods are required to evaluate an area for its abundance of CoTS (Fig. 7.7). Manta tows are a primary monitoring method, where a diver is pulled along a section of reef, counting the number of CoTS and their feeding scars (Moran et al. 1989; Bass and Miller 1996; Geoffroy et al. 2009; Geoffroy 2010). However, the efficacy of this method is contested, with reports estimating that it only detects 5% of CoTS due to their cryptic nature (Fernandes et al. 1990). The low accuracy of this method stems from the 3D habitat complexity of a coral reef, with CoTS often being highly camouflaged and subsisting in small crevices or hiding under corals. Furthermore, this method is highly subjective, with CoTS feeding scars being confused for *Drupella* scarring, coral disease or coral bleaching (Fernandes et al. 1990; GBRMPA 2017). Additionally due to the health and safety of the divers, manta tows occur during daylight hours. However, daytime monitoring probably underestimates the abundance of CoTS, as CoTS are a nocturnal species with most of their movements occurring at night, whilst they remain hidden during the day (Keesing 1995; Ling et al. 2020). Furthermore, CoTS can be found in deeper waters, up to 50 m (Beaman 2018; Byrne et al. 2024; Fig. 7.8); therefore, manta tows that are conducted at shallower depths are likely to miss these individuals. Although manta tows harbour several drawbacks, they currently remain a popular method for determining CoTS abundance due to their quick snapshot assessment of reefs for CoTS outbreaks. The high-throughput and low-cost nature of manta tows are attractive qualities for large scale reef systems such as the GBR (Miller et al. 2003). For instance, the Great Barrier Reef Marine Park Authority were able to survey 100 reefs between 2022 and 2023 to collect data (including CoTS abundance) for their long-term coral monitoring program using the manta tow method (GBRMPA 2023).

7.2.2 Management Strategies

After an area has been deemed threatened by CoTS through monitoring surveys of local abundance, control methods are deployed. Previously, control methods have involved both unsuccessful methods, e.g., underwater fences and the cutting of CoTS into small pieces (Fraser et al. 2000) and successful methods, e.g., manual CoTS removal and culling by local snorkellers, with CoTS killed either in situ or individuals being buried/burnt on land (Dumas et al. 2015). A seminal method using the injection of lime juice or white spirit vinegar into CoTS individuals resulted in 100% mortality (Fig. 7.5) (Dumas et al. 2015, 2018; Moutardier et al. 2015). Furthermore, the low-cost nature of this method makes it accessible to low-income Pacific Island nations (Moutardier et al. 2015). The bioactive route of the acidic agents used (lime juice, vinegar) are thought to kill CoTS by disrupting their physiology and causing cell death (Mendoza-Porras et al. 2023). Although control through manual injection is currently regarded as the most effective management strategy during an outbreak (Westcott et al. 2020; Matthews et al. 2024) and has support from the general public (Lockie et al. 2024), it has its drawbacks. For instance, manual control is very costly and can only be deployed in areas accessible to divers (e.g., <15–20 m depth due to health and safety) and even shallower for snorkellers, despite CoTS inhabiting much deeper areas of the reef (Babcock et al. 2020).

In addition to manual control, management strategies of CoTS population outbreaks have included water quality improvement and marine protected area (MPA) designation,

Fig. 7.7 Diver monitoring the abundance of CoTS in New Caledonia. (Image courtesy of Observatoire de l'environnement (OEIL)/ Matthieu Juncker)

Fig. 7.8 Photo of a deep-water CoTS taken at 41 m on the northern GBR (CSIRO Marine National Facility RV Investigator IN2022_V07). (Photo taken from Byrne et al. 2024))

the former with the assumption that decreased nutrient run-off decreases phytoplankton blooms and thereby food for CoTS larvae (Westcott et al. 2020). However, it is still unclear whether water quality measures would prove an effective management strategy in controlling CoTS outbreaks. Firstly, this is because local authorities in the GBR region have not successfully reduced terrestrial nutrients loads to the levels required to mitigate CoTS outbreaks, with nutrient loads currently needing to be reduced by 20–40% (Wooldridge and Brodie 2015; Westcott et al. 2020). Secondly, recent computer modelling of the GBR "initiation box" has shown that terrestrial nutrient loads are only weakly correlated to factors that encourage CoTS larval population booms, e.g., chlorophyll *a* concentrations, organic nitrogen concentrations and overall zooplankton concentrations (Kroon et al. 2023). Thus, water quality improvement is likely an inferior method for controlling primary outbreaks compared to other strategies.

Contrastingly, studies have shown MPAs to be an effective natural strategy to reduce CoTS densities through increasing the biomass of species that prey on and injure CoTS (Kroon et al. 2021; Caballes et al. 2022), leading to both a reduction in the frequency and an 86% reduction in the severity of CoTS outbreaks in no-take marine reserves (Sweatman 2008; Vanhatalo et al. 2017). There are contrasting results on MPA effectiveness in controlling CoTS out-

Fig. 7.9 *Charonia tritonis* predating on a CoTS individual in New Caledonia. (Image courtesy of Observatoire de l'environnement (OEIL)/ Matthieu Juncker)

breaks, with some studies on the GBR reporting no difference in CoTS presence, prevalence and population outbreaks and thus no difference in hard coral cover (Matthews et al. 2020a; Westcott et al. 2020). However, this is possibly due to the time taken to see benefits from MPAs on predator populations (and thus the benefits on CoTS abundances), which is around 10 years (Vanhatalo et al. 2017).

One CoTS predator species that will benefit through utilising MPAs is *Charonia tritonis*, also known as the giant triton and Triton's trumpet snail (Motti et al. 2022). *C. tritonis* is regarded as the primary predator of adult CoTS in the Pacific and is present throughout the Indo-Pacific in tropical waters, as well as the Red Sea and Indian Ocean, although their abundance is unclear in these latter areas (Motti et al. 2022; Fig. 7.9). Indo-Pacific populations of *C. tritonis* were decimated in the twentieth century with an estimated ~10,000 tritons removed from the GBR per year from 1947 to 1960 (Endean 1969), due to their ornamental shell and practical use (as a trumpet and food source) (Motti et al. 2022). The numbers of this species remain very low. Despite being protected, *C. tritonis* are still actively harvested and sold as tourist souvenirs in shops across the Pacific (Nijman and Lee 2016). These gastropods travel large distances (~234.24 m per d^{-1}) looking for prey, with most of their activity occurring at night (Schlaff et al. 2020). The feeding rate of *C. tritonis* on CoTS in the wild is unclear, with reports of *C. tritonis* in aquaria consuming one CoTS per week (Pearson and Endean 1969) and nine CoTS over 6 months when supplied with other echinoderm prey (Birkeland 1989b). Estimated times from hunt to ingestion by *C. tritonis* are approximately 4 h per CoTS (Bose et al. 2017). These predation rates are likely to be insufficient to curb CoTS outbreaks unless stakeholders plan to vastly increase the numbers of *C. tritonis* to pre-1950s levels (Campbell and Ormond 1970; Hall et al. 2017a, b). Importantly, *C. tritonis* is an echinoderm specialist predator that does not solely prey on CoTS, with reports that it prefers the blue starfish *Linckia laevigata* as opposed to CoTS when given a choice (Vine 1970; Motti et al. 2022). Thus, plans to increase the numbers of *C. tritonis* would need to consider the knock-on effects on other echinoderm species that may be adversely affected, especially outside of a CoTS outbreak period.

7.3 Emerging Methods in Studying CoTS Outbreaks

In the advent of *robotics and machine learning*, methods harnessing these technologies have been proposed to help in CoTS monitoring and management (Clement et al. 2005; Dayoub et al. 2015; Takemura et al. 2015; Matthews et al. 2020b; Pooloo et al. 2021; Heenaye-Mamode Khan et al. 2023). Computer algorithms can provide between 95% and 100% accuracy in detecting CoTS after training, providing an efficient method for determining the abundance of CoTS in real-time without human errors (Dayoub et al. 2015; Heenaye-Mamode Khan et al. 2023). After CoTS detection, robots can inject them, mitigating safety concerns for divers whilst culling (Dayoub et al. 2015; Takemura et al. 2015). Robots can operate day and night, in deeper waters inaccessible to divers and even under otherwise challenging sea conditions (Dayoub et al. 2015). The effectiveness of these technologies on a large scale in situ is still yet to be documented; however, in one experimental study (Takemura et al. 2015), a robot was able to conduct five acetic acid injections within 20 minutes in two CoTS individuals. Once optimised for large-scale deployment, these technologies will provide an attractive addition in the CoTS detection and control repertoire, enabling previously inaccessible areas of reefs to be reached across all hours of the day.

Although manta tows have previously (and are still currently) a widely used method to monitor CoTS populations, there is a push towards developing nuanced technologies that allow an increase in the depth and breadth of CoTS monitoring, e.g., genomic technologies. For instance, a seminal study using the universal barcoding gene cytochrome oxidase (COI) and polymerase chain reation (PCR) technology, allowed the detection of CoTS larvae at increasing distances from the site of the population outbreak (Uthicke et al. 2015). This method opened previously unattainable questions about the migration of CoTS larvae. Following this, the method was initially coupled with microscopy to quantify the number of CoTS larvae (Suzuki et al. 2016), and later it was coupled with environmental DNA (eDNA) and qPCR to quantify the density of CoTS larvae in the field (Doyle et al. 2017). This allowed the abundance of each larval stage to be quantified when samples were size-fractionated (Doyle et al. 2017) and allowed post-settlement CoTS across reefs to be quantified (Uthicke et al. 2018). The number of *reads* (copies of COI barcode) amplified with the eDNA qPCR approach is positively correlated with CoTS biomass on reefs, and its resolution was confirmed when CoTS eDNA was absent from both pre- and post-outbreak reefs (Uthicke et al. 2018). These advances have meant that monitoring systems can now detect both the presence/absence as well as the density of CoTS larvae and adults (Suzuki et al. 2016; Doyle et al. 2017; Uthicke et al. 2018; Wang et al. 2023) and calculate how the recruitment numbers of larval CoTS to reefs translates to adult CoTS densities over time (Uthicke et al. 2019), although, these estimates may be confounded by the ability of CoTS larvae to clone (Allen et al. 2019; Fig. 7.10). Of note, is that these genetic advances have sped up the detection of possible outbreak sites by an estimated 1–2 years compared with previous methods such as manta tows (Babcock et al. 2020). Thus, CoTS genetic barcoding should be carried out in addition to monitoring as a proactive complementary tool.

Ultimately, although monitoring methods and management strategies have become more nuanced and effective in the last decade, their effectiveness still lags behind what is needed, with research suggesting the critical density of CoTS needed to be targeted to stop an outbreak is <3 sea stars per ha^{-1} (Rogers et al. 2017). This concerningly low number highlights the need for more accurate monitoring methods and effective management strategies to aid in CoTS population biocontrol and suppress future outbreaks (Babcock et al.

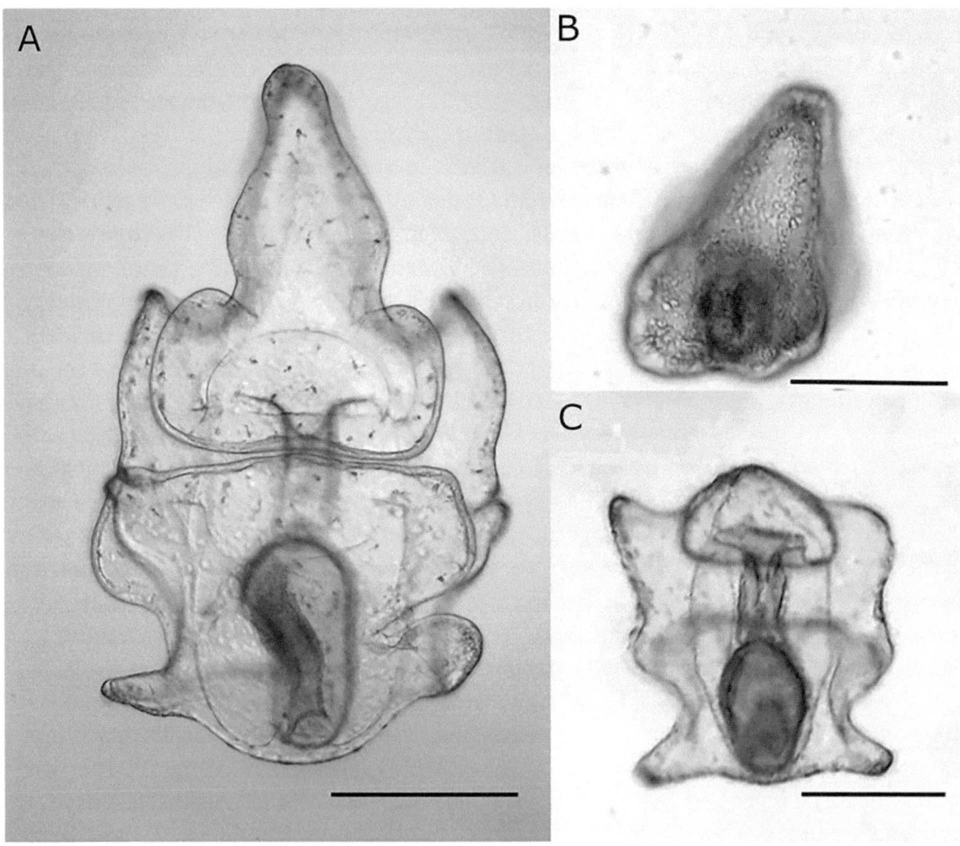

Fig. 7.10 Image of a CoTS bipinnaria larva (**a**) and recently formed anterior (**b**) and posterior (**c**) clones that can regenerate to form a new complete larva in a few days. Scale bars = 100 μm. (Image supplied by Maria Byrne)

2020). These strategies could build upon new discoveries in the fundamental concepts that underpin CoTS biology that have allowed the conception of new possible management strategies, which were previously inaccessible (Babcock et al. 2020; Pratchett et al. 2021). Over 40 years ago, Endean and Stablum (1973) stated that information is lacking about how CoTS invade reefs, the role in which larval and adult CoTS contribute to population outbreaks and the identity and biology of CoTS predators. Recent research shows that the same knowledge gaps can be applied to the early benthic CoTS juveniles (Deaker et al. 2020; Desbiens et al. 2023). Despite decades of research and considerable advances in the knowledge of CoTS biology and ecology, in addition to an overall increase in CoTS studies (Fig. 7.11), major knowledge gaps still exist in these areas (Pratchett et al. 2021). These knowledge gaps also create bottlenecks when using new technologies such as machine learning, as the models do not describe the data as accurately due to vague estimates on certain life-history parameters, e.g., larval predation and connectivity, rendering them less effective (Matthews et al. 2020a). Thus, research into the biology of CoTS is paramount to suppressing population outbreaks.

7.4 Emerging Technologies Addressing CoTS Knowledge Gaps

In the next three sections, we will discuss major knowledge gaps in the biology and ecology of CoTS that will be bridged by emerging technologies (Fig. 7.12). We also discuss how these technologies will aid in and create new management strategies and we highlight their current limitations. Firstly, we describe how advances in DNA technologies such as eDNA metabarcoding and qPCR can be used in better understanding both CoTS predators and CoTS larval recruitment and migration, a method vital for efficient control and management of outbreaks. We then explain the potential of genetic technologies to facilitate gene drives that spread genetic mutations throughout wild CoTS populations using bioengineering, which will reduce the fitness of CoTS and are subsequently passed onto the next CoTS generation. In addition, we discuss how bioengineering and assisted evolution of coral and CoTS biology, specifically venom, could be used to manage CoTS outbreaks by both increasing injuries to CoTS during corallivory or by making CoTS more palatable to predators. Secondly, we describe potential methods

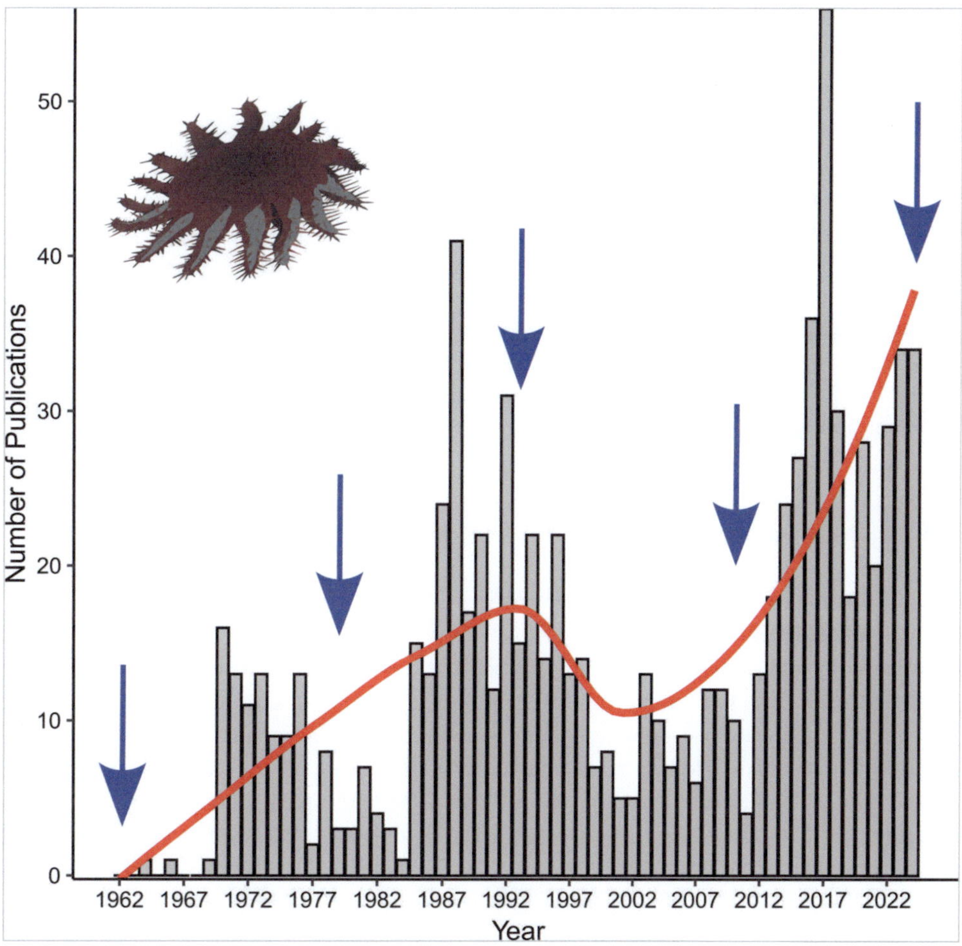

Fig. 7.11 Number of CoTS publications published overtime from 1962 to 2024. Search term: "allintitle: seastar OR starfish 'crown of thorns'" was used to search google scholar database for CoTS publications. A best fit locally estimated scatterplot smoothing (LOESS) curve was added to the data (red line). Blue arrows show when outbreaks of CoTS occurred on the GBR (Babcock et al. 2016a; Australian Institute of Marine Science 2024). CoTS illustration sourced from the Integration and Application Network, University of Maryland Center for Environmental Science (https://ian.umces.edu/media-library/)

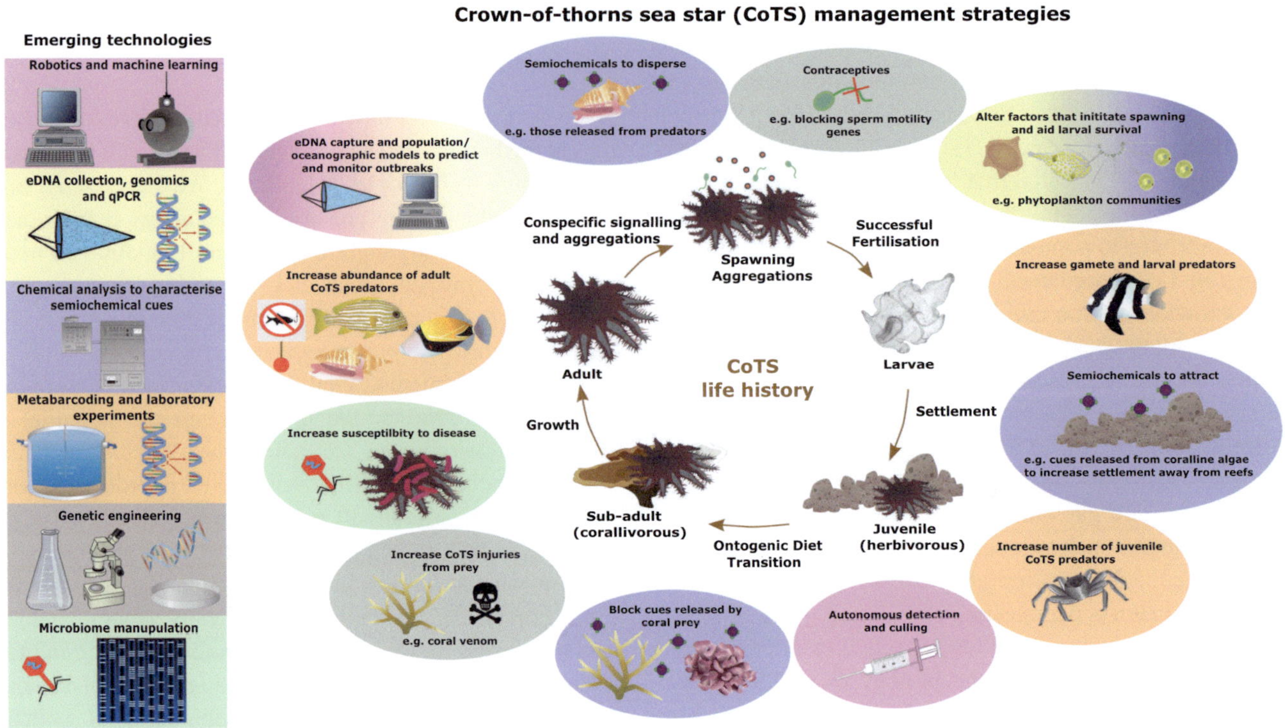

Fig. 7.12 Emerging technologies that are helping to answer knowledge gaps surrounding CoTS biology and their numbering as discussed in this chapter, with the subsequent management strategies they can aid at different CoTS life stages. (Figure produced in InkScape and illustrations downloaded from the Integration and Application Network, University of Maryland Center for Environmental Science (https://ian.umces.edu/media-library/))

that take advantage of chemical communication and how these methods can be utilised to both herd and disperse CoTS, allowing efficient culling and protecting vulnerable areas, e.g., reefs, respectively. Thirdly, we highlight how a better understanding of the CoTS microbiome could open the possibility of using probiotics to either reduce CoTS fecundity or their immunity. Finally, we end by discussing how these emerging technologies and management strategies need to be amiable to a changing climate and the ethical dilemmas behind the biocontrol of CoTS.

7.4.1 DNA Technologies

7.4.1.1 Using Metabarcoding to Redefine CoTS Predators

To effectively manage and control CoTS populations, it is important to understand the biology of their natural predators and to conserve their populations, as a natural tool to controlling outbreaks (Motti et al. 2022). Knowledge on the predators of CoTS gametes and/or larvae is limited due to the difficulty in field monitoring (Cowan et al. 2017). However, predation rates on the early life stages of CoTS (gametes, larvae and juveniles; Fig. 7.6) are postulated to have the largest influence on CoTS population outbreaks, as it directly influences the number of CoTS juveniles transitioning into adulthood (Keesing et al. 2018; Pratchett et al. 2021; Desbiens et al. 2023). For example, the predation rate on newly settled juveniles is reported to be up to 5% per day (Keesing and Halford 1992). In addition to knowledge gaps surrounding predation on early-life stages, there is currently also a lack of data on background mortality and natural predation rates on adult CoTS due to the lack of knowledge of known CoTS predators (Pratchett et al. 2021). However, new technologies such as metabarcoding of gut and faecal contents of marine fauna are increasing our knowledge of CoTS predators (Kroon et al. 2020).

So far, 80 species of coral reef organisms have been reported to feed on CoTS, including butterflyfish and triggerfish (Kroon et al. 2020). However, most of these observations have been in laboratory settings (Cowan et al. 2017; Motti et al. 2022), with field observations of predation on adult CoTS being isolated rare events and mostly documented by personal communication rather than in studies (Ormond et al. 1990; Sweatman 1995; Morello et al. 2014). Metabarcoding technology has facilitated the classification of nine potential fish predators of CoTS, including the painted sweetlips (*Diagramma pictum labiosum*) and Floral Maori Wrasse (*Cheilinus chlorourus*) by detecting CoTS DNA in faecal and gut samples of fish in the field (Kroon et al. 2020). Furthermore, metabarcoding has allowed confirmation of laboratory studies of CoTS predators, e.g., confirming that during the spawning season, some damselfish are common predators of CoTS gametes (Cowan et al. 2016b; Kroon et al. 2020).

Recently, mesocosm laboratory trials actively monitoring potential CoTS predators have identified 26 novel benthic invertebrate CoTS predators including species with large predatory impacts on CoTS juveniles (Desbiens et al. 2023). For instance, from the species tested, the red decorator crab *Schizophrys aspera* was the main predator of juvenile CoTS, consuming up to five CoTS per day (Desbiens et al. 2023). These findings show the value of multidisciplinary studies, such as using metabarcoding technologies combined with aquaria observations. These multidisciplinary studies can elucidate whether the organisms containing CoTS DNA are actively preying on live CoTS, scavenging dead CoTS, or if their position in the food chain is above that of a CoTS predator. Furthermore, these findings will highlight which species are important in controlling CoTS numbers during different life stages, e.g., damselfish for CoTS gametes, red decorator crabs for CoTS juveniles or fishes that prey on adults. This information is important in informing strategies to conserve active CoTS predators that can aid in curbing population outbreaks.

7.4.1.2 Genomics to Understand CoTS Larval Recruitment and Migration

Larval populations of CoTS have been extensively studied using indirect methods such as oceanographic simulation models to explain genetic connectivity due to the difficulty of tracking CoTS larvae in the field (Hock et al. 2014, 2017). In addition, identification of CoTS larvae is difficult due to their morphological similarities to the larvae of other asteroids and holothuroid species (Uthicke et al. 2015) and their small size (<1 mm; Horoiwa et al. 2022). However, the advent of recent genomic advances using eDNA and digital droplet qPCR has facilitated direct and more accurate measurements of CoTS larval abundance in the field (Doyle et al. 2017; Uthicke et al. 2019) and allowed larvae to be detected via eDNA, potentially indicating connectivity over large spatial scales (>500 km; Uthicke et al. 2015, 2018). Recently the assay has been optimised further by using a catalytic hairpin assembly (CHA) and *hybridisation chain reaction* (HCR) cyclic amplification strategy, allowing ultrasensitive and low-cost detection of CoTS eDNA when in the field (Wei et al. 2024). Understanding the abundance of CoTS larvae is critical as larval supply is a major predictor of CoTS outbreak intensity (Hock et al. 2017), and CoTS populations at early life stages may potentially be more easily managed. Thus, new technologies accurately monitoring the presence and abundance of early life stages are an important addition to the tools available to manage CoTS populations.

In addition to CoTS larva presence and abundance, understanding CoTS population connectivity between reefs is critical as it allows us to understand whether outbreaks are primary or secondary (Horoiwa et al. 2022). For instance, reefs experiencing both high larval retention and connectivity may be "superspreader reefs" that are the epicentres of CoTS population outbreaks (Hock et al. 2014). Genomic technologies such as genetic microsatellites (Yasuda et al. 2009) and whole-genome sequencing (Horoiwa et al. 2022) have allowed the gene flow between CoTS populations to be studied. Increasing the resolution of these genomic technologies is also allowing for more accurate population genealogies and abundances to be calculated. For example, it has recently been discovered that the Pacific CoTS population is composed of two distinct clades (Yasuda et al. 2022). Furthermore, the cloning rate in CoTS larval populations has started to be measured, allowing more accurate predictions of larval abundance and population connectivity (Allen et al. 2019; Hart et al. 2021; Uthicke et al. 2021). For the first time, qPCR has allowed CoTS to be classified into age groups by measuring the relative telomere length of different ages of CoTS (Kwong et al. 2023). Moreover, the increase in technological resolution has also facilitated the recent finding that CoTS larval presence is higher around full moons (Uthicke et al. 2019). These technologies are helping to explain CoTS abundance and the migration patterns found in the field and will ultimately help to determine times for more intense management intervention and to inform population models, helping to predict reefs under threat.

7.4.2 Bioengineering

7.4.2.1 Gene Drives to Reduce Fecundity

Gene drives to spread selfish genetic mutations throughout wild populations are an attractive avenue in pest control as these mutations reduce the fitness of the host and are passed onto the next generation (Burt 2003; Esvelt et al. 2014; McFarlane et al. 2018). Currently, two gene drives using clustered regularly interspaced short palindromic repeats (CRISPR) endonucleases (such as Cas9) are focused on inflicting sex distortion in wild populations of vertebrates; however, advances are also being made with these technologies in invertebrates such as mosquitos (McFarlane et al. 2018). This CRISPR-Cas9 technology targets genes in females that facilitate a male-dominated population (McFarlane et al. 2018). One of these gene drives targets a female-fertility gene using a homing gene, rendering the female offspring infertile (McFarlane et al. 2018). The second gene drive method uses an X-shredder that cuts the

X-chromosome during spermatogenesis, destroying the X-chromosome and therefore favouring sperm carrying the Y-chromosome to reach the mature oocyte (McFarlane et al. 2018). Female CoTS are notoriously fecund with high numbers of eggs being produced per spawning season (Babcock et al. 2016b). Therefore, methods rendering female CoTS infertile can have a dramatic effect on CoTS population sizes. That said, the mechanism of sex determination is not known and may be phenotypically plastic (Guerra et al. 2020).

Another gene that could be targeted in CoTS females is the peptide hormone, relaxin-like gonad-stimulating peptide (RGP), whose expression leads to oocyte maturation and ovulation in asteroids (Chaet 1959; Mita 2013). Recently, the expression of RGP has been investigated in CoTS, and the CoTS-specific RGP was inoculated in recombinant yeast (Smith et al. 2019). This recombinant CoTS RGP successfully initiated oocyte maturation in CoTS individuals (Smith et al. 2019). However, as RGPs can initiate reproduction across different sea star species (Mita 2016), any genetic manipulation of CoTS RGPs would have to ensure that it would not alter oocyte activity across other Asteroidea.

As males normally initiate the spawning cascade in CoTS due to pheromones released during the release of sperm (Caballes and Pratchett 2017), using gene editing to target male reproductive function and genes that inhibit/reduce sperm motility would greatly reduce the reproductive success of CoTS. Research is currently investigating genes involved in mammalian reproductive biology to find novel nonhormonal candidates for contraceptives that would stop conception but not affect overall fertility in mammals (Park et al. 2020). This has led to studies discovering genes both dispensable and indispensable for male fecundity (Miyata et al. 2015; Castaneda et al. 2017, 2020; Fujihara et al. 2019; Shimada et al. 2019; Oji et al. 2020; Park et al. 2020). As CoTS have diverged from the chordate line, gene editing therapies in CoTS should probably target well-studied evolutionarily conserved vertebrate genes.

A prime candidate to target for altering CoTS reproduction is the flagella of sperm, whose function is conserved across eukaryotes and contains genes that are also conserved (Carvalho-Santos et al. 2011). For instance, T-Complex-Associated–Testis-Expressed 1 (TCTE1) is a gene whose function helps use ATP to fuel motors required for flagella movement, and TCTE1 knockouts show male sterilisation due to abnormal sperm motility in mice (Castaneda et al. 2017). Other evolutionarily conserved genes in sperm flagella are radical spoke (RS) proteins (RSP1, RSP4, RSP6, RSP9 and RSP10) that regulate flagellar motility (Abbasi et al. 2018). Knockouts of the RS proteins cause infertility in both male mice (Tokuhiro et al. 2008; Abbasi et al. 2018) and the green alga *Chlamydomonas* (Witman et al. 1978; Huang et al. 1981).

In addition to inflicting a sex distortion in CoTS populations, targeting CoTS genes and proteins that are not required in reproduction but used in other functions, e.g., CoTS venom composition to make them more palatable to predators or chemo-detection to disrupt aggregation cues for spawning, could also be heritable modifications that reduce their population sizes. For instance, knockouts of the receptor coreceptor *Orco* gene, used for olfactory communication in the crop pest moth *Spodoptera littoralis* using CRISPR Cas9 technology, caused defects in the olfactory detection of both sex pheromones and host plants (Koutroumpa et al. 2016). Moreover, this genetic editing was heritable, with 70% passing this gene mutation onto the next generation. This method could be applied to CoTS individuals, targeting candidate genes hypothesised to be olfactory receptors in CoTS, e.g., chemosensory G-protein-coupled receptors (GPCRs) (Hall et al. 2017a, b), possibly disrupting the chemoreception of cues that induce spawning aggregations and/or the release of gametes.

Although CRISPR-Cas9 technology provides promise for targeting CoTS population, the drawbacks and ethical implications need to be considered (see Esvelt et al. (2014) for an in-depth review). For instance, gene drive-resistant alleles can occur in the wild population which could eliminate the desired gene drive (Esvelt et al. 2014). It is unknown how quickly resistant alleles will arise in CoTS populations, but ideally several gene drives would be constructed to avoid this bottleneck. Gene drives are not effective for species that undergo asexual or clonal reproduction (Esvelt et al. 2014). Although CoTS reproduce sexually, there is increasing evidence of hermaphroditism and cloning in CoTS populations (Allen et al. 2019; Guerra et al. 2020; Hart et al. 2021). The contribution of hermaphroditism and cloning rates in a natural CoTS population however is still unclear, but if significant, this may present a hurdle to eradicating CoTS outbreaks using genetic engineering technologies. Furthermore, gene drives may affect closely related species, which could harbour unintended devastating consequences (Esvelt et al. 2014). Thus, it would have to be determined whether gene drives developed in CoTS could be passed to other echinoderms. If genetic engineering is ultimately chosen as a strategy for CoTS outbreak suppression, scientists would have to develop these technologies in full transparency with local communities, stakeholders and organisations, due to the impact that this technology could have on those communities whose livelihood is dependent on reefs (Esvelt et al. 2014).

7.4.2.2 Assisted Evolution of Coral and CoTS Toxins

Toxins are produced by a wide variety of marine organisms acting as a deterrent to make them less palatable to predators and to inflict injury to predators/prey (Bornancin et al. 2017). CoTS and their predators/prey are no exception to this rule, all containing toxic peptides present in their venoms that aid them in their survival (Shiomi et al. 2004; Deaker et al. 2021; Zhang et al. 2022). Animal venoms are an area of huge interest to the pharmaceutical field due to their unique cocktail of specialised bioactive peptides, making them primary candidates for the biocontrol of pests (Moreau and Asgari 2015).

The biocontrol of CoTS outbreaks through the manipulation of coral/CoTS venom could include modern bioengineering technologies such as assisted evolution. Assisted evolution technologies are becoming increasingly popular methods to contribute to the management and restoration of coral reefs amid climate change (Suggett and van Oppen 2022). One of these technologies includes preconditioning, which is a bioengineering method where organisms can be exposed to sublethal stress to increase their overall tolerance to a particular stress and ultimately, increase their survival (Hackerott et al. 2021; Suggett and van Oppen 2022). In corals, preconditioning is a method proposed for increasing a coral's tolerance to bleaching and ocean acidification (Suggett and van Oppen 2022). Thus, assisted evolution and preconditioning methods could be employed to increase a coral's tolerance to CoTS outbreaks by manipulating their natural defence system—venom and toxin production. As CoTS themselves also synthesise venom which they use to defend themselves against predators, bioengineering technologies could also target CoTS venom to reduce its efficacy, potentially making CoTS more palatable to a wider range of predators.

7.4.2.2.1 CoTS Venom

CoTS are covered in spines (Fig. 7.13) of which the outer surface contains pores (2–10 μM diameter) joined by a porous inner network, allowing the venom to be secreted into the victims' tissues, inducing injury (Hillberg et al. 2023). The venom harbours at least two unique DNases—plancitoxin I and plancitoxin II (Shiomi et al. 2004); however, recent genome analysis has revealed a third plancitoxin gene (Hillberg et al. 2023). These toxins have been shown to induce apoptosis, oxidative stress and mitochondrial dysfunction in mammalian cells (Ota et al. 2006; Lee et al. 2015). In addition to plancitoxins, the venom contains the toxins phospholipase A2 and a peptidase inhibitor 16-like protein (PI16) in addition to an array of other proteins that may aid in CoTS envenomation (Hillberg et al. 2023). This venom may contribute to the low predation rates on CoTS. Molecular studies of the salivary glands of the CoTS predator *C. tritonis* have characterised peptides and toxins that are secreted by this species and which may be important in its efficacy in preying on CoTS (Bose et al. 2017; Zhang et al. 2022). When *C. tritonis* attacks CoTS, large amounts of mucus are produced, which may aid in sequestering and detoxifying CoTS saponin defences and may have the capacity to metabolise the saponins (Bose et al. 2017). In addition to blocking CoTS defences during predation, *C. tritonis* contains four homologous genes to cone snail conotoxin, including its own unique toxin—a Kunitz-type toxin, which causes muscle disintegration in CoTS and the destruction of tube feet tissue (Zhang et al. 2022; Jia et al. 2023). Whether assisted evolution can be utilised to lower the DNase activity of molecules in CoTS Plancitoxin or increase the sequestering activity of molecules in CoTS predators to render CoTS venom ineffective and thus make CoTS easier to predate upon, will require future investigation.

Fig. 7.13 Close-up of the venom apparatus (spines) of CoTS that harbour plancitoxins I and II in the secretory glands. (Image courtesy of Institut de Recherche pour le Développement (IRD), New Caledonia/Pascal Dumas)

7.4.2.2.2 Coral Venom

Although corals are the primary prey of CoTS, they are not submissive prey, with corals releasing defensive bioactive compounds that inflict sublethal and lethal wounds on CoTS (Deaker et al. 2020, 2021) from their toxin-producing cnidocysts (David et al. 2008; Fautin 2009). Juvenile CoTS have been reported to die after feeding on coral (Keesing and Halford 1992; Johansson et al. 2016; Deaker et al. 2021), with species such as *Echinopora lamellosa* causing 100% mortality post-predation (Johansson et al. 2016). Even the preferred prey genera of CoTS, *Acropora*, has been observed to inflict injury and death in CoTS (Deaker et al. 2020). Interestingly after injury from corals, CoTS have been shown to switch their diet to coralline algae to recover (Deaker et al. 2020). Furthermore, radial nerve cord bulb structures have been recently discovered in adult CoTS that have been hypothesised to develop prior to their ontogenetic diet transition and are postulated to aid in their defence against coral toxins during feeding (Smith et al. 2023). Interestingly, although not a scleractinian coral, the hydroid *Millepora platyphylla* has been shown to offer the preferred scleractinian coral prey genera of CoTS, *Acropora*, refuge from CoTS predation, as CoTS actively avoided areas where *M. platyphylla* was present (Kayal and Kayal 2017). *M. platyphylla* possesses haemolytic venom consisting of a phospholipase A_2 known as milleporin-1 (Radwan 2002; Radwan and Aboul-Dahab 2004). Whether the potency of *M. platyphylla* venom is strong enough to effect CoTS predation behaviour and movement and is the driver behind these coral refuges will require future work.

A recent study (Gorman et al. 2025) has utilised bioinformatics to start to investigate which species-specific coral toxins may cause injury in CoTS and how this differs between CoTS preferred (*Acropora* species) and non-preferred coral prey species (*Porites* species). Interestingly, several differences between the venom and toxin composition of these species were found including *Porites* species containing homologs to the lethal jellyfish JFT-1b toxins which were absent from *Acropora* species (Gorman et al. 2025). Both *Acropora* and *Porites* species contained several neurotoxic kunitztype proteins (Gorman et al. 2025). A kunitz-type protein isolated from the CoTS predator *Charonia tritonis* has previously been shown to degrade the tube feet of CoTS (Zhang et al. 2022). Whether these same proteins in *Acropora* and *Porites* species can damage CoTS requires future work. Furthermore, whether CoTS can become immune to these coral toxins over time through cell-based or humoral immunity (Smith et al. 2010) should be investigated. On the one hand, corals have been shown to increase the number of their nematocysts in response to predation e.g., *Porites compressa*, increased its number of nematocysts, thereby reducing its palatability to reef fishes (Gochfeld 2004). But on the other hand, after repeated feeding on a coral species, CoTS increase their readiness to feed on that species (Ormond et al. 1976). One explanation may be that of an evolutionary arms race, where CoTS develop immunity to the venom of a coral species, allowing the CoTS to continue to feed on the coral, and in turn the coral increases the potency of its venom. Once these toxins are characterised, studies can investigate ways to increase venom potency in CoTS prey/predators, e.g., preconditioning corals to CoTS predation, whilst decreasing venom potency in CoTS to facilitate predation upon CoTS. This would both increase their injury rates from prey species (possibly inhibiting feeding or changing their preferences to less suitable corals) and make them more palatable to predators.

7.4.3 Chemical Signals or Cues (on Life-History Traits)

Chemical cues are used by marine organisms for communication (e.g., Bornancin et al. 2017). CoTS are not exempt, using chemical communication in foraging (Ormond et al. 1976; Ling et al. 2020), between conspecifics to form spawning aggregations (Hall et al. 2017a, b), in larval settlement (Cowan et al. 2016a; Doll et al. 2023) and juvenile diet transition (Deaker et al. 2020; Neil et al. 2022). Thus, as CoTS rely on semiochemicals for most of their survival, utilising CoTS chemical cues may provide a more efficient way to manage CoTS numbers either by deterring or herding CoTS. Recent studies are starting to investigate potential semiochemical candidates (see Harris et al. 2025 for a review). Whichever semiochemical candidates are chosen for further research in controlling CoTS populations, they need to all harbour similar properties of being able to have long range dispersal, specific to CoTS, and effective in eliciting behavioural changes once CoTS have come into contact with the cue (Harris et al. 2025). Furthermore, it is imperative that it is understood whether the chemical cue attracts a specific, or multiple, CoTS life stages (Harris et al. 2025).

7.4.3.1 Foraging and Avoiding Predators

Movement of CoTS is mediated by chemoattractants released from coral prey (Ormond et al. 1976; Ling et al. 2020). Regardless of species, CoTS juveniles have shown feeding preferences for *Acropora* and pocilloporid corals (De Bruin 1972; Ormond et al. 1973; De'Ath and Moran 1998; Johansson et al. 2016; Keesing 2021), with *Porites* being the least preferred coral genus by CoTS (De'Ath and Moran 1998; Millican et al. 2024). *Acropora* and pocilloporid corals promote the fastest growth of CoTS (Keesing and Halford 1992; Keesing 2021), and CoTS still show preference for these corals, even at very low abundances, albeit with increased consumption of non-preferred coral taxa (Keesing et al. 2019). The density of preferred *Acropora* spp. in a local reef area has been shown to affect the movement of CoTS, with densities of >55% causing CoTS to remain localised, whilst <28% caused CoTS to emigrate (Ling et al. 2020). Furthermore, when stimuli from the coral *Fungia fungites* (previously considered a preferred species of CoTS) were extracted and implanted on plaster of Paris, they attracted CoTS individuals, which then attempted to feed on the artificial substrate (Ormond et al. 1976). This highlights the power of chemosensory attractants in influencing the movement of CoTS for foraging and provides a promising tool for herding CoTS to specific locations on reefs.

In addition to chemicals that attract CoTS to their prey, chemicals released by the CoTS predator, *C. tritonis*, alter CoTS behaviour causing them to disperse (Bose et al. 2017; Hall et al. 2017a, b). When CoTS were incubated in water containing the plumes released from *C. tritonis* CoTS acted rapidly and adversely (Hall et al. 2017a, b). This mechanism may be useful in disrupting CoTS aggregations and possibly reducing spawning success due to the negative relationship between spawning success and distance (Caballes and Pratchett 2014). The full mitochondrial genome of *C. tritonis* has been sequenced (Klein 2021), as well as a reference transcriptome (Klein et al. 2021), which may lend insight into both the chemical mechanisms used by *C. tritonis* to find/perceive CoTS and the chemicals released by *C. tritonis* that are perceived by CoTS which cause CoTS to disperse. These findings may help researchers to release chemoattractants into the water to attract CoTS predators into areas that are experiencing CoTS outbreaks, or to use chemorepellents to disperse CoTS aggregations.

7.4.3.2 Reproduction

Studies are starting to characterise the biomolecular signals that underpin CoTS reproduction (Hall et al. 2017a, b; Smith et al. 2017; Guerra et al. 2020; Mendoza-Porras et al. 2023). One class of molecules that may represent a significant role in CoTS reproduction is steroid oligoglycosides known as asterosaponins that have reproductive roles in Asteroidea (Ikegami et al. 1972). Asterosaponins inhibit oocyte maturation that allows the synchronisation of spawning in asteroids (Ikegami et al. 1972) and which synergistically react with sperm-activating peptides that induce the acrosome reaction, an essential reaction for successful fertilisation (Hoshi et al. 1994; Stonik et al. 2020). In addition to reproductive roles, asterosaponins are postulated to aid in both predator deterrence of CoTS gametes, making eggs and larvae unpalatable to certain planktivorous fish (Lucas et al. 1979; Tang et al. 2005; Malyarenko et al. 2012), and communication facilitating CoTS aggregation behaviour during spawning (Mendoza-Porras et al. 2023).

CoTS-specific asterosaponins have been characterised, with two asterosaponins—Regularoside B (Asterosaponin A) and Hillaside B present in male gonads—and over 34 asterosaponins detected in eggs, with Glycoside B2 and Thornasteroside A being the most abundant (Mendoza-Porras et al. 2023). In other Asteroidea, Regularoside B inhibits spawning (Ikegami 1976), Glycoside B and Asterosaponin 4 inhibit oocyte maturation (Ikegami 1976; Okano et al. 1985) and Thornasteroside A has a defensive role (Malyarenko et al. 2012). In addition to asterosaponins, two key metabolites found in high abundance in CoTS reproductive tissues may act as a pheromone complex to trigger spawning in CoTS—Inosine in male gonads and 2-hexyldecanoic acid in eggs (Mendoza-Porras et al. 2023). Future studies continuing this work to characterise CoTS asterosaponins/spawning metabolites and to elucidate their functions may aid in creating management strategies to disrupt their signalling and prevent spawning synchronicity (Mendoza-Porras et al. 2023).

7.4.3.3 Spawning Aggregations

Spawning aggregations are a primary mechanism to be targeted for CoTS population control as they cause mass predation on coral at local scales, decreasing the ability of a reef to recover and promote the formation of local aggregations for spawning and fertilisation success (Pratchett et al. 2014). When naïve CoTS were incubated in water containing CoTS aggregation cues, they became highly active and moved in the direction of the aggregation factors (Hall et al. 2017a, b). Subsequently, the genome of CoTS was sequenced to characterise gene products that may allow chemical communication between conspecifics to form aggregations (Hall et al. 2017a, b). The main communication proteins secreted by aggregating CoTS were CoTS-specific ependymin-related proteins (EPDRs), which also function as signalling molecules in vertebrates (Hall et al. 2017a, b). EPDRs have also been recently discovered in CoTS venom extracts (Hillberg et al. 2023). The candidate genes for communication receptors that would receive these EPDR signals in CoTS are genes in the large family of chemosensory G-protein-coupled receptors (GPCRs), confined to CoTS sensory tissues (Hall et al. 2017a, b). Studies investigating CoTS tissue-specific secretory proteins and sensors are still in their infancy; however, recent proteomic analysis has started to characterise putative secretory proteins in the CoTS radial nerve (Smith et al. 2023). Future work should focus on investigating the function of these hypothesised spawning aggregation factors and communication receptors, as these could prove useful mechanisms to concentrate CoTS into an area away from target reefs for their eradication.

7.4.3.4 Spawning

As for most broadcast spawners, CoTS spawning is synchronised, in which the release of male gametes initiates a spawning cascade in conspecifics (Caballes and Pratchett 2017). The synchronisation of this spawning cascade is induced by pheromones released from the sperm and gonads/ovaries of spawning individuals (Beach et al. 1975; Caballes and Pratchett 2017). This sperm pheromone also causes CoTS to aggregate, migrating towards the spawning individual (Beach et al. 1975). These aggregations increase fertilisation success with 100% fertilisation in individuals <1 m apart (Pratchett et al. 2014). However, even at larger distances (>20 m apart) CoTS have high fertilisation success, significantly greater than that reported for other marine invertebrates (Caballes and Pratchett 2014; Babcock et al. 2016b), due to the high fecundity of CoTS and mass of sperm produced (Babcock et al. 2016b). A mature female can produce >100 million eggs and an average-sized male produces 1.1×10^{13} sperm per spawning season (Babcock et al. 2016b). The high fertilisation success of isolated females may also be due to the presence of cryptic hermaphroditism, as seen in a CoTS population in Japan (Guerra et al. 2020; Hart et al. 2021). In contrast to other dioecious sea stars, the ovaries of CoTS express male genes, and the testes express female genes (Guerra et al. 2020).

In a laboratory setting, spawning of male CoTS is linked to cues such as an abrupt temperature increase, the presence of the phytoplankton *Skeletonema* and the presence of conspecific sperm (Caballes and Pratchett 2017). Contrastingly, female CoTS were only induced to spawn by the presence of male sperm in the water column, being insensitive to other environmental triggers (Caballes and Pratchett 2017). Male and female CoTS have different expressions of chemosensory receptors (Hall et al. 2017a, b). Whether this difference is linked to the difference in response to spawning cues is unknown. Male CoTS release their gametes 240 minutes after an increase in temperature (Caballes and Pratchett 2017), highlighting their response to environmental cues. Not the temperature itself, but the abrupt increase of temperature, triggers spawning in male CoTS (Caballes and Pratchett 2017). CoTS spawning occurs from spring to summer when the oceans begin to warm. This timing differs with location, with spawning occurring in December and January on the GBR, New Caledonia and Vanuatu, whereas in Hawaii spawning occurs during April to June (Pratchett et al. 2014; Dumas et al. 2016; Hue et al. 2020). Neither phytoplankton abundance nor nutrient levels were linked with spawning in CoTS; however, when CoTS were exposed to the diatom genus *Skeletonema*, 62.5% spawned (Caballes and Pratchett 2017). The specific chemosensory cues released from CoTS gametes and *Skeletonema* that induce spawning are unknown (Pratchett et al. 2021); however, they could prove vital in the fight against CoTS, with their inhibition possibly desynchronising spawning between conspecifics.

7.4.3.5 Larval Feeding and Settlement

As is typical of most sea stars, CoTS have a biphasic planktonic-benthic life cycle (Fig. 7.6). Whilst it is not widely known which phytoplankton the larvae eat in the field, in the laboratory they are resilient to low food conditions (typical of reefs), as well as developing faster in more enriched food conditions (Wolfe et al. 2015a, 2017). A recent study of phytoplankton community structure in the South China Sea has discovered that during a CoTS outbreak, Bacillariophyta were the dominant phytoplankton (Peng et al. 2023). However, it is still unclear the extent to which CoTS larvae feed on Bacillariophyta and other phytoplankton in situ and which species can support larval CoTS outbreak densities in either abundance, nutrition or both. In low food conditions, CoTS larvae can adjust the profile of their body to increase their ciliary band length to enhance particle capture, exhibiting the phenotypic plasticity typical of many echinoderm feeding larvae (Wolfe et al. 2015b). Furthermore, CoTS larvae harbour a symbiosis with resident photobacteria in their gut microbiome, which also points to the possibility of a source of nutrients from symbiotic bacteria (Carrier et al. 2018). The plasticity of CoTS larvae to extreme food conditions has been shown by larvae able to complete their larval phase only feeding on energetically poor cyanobacteria (Mos et al. 2024). In laboratory cultures, peak settlement occurs after 16–22 days post-fertilisation under optimum food conditions (Wolfe et al. 2015a; Pratchett et al. 2017b). However in the field, settlement occurred as early as 14 days (Olson 1987). It appears that CoTS larvae primarily settle in coral rubble habitats at intermediate depths and in response to the presence of certain crustose coralline algae (CCA) species in the field (Wilmes et al. 2020a; Doll et al. 2023). Moreover, the loss of live coral (facilitated by corallivorous CoTS adults) creates an environment that increases the settlement of juveniles, invoking a negative feedback loop known as the "degraded reef hypothesis" (Wolfe and Byrne 2024).

The early CoTS juveniles are subject to mortality and injury (Wilmes et al. 2019). This may be due to predators in the rubble such as crabs, as well as injuries from coral venom (Deaker et al. 2021; Desbiens et al. 2023). CoTS larvae are able to detect chemosensory signals in the water that impede or induce this settlement (Cowan et al. 2016a; Doll et al. 2023). In the laboratory CoTS larvae have a strong settlement response to several species of coralline algae (Deaker et al. 2020; Byrne et al. 2023; Doll et al. 2023) probably due to the diet of juvenile CoTS being primarily CCA (Zann et al. 1987). Settling in areas with high prey is an advantageous trait for CoTS juveniles, especially due to their high mortality rates (>99%; Zann et al. 1987). In addition to CCA, the strongest settlement cues in CoTS larvae are from adult conspecifics (Cowan et al. 2016a). However, CoTS can also settle in the absence of a cue (Wolfe et al. 2015a), albeit at low rates.

The larvae of other echinoderms (sea stars and sea urchins) as well as corals also undergo metamorphosis in the presence of a range of CCA species (Huggett et al. 2006; Swanson et al. 2012). Like CoTS, coral larval settlement induced by CCA depends on both the CCA species and the community composition of the CCA microbiota (Gómez-Lemos et al. 2018; Whitman et al. 2020; Jorissen et al. 2021; Petersen et al. 2021; Yang et al. 2021). For instance, the settlement rates of CoTS larvae varied markedly with CCA species, with *Melyvonnea* cf. *madagascariensis* promoting the settlement >90% of CoTS larvae, whereas *Lithophyllum* and *Porolithon* induced less than 2% settlement (Doll et al. 2023). The addition of antibiotics to the CCA surface inhibited CoTS larvae metamorphosis (Johnson and Sutton 1994), suggesting a strong influence of the CCA microbiome on CoTS settlement as witnessed for certain coral larvae. For example, CCA species with markedly distinct microbiomes and metabolomes induced high rates of settlement in *Acropora cytherea* larvae, suggesting that several different classes of biomolecules may induce settlement (Jorissen et al. 2021). Therefore, it is likely that both the CCA and its microbiome induce settlement in CoTS larvae; however, the specific chemical cues/molecules that initiate settlement and which CCA species/CCA-associated bacteria produce these cues are still unknown.

Using modern techniques such as bioassay-guided fractionation and 16S metabarcoding, studies have been able to characterise classes of molecules in CCA and their microbial communities that may be the chemical inducers of marine invertebrate settlement, e.g., histamine, glycoglycerolipids, 11-deoxyfistularin-3, polysaccharides and betaine lipids (Kitamura et al. 2007; Swanson et al. 2012; Tebben et al. 2015; Jorissen et al. 2021). Future research should use these techniques to investigate the classes of molecules that induce settlement in CoTS. These molecules could then be synthesised to promote settlement to herd CoTS in areas designated for culling, or these molecules could be scavenged/blocked in areas to inhibit settlement.

In addition to settlement induction, areas with higher benthic predators such as polychaetes, trapeziid crabs and scleractinian corals, elicited a decrease in the settlement of competent CoTS brachiolaria (Cowan et al. 2016a). Furthermore, if CoTS larvae settled in the presence of these predators, none of them metamorphosed, compared with 55% in areas devoid of benthic predators (Cowan et al. 2016a). This is unsurprising, due to the significant effect of predation on the population of newly settled CoTS (Keesing and Halford 1992). Future research should prioritise characterising the molecules that induce the settlement of CoTS larvae by comparing the metabolome and microbiome of a high-CoTS settlement inducer (e.g., *Melyvonnea* cf. *madagascariensis*) with a low-CoTS settlement inducer (e.g., *Lithophyllum* cf. *kotschyanum*), as these molecules will ultimately determine the recruitment potential onto reefs (Cowan et al. 2016a; Doll et al. 2023). Furthermore, these molecules could be used to spike water with olfactory cues released from benthic predators (Desbiens et al. 2023) to inhibit settlement or olfactory cues from CCA that induce settlement, allowing CoTS to be herded for efficient culling or biocontrol. Although it should be investigated what effect this chemical stimulation or inhibition would have on the behaviour of other marine species, especially since chemicals released from CCA are a vital cue for the settlement of the larvae of several coral reef species.

7.4.3.6 Diet Transition of Juveniles from Herbivory to Corallivory

Post-settlement juvenile CoTS are herbivorous, feeding on CCA and other algae (Zann et al. 1987; Deaker et al. 2021). In optimal conditions once CoTS juveniles reach a certain size (>8 mm; 13–15 months of age), they are competent to switch their diet to feed on scleractinian corals (Yamaguchi 1974; Zann et al. 1987). After transitioning to a corallivorous diet, the juveniles experience exponential growth and subsequently transition into adulthood (Zann et al. 1987; Deaker et al. 2020; Wilmes et al. 2020b). The presence of coral prey determines this ontogenetic diet transition (Deaker et al. 2020). In the absence of coral prey, the juveniles can remain as cryptic herbivores feeding on CCA and algal biofilms (Deaker et al. 2020; Neil et al. 2022). This delay in metamorphosis poses an alarming risk, allowing the build-up of recruitment pulses over years to generate high background populations of CoTS juveniles in the reef infrastructure (Deaker et al. 2020). This disconnect between the age of CoTS and their ontogenetic diet transition also explains the disconnect between conditions leading to CoTS larval blooms and adult population outbreaks (Deaker et al. 2020). This makes the boom-and-bust cycle of CoTS even more erratic and highlights the inherent plasticity of CoTS (Deaker and Byrne 2022).

How a CoTS individual perceives the prospective environment and determines whether they shift their diet from CCA to coral is unknown. Whether the presence of any coral will cause this shift or whether the presence of a preferred coral species will elicit this shift requires further study. Presumably juvenile CoTS would use sensory cues to perceive the environment to decide whether they make the switch from CCA in the coral rubble to coral on a reef habitat (Wilmes et al. 2020b). Corals and coral reefs release sensory cues including olfactory (Lecchini et al. 2014), visual (Foster and Gilmour 2016) and auditory (Lillis et al. 2016) cues that attract larvae to the reef (Barth et al. 2015). All three cues have different signatures from healthy reefs compared with degraded reefs (Lecchini et al. 2014; Piercy et al. 2014; Bertucci et al. 2016), and both olfactory and audial cues can be transported kilometres away from reefs (Lecchini et al.

2014; Piercy et al. 2014). Whether CoTS use these cues to help inform their decision to change diet is unknown. CoTS have been shown to be sensitive to olfactory cues (Hall et al. 2017a, b), and echinoderms can respond to sound (Vazzana et al. 2020), although very little is known about noise stimulus on echinoderms. Although vision in CoTS is reported to be limited to close range and a narrow horizontal field of view, it is important to adult CoTS who use eye spots at the end of each arm to navigate towards reef structures (Garm and Nilsson 2014; Petie et al. 2016). These eyespots function as simple compound eyes and are highly sensitive to visible light, with a photoreception wavelength range of 425–580 nm (Garm and Nilsson 2014; Petie et al. 2016; Hall et al. 2017a, b). Visual cues inform settlement decisions of certain coral reef organisms such as fish larvae (Chaput et al. 2019), and thus, may also be important for CoTS larvae.

If the sensory signals produced by corals and coral reefs that elicit a change in diet can be scavenged or blocked, this would mean that even if there are high background populations of cryptic juvenile CoTS, adult scleractinian corals would not be threatened. Furthermore, as CoTS can transition back to a CCA diet in unfavourable conditions (Deaker et al. 2020), it may force adult CoTS to revert to a herbivorous diet. However, large populations of CoTS feeding on CCA would potentially reduce the biomass of CCA on reefs. As certain CCA species induce the settlement of scleractinian corals, this in turn could create a negative feedback loop that reduces coral larvae recruitment and subsequently reduces hard coral cover on reefs.

7.4.4 Microbiome Manipulation

7.4.4.1 Infectious Diseases

Microbiomes play an important part in innate immunity for eukaryotes, with a dysbiosis in the bacterial community leading to the spread of disease (Round and Mazmanian 2009; Vega Thurber et al. 2020). Diseased CoTS are rarely encountered in the wild (Pratchett 1999); however, CoTS can experience disease, showing symptoms of lesions, tissue disintegration, discoloured/necrotic skin, accumulation of mucus on spines and collapsed spines (Rivera-Posada et al. 2011a, b; Høj et al. 2018). Currently, there are only a handful of studies (Carrier et al. 2018; Høj et al. 2018) that have produced work on culture-independent microbial communities in CoTS (Wada et al. 2020). When injected with a culture medium that promotes the exponential growth of the natural bacterium, *Vibrionales*, CoTS showed signs of dysbiosis followed by diseased and mortality in 100% of individuals injected (Rivera-Posada et al. 2011a, b). Furthermore, on encountering infected CoTS, this disease spread to healthy CoTS individuals which died within 24 h (Rivera-Posada et al. 2011a). This agrees with previous studies that have shown CoTS are capable of succumbing to infectious diseases or infectious maladies (Pratchett 1999), despite upregulating their immune response during outbreaks (Mills 2012). Thus, there is potential to identify and synthesise a pathogen that could be used as a biocontrol agent for CoTS that elicits a wave of infectious disease or systemic illness that targets life-history traits of CoTS, e.g., fecundity (Smith et al. 2022). For instance, the ciliate *Orchitophyra stellarum* has been shown to cause castration in male echinoderms in the Asteriidae family once infected, and it was considered to be a potential biocontrol for the invasive Japanese sea star *Asterias amurensis* in Australia (Byrne et al. 1997; Smith et al. 2022). However, the release of biocontrol agents in the sea is problematic, as this may have unanticipated consequences such as the spreading of harmful microbes to other species. For instance, the parasitic ciliate *O. stellarum* infects several other asteroid species, and so was not used for biocontrol as its application was deemed to have a risk of infecting native species (Byrne et al. 1997).

7.4.4.2 Specific Bacteria Affecting Sperm Production and Larval Development

Electron microscopy and metabarcoding using 16 s rRNA were used to examine the microbiome from healthy and diseased CoTS across different tissues (Høj et al. 2018). Across all tissue types, healthy CoTS had a significantly higher presence of a *Flavobacterium*-related operational taxonomic unit (OTU), an OTU related to a marine spirochete and an OTU belonging to the phylum *Bacteroidetes*. The dominant bacteria in specific CoTS tissue types were documented, with *Mollicutes* dominating the male gonad microbiome, *Spirochaetales* dominating the body wall microbiome, *Hyphomonadaxeae* dominating the tube feet microbiome and *Oceanospirillales* dominating microbiomes in all tissue types (Høj et al. 2018). Although present in healthy individuals, diseased CoTS microbiomes showed significant increases in *Oceanospirillales*- and *Endozoicomonas*-related OTUs and *Vibrionaceae* species. Furthermore, a bacterial species only associated with diseased individuals (an OTU related to the genus *Arcanobacterium*) was identified (Høj et al. 2018). Bacteria associated with diseased CoTS gonads (OTUs related to *Spiroplasma*) were of particular interest, with infections of related bacteria shown to affect sperm production in male crayfish (Longshaw et al. 2012) and kill male progeny in *Drosophila* (Regassa and Gasparich 2006). A long-term study conducted over 13 years investigated CoTS-associated bacteria and found that a single bacterium related to the *Spirochaetes* phylum, named CoTS27, was universally present in, and dominated, the subcuticular space throughout CoTS in the Indo-Pacific (Wada et al. 2020). It was concluded that CoTS27 formed a monospecific symbiotic relationship with CoTS in the subcuticular space (Wada et al. 2020). However, subcuticular bacteria are a common feature of sea stars, and although their exact function is unknown, they are postulated to be symbiotic and serve to aid larval development (Cerra et al. 1997; Carrier and Reitzel 2020). Our understanding of the CoTS microbiome is in its infancy, with most studies focusing on classifying CoTS-associated bacteria, with knowledge gaps remaining on the function of these specific bacterial species in CoTS health. Future work on probiotics should investigate the potential to manipulate the CoTS microbiome to alter CoTS physiology, especially focusing on microbes (or functions) that have been hypothesised to alter fecundity and immunity or aid CoTS health (e.g., CoTS27 that is hypothesised to aid CoTS larval development), to reduce the abundance of CoTS.

7.5 Conclusion

This chapter has reviewed recent research on the life history and biological traits of CoTS and highlighted five avenues of future research using emerging technologies that may contribute to promising management strategies in CoTS biocontrol. The ethical implications of these possible management strategies must also be assessed prior to deployment. For instance, will increasing the abundance of certain CoTS predators at the reef-level reduce the abundance of their other prey species, initiating cascading effects? Would bioengineering coral venom to become more potent to CoTS cause CoTS to rapidly evolve their own biological defences for sequestering coral venom more efficiently, making them become more prolific corallivores? Furthermore, will using chemosensory cues to facilitate or disperse CoTS aggregations interfere with the chemosensory cues of other species, e.g., coral larval recruitment to reefs? Will releasing pathogens that initiate diseases that spread through CoTS populations also be transmitted to other echinoderm species causing population collapses? All these areas require careful consideration from scientists and stakeholders before being deployed into the field, to evaluate the "butterfly effect" of a proposed CoTS management strategy on the rest of the reef.

In addition to ethical implications, management strategies must also be designed to be able to adapt to the changing environmental conditions (e.g., increased sea surface temperatures, ocean acidification, increased terrestrial run-off events from cyclones and increased sea-levels) in the advent of climate change, undoubtedly adding another layer of complexity to managing CoTS outbreaks (Pratchett et al. 2021). Currently, studies are lacking that investigate the compounding effects of all future changes in the environment (temperature, ocean acidification, salinity, etc.) on CoTS populations (Pratchett et al. 2021). Troublingly, the future environmental predictions of temperature and ocean acidification indicate that CoTS outbreaks will pose a continuing indefinite threat to reefs, with conditions favouring outbreaks and increasing their intensity.

For instance, studies have shown that CoTS larvae have a broad thermal optimum with high survival and normal development across 25–32 °C (Lamare et al. 2014; Uthicke et al. 2013; although see Kamya et al. 2014), and recently, the thermal limit of juvenile CoTS has been recorded at 34–36 °C (Byrne et al. 2023). However, under ocean acidification conditions, larvae have shown slowed growth and attained more developmental abnormalities (Mos et al. 2023). Contrastingly to larvae, increased ocean temperatures and acidification have elicited positive effects on the growth of juvenile CoTS (Kamya et al. 2016, 2017, 2018). Furthermore, these juveniles have been observed to tolerate heatwave conditions over a prolonged period which has been shown to kill their coral prey (Byrne et al. 2023). Increased temperatures caused higher activity, movement and metabolic rates in juvenile and adult CoTS, probably leading to an increase in their feeding rate to compensate for their increased energy expenditure (Kamya et al. 2018; Lang et al. 2022; Byrne et al. 2023). Reproductive success in adult CoTS exposed to increased temperatures for 4 months was reduced around three-fold, whereas acidification exposure had no effect on reproductive success (Hue et al. 2020). However, even though less offspring were produced, the ones that were produced in these conditions had no significant mortality (Hue et al. 2022), highlighting the potential for transgenerational plasticity in response to changing environmental conditions. In addition, the increase in the availability of coral rubble habitat (required by CoTS juveniles) post bleaching is also a concern (Wolfe and Byrne 2024). Thus, it is still not clear whether the combined effects of increased temperature and acidification will elicit a negative or positive impact on CoTS populations over generations and whether these conditions will inhibit or generate outbreaks. However, what is becoming clear is that CoTS life stage will have a large influence in determining this impact.

Although storms and cyclones are becoming more prevalent with climate change (Heron et al. 2005), whether this would aid, be negligible, or hinder CoTS outbreaks is yet to be determined. Storms and cyclones would increase terrestrial run-off (and therefore nutrient loads) to oceans and there is a correlation between the increase in nutrient loads and CoTS larval development and survival (Brodie et al. 2005; Fabricius et al. 2010; but see Wolfe et al. 2017). For instance, increased nutrient loads leads to high phytoplankton abundance which has been shown to aid CoTS larvae in overcoming the negative effects of increased sea temperatures (Mos et al. 2023). Furthermore, the presence of certain phytoplankton species (*Skeletonema*) triggers spawning in male CoTS which subsequently triggers a spawning cascade (Caballes and Pratchett 2017). However, with increased terrestrial run-off also comes lower salinities, and this decrease in salinity has been shown to decrease survival of CoTS larvae, despite providing enriched nutrients (Clements et al. 2022). This deleterious effect of salinity coupled with the physical damage inflicted by a cyclone may cause widespread mortality in CoTS populations and potentially explains the absence of a temporal link between terrestrial run-off and CoTS larval populations to reefs to date (Clements et al. 2022). Future research should aim to elucidate CoTS population dynamics following storms and cyclones.

What is clear is that the diversity and abundance of coral reefs are decreasing and will continue to decrease, causing a reduction in the total hard coral cover on reefs (Hoegh-Guldberg et al. 2007; Hughes et al. 2017). Mass die-offs of coral species sensitive to temperature anomalies are permanently transforming the reef habitat (Hughes et al. 2018). This reduction in the availability of coral prey species on reefs changes the behaviour of CoTS, causing aggregations of individuals around remaining coral, indirectly facilitating population aggregations and ultimately reducing the resilience of a reef to recover following bleaching (Haywood et al. 2019). The decline in corals combined with the response of CoTS across different life stages to climate change stressors for coral reefs is concerning. For instance, the robustness of CoTS larvae at ≤32 °C, the positive response of CoTS juveniles to heatwave and ocean acidification conditions and adult CoTS not exceeding their biological optimum until temperatures of 32 °C (the predicted 2100 summer maximum temperature) are reached (Lucas 1973; Lang et al. 2022). By 2100, CoTS will have already carried out a further 70 years of damage to reefs in conditions that do not exceed their biological optimum.

The fifth documented CoTS outbreak is occurring at this very moment across the GBR (Australian Institute of Marine Science 2024), and the fourth documented CoTS outbreak is occurring across French Polynesia (pers. comm. LMG & SCM). Effective management strategies to preserve coral reefs by shielding them from the nefarious effects of CoTS outbreaks need to be employed sooner rather than later. Climatic changes such as current summer maximum temperatures of 28 °C, shortly becoming average summer temperatures, will cause massive bleaching whilst simultaneously increasing CoTS feeding rates (to compensate for their increased metabolism), further exacerbating the effects of CoTS outbreaks. That said, the demise of their coral prey due to bleaching would prevent the generation of outbreaks.

The authors of this chapter include global coral reef and CoTS experts, and our consensus is that management strategies (including the ones mentioned in this chapter) need to be prioritised to help combat the threats that CoTS pose to coral reefs, agreeing with recent calls for rapid prioritisation from other lead experts (Babcock et al. 2020; Pratchett et al. 2021). We also call for more stringent enforcement on the illegal trade of *C. tritonis* shells and increased education of tourists on the importance of this keystone coral reef species, greater designation of MPAs across reef habitats to help increase the biomass of the plethora of CoTS predators, and governments to make more concentrated efforts to reduce carbon emissions to combat climate change. All these factors will work synergistically with CoTS management strategies to help mitigate the impact of CoTS outbreaks and aid in the conservation of coral reefs.

Acknowledgements We would like to thank Daphne Cortese for creating Fig. 7.4 as well as Yannick Chancerelle, Gilles Siu, Bernard Salvat, René Galzin, Claude Bouchon and other members of CRIOBE for providing the data for the long-term monitoring of live coral cover around Moorea (SNO CORAIL) for Fig. 7.4. In addition, we would like to thank Jean-Michel Boré for supplying Fig. 7.3, Dione Deaker and Matthew Clements for supplying images (A–E and H and I, respectively) for Fig. 7.6 and Matthieu Juncker for supplying Figs. 7.7 and 7.9. Finally, we would like to thank both the Laboratoire d'Excellence (LabEx) Coral reefs facing global change (CORAIL) and École Pratique des Hautes Études (EPHE)-PSL ATER for salaries for LMG.

References

Abbasi F, Miyata H, Shimada K et al (2018) RSPH6A is required for sperm flagellum formation and male fertility in mice. J Cell Sci 131(19):jcs221648

Abou Zaid MM, Kotb MA (2000) Human and natural induced impacts to the Egyptian Red Sea coral reefs. In: International symposium on the extent of coral reef bleaching, Riyadh, Saudi Arabia, 5–9 February 2000

Adjeroud M, Kayal M, Iborra-Cantonnet C et al (2018) Recovery of coral assemblages despite acute and recurrent disturbances on a South-Central Pacific reef. Sci Rep 8(1):1–8. https://doi.org/10.1038/s41598-018-27891-3

Allen JD, Richardson EL, Deaker D et al (2019) Larval cloning in the crown-of-thorns sea star, a keystone coral predator. Mar Ecol Prog Ser 609:271–276. https://doi.org/10.3354/meps12843

Australian Institute of Marine Science (AIMS) (2024) Understanding crown-of-thorns starfish outbreaks. Available at: https://www.aims.gov.au/research-topics/featured-projects/reef-spawning-research-aims/understanding-crown-thorns-starfish-outbreaks. Accessed 21 Feb 2025

Babcock RC, Dambacher JM, Morello EB et al (2016a) Assessing different causes of crown-of-thorns starfish outbreaks and appropriate responses for management on the Great Barrier Reef. PLoS One 11(12):e0169048

Babcock RC, Milton DA, Pratchett MS (2016b) Relationships between size and reproductive output in the crown-of-thorns starfish. Mar Biol 163(11):234. https://doi.org/10.1007/s00227-016-3009-5

Babcock RC, Plagányi ÉE, Condie SA et al (2020) Suppressing the next crown-of-thorns outbreak on the Great Barrier Reef. Coral Reefs 39(5):1233–1244

Barnes J, Endean R (1964) A dangerous starfish – *Acanthaster planci* (Linné). Med J Aust 1(16):592–593

Barth P, Berenshtein I, Besson M et al (2015) From the ocean to a reef habitat: how do the larvae of coral reef fishes find their way home? A state of art on the latest advances. Vie et Milieu 65(2):91–100

Bass DK, Miller IR (1996) Crown-of-thorns starfish and coral surveys using the manta tow and scuba search techniques. Australian Institute of Marine Science

Beach DH, Hanscomb NJ, Ormond RFG (1975) Spawning pheromone in crown-of-thorns starfish. Nature 254(5496):135–136. https://doi.org/10.1038/254135a0

Beaman RJ (2018) Assessment of deep-water habitat for crown-of-thorns starfish (COTS) in the Great Barrier Reef. Report to the National Environmental Science Program, Reef and Rainforest Research Centre Limited, Cairns

Bellwood DR, Brandl SJ, McWilliam M et al (2024) Studying functions on coral reefs: past perspectives, current conundrums, and future potential. Coral Reefs 43(2):281–297

Bertucci F, Parmentier E, Lecellier G et al (2016) Acoustic indices provide information on the status of coral reefs: an example from Moorea Island in the South Pacific. Sci Rep 6:1–9. https://doi.org/10.1038/srep33326

Berumen ML, Pratchett MS (2006) Recovery without resilience: persistent disturbance and long-term shifts in the structure of fish and coral communities at Tiahura Reef, Moorea. Coral Reefs 25:647–653. https://doi.org/10.1007/s00338-006-0145-2

Birkeland C (1982) Terrestrial run-off as a cause of outbreaks of *Acanthaster planci* (Echinodermata: Asteroidea). Mar Biol 69(2):175–185. https://doi.org/10.1007/BF00396897

Birkeland C (1989a) The Faustian traits of the crown-of-thorns starfish. Am Sci 77(2):154–163

Birkeland C (1989b) The influence of echinoderms on coral-reef communities. Echinoderm Stud 3:1–79

Bornancin L, Bonnard I, Mills SC et al (2017) Chemical mediation as a structuring element in marine gastropod predator-prey interactions. Nat Prod Rep 34(6):644–676. https://doi.org/10.1039/c6np00097e

Bose U, Wang T, Zhao M et al (2017) Multiomics analysis of the giant triton snail salivary gland, a crown-of-thorns starfish predator. Sci Rep 7(1):1–14. https://doi.org/10.1038/s41598-017-05974-x

Brodie J, Fabricius K, De'ath G et al (2005) Are increased nutrient inputs responsible for more outbreaks of crown-of-thorns starfish? An appraisal of the evidence. Mar Pollut Bull 51(1):266–278. https://doi.org/10.1016/j.marpolbul.2004.10.035

Burt A (2003) Site-specific selfish genes as tools for the control and genetic engineering of natural populations. Proc R Soc London Ser B: Biol Sci 270(1518):921–928

Byrne M, Cerra A, Nishigaki T et al (1997) Infestation of the testes of the Japanese sea star *Asterias amurensis* by the ciliate *Orchitophyra stellarum*: a caution against the use of this ciliate for biological control. Dis Aquat Org 28(3):235–239

Byrne M, Deaker DJ, Gibbs M et al (2023) Juvenile waiting stage crown-of-thorns sea stars are resilient in heatwave conditions that bleach and kill corals. Glob Chang Biol 29:6493. https://doi.org/10.1111/gcb.16946

Byrne M, Foo SA, Vila-Concejo A et al (2024) Impacts of climate change stressors on the Great Barrier Reef. In: Oceanographic processes of coral reefs. CRC Press, pp 323–340

Byrne M, Waller A, Clements M et al (2025) Catastrophic bleaching in protected reefs of the Southern Great Barrier Reef. Limnol Oceanogr Lett 10:340. https://doi.org/10.1002/lol2.10456

Caballes CF, Pratchett MS (2014) Reproductive biology and early life history of the crown-of-thorns starfish. In: Whitmore E (ed) Echinoderms: ecology, habitats and reproductive biology. Nova Science Publishers, New York, pp 101–146

Caballes CF, Pratchett MS (2017) Environmental and biological cues for spawning in the crown-of-thorns starfish. PLoS One 12(3):1–18. https://doi.org/10.1371/journal.pone.0173964

Caballes CF, Messmer V, Raymundo ML et al (2022) Prevalence and severity of sublethal injuries in crown-of-thorns starfish relative to marine reserves in the Great Barrier Reef. Aquat Conserv Mar Freshwat Ecosyst 32(6):993–1004

Campbell AC, Ormond RF (1970) The threat of the 'crown-of-thorns' starfish (*Acanthaster planci*) to coral reefs in the Indo-Pacific area: observations on a normal population in the Red Sea. Biol Conserv 2(4):246–251

Carrier TJ, Reitzel AM (2020) Symbiotic life of echinoderm larvae. Front Ecol Evol 7:1–7. https://doi.org/10.3389/fevo.2019.00509

Carrier TJ, Wolfe K, Lopez K et al (2018) Diet-induced shifts in the crown-of-thorns (*Acanthaster* sp.) larval microbiome. Mar Biol 165(10):157. https://doi.org/10.1007/s00227-018-3416-x

Carvalho-Santos Z, Azimzadeh J, Pereira-Leal JB et al (2011) Tracing the origins of centrioles, cilia, and flagella. J Cell Biol 195(2):341. https://doi.org/10.1083/jcb.201011521952c

Castaneda JM, Hua R, Miyata H et al (2017) TCTE1 is a conserved component of the dynein regulatory complex and is required for motility and metabolism in mouse spermatozoa. Proc Natl Acad Sci 114(27):E5370–E5378

Castaneda JM, Miyata H, Archambeault DR et al (2020) Mouse t-complex protein 11 is important for progressive motility in sperm. Biol Reprod 102(4):852–862

Celliers L, Schleyer MH (2006) Observations on the behaviour and the character of an *Acanthaster planci* (L.) Aggregation in a high latitude coral community in South Africa. West Indian Ocean J Mar Sci 5(1):105–113

Cerra A, Byrne M, Hoegh-Guldberg O (1997) Development of the hyaline layer around the planktonic embryos and larvae of the asteroid *Patiriella calcar* and the presence of associated bacteria. Invertebr Reprod Dev 31(1–3):337–343

Chaet AB (1959) Physiologic activity of nerve extracts. Biol Bull 117:407–408

Chaput R, Paris CB, Smith SG et al (2019) Reef fish larvae visually discriminate coral diversity during settlement. Bull Mar Sci 95(3):449–458

Chesher RH (1969) Destruction of pacific corals by the sea star *Acanthaster planci*. Science 165(3890):280–283. https://doi.org/10.1126/science.165.3890.280

Clement R, Dunbabin M, Wyeth G (2005) Toward robust image detection of crown-of-thorns starfish for autonomous population monitoring. In: *Proceedings of the 2005 Australasian conference on robotics and automation*, pp 1–8

Clements M, Selvakumaraswamy P, Deaker D et al (2022) Freshening of Great Barrier Reef waters is deleterious for larval crown-of-thorns starfish, counter to the terrestrial run-off hypothesis. Mar Ecol Prog Ser 696:1–14. https://doi.org/10.3354/meps14150

Cowan ZL, Dworjanyn SA, Caballes CF et al (2016a) Benthic predators influence microhabitat preferences and settlement success of crown-of-thorns starfish (*Acanthaster* cf. solaris). Diversity 8(4). https://doi.org/10.3390/d8040027

Cowan ZL, Dworjanyn SA, Caballes CF, Pratchett MS (2016b) Predation on crown-of-thorns starfish larvae by damselfishes. Coral Reefs 35(4):1253–1262. https://doi.org/10.1007/s00338-016-1491-3

Cowan ZL, Pratchett MS, Messmer V et al (2017) Known predators of crown-of-thorns starfish (*Acanthaster* spp.) and their role in mitigating, if not preventing, population outbreaks. Diversity 9(1):1–19. https://doi.org/10.3390/d9010007

Dana TF (1970) *Acanthaster*: a rarity in the past? Science 169(3948):894. https://doi.org/10.1126/science.169.3948.894

Dana TF, Wolfson A (1970) Eastern Pacific crown-of-thorns starfish populations in the lower gulf of California. Trans San Diego Soc Nat Hist 16:83–90. https://doi.org/10.5962/bhl.part.15455

Dana TF, Newman WA, Fager EW (1972) *Acanthaster* aggregations: interpreted as primarily responses to natural phenomena. Pac Sci 26:1–18

David CN, Özbek S, Adamczyk P et al (2008) Evolution of complex structures: minicollagens shape the cnidarian nematocyst. Trends Genet 24(9):431–438. https://doi.org/10.1016/j.tig.2008.07.001

Dayoub F, Dunbabin M, Corke P (2015) Robotic detection and tracking of crown-of-thorns starfish. In: 2015 IEEE/RSJ International

Conference on Intelligent Robots and Systems (IROS), pp 1921–1928. https://doi.org/10.1109/IROS.2015.7353629

De Bruin GHP (1972) The crown-of-thorns starfish *Acanthaster planci* (Linne') in Ceylon. Bull Fish Res Station Sri Lanka Ceylon 23(1 & 2):37–41

De'Ath G, Moran PJ (1998) Factors affecting the behaviour of crown-of-thorns starfish (*Acanthaster planci* L.) on the Great Barrier Reef: 2: feeding preferences. J Exp Mar Biol Ecol 220(1):107–126. https://doi.org/10.1016/S0022-0981(97)00100-7

De'Ath G, Fabricius KE, Sweatman H et al (2012) The 27-year decline of coral cover on the great barrier reef and its causes. Proc Natl Acad Sci USA 109(44):17995–17999. https://doi.org/10.1073/pnas.1208909109

Deaker DJ, Byrne M (2022) Crown-of-thorns starfish life-history traits contribute to outbreaks, a continuing concern for coral reefs. Emerging Top Life Sci 6(1):67–79. https://doi.org/10.1042/ETLS20210239

Deaker DJ, Agüera A, Lin H et al (2020) The hidden army: corallivorous crown-of- thorns sea stars can spend years as herbivorous juveniles. Biol Lett 16:20190849. https://doi.org/10.1098/rsbl.2019.0849

Deaker DJ, Mos B, Lawson C et al (2021) Coral defences: the perilous transition of juvenile crown-of-thorns starfish to corallivory. Mar Ecol Prog Ser 665:115–125

Desbiens AA, Mumby PJ, Dworjanyn S et al (2023) Novel rubble dwelling predators of herbivorous juvenile crown-of-thorns starfish (*Acanthaster* sp.). Coral Reefs 42(2):579–591. https://doi.org/10.1007/s00338-023-02364-w

Dissanayak KN, Terney Pradeep Kumara PB (2015) Impacts of *Acanthaster planci* (Crown-of-Thorns Starfish) population expansion on the Coral ecosystem of Pigeon Island Marine National Park, Sri Lanka. In: Proceedings of the National Aquatic Resources Research and Development Agency (NARA). Scientific Sessions

Doll PC, Uthicke S, Caballes CF et al (2023) Settlement cue selectivity by larvae of the destructive crown-of-thorns starfish. Biol Lett 19(1):20220399. https://doi.org/10.1098/rsbl.2022.0399

Domantay JS, Roxas HA (1938) The littoral Asteroidea of Port Galera Bay and adjacent waters. Philipp J Sci 65:203–237

Done TJ, Potts DC (1992) Influences of habitat and natural disturbances on contributions of massive *Porites* corals to reef communities. Mar Biol 114(3):479–493. https://doi.org/10.1007/BF00350040

Doyle JR, McKinnon AD, Uthicke S (2017) Quantifying larvae of the corallivorous seastar *Acanthaster* cf. *solaris* on the Great Barrier Reef using qPCR. Mar Biol 164(8). https://doi.org/10.1007/s00227-017-3206-x

Dumas P, Gereva S, Moutardier G et al (2015) Collective action and lime juice fight crown-of-thorns starfish outbreaks in Vanuatu. SPC Fis Newsl 146:47–52

Dumas P, Moutardier G, Ham J et al (2016) Timing within the reproduction cycle modulates the efficiency of village-based crown-of-thorns starfish removal. Biol Conserv 204:237–246. https://doi.org/10.1016/j.biocon.2016.10.027

Dumas P, Mills SC, Moutardier G et al (2018) The chaotic history of using vinegar injections to control *Acanthaster* spp. populations. A comment to Boström-Einarsson L., Bonin M. C., Moon S. & Firth S. (2018). Environmental impact monitoring of household vinegar-injections to cull crown-of-thorns. Ocean Coast Manag 165:434–435. https://doi.org/10.1016/j.ocecoaman.2018.05.017

Dumas P, Fiat S, Durbano A et al (2020) Citizen science, a promising tool for detecting and monitoring outbreaks of the crown-of-thorns starfish *Acanthaster* spp. Sci Rep 10(1):1–10. https://doi.org/10.1038/s41598-019-57251-8

Dumas P, Peignon C, Dumas M et al (2022) Erratum: destructive outbreaks of the corallivorous starfish *Acanthaster* cf. *solaris* spare coral assemblages in the shallowest reef flat areas in New Caledonia. ICES J Mar Sci 79(7):2161. https://doi.org/10.1093/icesjms/fsac031

Endean R (1969) Report on Investigations Made into Aspects of the Current Acanthaster planci (Crown-of-Thorns Starfish) Infestations on Certain Reefs of the Great Barrier Reef. Queensland Department of Primary Industries (Fisheries Branch), Brisbane, pp 1–35

Endean R, Stablum W (1973) A study of some aspects of the crown-of-thorns starfish (*Acanthaster planci*) infestations of reefs of Australia's Great Barrier Reef, Atoll research bulletin. The Smithsonian Institution, Washington, DC

Esvelt KM, Smidler AL, Catteruccia F et al (2014) Concerning RNA-guided gene drives for the alteration of wild populations. elife 3:1–21. https://doi.org/10.7554/eLife.03401

Fabricius KE, Okaji K, De'ath G (2010) Three lines of evidence to link outbreaks of the crown-of-thorns seastar *Acanthaster planci* to the release of larval food limitation. Coral Reefs 29(3):593–605. https://doi.org/10.1007/s00338-010-0628-z

Fagoonee I (1990) Coastal marine ecosystems of Mauritius. Hydrobiologia 208:55–62. https://doi.org/10.1007/bf00008443

Fautin DG (2009) Structural diversity, systematics, and evolution of cnidae. Toxicon 54(8):1054–1064

Fernandes L, Marsh H, Moran PJ et al (1990) Bias in manta tow surveys of *Acanthaster planci*. Coral Reefs 9(3):155–160. https://doi.org/10.1007/BF00258228

Foo SA, Millican HR, Byrne M (2024) Crown-of-thorns seastar (*Acanthaster* spp.) feeding ecology across species and regions. Sci Total Environ 930:172691

Foster T, Gilmour JP (2016) Seeing red: coral larvae are attracted to healthy-looking reefs. Mar Ecol Prog Ser 559:65–71. https://doi.org/10.3354/meps11902

Fraser N, Crawford BR, Kusen J et al (2000) Best practices guide for crown-of-thorns clean-ups, Proyek Pesisir special publication. Coastal resources center coastal management report #2225. Coastal Resources Center, University of Rhode Island Narragansett, Rhode Island, pp 1–38

Fujihara Y, Noda T, Kobayashi K et al (2019) Identification of multiple male reproductive tract-specific proteins that regulate sperm migration through the oviduct in mice. Proc Natl Acad Sci 116(37):18498–18506

Garm A, Nilsson DE (2014) Visual navigation in starfish: first evidence for the use of vision and eyes in starfish. Proc R Soc B Biol Sci 281(1777):20133011

GBRMPA (2017) Crown-of-thorns starfish control guidelines, 2nd edn. Great Barrier Reef Marine Park Authority, Townsville

GBRMPA, Australian Institute of Marine Science, and CSIRO (2023) Reef snapshot: summer 2022–23. Great Barrier Reef Marine Park Authority, Townsville

Geoffroy S (2010) Évaluation spatio-temporelle de la prolifération des populations d'étoiles de mer corallivores, Acanthaster planci (L.) en Polynésie française (Moorea, Raiatea-Tahaa, Bora-Bora). Report. Centre de Recherches Insulaires et Observatoire de l'Environnemet, Ecole Pratique des Hautes Etudes, French Polynesia

Geoffroy S, Petit M, Lison de Loma T (2009) Suivi de l'invasion des Acanthaster plancii dans les iles de Moorea, Raiatea, Tahaa, *Bora-Bora*. Report. Centre de Recherches Insulaires et Observatoire de l'Environnemet, Ecole Pratique des Hautes Etudes, French Polynesia

Gochfeld DJ (2004) Predation-induced morphological and behavioral defenses in a hard coral: implications for foraging behavior of coral-feeding butterflyfishes. Mar Ecol Prog Ser 267:145–158. https://doi.org/10.3354/meps267145

Gómez-Lemos LA, Doropoulos C, Bayraktarov E et al (2018) Coralline algal metabolites induce settlement and mediate the inductive effect of epiphytic microbes on coral larvae. Sci Rep 8(1):1–11. https://doi.org/10.1038/s41598-018-35206-9

Gorman LM, Huffmyer AS, Byrne M et al (2025) Coral venom and toxins as protection against crown-of-thorns sea star attack. bioRxiv, 2025-02. https://doi.org/10.1101/2025.02.10.637440

Guerra V, Haynes G, Byrne M et al (2020) Nonspecific expression of fertilization genes in the crown-of-thorns *Acanthaster* cf. *solaris*: unexpected evidence of hermaphroditism in a coral reef predator. Mol Ecol 29(2):363–379

Hackerott S, Martell HA, Eirin-Lopez JM (2021) Coral environmental memory: causes, mechanisms, and consequences for future reefs. Trends Ecol Evol 36(11):1011–1023. https://doi.org/10.1016/j.tree.2021.06.014

Haguenauer A, Zuberer F, Siu G et al (2021) Deep heat: a comparison of water temperature, anemone bleaching, anemonefish density and reproduction between shallow and mesophotic reefs. Fishes 6(3). https://doi.org/10.3390/fishes6030037

Hall MR, Kocot KM, Baughman KW et al (2017a) The crown-of-thorns starfish genome as a guide for biocontrol of this coral reef pest. Nature 544(7649):231–234. https://doi.org/10.1038/nature22033

Hall MR, Motti CA, Kroon F (2017b) The potential role of the giant triton snail, *Charonia tritonis* (Gastropoda: Ranellidae) in mitigating populations of the crown-of-thorns starfish. In: Integrated Pest management of crown-of-thorns starfish. report to the national environmental science programme. Reef and Rainforest Research Centre Limited, Cairns, pp 1–58

Harris RJ, d'Artagnan LB, Paxton G et al (2025) The future of utilising semiochemical pest control methods to manage the destructive crown-of-thorns starfish outbreaks on coral reefs. Biol Conserv 302:110984

Hart MW, Guerra VI, Allen JD et al (2021) Cloning and selfing affect population genetic variation in simulations of outcrossing, sexual sea stars. Biol Bull 241(3):286–302. https://doi.org/10.1086/717293

Haszprunar G, Vogler C, Wörheide G (2017) Persistent gaps of knowledge for naming and distinguishing multiple species of crown-of-thorns-seastar in the *Acanthaster planci* species complex. Diversity 9(2):1–10. https://doi.org/10.3390/d9020022

Hayashi R (1938) Sea-stars of the Caroline islands. Palao Trop Biol Res Station Stud 1:418–446

Haywood MDE, Thomson DP, Babcock RC et al (2019) Crown-of-thorns starfish impede the recovery potential of coral reefs following bleaching. Mar Biol 166(7):99. https://doi.org/10.1007/s00227-019-3543-z

Heenaye-Mamode Khan M, Makoonlall A, Nazurally N et al (2023) Identification of crown-of-thorns starfish (COTS) using convolutional neural network (CNN) and attention model. PLoS One 18(4):e0283121. https://doi.org/10.1371/journal.pone.0283121

Heron S, Morgan J, Eakin M et al (2005) Hurricanes and their effects on coral reefs. In: Wilkinson C, Souter D (eds) Status of Caribbean Coral Reefs after Bleaching and Hurricanes in 2005. Global Coral Reef Monitoring Network, and Reef and Rainforest Research Centre, Townsville, pp 31–36

Hillberg AK, Smith MK, Lausen BS et al (2023) Crown-of-thorns starfish spines secrete defence proteins. PeerJ 11:e15689. https://doi.org/10.7717/peerj.15689

Hobbs JP, Hender J, Gilligan JJ (2005) The status of the coral reef community at the Cocos (Keeling) Islands, Eastern Indian Ocean, 1997–2005. Commonwealth of Australia

Hock K, Wolff NH, Condie SA et al (2014) Connectivity networks reveal the risks of crown-of- thorns starfish outbreaks on the Great Barrier Reef. J Appl Ecol 51(1):1188–1196. https://doi.org/10.1111/1365-2664.12320

Hock K, Wolff NH, Ortiz JC et al (2017) Connectivity and systemic resilience of the Great Barrier Reef. PLoS Biol:1–23. https://doi.org/10.5281/zenodo.1017925

Hoegh-Guldberg O, Mumby PJ, Hooten AJ et al (2007) Coral reefs under rapid climate change and ocean acidification. Science 318(5857):1737–1742. https://doi.org/10.1126/science.1152509

Høj L, Levy N, Baillie BK et al (2018) Crown-of-thorns sea star *Acanthaster* cf. *solaris* has tissue-characteristic microbiomes with potential roles in health and reproduction. Appl Environ Microbiol 84(13):1–18. https://doi.org/10.1128/AEM.00181-18

Horoiwa M, Nakamura T, Yuasa H et al (2022) Integrated population genomic analysis and numerical simulation to estimate larval dispersal of *Acanthaster* cf. *solaris* between Ogasawara and other Japanese regions. Front Mar Sci 8:1–12. https://doi.org/10.3389/fmars.2021.688139

Hoshi M, Nishigaki T, Ushiyama A et al (1994) Egg-jelly signal molecules for triggering the acrosome reaction in starfish spermatozoa. Int J Dev Biol 38(2):167–174

Huang B, Piperno G, Ramanis Z et al (1981) Radial spokes of *Chlamydomonas* flagella: genetic analysis of assembly and function. J Cell Biol 88(1):80–88. https://doi.org/10.1083/jcb.88.1.80

Hue T, Chateau O, Lecellier G et al (2020) Temperature affects the reproductive outputs of coral-eating starfish *Acanthaster* spp. after adult exposure to near-future ocean warming and acidification. Mar Environ Res 162:105164

Hue T, Chateau O, Lecellier G et al (2022) Impact of near-future ocean warming and acidification on the larval development of coral-eating starfish *Acanthaster* cf. *solaris* after parental exposure. J Exp Mar Biol Ecol 548:151685. https://doi.org/10.1016/j.jembe.2021.151685

Huggett MJ, Williamson JE, De Nys R et al (2006) Larval settlement of the common Australian sea urchin *Heliocidaris erythrogramma* in response to bacteria from the surface of coralline algae. Oecologia 149:604–619

Hughes TP, Kerry JT, Álvarez-Noriega M et al (2017) Global warming and recurrent mass bleaching of corals. Nature 543(7645):373–377. https://doi.org/10.1038/nature21707

Hughes TP, Kerry JT, Baird AH et al (2018) Global warming transforms coral reef assemblages. Nature 556:492–496. https://doi.org/10.1038/s41586-018-0041-2

Ikegami S (1976) Role of asterosaponin A in starfish spawning induced by gonad-stimulating substance and 1-methyladenine. J Exp Zool 198(3):359–366

Ikegami S, Kamiya Y, Tamura S (1972) Isolation and characterization of spawning inhibitors in ovary of the starfish, Asterias amurensis. Agric Biol Chem 36(11):2005–2011

Jia H, Zhang G, Zhang C et al (2023) Identification and expression of the Conotoxin homologous genes in the giant triton snail (*Charonia tritonis*). J Ocean Univ China 22(1):213–220

Johansson CL, Francis DS, Uthicke S (2016) Food preferences of juvenile corallivorous crown-of-thorns (*Acanthaster planci*) sea stars. Mar Biol 163(3):49. https://doi.org/10.1007/s00227-016-2823-0

Johnson CR, Sutton DC (1994) Bacteria on the surface of crustose coralline algae induce metamorphosis of the crown-of-thorns starfish *Acanthaster planci*. Mar Biol 120(2):305–310. https://doi.org/10.1007/BF00349692

Jorissen H, Galand PE, Bonnard I et al (2021) Coral larval settlement preferences linked to crustose coralline algae with distinct chemical and microbial signatures. Sci Rep 11(1):1–11. https://doi.org/10.1038/s41598-021-94096-6

Kamya PZ, Dworjanyn SA, Hardy N et al (2014) Larvae of the coral eating crown-of-thorns starfish, *Acanthaster planci* in a warmer-high CO_2 Ocean. Glob Chang Biol 20(11):3365–3376

Kamya PZ, Byrne M, Graba-Landry A et al (2016) Near-future ocean acidification enhances the feeding rate and development of the herbivorous juveniles of the crown-of-thorns starfish, Acanthaster planci. Coral Reefs 35(4):1241–1251

Kamya PZ, Byrne M, Mos B et al (2017) Indirect effects of ocean acidification drive feeding and growth of juvenile crown-of-thorns starfish, *Acanthaster planci*. Proc R Soc B Biol Sci 284(1856):20170778

Kamya PZ, Byrne M, Mos B et al (2018) Enhanced performance of juvenile crown-of-thorns starfish in a warm-high CO_2 Ocean exacerbates poor growth and survival of their coral prey. Coral Reefs 37(3):751–762

Kayal M, Kayal E (2017) Colonies of the fire coral *Millepora platyphylla* constitute scleractinian survival oases during *Acanthaster* outbreaks in French Polynesia. Mar Biodivers 47:255–258. https://doi.org/10.1007/s12526-016-0465-6

Kayal M, Vercelloni J, Lison de Loma T et al (2012) Predator crown-of-thorns starfish (*Acanthaster planci*) outbreak, mass mortality of corals, and cascading effects on reef fish and benthic communities. PLoS One 7(10):e47363. https://doi.org/10.1371/journal.pone.0047363

Keesing JK (1995) Temporal patterns in the feeding and emergence behaviour of the crown-of-thorns starfish *Acanthaster planci*. Mar Freshw Behav Physiol 25(4):209–232. https://doi.org/10.1080/10236249509378919

Keesing JK (2021) Optimal foraging theory explains feeding preferences in the Western Pacific crown-of-Thorns Sea star *Acanthaster* sp. Biol Bull 241(3):303–329. https://doi.org/10.1086/718141

Keesing JK, Halford AR (1992) Importance of post-settlement processes for the population dynamics of *Acanthaster planci* (l.). Mar Freshw Res 43(3):635–651. https://doi.org/10.1071/MF9920635

Keesing JK, Halford AR, Hall KC (2018) Mortality rates of small juvenile crown-of-thorns starfish *Acanthaster planci* on the Great Barrier Reef: implications for population size and larval settlement thresholds for outbreaks. Mar Ecol Prog Ser 597:179–190

Keesing JK, Thomson DP, Haywood MDE et al (2019) Two-time losers: selective feeding by crown-of-thorns starfish on corals most affected by successive coral-bleaching episodes on western Australian coral reefs. Mar Biol 166(6):72. https://doi.org/10.1007/s00227-019-3515-3

Keesing JK, Mortimer N, Hellmrich L et al (2023) The short spined crown-of-thorns starfish *Acanthaster brevispinus* is a corallivore too. Coral Reefs 42(2):399–404

Kenchington RA (1977) Growth and recruitment of *Acanthaster planci* (L.) on the Great Barrier Reef. Biol Conserv 11(2):103–118. https://doi.org/10.1016/0006-3207(77)90032-5

Kitamura M, Koyama T, Nakano Y et al (2007) Characterization of a natural inducer of coral larval metamorphosis. J Exp Mar Biol Ecol 340(1):96–102. https://doi.org/10.1016/j.jembe.2006.08.012

Klein AH (2021) Establishment of multi-omics resources for the giant triton snail and the silver-lipped pearl oyster. Dissertation. University of the Sunshine Coast, Australia. https://doi.org/10.25907/00132

Klein AH, Motti CA, Hillberg AK et al (2021) Development and interrogation of a transcriptomic resource for the Giant triton snail (*Charonia tritonis*). Mar Biotechnol 23(3):501–515. https://doi.org/10.1007/s10126-021-10042-7

Koutroumpa FA, Monsempes C, François MC et al (2016) Heritable genome editing with CRISPR/Cas9 induces anosmia in a crop pest moth. Sci Rep 6(1):1–9. https://doi.org/10.1038/srep29620

Kroon FJ, Lefèvre CD, Doyle JR et al (2020) DNA-based identification of predators of the corallivorous Crown-of-Thorns Starfish (*Acanthaster* cf. *solaris*) from fish faeces and gut contents. Sci Rep 10(1):1–14. https://doi.org/10.1038/s41598-020-65136-4

Kroon FJ, Barneche DR, Emslie MJ (2021) Fish predators control outbreaks of crown-of-thorns starfish. Nat Commun 12(1). https://doi.org/10.1038/s41467-021-26786-8

Kroon FJ, Crosswell JR, Robson BJ (2023) The effect of catchment load reductions on water quality in the crown-of-thorn starfish outbreak initiation zone. Mar Pollut Bull 195:115255. https://doi.org/10.1016/j.marpolbul.2023.115255

Kwong SLT, Villacorta-Rath C, Pratchett M et al (2023) Telomere dynamics in the Pacific crown-of-thorns seastar (*Acanthaster* cf. *solaris*): effect of age, diet, and tissue type. Coral Reefs 42:1–9. https://doi.org/10.1007/s00338-023-02405-4

Lamare M, Pecorino D, Hardy N et al (2014) The thermal tolerance of crown-of-thorns (*Acanthaster planci*) embryos and bipinnaria larvae: implications for spatial and temporal variation in adult populations. Coral Reefs 33:207–219

Lamy T, Galzin R, Kulbicki M et al (2016) Three decades of recurrent declines and recoveries in corals belie ongoing change in fish assemblages. Coral Reefs 35(1):293–302. https://doi.org/10.1007/s00338-015-1371-2

Lang BJ, Donelson JM, Caballes CF et al (2022) Effects of elevated temperature on the performance and survival of pacific crown-of-thorns starfish (*Acanthaster* cf. *solaris*). Mar Biol 169(4):1–13. https://doi.org/10.1007/s00227-022-04027-w

Lang BJ, Caballes CF, Uthicke S et al (2023) Impacts of ocean warming on the settlement success and post-settlement survival of Pacific crown-of-thorns starfish (*Acanthaster* cf. *solaris*). Coral Reefs 42(1):143–155

Lecchini D, Miura T, Lecellier G et al (2014) Transmission distance of chemical cues from coral habitats: implications for marine larval settlement in context of reef degradation. Mar Biol 161(7):1677–1686. https://doi.org/10.1007/s00227-014-2451-5

Lee CC, Hsieh HJ, Hsieh CH et al (2015) Plancitoxin I from the venom of crown-of-thorns starfish (*Acanthaster planci*) induces oxidative and endoplasmic reticulum stress associated cytotoxicity in A375. S2 cells. Exp Mol Pathol 99(1):7–15. https://doi.org/10.1016/j.yexmp.2015.05.001

Leray M, Béraud M, Anker A et al (2012) *Acanthaster planci* outbreak: decline in coral health, coral size structure modification and consequences for obligate decapod assemblages. PLoS One 7(4):e35456. https://doi.org/10.1371/journal.pone.0035456

Lillis A, Bohnenstiehl DW, Peters JW et al (2016) Variation in habitat soundscape characteristics influences settlement of a reef-building coral. PeerJ 2016(10):1–16. https://doi.org/10.7717/peerj.2557

Ling SD, Cowan ZL, Boada J et al (2020) Homing behaviour by destructive crown-of-thorns starfish is triggered by local availability of coral prey. Proc R Soc B Biol Sci 287(1938):20201341. https://doi.org/10.1098/rspb.2020.1341

Lockie S, Bartelet HA, Ritchie BW et al (2024) Quantifying public support for culling crown-of-thorns starfish (*Acanthaster* spp.) on the Great Barrier Reef. Conserv Sci Pract 6(11):e13252. https://doi.org/10.1111/csp2.13252

Longshaw M, Bateman KS, Stebbing P et al (2012) Disease risks associated with the importation and release of non-native crayfish species into mainland Britain. Aquat Biol 16(1):1–15. https://doi.org/10.3354/ab00417

Lucas JS (1973) Reproductive and larval biology of *Acanthaster planci* (L.) in Great Barrier Reef waters. Micronesica 9(2):197–203

Lucas JS, Hart RJ, Howden ME et al (1979) Saponins in eggs and larvae of *Acanthaster planci* (L.) (Asteroidea) as chemical defences against planktivorous fish. J Exp Mar Biol Ecol 40(2):155–165. https://doi.org/10.1016/0022-0981(79)90042-X

Malyarenko TV, Kicha AA, Ivanchina NV et al (2012) Asteropsiside A and other asterosaponins from the starfish *Asteropsis carinifera*. Russ Chem Bull 61(10):1986–1991. https://doi.org/10.1007/s11172-012-0275-3

Matthews SA, Mellin C, Pratchett MS (2020a) Chapter Nine: Larval connectivity and water quality explain spatial distribution of crown-of-thorns starfish outbreaks across the Great Barrier Reef. Adv Mar Biol 87(1):223–258. https://doi.org/10.1016/bs.amb.2020.08.007

Matthews SA, Shoemaker K, Pratchett MS et al (2020b) Chapter Ten: COTSMod: a spatially explicit metacommunity model of outbreaks of crown-of-thorns starfish and coral recovery. Adv Mar Biol 87(1):259–290. https://doi.org/10.1016/bs.amb.2020.09.001

Matthews SA, Williamson DH, Beeden R et al (2024) Protecting Great Barrier Reef resilience through effective management of crown-of-thorns starfish outbreaks. PLoS One 19(4):e0298073

McFarlane GR, Whitelaw CBA, Lillico SG (2018) CRISPR-based gene drives for Pest control. Trends Biotechnol 36(2):130–133. https://doi.org/10.1016/j.tibtech.2017.10.001

Mendonça VM, Al Jabri MM, Al Ajmi I et al (2010) Persistent and expanding population outbreaks of the Corallivorous starfish

Acanthaster planci in the North-Western Indian Ocean: are they really a consequence of unsustainable starfish predator removal through overfishing in coral reefs, or a response to a changing environment? Zool Stud 49(1):108–123

Mendoza-Porras O, Nguyen TV, Shah RM et al (2023) Biochemical metabolomic profiling of the crown-of-thorns starfish (*Acanthaster*): new insight into its biology for improved pest management. Sci Total Environ 861:160525. https://doi.org/10.1016/j.scitotenv.2022.160525

Miller IR, Jonker M, Coleman G (2003) Crown-of-thorns starfish and coral surveys using the manta tow and SCUBA search techniques. In: Long-term monitoring of the great barrier reef standard operation procedure Number 9 Edition 3. Australian Institute of Marine Science, Townsville

Millican HR, Byrne M, Keesing J et al (2024) Feeding biology of crown-of-thorns seastars across sites differing in *Acropora* availability. Mar Environ Res 200:106655

Mills SC (2012) Density-dependent prophylaxis in the coral-eating crown-of-thorns sea star, Acanthaster planci. Coral Reefs 31(2):603–612

Milne R, Anand M, Bauch CT (2023) Preparing for and managing crown-of-thorns starfish outbreaks on reefs under threat from interacting anthropogenic stressors. Ecol Model 484:110443. https://doi.org/10.1016/j.ecolmodel.2023.110443

Mita M (2013) Relaxin-like gonad-stimulating substance in an echinoderm, the starfish: a novel relaxin system in reproduction of invertebrates. Gen Comp Endocrinol 181:241–245. https://doi.org/10.1016/j.ygcen.2012.07.015

Mita M (2016) Starfish gonadotropic hormone: Relaxin-like gonad-stimulating peptides. Gen Comp Endocrinol 230–231:166–169. https://doi.org/10.1016/j.ygcen.2016.04.016

Miura C, Ohta T, Miyake K et al (2023) Analysis of the positive effects of crown-of-thorns starfish on the growth and immune system of red sea bream. Res Square. https://doi.org/10.21203/rs.3.rs-3114497/v1

Miyata H, Satouh Y, Mashiko D et al (2015) Sperm calcineurin inhibition prevents mouse fertility with implications for male contraceptive. Science 350(6259):442–445

Moran PJ, Johnson DB, Miller-Smith BA et al (1989) A guide to the AIMS manta tow technique. Australian Institute of Marine Science

Moreau SJM, Asgari S (2015) Venom proteins from parasitoid wasps and their biological functions. Toxins 7(7):2385–2412

Morello EB, Plagányi ÉE, Babcock RC et al (2014) Model to manage and reduce crown-of-thorns starfish outbreaks. Mar Ecol Prog Ser 512:167–183

Mortensen T (1931) Contribution to the study of the development and larval forms of some echinoderms. Kongelie Danske Videnskabernes Selskab Skrifter Naturvidenskabelig Og Mathematisk 4:39

Mos B, Mesic N, Dworjanyn SA (2023) Variable food alters responses of larval crown-of-thorns starfish to ocean warming but not acidification. Commun Biol 6:639. https://doi.org/10.1038/s42003-023-05028-1

Mos B, Erler D, Lawson C et al (2024) Crown-of-thorns starfish complete their larval phase eating only nitrogen-fixing *Trichodesmium cyanobacteria*. Sci Adv 10(29):eado2682

Motti CA, Cummins SF, Hall MR (2022) A review of the Giant triton (*Charonia tritonis*), from exploitation to coral reef protector? Diversity 14(11):961. https://doi.org/10.3390/d14110961

Moutardier G, Gereva S, Mills SC et al (2015) Lime juice and vinegar injections as a cheap and natural alternative to control COTS outbreaks. PLoS One 10(9):1–16. https://doi.org/10.1371/journal.pone.0137605

Neil RC, Gomez Cabrera M, Uthicke S (2022) Juvenile age and available coral species modulate transition probability from herbivory to corallivory in *Acanthaster* cf. *solaris* (crown-of-thorns sea star). Coral Reefs 41(4):843–848. https://doi.org/10.1007/s00338-022-02255-6

Nijman V, Lee PB (2016) Trade in nautilus and other large marine molluscs as ornaments and decorations in Bali, Indonesia. Raffles Bull Zool 64:368–373

Obura D, Gudka M, Rabi FA et al (2017) Coral reef status report for the Western Indian Ocean. Global Coral Reef Monitoring Network (GCRMN)/International Coral Reef Initiative (ICRI), p 144

Oji A, Isotani A, Fujihara Y et al (2020) Tesmin, metallothionein-like 5, is required for spermatogenesis in mice. Biol Reprod 102(4):975–983

Okano K, Ohkawa N, Ikegami S (1985) Structure of ovarian Asterosaponin-4, an inhibitor of spontaneous oocyte maturation from the starfish *Asterias amurensis*. Agric Biol Chem 49(9):2823–2826

Olson RR (1987) *In situ* culturing as a test of the larval starvation hypothesis for the crown-of-thorns starfish, Acanthaster planci. Limnol Oceanogr 32(4):895–904. https://doi.org/10.4319/lo.1987.32.4.0895

Ormond RFG, Campbell AC, Head SH et al (1973) Formation and breakdown of aggregations of the crown-of-thorns starfish, *Acanthaster planci* (L.). Nature 246(5429):167–169

Ormond RFG, Hanscomb NJ, Beach DH (1976) Food selection and learning in the crown-of-thorns starfish, *Acanthaster planci* (L). Mar Behav Physiol 4(2):93–105. https://doi.org/10.1080/10236247609386944

Ormond R, Bradbury R, Bainbridge S et al (1990) Test of a model of regulation of crown-of-thorns starfish by fish predators. In: Acanthaster and the coral reef: a theoretical perspective. Springer-Verlag, Berlin, pp 189–207

Ota E, Nagashima Y, Shiomi K et al (2006) Caspase-independent apoptosis induced in rat liver cells by plancitoxin I, the major lethal factor from the crown-of-thorns starfish *Acanthaster planci* venom. Toxicon 48(8):1002–1010. https://doi.org/10.1016/j.toxicon.2006.08.005

Park S, Shimada K, Fujihara Y et al (2020) CRISPR/Cas9-mediated genome-edited mice reveal 10 testis-enriched genes are dispensable for male fecundity. Biol Reprod 103(2):195–204. https://doi.org/10.1093/biolre/ioaa084

Pearson RG, Endean R (1969) A preliminary study of the coral predator *Acanthaster planci* (L.) (Asteroidea) on the Great Barrier Reef, vol 3. Queensland Department Harbour Marine Fish Notes, pp 27–55

Peng C, Wang K, Wang W et al (2023) Phytoplankton community structure and environmental factors during the outbreak of Crown-of-Thorns Starfish in Xisha Islands, South China Sea. Environ Res 235:116568

Petersen LE, Moeller M, Versluis D et al (2021) Mono- and multispecies biofilms from a crustose coralline alga induce settlement in the scleractinian coral *Leptastrea purpurea*. Coral Reefs 40(2):381–394. https://doi.org/10.1007/s00338-021-02062-5

Petie R, Hall MR, Hyldahl M et al (2016) Visual orientation by the crown-of-thorns starfish (*Acanthaster planci*). Coral Reefs 35:1139–1150

Piercy JJB, Codling EA, Hill AJ et al (2014) Habitat quality affects sound production and likely distance of detection on coral reefs. Mar Ecol Prog Ser 516:35–47. https://doi.org/10.3354/meps10986

Plass-Johnson JG, Schwieder H, Heiden J et al (2015) A recent outbreak of crown-of-thorns starfish (*Acanthaster planci*) in the Spermonde Archipelago, Indonesia. Reg Environ Chang 15:1157–1162

Pooloo N, Aumeer W, Khoodeeram R (2021) Monitoring coral reefs death causes with artificial intelligence. In: 2021 IST-Africa Conference (IST-Africa), pp 1–9

Pratchett MS (1999) An infectious disease in crown-of-thorns starfish on the Great Barrier Reef. Coral Reefs 18:272

Pratchett MS, Caballes CF, Rivera-Posada JA et al (2014) Limits to understanding and managing outbreaks of crown-of-thorns starfish (*Acanthaster* spp.). Oceanogr Mar Biol 52:133–200. https://doi.org/10.1201/b17143-4

Pratchett MS, Caballes CF, Wilmes JC et al (2017a) Thirty years of research on crown-of-thorns starfish (1986–2016): scientific advances and emerging opportunities. Diversity 9(4). https://doi.org/10.3390/d9040041

Pratchett MS, Dworjanyn S, Mos B et al (2017b) Larval survivorship and settlement of crown-of-thorns starfish (*Acanthaster* cf. *solaris*) at varying algal cell densities. Diversity 9(1). https://doi.org/10.3390/d9010002

Pratchett MS, Caballes CF, Cvitanovic C et al (2021) Knowledge gaps in the biology, ecology, and management of the Pacific crown-of-thorns sea star *Acanthaster* sp. on Australia's Great Barrier Reef. Biol Bull 241(3):330–346. https://doi.org/10.1086/717026

Radwan FF (2002) Comparative toxinological and immunological studies on the nematocyst venoms of the Red Sea fire corals *Millepora dichotoma* and *M. platyphylla*. Comp Biochem Physiol Part C: Toxicol Pharmacol 131(3):323–334. https://doi.org/10.1016/S1532-0456(02)00017-0

Radwan FF, Aboul-Dahab HM (2004) Milleporin-1, a new phospholipase A2 active protein from the fire coral *Millepora platyphylla* nematocysts. Comp Biochem Physiol Part C: Toxicol Pharmacol 139(4):267–272. https://doi.org/10.1016/j.cca.2004.12.002

Regassa LB, Gasparich GE (2006) Spiroplasmas: evolutionary relationships and biodiversity. Front Biosci 11(3):2983–3002. https://doi.org/10.2741/2027

Reichelt RE, Bradbury RH, Moran PJ (1990) The crown-of-thorns starfish, *Acanthaster planci*, on the Great Barrier Reef. Math Comput Model 13(6):45–60. https://doi.org/10.1016/0895-7177(90)90008-B

Rivera-Posada JA, Pratchett MS, Cano-Gómez A et al (2011a) Injection of *Acanthaster planci* with thiosulfate-citrate-bile-sucrose agar (TCBS). I. Disease induction. Dis Aquat Org 97(2):85–94. https://doi.org/10.3354/dao02401

Rivera-Posada JA, Pratchett MS, Owens L (2011b) Injection of *Acanthaster planci* with thiosulfate-citrate-bile-sucrose agar (TCBS). II. Histopathological changes. Dis Aquat Org 97(2):95–102. https://doi.org/10.3354/dao02400

Roche RC, Pratchett MS, Carr P et al (2015) Localized outbreaks of *Acanthaster planci* at an isolated and unpopulated reef atoll in the Chagos archipelago. Mar Biol 162:1695–1704. https://doi.org/10.1007/s00227-015-2708-7

Rogers JGD, Plágányi ÉE, Babcock RC (2017) Aggregation, allee effects and critical thresholds for the management of the crown-of-thorns starfish *Acanthaster planci*. Mar Ecol Prog Ser 578:99–114. https://doi.org/10.3354/meps12252

Round JL, Mazmanian SK (2009) The gut microbiota shapes intestinal immune responses during health and disease. Nat Rev Immunol 9(5):313–323

Salvat B (1992) The 1991 bleaching event in the society islands, French Polynesia. In: *Proceedings of the 7th international coral reef symposium*, vol 1, p 73

Saponari L, Montano S, Seveso D et al (2015) The occurrence of an *Acanthaster planci* outbreak in Ari Atoll, Maldives. Mar Biodivers 45:599–600. https://doi.org/10.1007/s12526-014-0276-6

Schlaff A, Menéndez P, Hall M et al (2020) Acoustic tracking of a large predatory marine gastropod, *Charonia tritonis*, on the Great Barrier Reef. Mar Ecol Prog Ser 642:147–161. https://doi.org/10.3354/meps13291

Seveso D, Stahl JH, Landes A (2024) An outbreak of *Acanthaster* spp. in the Gulf of Oman (United Arab Emirates). Mar Biodivers 54:62. https://doi.org/10.1007/s12526-024-01456-2

Shimada K, Kato H, Miyata H et al (2019) Glycerol kinase 2 is essential for proper arrangement of crescent-like mitochondria to form the mitochondrial sheath during mouse spermatogenesis. J Reprod Dev 65(2):155–162

Shiomi K, Midorikawa S, Ishida M et al (2004) Plancitoxins, lethal factors from the crown-of-thorns starfish *Acanthaster planci*, are deoxyribonucleases II. Toxicon 44(5):499–506. https://doi.org/10.1016/j.toxicon.2004.06.012

Smith CL, Ghosh J, Buckley KM et al (2010) Echinoderm immunity. In: Invertebrate immunity, pp 260–301

Smith MK, Wang T, Suwansa-ard S et al (2017) The neuropeptidome of the crown-of-thorns starfish, *Acanthaster planci*. J Proteome 165:61–68. https://doi.org/10.1016/j.jprot.2017.05.026

Smith MK, Chieu HD, Aizen J et al (2019) A crown-of-thorns sea star recombinant relaxin-like gonad-stimulating peptide triggers oocyte maturation and ovulation. Gen Comp Endocrinol 281:41–48. https://doi.org/10.1016/j.ygcen.2019.05.009

Smith LC, Boettgerm SA, Byrne M et al (2022) Echinoderm diseases and pathologies. In: Invertebrate pathology. Oxford University Press. https://doi.org/10.1093/oso/9780198853756.003.0018

Smith MK, Rotgans BA, Lang T et al (2023) Structure and proteomic analysis of the crown-of-thorns starfish (*Acanthaster* sp.) radial nerve cord. Sci Rep 13(1):3349. https://doi.org/10.1038/s41598-023-34881-7

Stonik VA, Kicha AA, Malyarenko TV et al (2020) Asterosaponins: structures, taxonomic distribution, biogenesis and biological activities. Mar Drugs 18(12):584. https://doi.org/10.3390/MD18120584

Suggett DJ, van Oppen MJ (2022) Horizon scan of rapidly advancing coral restoration approaches for 21st century reef management. Emerging Top Life Sci 6(1):125–136. https://doi.org/10.1042/ETLS20210240

Suzuki G, Yasuda N, Ikehara K et al (2016) Detection of a high-density brachiolaria-stage larval population of crown-of-thorns sea star (*Acanthaster planci*) in Sekisei Lagoon (Okinawa, Japan). Diversity 8(2):9. https://doi.org/10.3390/d8020009

Swanson RL, Byrne M, Prowse TA et al (2012) Dissolved histamine: a potential habitat marker promoting settlement and metamorphosis in sea urchin larvae. Mar Biol 159:915–925. https://doi.org/10.1007/s00227-011-1869-2

Sweatman HPA (1995) A field study of fish predation on juvenile crown-of-thorns starfish. Coral Reefs 14:47–53. https://doi.org/10.1007/BF00304071

Sweatman H (2008) No-take reserves protect coral reefs from predatory starfish. Curr Biol 18(14):R598–R599. https://doi.org/10.1016/j.cub.2008.05.033

Takemura F, Kobashigawa S, Hirayama K et al (2015) Development of an acetic acid injection device for crown-of-thorns starfish controlled by a remotely operated underwater robot. J Rob Mechatronics 27(5):571–578. https://doi.org/10.20965/jrm.2015.p0571

Tang HF, Yi YH, Li L et al (2005) Three new asterosaponins from the starfish *Culcita novaeguineae* and their bioactivity. Planta Med 71(5):458–463. https://doi.org/10.1055/s-2005-871215

Tebben J, Motti CA, Siboni N et al (2015) Chemical mediation of coral larval settlement by crustose coralline algae. Sci Rep 5:1–11. https://doi.org/10.1038/srep10803

Timmers MA, Andrews KR, Bird CE et al (2011) Widespread dispersal of the crown-of-thorns sea star, *Acanthaster planci*, across the Hawaiian Archipelago and Johnston Atoll. J Marine Biol 2011:934269. https://doi.org/10.1155/2011/934269

Tokuhiro K, Hirose M, Miyagawa Y et al (2008) Meichroacidin containing the membrane occupation and recognition nexus motif is essential for spermatozoa morphogenesis. J Biol Chem 283(27):19039–19048. https://doi.org/10.1074/jbc.M708590200

Tourrand C, Naim O, Bigot L et al (2013) Fringing reefs of Reunion Island and eutrophication effects-part 1: long-term monitoring of two shallow coral reef communities. Atoll Res Bull 596:1–35. https://doi.org/10.5479/si.00775630.596

Uthicke S, Schaffelke B, Byrne M (2009) A boom–bust phylum? Ecological and evolutionary consequences of density variations in echinoderms. Ecol Monogr 79(1):3–24

Uthicke S, Pecorino D, Albright R et al (2013) Impacts of ocean acidification on early life-history stages and settlement of the coral-eating sea star *Acanthaster planci*. PLoS One 8(12):e82938

Uthicke S, Doyle J, Duggan S et al (2015) Outbreak of coral-eating crown-of-thorns creates continuous cloud of larvae over 320 km of the Great Barrier Reef. Sci Rep 5. https://doi.org/10.1038/srep16885

Uthicke S, Lamare M, Doyle JR (2018) eDNA detection of corallivorous sea star (*Acanthaster* cf. *solaris*) outbreaks on the Great Barrier Reef using digital droplet PCR. Coral Reefs 37(4):1229–1239. https://doi.org/10.1007/s00338-018-1734-6

Uthicke S, Fisher EE, Patel F et al (2019) Spawning time of *Acanthaster* cf. *solaris* on the Great Barrier Reef inferred using qPCR quantification of embryos and larvae: do they know it's Christmas? Mar Biol 166(10):1–10. https://doi.org/10.1007/s00227-019-3582-5

Uthicke S, Pratchett MS, Messmer V et al (2021) Limited genetic signal from potential cloning and selfing within wild populations of coral-eating crown-of-thorns sea stars (*Acanthaster* cf. *solaris*). Coral Reefs 40(1):131–138. https://doi.org/10.1007/s00338-020-02022-5

Vanhatalo J, Hosack GR, Sweatman H (2017) Spatiotemporal modelling of crown-of-thorns starfish outbreaks on the Great Barrier Reef to inform control strategies. J Appl Ecol 54(1):188–197

Vazzana M, Mauro M, Ceraulo M et al (2020) Underwater high frequency noise: biological responses in sea urchin *Arbacia lixula* (Linnaeus, 1758). Comp Biochem Physiol A Mol Integr Physiol 242:110650. https://doi.org/10.1016/j.cbpa.2020.110650

Vega Thurber R, Mydlarz LD, Brandt M et al (2020) Deciphering coral disease dynamics: integrating host, microbiome, and the changing environment. Front Ecol Evol 8:1–18. https://doi.org/10.3389/fevo.2020.575927

Vercelloni J, Caley MJ, Mengersen K (2017) Crown-of-thorns starfish undermine the resilience of coral populations on the Great Barrier Reef. Glob Ecol Biogeogr 26(7):846–853. https://doi.org/10.1111/geb.12590

Vine PJ (1970) Field and laboratory observations of the crown-of-thorns starfish, *Acanthaster planci*: densities of *Acanthaster planci* in the Pacific Ocean. Nature 228(5269):341–342. https://doi.org/10.1038/228341a0

Vine PJ (1973) Crown of thorns (*Acanthaster planci*) plagues: the natural causes theory, Atoll research bulletin, 166. The Smithsonian Institution, Washington, DC

Vogler C, Benzie J, Lessios H et al (2008) A threat to coral reefs multiplied? Four species of crown-of-thorns starfish. Biol Lett 4(6):696–699. https://doi.org/10.1098/rsbl.2008.0454

Vogler C, Benzie J, Barber PH et al (2012) Phylogeography of the crown-of-thorns starfish in the Indian ocean. PLoS One 7(8):1–10. https://doi.org/10.1371/journal.pone.0043499

Wada N, Yuasa H, Kajitani R et al (2020) A ubiquitous subcuticular bacterial symbiont of a coral predator, the crown-of-thorns starfish, in the Indo-Pacific. Microbiome 8(1):1–14. https://doi.org/10.1186/s40168-020-00880-3

Wang L, Xu J, Liu H et al (2023) Ultrasensitive and on-site eDNA detection for the monitoring of crown-of-thorns starfish densities at the pre-outbreak stage using an electrochemical biosensor. Biosens Bioelectron 230:115265. https://doi.org/10.1016/j.bios.2023.115265

Wei Z, Zhang X, Chen Y et al (2024) A new strategy based on a cascade amplification strategy biosensor for on-site eDNA detection and outbreak warning of crown-of-thorns starfish. Sci Total Environ 927:172258

Westcott DA, Fletcher CS, Kroon FJ et al (2020) Relative efficacy of three approaches to mitigate crown-of-thorns starfish outbreaks on Australia's Great Barrier Reef. Sci Rep 10(1):1–13. https://doi.org/10.1038/s41598-020-69466-1

Whitman TN, Negri AP, Bourne DG et al (2020) Settlement of larvae from four families of corals in response to a crustose coralline alga and its biochemical morphogens. Sci Rep:1–10. https://doi.org/10.1038/s41598-020-73103-2

Wilmes JC, Hoey AS, Messmer V et al (2019) Incidence and severity of injuries among juvenile crown-of-thorns starfish on Australia's Great Barrier Reef. Coral Reefs 38(6):1187–1195. https://doi.org/10.1007/s00338-019-01845-1

Wilmes JC, Schultz DJ, Hoey AS et al (2020a) Habitat associations of settlement-stage crown-of-thorns starfish on Australia's Great Barrier Reef. Coral Reefs 39(4):1163–1174. https://doi.org/10.1007/s00338-020-01950-6

Wilmes JC, Hoey AS, Pratchett MS (2020b) Contrasting size and fate of juvenile crown-of-thorns starfish linked to ontogenetic diet shifts. Proc R Soc B Biol Sci 287(1931):20201052. https://doi.org/10.1098/rspb.2020.1052

Witman GB, Plummer J, Sander G (1978) *Chlamydomonas* flagellar mutants lacking radial spokes and central tubules. Structure, composition, and function of specific axonemal components. J Cell Biol 76(3):729–747. https://doi.org/10.1083/jcb.76.3.729

Wolfe K, Byrne M (2024) Dead foundation species create coral rubble habitat that benefits a resilient pest species. Mar Environ Res 202:106740

Wolfe K, Graba-Landry A, Dworjanyn SA et al (2015a) Larval starvation to satiation: influence of nutrient regime on the success of *Acanthaster planci*. PLoS One 10(3):1–17. https://doi.org/10.1371/journal.pone.0122010

Wolfe K, Graba-Landry A, Dworjanyn SA et al (2015b) Larval phenotypic plasticity in the boom-and-bust crown-of-thorns sea star, *Acanthaster planci*. Mar Ecol Prog Ser 539:179–189

Wolfe K, Graba-Landry A, Dworjanyn SA et al (2017) Superstars: assessing nutrient thresholds for enhanced larval success of *Acanthaster planci*, a review of the evidence. Mar Pollut Bull 116(1):307–314. https://doi.org/10.1016/j.marpolbul.2016.12.079

Wooldridge SA, Brodie JE (2015) Environmental triggers for primary outbreaks of crown-of-thorns starfish on the Great Barrier Reef, Australia. Mar Pollut Bull 101(2):805–815. https://doi.org/10.1016/j.marpolbul.2015.08.049

Wörheide G, Kaltenbacher E, Cowan ZL et al (2022) A new species of crown-of-thorns sea star, *Acanthaster benziei* sp. nov. (Valvatida: Acanthasteridae), from the Red Sea. Zootaxa 5209(3):379–393. https://doi.org/10.11646/zootaxa.5209.3.7

Xiao J, Wang W, Wang X et al (2022) Recent deterioration of coral reefs in the South China Sea due to multiple disturbances. PeerJ 10:e13634. https://doi.org/10.7717/peerj.13634

Yamaguchi M (1974) Growth of juvenile *Acanthaster planci* (L.) in the laboratory. Pac Sci 28(2):123–138

Yang F, Mo J, Wei Z et al (2021) Calcified macroalgae and their bacterial community in relation to larval settlement and metamorphosis of reef-building coral *Pocillopora damicornis*. FEMS Microbiol Ecol 97(1):1–11. https://doi.org/10.1093/femsec/fiaa215

Yasuda N (2018) Distribution expansion and historical population outbreak patterns of crown-of-thorns starfish, *Acanthaster planci sensu lato*, in Japan from 1912 to 2015. In: Coral Reef Studies of Japan, pp 125–148

Yasuda N, Nagai S, Hamaguchi M et al (2009) Gene flow of *Acanthaster planci* (L.) in relation to ocean currents revealed by microsatellite analysis. Mol Ecol 18(8):1574–1590. https://doi.org/10.1111/j.1365-294X.2009.04133.x

Yasuda N, Inoue J, Hall MR et al (2022) Two hidden mtDNA-clades of crown-of-thorns starfish in the Pacific Ocean. Front Mar Sci 9. https://doi.org/10.3389/fmars.2022.831240

Yuasa H, Higashimura Y, Nomura K et al (2017) Diet of *Acanthaster brevispinus*, sibling species of the coral-eating crown-of-thorns starfish, *Acanthaster planci sensu lato*. Bull Mar Sci 93(4):1009–1010

Yuasa H, Kajitani R, Nakamura Y et al (2021) Elucidation of the speciation history of three sister species of crown-of-thorns starfish (*Acanthaster* spp.) based on genomic analysis. DNA Res 28(4). https://doi.org/10.1093/dnares/dsab012

Zann L, Brodie J, Berryman C et al (1987) Recruitment, ecology, growth, and behavior of juvenile *Acanthaster planci* (L.) (Echinodermata: Asteroidea). Bull Mar Sci 41(2):561–575

Zhang G, Jia H, Luo L et al (2022) Molecular and functional characterization of a novel Kunitz-type toxin-like peptide in the giant triton snail *Charonia tritonis*. Mar Drugs 20(11). https://doi.org/10.3390/md20110686

Coral Heterotrophy in an Era of Climate Change

8

Valentine Meunier, Christine Ferrier-Pagès, Charlotte R. Dromard, and Fanny Houlbrèque

Abstract

Reef-building corals are mixotrophic organisms that obtain photosynthetically fixed carbon from their symbiotic association with algae of the family Symbiodiniaceae (from autotrophy) and assimilation of externally sourced organic compounds (heterotrophic uptake). Heterotrophic nutrition is highly beneficial to many coral species, particularly under environmental stress. Numerous experimental works have demonstrated that heterotrophy can enhance coral growth and fecundity, reduce mortality during a bleaching event, and accelerate post-bleaching recovery. Yet, the extent to which different coral species rely on heterotrophy remains largely complex and challenging to quantify. Understanding the importance of heterotrophy in coral resilience for the long-term survival of reef environments is critical. In an era largely focused on finding ways to limit the loss of coral reefs to global environmental change, the study of how heterotrophy can mitigate coral bleaching is timely. Here, we summarize our current understanding of the importance of heterotrophy for coral health and its contribution in terms of macro- and micronutrients on the survival of adult coral and early-life stages. This review highlights the urgency to further investigate the role of heterotrophy in coral resistance and resilience, and emphasizes the potential for identifying trophic niches and geographical areas that can serve as refuges from climate change.

Keywords

Coral · Heterotrophy · Climate change · Thermal stress · Review

8.1 Introduction

The coral-algae symbiosis drives reef accretion, formation, and functioning as we know it, but corals are also associated with a variety of prokaryotic and eukaryotic microorganisms (Rohwer et al. 2002; Rosenberg et al. 2007). They form a meta-organism called holobiont (Bang et al. 2018; Jaspers et al. 2019). Some of these partners within the holobiont are known to contribute to nutrient acquisition and cycling and allows corals to thrive in oligotrophic waters (Muscatine et al. 1984; Camp et al. 2018). Scleractinian corals, which live in symbiosis with dinoflagellate algal symbionts of the family Symbiodiniaceae (LaJeunesse et al. 2018), are considered as mixotrophic organisms. Symbiodiniaceae constitutes the main source of fixed carbon for the coral and produces photosynthates, carbon-rich organic compounds, fulfilling a significant proportion of the energy requirements of their coral host (i.e., autotrophy) (Muscatine et al. 1984; Grottoli et al. 2006). They thus provide sufficient energy to build the corals' calcium carbonate skeleton and thus the reef habitat (Muscatine and Porter 1977; Muscatine et al. 1984; Hughes et al. 2010; Tremblay et al. 2012a, b). In return, the coral provides the symbionts with habitat and nutrients derived from its catabolism. Additionally, corals exhibit trophic plasticity by feeding on particulate organic matter and plankton (i.e., heterotrophy; see review by Houlbrèque and Ferrier-Pagès 2009). Corals therefore covary across the resource acquisition spectrum and their high trophic plasticity underlie their evolutionary success (Goreau et al. 1971; Houlbrèque and Ferrier-Pagès 2009; Goldberg 2018).

V. Meunier (✉) · F. Houlbrèque
ENTROPIE UMR 9220 (CNRS, IRD, UR, UNC, IFREMER), Institut de Recherche pour le Développement, Nouméa, New Caledonia

Laboratoire d'excellence « CORAIL », Paris, France

C. Ferrier-Pagès
Laboratoire d'excellence « CORAIL », Paris, France

Coral Ecophysiology Team, Centre Scientifique de Monaco, Monte Carlo, Monaco

C. R. Dromard
Laboratoire d'excellence « CORAIL », Paris, France

UMR 8067 BOREA, Univ Antilles, MNHN, SU, UCN, CNRS, IRD, Pointe-à-Pitre, Guadeloupe, France

S. Planes (ed.), *The Future of Coral Reefs*, Coral Reefs of the World 21, https://doi.org/10.1007/978-3-031-98584-3_8

Scleractinian corals are most common in the tropics but also occur in subtropical areas at high latitudes and in the deep sea, where they show adaptability to fluctuating environmental conditions (albeit no photosynthesis) (Goldberg 2018; Martinez et al. 2020). Feeding strategies are species-specific and depend on various factors: geographic location, feeding history of the host, thermal conditions and tidal variability, local pollution, etc. (Ferrier-Pagès et al. 1998; Grover et al. 2002; Fujii et al. 2020; Sturaro et al. 2021). Corals are able to shift from predominantly autotrophic to heterotrophic and vice versa (Grottoli et al. 2006; Houlbrèque and Ferrier-Pagès 2009; Fox et al. 2019). In shallow, well-lit areas, corals can be mostly autotrophic and photosynthetically fixed carbon meeting up to 90% of their metabolic requirements (Falkowski et al. 1984; Muscatine 1990; Tremblay et al. 2012a, b), while in deeper or light-limited areas such as mesotrophic reefs, corals can be much more heterotrophic (Houlbrèque and Ferrier-Pagès 2009; Goldberg 2018; Kahng et al. 2019; Martinez et al. 2020). The uptake of externally sourced organic compounds (from live planktonic matter, suspended particulate, or dissolved organic matter; POM and DOM respectively) allows corals to meet up to 66% of the fixed carbon incorporated into coral skeletons (Houlbrèque and Ferrier-Pagès 2009; Martinez et al. 2020).

The main threat that corals are facing today are the effects of global warming as reflected in anomalously high seawater temperatures (Hoegh-Guldberg 1999; Hughes et al. 2018; Sully et al. 2019) leading to increasingly severe and frequent massive bleaching events (Shlesinger and van Woesik 2023). Bleaching results in the loss of Symbiodiniaceae from the coral tissues (or the loss of photosynthetic pigment from the algae) and as such deprives coral hosts of their main energy source, even leading to coral death if the stress persists (Hoegh-Guldberg 1999). Now more than ever, understanding the drivers of and improving coral health and resilience is a priority. Knowledge of molecular, genetic, and phenotypic traits that determine coral resistance and resilience is still in its infancy. In the last 20 years, the role of Symbiodiniaceae under thermal stress has been much more extensively studied than that of heterotrophy. However, heterotrophy promotes resistance and resilience by building up energy reserves to better withstand the energetic losses associated with thermal stress (Grottoli et al. 2006; Palardy et al. 2008; Goldberg 2018) understanding coral trophic plasticity (e.g., Chei et al. 2025) and identifying geographic areas that facilitate higher coral heterotrophy that might improve protection and preservation measures for coral ecosystems in the face of our escalating climate emergency (van Woesik et al. 2022).

In this chapter, we briefly review the mechanisms of heterotrophy used by scleractinian corals and their various food sources. The benefits of heterotrophy in the coral holobiont and its performance under thermal stress are described. Furthermore, the chapter emphasizes the importance of heterotrophically acquired energy reserves and their role in physiology in mature corals and their early-life stages. Additionally, we highlight knowledge gaps and recommend future research to identify the trophic niches of corals and geographical areas as possible thermal refugia.

8.2 Part I | The Benefits of Heterotrophy Under Non-stressful Growth Conditions

8.2.1 The Mixotrophic Nature of Corals

Scleractinian corals are the main ecosystem engineers of tropical coral reefs (Knowlton and Rohwer 2003; Ziegler et al. 2019). They are complex organisms, in which the coral host is associated with a variety of microbes (i.e., protists, bacteria, archaea, and fungi) as well as viruses (Rohwer et al. 2002; Rosenberg et al. 2007). This assemblage, also known as the coral holobiont (Rohwer et al. 2002), is highly dynamic, and some partners are known to play major roles in nutrient acquisition. As such, efficient nutrient cycling within the holobiont (Cardini et al. 2015; Rädecker et al. 2015) ensures the ecological success of scleractinian corals in nutrient-poor waters. Unicellular endosymbiotic dinoflagellates of the family Symbiodiniaceae (Muscatine et al. 1984; LaJeunesse et al. 2018; Hume et al. 2020) are photosynthetically competent, fixing carbon dioxide and producing organic compounds commonly called photosynthates. These photosynthates are carbon-rich, low-molecular-weight compounds, such as glycerol, organic acids, glucose, amino acids, and lipids (Trench 1979), of which the vast majority are translocated to the animal host (e.g., Muscatine and Porter 1977; Davies 1984; Muscatine 1990; Sutton and Hoegh-Guldberg 1990; Godinot et al. 2011). This large supply of autotrophic energy source is respired, released as mucus (see 2.2 below) or stored, providing a large fraction of the holobiont's daily metabolic energy requirements for tissue and skeletal growth (Muscatine 1990; Grottoli et al. 2006, 2014; Comeau et al. 2017). In return, the animal host provides the symbionts with habitat and nutrients derived from prey digestion. In addition, bacterial assemblages can contribute to nutrient acquisition (Rädecker et al. 2015; Bourne et al. 2016; Meunier et al. 2021; Saper et al. 2025) and, potentially nutrient synthesis (Pogoreutz et al. 2022; Maire et al. 2022) as well as the provisioning with antibiotic compounds (Ritchie 2006). Most scleractinian species take up substantial amounts of nutrients by heterotrophic on particulate organic matter and plankton and are therefore considered as mixotrophs (Houlbrèque and Ferrier-Pagès 2009; Goldberg 2018). Their trophic plasticity has been poorly understood and oversimplified for years using impaired coral nutrition proxies and neglecting intraspecific variability. As photosynthates are considered as junk food, due to their high C/N ratios, it has been suggested that corals would preferentially use photosynthates to meet their immediate energy needs, while they would need additional N through feeding to sustain their growth (Falkowski et al. 1984; Bachar et al. 2007).

8.2.2 Brief Overview of Heterotrophy Mechanisms

Prey capture mechanisms are diverse and well documented, but the two most common are the use of tentacles and mucociliary trapping (Lewis and Price 1975; Ferrier-Pagès et al. 1998; Houlbrèque and Ferrier-Pagès 2009; Goldberg 2018). Tentacles are covered with stinging vesicles called nematocysts, which are discharged upon contact with a mobile prey, paralyzing it (Houlbrèque and Ferrier-Pagès 2009; Goldberg 2018). Prey is then transported by the tentacles into the oral disk and wrapped in a mucus net, before being ingested through the polyp mouth (Figs. 8.1a and 8.1b). Mucus nets (Fig. 8.2), produced by mucocyte cells, are a complex of lipids, polysaccharides, and proteins. In addition to its role in trapping particles, this viscous and elastic structure lubricates and protects the coral's outer epithelial layer (e.g., Carpenter 1910; Lewis and Price 1975; Bythell and Wild 2011; Goldberg 2018). The mucus, highly enriched in organic carbon and nitrogen, also represents by itself an important food source for bacteria (e.g., Ferrier-Pagès et al. 1998; Wild et al. 2005; Huettel et al. 2006; Glasl et al. 2016; Peixoto et al. 2017; Hadaidi et al. 2018; Bhagwat et al. 2024). Once the prey has been captured, the digestive process (extra- and intra-digestion) of particulate food starts within a time frame of several hours (Wijgerde et al. 2011). It can be considerably longer than previously reported (Leal et al. 2014), depending on the type of prey, coral species, and digestion process (Wijgerde et al. 2011; Raz-Bahat et al. 2016; Goldberg 2018).

8.2.3 Heterotrophic Sources

Corals are opportunistic feeders, able to ingest a wide variety of food sources, ranging from DOM (Grover et al. 2008; Levas et al. 2015, 2016) to POM, represented by detritus (Rosenfeld et al. 1999; Anthony 2000; Ferrier-Pagès et al. 2003; Houlbrèque et al. 2004a; Levas et al. 2015, 2016), pico-nanoplankton (with a size <5 μm; Houlbrèque et al. 2004b), and larger particles such as meso- and macrozooplankton (Anthony 1999; Anthony and Fabricius 2000; Grottoli et al. 2006; Houlbrèque and Ferrier-Pagès 2009; Levas et al. 2016; Goldberg 2018). In turbid nearshore areas, sediment can provide up to 80% of the total POM ingested by corals (Anthony 2000; Houlbrèque and Ferrier-Pagès 2009; Lesser et al. 2009; Goldberg 2018). Rightly so, corals are considered voracious predators. Within the DOM pool, amino acids are present at concentrations between 0.02 and 1 μM (Palenik and Morel 1990; Ferrier 1991; Hansell and Carlson 2014) and provide up to 21% of total tissue N requirements (Grover et al. 2008). DOM is also a source of dissolved organic phosphorus compounds, which have to be hydrolyzed to dissolved inorganic phosphorus, via the activity of phosphatase enzymes localized in coral tissue, before being absorbed (Jackson et al. 1989; Anthony and Fabricius 2000; Annis and Cook 2002; Ferrier-Pagès et al. 2016). Within the living particulate matter pool, zooplankton is recognized as one of the primary food sources in a coral diet; Fig. 8.3). Corals are able to feed opportunistically on zooplankton (Coles 1969), but ingestion rates generally increase with the zooplankton abundance in the environment and the

a b

Fig. 8.1 Expended polyps of the scleractinian coral (**a**) *Stylophora pistillata* and (**b**) *Galaxea fascicularis* (Photos by Tom Schlesinger, Tel Aviv University)

Fig. 8.2 Massive *Porites* colony coated with a wide mucus net used, among other mechanisms, to capture organic matter in the surrounding waters (Photo by Tom Schlesinger, Tel Aviv University)

Fig. 8.3 Lateral view of a fluorescent solitary scleractinian *Fungia* feeding on shrimps © Institut Océanographique de Monaco, Frédéric Pacorel

depth at which corals live. Seawater temperature and the coral health status can have different effects on ingestion rates depending on the coral species (Palardy et al. 2005, 2006; Ferrier-Pagès et al. 2010; McLachlan et al. 2022). The prey found in coral coelenteron (mostly Crustacea: copepods, isopods, amphipods, crab zoeae, but also rarer and less motile prey) depend on coral species and the local relative abundance of prey types (Johnson and Sebens 1993; Leichter et al. 1998; Palardy et al. 2006, 2008). Coral do not eat prey in proportion to their availability in the environment because they either can't catch all prey types equally or because they chose some prey types (Palardy et al. 2006, 2008). Also, since the prey are rapidly digested, the analyses of the stomach content of the corals do not always reflect the true ingestion rates (Johannes and Tepley 1974; Johnson and Sebens 1993; Sebens et al. 1996; Palardy et al. 2005; Grottoli et al. 2006; Osinga et al. 2011).

8.2.4 Heterotrophic Energy Reserve and Role in Adult Coral Physiology

The very first evidence of the heterotrophic nature of corals was provided by Edmondson (1928) and Yonge and Nicholls (1931) during an expedition to the Great Barrier Reef in Australia. Corals were kept in the dark and fed at repletion with zooplankton. These studies demonstrated that corals could survive more than 45 days without autotrophic input. Since then, an increasing number of studies have confirmed the importance of heterotrophy for coral metabolism (e.g., Sebens et al. 1996; Houlbrèque et al. 2003, 2004a, b; Palardy et al. 2005, 2008; Grottoli et al. 2006; Houlbrèque and Ferrier-Pagès 2009; Levas et al. 2016). Prey capture and organic matter ingestion can be essential for coral growth, even for coral colonies located in shallow areas (under favorable light conditions) or with high Symbiodiniaceae densities (Houlbrèque and Ferrier-Pagès 2009; Martinez et al. 2020). Heterotrophy contributions are even more essential in low-light environments when photosynthetic inputs are limited (e.g., deep or turbid seawater) (Falkowski et al. 1984; Anthony and Fabricius 2000; Travaglione et al. 2023). The reliance on autotrophic versus heterotrophic diet is however species- and context-dependent (Lewis and Price 1975; Houlbrèque and Ferrier-Pagès 2009; Goldberg 2018; Conti-Jerpe et al. 2020; Price et al. 2021; McLachlan et al. 2022; Chei et al. 2025).

Heterotrophy provides corals with additional carbon and essential nutrients such as nitrogen and phosphorus (Falkowski et al. 1984), not supplied through symbiont photosynthates, and allows them to synthesize greater amounts of proteins, lipids and carbohydrates. Proteins are a central component of coral cellular physiology (i.e., enzymatic catalysis, cellular transportation, immunity, and growth) (Ferrier-Pagès et al. 2003; Houlbrèque et al. 2004a; Borell et al. 2008; Houlbrèque and Ferrier-Pagès 2009). Lipids account for at least 30% of the coral's energy reserves (Rodrigues and Grottoli 2007; Parrish 2013; Imbs 2013) and the composition of cell membranes and subcellular organelles (Liu et al. 2020). Carbohydrates are involved in the production of energetic metabolites such as ATP (Hardie et al. 2003). They can also be released externally as a mucoid matrix (Hadaidi et al. 2018) which serves as a mechanism to discard excess fixed carbon (Levas et al. 2015) and as an attractive layer to capture plankton and bacteria. Mucus also constitutes a physical barrier with distinct microbial communities, which may help to protect the holobiont from external pathogens and viruses (Wild et al. 2005; Glasl et al. 2016; Hadaidi et al. 2017). Hence, tissue contents of these different molecules are considered strong indicators of coral health status, since they tend to decrease in bleached and heat-stressed corals (Hughes et al. 2017; Kochman et al. 2021). Fatty acid and sterol compositions are often species-specific and different from one plankton source to another. They are therefore used as indicators of dietary sources in corals (Treignier et al. 2008; Tolosa et al. 2011; Naumann et al. 2015). Enhanced tissue growth often parallels skeletal growth (linear extension). Several hypotheses have been proposed to explain coral skeletal growth stimulation by heterotrophy (reviewed in Houlbrèque and Ferrier-Pagès 2009). Heterotrophy can increase (*i*) tissue growth, thus providing additional inorganic carbon for calcification (Furla et al. 2000); (*ii*) energy reserves and photosynthetic rates which increase the supply of Ca^{2+} required for the calcification process by stimulating ATP synthesis needed to activate proton pumps, which would promote carbonate precipitation (McConnaughey 1989; Borell et al. 2008; Ferrier-Pagès et al. 2010; Allemand et al. 2011); and (*iii*) organic matrix synthesis (Houlbrèque et al. 2004a), considered as an essential precondition for the formation of the biomineral structure (Allemand et al. 1998).

Box 8.1. The Perfect Microfood Source for Scleractinian Corals

Despite the fact that heterotrophy in corals has started to be explored more intensively since these last decades, there are still unresolved questions such as (*i*) how does unavailability of preferred prey items affect coral feeding behavior and, subsequently, health/resilience to stress? (*ii*) How this trophic mismatch is going to affect coral reef benthic-pelagic coupling in the future? (*iii*) How does prey selection work in the environment? (*iii*) Can wild corals adapt their feeding rates to the availability of a specific plankton type? For example, coral morphology directly affects the composition of captured plankton (Houlbrèque et al. 2004b; Palardy et al. 2005). Wild corals exhibit higher predation rates at night, when zooplankton density is greater, while in captivity corals feed just as well during the day (Houlbrèque and Ferrier-Pagès 2009; Tagliafico et al. 2018). Scleractinian corals rely on chemosensory clues to select and capture prey (i.e., through their ability to react to specific metabolites present on the surface or inside the prey) (Lenhoff and Howard 1977). However, this selection is still poorly understood, and additional experiments are needed to identify the chemosensory cues; (*iv*) can scleractinian corals actively select feeding on high-nutrient or high-energy prey? It remains unclear how corals feed selectively upon certain plankton groups (e.g., zooplankton and picoplankton; (Goldberg 2018; Meunier et al. 2019; Hoadley et al. 2021; Joppien et al. 2022). Most studies have focused on capture and predation of large planktonic organisms (> 100 μm) such as zooplankton, but some research has also investigated particulate and dissolved organic matter as heterotrophic sources (Houlbrèque and Ferrier-Pagès 2009; Levas et al. 2015, 2016; Goldberg 2018). Numerous coral feeding experiments using *Artemia salina* nauplii have highlighted the importance of zooplankton/microcrustaceans/small planktonic fraction/heterotrophic feeding in coral physiology, in particular with regard to tissue growth, calcification (Coles 1969; Johnson and Sebens 1993; Sebens et al. 1996; Grottoli 2002; Ferrier-Pagès et al. 2003; Houlbrèque et al. 2004a; Houlbrèque and Ferrier-Pagès 2009; Dobson et al. 2021). Coral feeding on the small planktonic fraction (i.e., nano- and pico-plankton) has long been neglected. However, this community contributes up to 50% of the chlorophyll biomass in the coral reef water column (Legendre et al. 1988; Sorokin 1991; Ferrier-Pages and Gattuso 1998). Hoadley et al. 2021 demonstrated the capacity of three Caribbean scleractinian coral species (*Madracis mirabilis*, *Porites astreoides*, and *Stephanocoenia intersepta*) to actively consume free-living *Prochlorococcus* and *Synechococcus* populations. Planktonic diazotrophs have also been identified as an important nitrogen source for various common coral species (Benavides et al. 2016; Meunier et al. 2021). Such selective feeding on specific microbial functional groups may be crucial for corals under environmentally stressful conditions, when heterotrophically derived carbon demand is particularly high.

8.2.5 Effects of Heterotrophy on Coral Sexual Reproduction and Early Life Stage

The relative importance of heterotrophy in coral sexual reproduction is still unclear, as neither qualitative nor quantitative data are available. The complex reproduction of scleractinian corals makes this type of study even more difficult. Photosynthates derived from Symbiodiniaceae are low in nitrogen and lack phosphorus and amino acids that are essential for gamete production (Battey and Patton 1984, 1987. Hence, heterotrophy may play a key role in coral reproductive success by providing these valuable nutrients (Ferrier-Pagès et al. 2003; Houlbrèque et al. 2003). An increase in the number and size of oocytes, leading to a better viability of the planula, due to a greater organic nutrient availability, has been already highlighted (Ward and Harrison 2000; Bongiorni et al. 2003; Amar et al. 2007; Petersen et al. 2008). One study conducted on *Pocillopora verrucosa* observed larger oocytes in fed corals compared to unfed ones but showed, on the contrary, a reduced fertility success and a gamete decrease when available prey concentration increased (Séré et al. 2010). Similarly, little is known about the heterotrophy role in the earliest life stages. Recent research has however shown that healthy new coral recruits of nine species of Caribbean corals all initiate zooplanktivory within 6 days following metamorphosis (Geertsma et al. 2022). Considering that most coral larvae acquire symbionts horizontally (i.e., taken up from the environment; Baird et al. 2009) after a few days/weeks, their only food source during this initial step should come from heterotrophy. As early-life stages, especially during metamorphosis, require a large amount of nutrients and energy reserves for development, growth, and survival (Lewis 1974; Petersen et al. 2008; Edmunds et al. 2013) and are vulnerable, they experience high mortality rates during stress events. As such, a positive effect of heterotrophy on coral gametogenesis, recruit growth, and survival rates as well as a boosted development have been demonstrated (e.g., Petersen et al. 2008; Drenkard et al. 2013; Jaffe et al. 2023; Viladrich et al. 2024). Food intake begins early, during the first 6 days after metamorphosis and the formation of polyp and tentacle structures that enable particle capture, even in coral species with vertical transmission of symbionts (from parents to offspring) (Cumbo et al. 2012; Toh et al. 2013; Geertsma et al. 2022). For example, feeding increased growth and resulted in thicker tissues and elevated symbiont fluorescence in *Pocillopora acuta* juveniles (Huffmyer et al. 2021).

8.3 Part II | The Importance of Heterotrophy Under Climate Change Conditions

8.3.1 Elevated Seawater Temperatures are the Main Drivers of Coral Bleaching

Coral reefs and their associated diversity have declined drastically around the world since the 1950s, due to the increased intensity and frequency of marine heat-waves (MHWs) (Hoegh-Guldberg 1999; Sully et al. 2019, 2022; Eddy et al. 2021; Starko et al. 2023). When ocean temperatures are anomalously high, the relationship between the animal host and its endosymbionts becomes dysfunctional (e.g., Glynn 1984; Gates et al. 1992; Brown 1997; Hughes et al. 2017) with millions of Symbiodiniaceae being expelled from coral tissue (i.e., bleaching; Warner et al. 1999; Fitt et al. 2001; Douglas 2003; Suggett and Smith 2011; Leggat et al. 2019). Thermal stress leads to increased oxidative stress and affects a variety of processes in both coral host and Symbiodiniaceae, resulting in reduced photosynthesis and calcification (Rädecker et al. 2021; Weis 2008; Leggat et al. 2011). Early developmental stages are the most sensitive and vulnerable points in coral life cycle, and the impacts of MHWs on coral larvae and recruits are more severe than those on adult colonies (Putnam et al. 2010). Determining the factors that modulate this coral response/resilience to thermal stress is essential to predict reef trajectory in the coming years and protect them more effectively.

8.3.2 Heterotrophy: The Winning Strategy for Scleractinian Corals to Cope with Thermal Stress?

During bleaching events, due to the loss of Symbiodiniaceae, coral hosts are deprived of photosynthates, their main food source (Muscatine 1990), and suffer from starvation. They may die unless they receive sufficient heterotrophic compensation or possess high energy reserves (Grottoli et al. 2006; Anthony et al. 2009; Baker et al. 2018; Palardy et al. 2008; Levas et al. 2016; Conti-Jerpe et al. 2020). Scleractinian corals with high heterotrophic feeding rates have higher survival and recovery rates after bleaching than those with low feeding rates and limited heterotrophic plasticity. Adult colonies of some coral species are able to increase their meso- and macrozooplankton consumption when exposed to thermal stress, potentially sustaining a critical nutrient supply needed for their main metabolic functions and recovery following bleaching (Grottoli et al. 2006, 2014; Palardy et al. 2008; Goldberg 2018; Saper et al. 2025). For example, bleached and recovered *Montipora capitata* were able to meet more than 100% of their daily metabolic energy requirement by increasing their feeding rates on zooplankton (Grottoli et al. 2006; Palardy et al. 2008). Other coral species, such as *Pocillopora damicornis*, *Stylophora pistillata*, and *Galaxea fascicularis*, are also able to maintain and/or adapt their metabolisms through heterotrophic feeding during heat stress (e.g., Borell et al. 2008; Lyndby et al. 2020; Martinez et al. 2024). Nevertheless, feeding behavior is highly species-specific and may be sensitive to prey availability and environmental factors such as water flow, turbidity, seasonality, and temperature (Price et al. 2021; Chapron et al. 2023; Wang et al. 2024a, b; Chei et al. 2025). Nutrition enhancement also increases coral recruit survival as well as larvae and juvenile growth (Toh et al. 2014; Conlan et al. 2017). Indeed, reproduction, as an energy-intensive process (through the production of lipid-rich gametes), is closely associated with the photosynthetic activities of Symbiodiniaceae or the energetic content of coral tissues (Ward 1995; Leuzinger et al. 2003). Scleractinian corals relying more on heterotrophy and able to maintain or replenish their energy reserves (lipid stores) between two bleaching events may thus have a higher reproductive fitness (Cox 2007), as a lipidic stock would be immediately available for gametogenesis. In addition, heterotrophy can help generate healthy recruits (Hughes and Tanner 2000; Santiago-Valentín et al. 2023), by providing additional energy and promoting biomass storage (Huffmyer et al. 2021). Feeding capacities begin very early at the larval stage (6 days after metamorphosis), but the heterotrophic influence on coral larvae or post-settlement juveniles stages during thermal stress is still poorly understood (Cumbo et al. 2012; Toh et al. 2013, 2014; Conlan et al. 2017; Geertsma et al. 2022).

Box 8.2. Pico- and Nanoplankton as a Prey Source for Bleached Corals

For many corals, predation on pico- and nanoplankton is influenced by both prey availability and prey physiology, including factors such as cell size and growth rate and nutrient content (Houlbrèque et al. 2004b; Picciano and Ferrier-Pagès 2007; Orejas et al. 2016; Goldberg 2018). Consequently, corals tend to adopt a selective feeding strategy for specific types of prey, whose growth and cell division are in turn maximized by feeding pressure. For example, corals such as *Porites astreoides* and *Stephanocoenia intersepta* feed only during the peak of cell division and primarily on *Synechococcus* sp., a cyanobacterium with the highest growth rates among other pico- and nanoplankton prey (Hoadley et al. 2021). This preference for cyanobacteria has also been observed in other scleractinian coral species (McNally et al. 2017; Sangmanee et al. 2020). Additionally, cyanobacteria capable of fixing dinitrogen are nitrogen-rich (Bertilsson et al. 2003; Jacquet et al. 2006) and may be selectively grazed upon, especially during thermal stress. Indeed, when corals bleach, they lose some of their photosynthetic symbionts that provide them with nitrogen (Muscatine and D'Elia 1978), and warming seawater also reduces the ability of corals to take up nitrogen (Godinot et al. 2011). By ingesting *Synechococcus* sp., corals can obtain an alternative source of bioavailable nitrogen, which may enhance their ability to recover from bleaching (Meunier et al. 2019). Consequently, bleached *Stylophora pistillata* exhibits a preference for feeding on *Synechococcus* sp. (1–2 μm) over *Prochlorococcus* (0.6 μm) and picoeukaryotes (0.2–2 μm), as the former has higher cellular nitrogen content (Meunier et al. 2019). This feeding strategy, targeting low-motile and immobile prey, requires less energy compared to capturing motile plankton like copepods (Tremblay et al. 2012a, b; Bednarz et al. 2017). Considering the high abundance of unicellular cyanobacteria, such as *Synechococcus* sp., in nutrient-poor waters, corals capable of exploiting this nutrient source may exhibit greater resistance and resilience to climate change. Numerous studies indicate that *Synechococcus* sp. will thrive and become more abundant in future oceans (Fu et al. 2007; Chen et al. 2014; Pittera et al. 2014). This nutritional advantage could play a significant role in the adaptation of scleractinian corals to climate change. The ability to selectively feed on *Synechococcus* sp., coupled with the anticipated increase in its activity and distribution due to rising sea surface temperatures, may be a crucial strategy for the recovery of corals affected by bleaching. Nevertheless, since the contribution of pico- and nanoplankton is small, the role of this food source may be more valuable as a nitrogen-fixing mechanism than for meeting metabolic demand. Further research is needed to understand the extent to which selective feeding and the assembly of picoplankton can be adjusted across coral holobionts as a means of acclimation to a high-temperature environment.

8.3.3 Prey Quality and Availability in a Warmer Climate

Composition, abundance, and trophic efficiency of planktonic communities are closely linked to water temperature (Lewandowska et al. 2014; Gahan et al. 2023; Hays et al. 2024; Albini et al. 2025). Thermal stress alters food quantity and quality (in terms of composition and energy storage of the plankton), by changing the seasonal dynamics of plankton communities and increasing the abundance of species adapted to highly variable environments (Mckinnon et al. 2007). In addition, ocean stratification decreases nutrient concentrations and zooplankton abundance in surface waters (Tada et al. 2003; Mckinnon et al. 2007; Bellantuono et al. 2012; Garzke et al. 2017; Hays et al. 2024) and intensifies fish predation on large zooplankton communities, resulting in sustained low biomass and smaller zooplankton (Dam and Baumann 2017; Heneghan et al. 2023). This significant decrease in zooplankton biomass and changes in plankton communities may also occur in coral reef areas (Stock et al. 2014; Chust et al. 2014; Kwiatkowski et al. 2019) and may hence impact coral diet, although few studies have been performed in coral reef areas (Mckinnon et al. 2007; Bouvy et al. 2022). Ocean stratification should promote picoplankton growth (Mckinnon et al. 2007). Within this picoplankton group, the genus *Synechococcus* (see Box 8.2) is widely distributed from the equator to the poles and shows a strong adaptation to temperature variations. Its availability and geographic distribution are likely to increase with global warming (Callieri 2017) and as such could play a role in the fate of coral reefs.

8.4 Part III | Lack of Knowledge and Future Research

Heterotrophy is more important to scleractinian coral ecology than previously considered but many grey areas remain, as heterotrophic contribution is highly variable and difficult to quantify. Future research is crucial to improve our understanding on trophic variation in survival and recovery processes following bleaching events. Here, we highlight some of the important knowledge gaps to be explored in the future.

8.4.1 Heterotrophic Nutrition Must Be Investigated in the Context of the Coral Holobiont

The trophic status of individual coral species and the respective contributions of auto- and heterotrophy to the coral energetic demand are not well defined under present and future conditions. In addition, only a few coral species (around 20) have been studied until now (Palardy et al. 2005, 2006, 2008; Grottoli et al. 2006, 2014; Houlbrèque and Ferrier-Pagès 2009; Levas et al. 2015, 2016; Goldberg 2018; Conti-Jerpe et al. 2020; Price et al. 2021; Dobson et al. 2021; McLachlan et al. 2022). These species represent a small percentage of all coral species (more than 800 species worldwide have been identified), making it difficult to generalize conclusions to the reef as a whole (McWilliam et al. 2018; Kramer et al. 2020). Corals have varying degrees of heterotrophic capacity to supplement their diet. Other important parameters affecting coral physiology, such as the generation of reactive oxygen species (ROS) and their possible effect on feeding capacity, must be considered when corals are grown under high light conditions, which generates high ROS levels in coral tissue (Lesser 1997, 2011; Lesser and Farrell 2004). Given the growing interest in holobiont studies (Grottoli et al. 2021; Saper et al. 2025), it is also essential to correlate the trophic behavior of corals with their associated microorganisms including their Symbiodiniaceae. Recent papers have shown that most of the amino acids or some essential metals acquired through heterotrophy are directly transferred to the Symbiodiniaceae and as such may influence coral tolerance to thermal stress (Ferrier-Pagès et al. 2018; Martinez et al. 2022; Martinez et al. 2024). Finally, symbionts' genus and their role in the assimilation of heterotrophically derived nutrients should be considered as a trophic strategy and thermotolerance trait of future corals (Thompson and van Woesik 2009; Peixoto et al. 2017; Camp et al. 2018; Wall et al. 2020; Jaffe et al. 2023; Wang et al. 2024a).

8.4.2 Specific Prey Selection: A Strategy with a Promising Future in the Resilience of Scleractinian Corals to Environmental Stress

Feeding rates and the type of prey ingested are largely influenced by prey availability, water flow, and morphology (Sebens et al. 1996, 1997; Palardy et al. 2005). We suggest that some prey may be more "appealing" to corals, and additional studies are needed to evaluate the role of chemotropism mechanisms in feeding. For example, it may be interesting to use fluorescent markers that can be proxies to further develop optimal dietary supplementation for corals. Zooplankton species readily ingest/take up microbeads (Setälä et al. 2014) which could provide visual markers (i.e., fluorescent microbeads or dyes) for assessing feeding rates and identifying preferential feeding behaviors (Saper 2022). Recent work has shown that corals with greater capacity for heterotrophy or evolving in plankton-rich areas would have greater resistance and/or resilience to bleaching (Fox et al. 2018; Krueger et al. 2018; Meunier et al. 2022). Regions with high primary production (i.e., plankton biomass) should be related to the resistance and persistence of corals (Fox et al. 2018; Keighan et al. 2023) and have to be taken into account.

8.4.3 Heterotrophy Must Be Integrated into Restoration Methods

Heterotrophic carbon sources are key to maintaining carbon budgets and facilitating recovery from bleaching until symbiosis is restored (Hughes et al. 2010; Hughes and Grottoli 2013). Nevertheless, in situ studies need to be extended with more coral species, as fewer than 20 coral species have been studied (mature corals; even fewer species were used for assessments of early life stages) (Grottoli et al. 2006, 2014; Palardy et al. 2008). The positive effects of heterotrophy on early life stages, post-settlement processes, and juvenile growth need to be studied in more detail to improve conservation and restoration efforts (Toh et al. 2013, 2014; Conlan et al. 2017). Outplanting of healthy individuals could be greatly enhanced by upstream work on fed young recruits raised in nurseries (Ladd and Shantz 2020; Lachs et al. 2023). Experiments observing plankton prey communities during ingestion and digestion by scleractinian corals could be conducted to quantify differences in nutrient profiles between natural (i.e., plankton) and cultured (e.g., *Artemia* nauplii) food sources. The next step would be to scale up lab experiments by using mesocosms set up on the reef, which would provide us rates, dynamics, and interactions that are virtually impossible to observe with ex situ or aquarium approaches.

Box 8.3. Mesophotic Reefs

Mesophotic scleractinian corals evolve at depths of 30–40 m (extending to 150 m deep in some locations (Hinderstein et al. 2010) with reduced light availability compared to shallow-water corals (Pyle and Copus 2019). Thus, they may overcome light-limited conditions either through photoadaptation processes, as photosynthates production slows over depth, or through increased heterotrophic nutrient acquisition (Kahng et al. 2019; Carmignani et al. 2023). However, until now, coral nutrition and physiology studies have predominantly used common shallow-water scleractinian coral species. This is due to the fact that mesophotic ecosystems are generally located in remote areas and in very deep waters, making them difficult to access and explore (Turner et al. 2017). Therefore, for mesophotic species, trophic strategy and adaptive limits are unknown, and research on this topic is still in its infancy (Loya et al. 2016; Pyle and Copus 2019). For many years, these reefs were considered to be the continuation of shallow reef communities, with a similarly diverse range of taxa and physiological mechanisms (Lesser et al. 2009). However, even the limited number of related studies suggests that major biogeographic patterns described for shallow reef organisms may not apply to mesophotic corals (Turner et al. 2017; Pyle and Copus 2019). On the Great Barrier Reef in Australia, a study based on natural selection across environmental gradients, demonstrated that *Seriatopora hystrix* is better able to utilize heterotrophic resources in mesophotic habitats when adapted to low-light conditions (Bongaerts et al. 2011). Overall, no patterns were reported between $\delta^{15}N$ values and heterotrophic behavior in mesophotic corals, likely due to the complexity of available nitrogen sources (Carmignani et al. 2023). This latest study provides new insights into the possible links between heterotrophic nitrogen benefits and species-specific nitrogen requirements at mesophotic depths. These results are promising for the potential benefits of a heterotrophic food source in the context of rising ocean temperatures. Important indicators of increased nutrient and plankton availability in the deeper habitat have also already been demonstrated (Wolanski et al. 1988; Bongaerts et al. 2011). We hypothesize that repeated thermal stress episodes could lead to the migration of shallow-water corals to deeper waters. As such, mesophotic areas might play an essential role as ecological refugia providing protection for coral reefs facing climate change.

8.4.4 Trophic Niches and Future Coral Refugia

Today, the primary challenge is to identify key patterns in coral trophic ecology across varying environmental stress gradients. Isotopic indicators (stable isotope analysis; SIA) have emerged as powerful tools to investigate coral niche and resource selectivity of scleractinian corals (e.g., Minagawa and Wada 1984; Conti-Jerpe et al. 2020; Price et al. 2021; Thibault et al. 2021; Chei et al. 2025). Global warming and recurrent MHWs have resulted in a significant reduction in the ecological niche of many corals, predominantly autotrophic (Dutkiewicz et al. 2013; Sully et al. 2019). Conversely, coral species capable of increasing the proportionate contribution of heterotrophy in response to environmental conditions, or with naturally high baseline

Fig. 8.4 Massive and branching corals air-exposed during low tide at the semi-enclosed bay of Bouraké, New Caledonia (Photo by Riccardo Rodolfo-Metalpa, IRD, UMR Entropie)

heterotrophic capacity, may benefit from niche expansion in the future compared to species lacking this capability (Loya et al. 2001; Grottoli et al. 2006; Rodrigues and Grottoli 2007; Conti-Jerpe et al. 2020; Wang et al. 2024b; Chei et al. 2025). This might lead to niche overlaps and competition for heterotrophic resources (Dixon et al. 2022; Skinner et al. 2022). Changes in coral niche space and distribution may form new "refugia" created by environmental stressors (Nakabayashi et al. 2019; Price et al. 2019). Refugia are defined as "habitats that components of biodiversity retreat to, persist in and can potentially expand from under changing environmental conditions" (Mertz and McDonald 2021). To date, a limited number of long-term and climate refugia (e.g., mangroves, upwelling regions) have been identified, and the coral resilience in these environments needs to be further studied (Cacciapaglia and van Woesik 2015; Kavousi and Keppel 2017; Dixon et al. 2022). It is likely that these refugia will become increasingly rare as climate change accelerates (Perry et al. 2011; Muir et al. 2015; Hughes et al. 2018). Mangrove coral habitats have shown promise as prospective refuges for corals that can show high heterotrophic capacity (Yates et al. 2014). Although these ecosystems are characterized by low-light conditions, high pCO_2, particulate sediment, and exceptionally high temperature (Anthony and Fabricius 2000; Tanvet et al. 2023), they shelter a rich variety of corals (Fig. 8.4) (Maggioni et al. 2023; Yates et al. 2014). Higher abundances of living and detrital organic matter could largely explain the greater resistance/resilience of corals in these ecosystems (Camp et al. 2017; Tanvet et al. 2023). Some of the coral species evolving in mangrove ecosystems, such as *Porites* sp., are able to easily shift from autotrophy to an increased heterotrophic feeding regime, helping them withstand these extreme conditions (Lopez and Levinton 1987; Camp et al. 2017, 2018; Haydon et al. 2023). Identification of potential refugia is crucial to safeguarding the future survival of coral reefs and associated reef species. As such, considering the potential for heterotrophic food availability may be one key criterium in the selection of

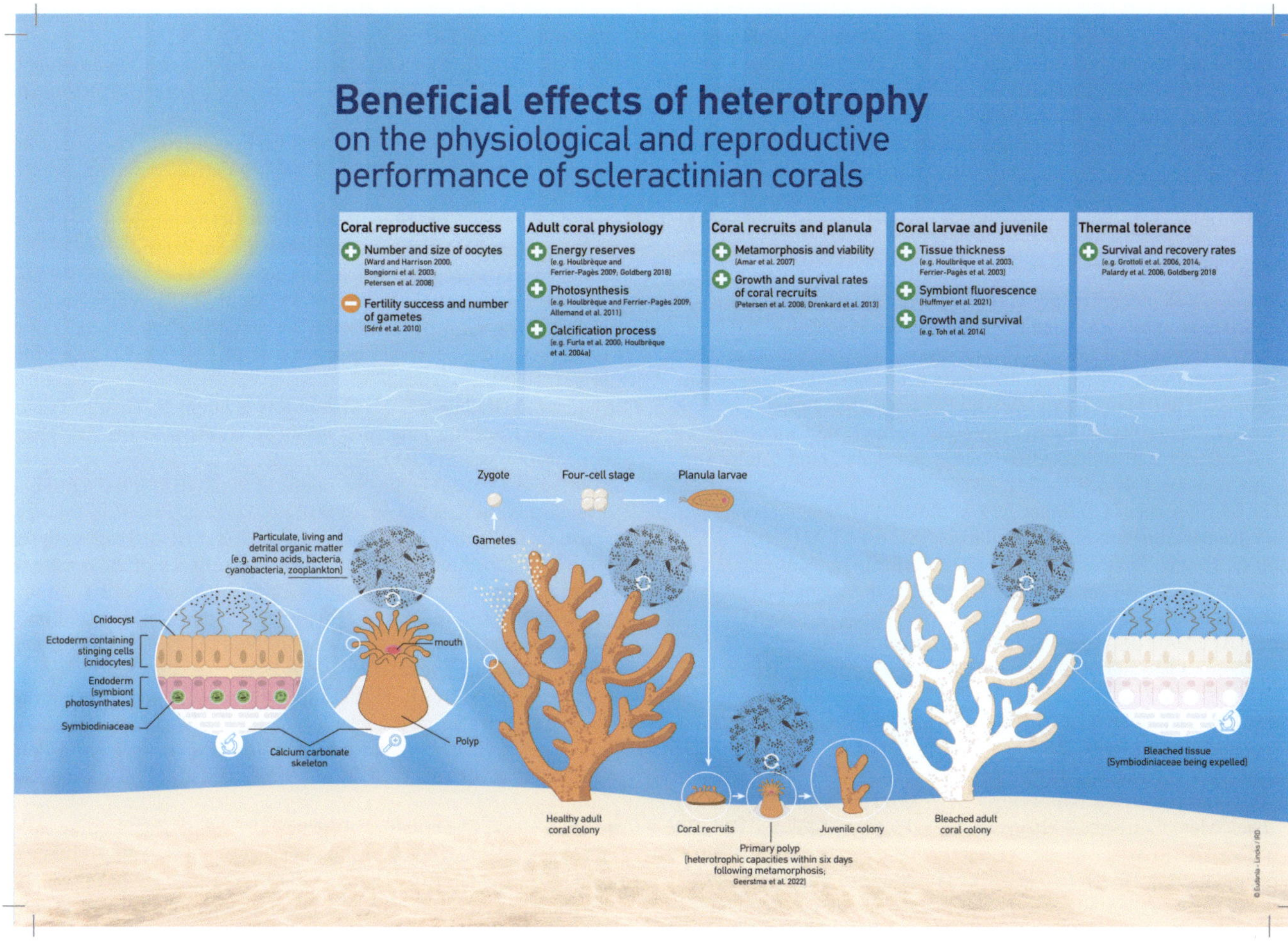

Fig. 8.5 Beneficial effects of heterotrophy on the physiological and reproductive performance of scleractinian corals. Overview of the critical role of heterotrophy in maintaining coral health across various life stages, including both adult corals and early-life stages. Emphasizing the necessity for further investigation into the heterotrophic mechanisms that drive coral resistance and resilience, we underline the importance of identifying trophic refugia. A thorough understanding of coral nutritional ecology is considered essential to develop effective strategies to mitigate coral reef decline and ensure their survival and persistence for future generations

future refugia. Furthermore, coral reefs that have previously encountered thermal stress and successfully acclimated could also serve as refugia (Thompson and van Woesik 2009; Mumby et al. 2011; Wall et al. 2018). By pinpointing these natural refugia, restoration initiatives can be better focused, ensuring the protection and longevity of corals and other reef-dwelling species in the face of future challenges (Fig. 8.5).

Authors' information and contributions V.M: conceptualization, investigation, resources, writing—original draft, and writing—review and editing (valentine0meunier@gmail.com); C.F-P: conceptualization and writing—review and editing (ferrier@centrescientifique.mc); C.R.D: conceptualization and writing—review and editing (Charlotte.Dromard@univ-antilles.fr); F.H: conceptualization, supervision and writing—review and editing (fanny.houlbreque@ird.fr).

Declaration of Competing Interest The authors declare no conflict of interest.

References

Albini D, Ransome E, Dumbrell AJ et al (2025) Warming alters plankton body-size distributions in a large field experiment. Commun Biol 8:162. https://doi.org/10.1038/s42003-024-07380-2

Allemand D, TambuttE E, Girard JP, Jaubert J (1998) Organic matrix synthesis in the scleractinian coral stylophora pistillata: role in biomineralization and potential target of the organotin tributyltin. J Exp Biol 201(Pt 13):2001–2009

Allemand D, Tambutté É, Zoccola D, Tambutté S (2011) Coral calcification, cells to reefs. In: Dubinsky Z, Stambler N (eds) Coral reefs: an ecosystem in transition. Springer, Dordrecht, pp 119–150

Amar KO, Chadwick NE, Rinkevich B (2007) Coral planulae as dispersion vehicles: biological properties of larvae released early and late in the season. Mar Ecol Prog Ser 350:71–78

Annis ER, Cook CB (2002) Alkaline phosphatase activity in symbiotic dinoflagellates (zooxanthellae) as a biological indicator of environmental phosphate exposure. Mar Ecol Prog Ser 245:11–20
Anthony KRN (1999) Coral suspension feeding on fine particulate matter. J. Exp. Mar. Bio. Ecol. 232:85–106
Anthony KRN (2000) Enhanced particle-feeding capacity of corals on turbid reefs (Great Barrier Reef, Australia). Coral Reefs 19:59–67
Anthony KR, Fabricius KE (2000) Shifting roles of heterotrophy and autotrophy in coral energetics under varying turbidity. J Exp Mar Bio Ecol 252:221–253
Anthony KRN, Hoogenboom MO, Maynard JA et al (2009) Energetics approach to predicting mortality risk from environmental stress: a case study of coral bleaching. Funct Ecol 23:539–550
Bachar A, Achituv Y, Pasternak Z, Dubinsky Z (2007) Autotrophy versus heterotrophy: the origin of carbon determines its fate in a symbiotic sea anemone. J Exp Mar Bio Ecol 349:295–298
Baird AH, Guest JR, Willis BL (2009) Systematic and biogeographical patterns in thereproductive biology of scleractinian corals. Annu Rev Ecol Evol Syst 40(1):551–571
Baker DM, Freeman CJ, Wong JCY et al (2018) Climate change promotes parasitism in a coral symbiosis. ISME J 12:921–930
Bang C, Dagan T, Deines P et al (2018) Metaorganisms in extreme environments: do microbes play a role in organismal adaptation? Zoology 127:1–19
Battey JF, Patton JS (1984) A reevaluation of the role of glycerol in carbon translocation in zooxanthellae-coelenterate symbiosis. Mar Biol 79:27–38
Battey JF, Patton JS (1987) Glycerol translocation in Condylactis gigantea. Mar Biol 95:37–46
Bednarz VN, Grover R, Maguer J-F et al (2017) The Assimilation of Diazotroph-Derived Nitrogen by Scleractinian Corals Depends on Their Metabolic Status. MBio 8:e02058-16. https://doi.org/10.1128/mBio.02058-16
Bellantuono AJ, Hoegh-Guldberg O, Rodriguez-Lanetty M (2012) Resistance to thermal stress in corals without changes in symbiont composition. Proc R Soc B Biol Sci 279:1100–1107
Benavides M, Houlbrèque F, Camps M et al (2016) Diazotrophs: a non-negligible source of nitrogen for the tropical coral Stylophora pistillata. J Exp Biol 219:2608. https://doi.org/10.1242/jeb.139451
Bertilsson S, Berglund O, Karl DM, Chisholm SW (2003) Elemental composition of marine Prochlorococcus and Synechococcus: Implications for the ecological stoichiometry of the sea. Limnol Oceanogr 48:1721–1731. https://doi.org/10.4319/lo.2003.48.5.1721
Bhagwat PV, Ravindran C, Irudayarajan L (2024) Beneficial properties of mucus in coral adaptations and ecological interactions. Mar Biol 171(2):46
Bongaerts P, Riginos C, Hay KB et al (2011) Adaptive divergence in a scleractinian coral: physiological adaptation of Seriatopora hystrix to shallow and deep reef habitats. BMC Evol Biol 11:303
Bongiorni L, Shafir S, Rinkevich B (2003) Effects of particulate matter released by a fish farm (Eilat, Red Sea) on survival and growth of Stylophora pistillata coral nubbins. Mar Pollut Bull 46:1120–1124
Borell EM, Yuliantri AR, Bischof K, Richter C (2008) The effect of heterotrophy on photosynthesis and tissue composition of two scleractinian corals under elevated temperature. J Exp Mar Bio Ecol 364:116–123
Bourne DG, Morrow KM, Webster NS (2016) Insights into the coral microbiome: underpinning the health and resilience of reef ecosystems. Ann Rev Microbiol 70:317–340
Bouvy M, Bélières A, Carré C et al (2022) Do microbial planktonic communities reflect the ecological changes of Glorieuses coral reefs (Iles Eparses, Western Indian Ocean)? Mar Pollut Bull 174:113218
Brown BE (1997) Coral bleaching: causes and consequences. Coral Reefs 16:S129–S138
Bythell JC, Wild C (2011) Biology and ecology of coral mucus release. J Exp Mar Bio Ecol 408:88–93
Cacciapaglia C, van Woesik R (2015) Reef-coral refugia in a rapidly changing ocean. Glob Chang Biol 21:2272–2282
Callieri C (2017) Synechococcus plasticity under environmental changes. FEMS Microbiol Lett 364. https://doi.org/10.1093/femsle/fnx229
Camp EF, Nitschke MR, Rodolfo-Metalpa R et al (2017) Reef-building corals thrive within hot-acidified and deoxygenated waters. Sci Rep 7:1–9
Camp EF, Schoepf V, Mumby PJ et al (2018) The future of coral reefs subject to rapid climate change: lessons from natural extreme environments. FrontMar Sci 5:4
Cardini U, Bednarz VN, Naumann MS et al (2015) Functional significance of dinitrogen fixation in sustaining coral productivity under oligotrophic conditions. Proc B 282:20152257. https://doi.org/10.1098/rspb.2015.2257
Carmignani A, Radice VZ, McMahon KM et al (2023) Levels of autotrophy and heterotrophy in mesophotic corals near the end photic zone. Front Mar Sci 10:1089746
Carpenter FW (1910) Feeding reactions of the Rose Coral (Isophyllia). Proc Am Acad Arts Sci 46:149–162
Chapron L, Kuffner IB, Kemp DW et al (2023) Heterotrophy, microbiome, and location effects on restoration efficacy of the threatened coral Acropora palmata. Commun Earth Environ 4:1–12
Chei E, Conti-Jerpe IE, Pons L, Baker DM (2025) Changes within the coral symbiosis underpin seasonal trophic plasticity in reef corals. ISME Commun 5:ycae162
Chen B, Liu H, Huang B, Wang J (2014) Temperature effects on the growth rate of marine picoplankton. Mar Ecol Prog Ser 505:37–47
Chust G, Allen JI, Bopp L et al (2014) Biomass changes and trophic amplification of plankton in a warmer ocean. Glob Chang Biol 20:2124–2139
Coles SL (1969) Quantitative estimates of feeding and respiration for three scleractinian corals. Limnol Oceanogr 14:949–953
Comeau S, Carpenter RC, Edmunds PJ (2017) Effects of pCO2 on photosynthesis and respiration of tropical scleractinian corals and calcified algae. ICES J Mar Sci 74:1092–1102
Conlan JA, Humphrey CA, Severati A, Francis DS (2017) Influence of different feeding regimes on the survival, growth, and biochemical composition of Acropora coral recruits. PLoS One 12:e0188568
Conti-Jerpe IE, Thompson PD, Wai C et al (2020) Trophic strategy and bleaching resistance in reef-building corals. Sci Adv 6(15):eaaz5443
Cox EF (2007) Continuation of sexual reproduction in Montipora capitata following bleaching. Coral Reefs 26:721–724
Cumbo VR, Fan TY, Edmunds PJ (2012) Scleractinian corals capture zooplankton within days of settlement and metamorphosis. Coral Reefs 31:1155
Dam HG, Baumann H (2017) Climate change, zooplankton and fisheries. In: Climate change impacts on fisheries and aquaculture. Wiley, Chichester, pp 851–874
Davies PS (1984) The role of zooxanthellae in the nutritional energy requirements of Pocillopora eydouxi. Coral Reefs 2(4):181–186
Dixon AM, Forster PM, Heron SF et al (2022) Future loss of local-scale thermal refugia in coral reef ecosystems. PLOS Climate 1:e0000004
Dobson KL, Ferrier-Pagès C, Saup CM, Grottoli AG (2021) The effects of temperature, light, and feeding on the physiology of Pocillopora damicornis, Stylophora pistillata, and Turbinaria reniformis corals. Water 13:2048
Douglas AE (2003) Coral bleaching—how and why? Mar Pollut Bull 46:385–392
Drenkard EJ, Cohen AL, McCorkle DC et al (2013) Calcification by juvenile corals under heterotrophy and elevated CO2. Coral Reefs 32:727–735
Dutkiewicz S, Scott JR, Follows MJ (2013) Winners and losers: ecological and biogeochemical changes in a warming ocean. Global Biogeochem Cycles 27:463–477

Eddy TD, Lam VWY, Reygondeau G et al (2021) Global decline in capacity of coral reefs to provide ecosystem services. One Earth 4:1278–1285

Edmondson CH (1928) The ecology of an Hawaiian coral reef. Bull Bernice P Bishop Mus 45:1–64

Edmunds PJ, Cumbo VR, Fan T-Y (2013) Metabolic costs of larval settlement and metamorphosis in the coral Seriatopora caliendrum under ambient and elevated pCO2. J Exp Mar Bio Ecol 443:33–38

Falkowski PG, Dubinsky Z, Muscatine L, Porter JW (1984) Light and the bioenergetics of a symbiotic coral. Sciences-New York 34:705–709

Ferrier MD (1991) Net uptake of dissolved free amino acids by four scleractinian corals. Coral Reefs 10:183–187

Ferrier-Pages C, Gattuso J (1998) Biomass, production and grazing rates of Pico- and Nanoplankton in coral reef waters (Miyako Island, Japan). Microb Ecol 35:46–57

Ferrier-Pagès C, Gattuso JP, Cauwet G et al (1998) Release of dissolved organic carbon and nitrogen by the zooxanthellate coral Galaxea fascicularis. Mar Ecol Prog Ser 172:265–274

Ferrier-Pagès C, Witting J, Tambutté E, Sebens KP (2003) Effect of natural zooplankton feeding on the tissue and skeletal growth of the scleractinian coral Stylophora pistillata. Coral Reefs 22:229–240

Ferrier-Pagès C, Rottier C, Beraud E, Levy O (2010) Experimental assessment of the feeding effort of three scleractinian coral species during a thermal stress: effect on the rates of photosynthesis. J Exp Mar Bio Ecol 390:118–124

Ferrier-Pagès C, Godinot C, D'Angelo C et al (2016) Phosphorus metabolism of reef organisms with algal symbionts. Ecol Monogr 86:262–277

Ferrier-Pagès C, Sauzéat L, Balter V (2018) Coral bleaching is linked to the capacity of the animal host to supply essential metals to the symbionts. Glob Chang Biol 24:3145–3157

Fitt W, Brown B, Warner M, Dunne R (2001) Coral bleaching: interpretation of thermal tolerance limits and thermal thresholds in tropical corals. Coral Reefs 20:51–65

Fox MD, Williams GJ, Johnson MD et al (2018) Gradients in primary production predict trophic strategies of Mixotrophic corals across spatial scales. Curr Biol 28:3355–3363.e4

Fox MD, Elliott Smith EA, Smith JE, Newsome SD (2019) Trophic plasticity in a common reef-building coral: insights from δ13C analysis of essential amino acids. Funct Ecol 33:2203–2214

Fu F-X, Warner ME, Zhang Y et al (2007) Effects of increased temperature and co2on photosynthesis, growth, and elemental ratios in mar inesynechococcusandprochlorococcus(Cyanobacteria). J Phycol 43:485–496

Fujii T, Tanaka Y, Maki K et al (2020) Organic carbon and nitrogen Isoscapes of reef corals and algal symbionts: relative influences of environmental gradients and heterotrophy. Microorganisms 8:1221. https://doi.org/10.3390/microorganisms8081221

Furla P, Galgani I, Durand I, Allemand D (2000) Sources and mechanisms of inorganic carbon transport for coral calcification and photosynthesis. J Exp Biol 203:3445–3457

Gahan J, Bellwood DR, Nankervis L, Tebbett SB (2023) Spatial and temporal variability in tropical off-reef zooplankton across broad spatial and temporal scales. Mar Environ Res 191:106169

Garzke J, Sommer U, Ismar SMH (2017) Is the chemical composition of biomass the agent by which ocean acidification influences on zooplankton ecology? Aquat Sci 79:733–748

Gates RD, Baghdasarian G, Muscatine L (1992) Temperature stress causes host cell detachment in symbiotic cnidarians: implications for coral bleaching. Biol Bull 182:324–332

Geertsma RC, Wijgerde T, Latijnhouwers KRW, Chamberland VF (2022) Onset of zooplanktivory and optimal water flow rates for prey capture in newly settled polyps of ten Caribbean coral species. Coral Reefs 41:1651–1664

Glasl B, Herndl GJ, Frade PR (2016) The microbiome of coral surface mucus has a key role in mediating holobiont health and survival upon disturbance. ISME J 10:2280–2292

Glynn PW (1984) Widespread coral mortality and the 1982–83 El Niño warming event. Environ Conserv 11:133–146

Godinot C, Grover R, Allemand D, Ferrier-Pagès C (2011) High phosphate uptake requirements of the scleractinian coral Stylophora pistillata. J Exp Biol 214:2749–2754

Goldberg WM (2018) Coral food, feeding, nutrition, and secretion: a review. In: Results and problems in cell differentiation. Springer Verlag, pp 377–421

Goreau TF, Goreau NI, Yonge CM (1971) Reef corals: autotrophs or heterotrophs? Biol Bull 141

Grottoli AG (2002) Effect of light and brine shrimp on skeletal δ13C in the Hawaiian coral Porites compressa: a tank experiment. Geochim Cosmochim Acta 66:1955–1967

Grottoli AG, Rodrigues LJ, Palardy JE (2006) Heterotrophic plasticity and resilience in bleached corals. Nature 440:1186–1189

Grottoli AG, Warner ME, Levas SJ et al (2014) The cumulative impact of annual coral bleaching can turn some coral species winners into losers. Glob Chang Biol 20:3823–3833

Grottoli AG, Toonen RJ, van Woesik R et al (2021) Increasing comparability among coral bleaching experiments. Ecol Appl 31:e02262

Grover R, Maguer J-F, Reynaud-Vaganay S, Ferrier-Pagè C (2002) Uptake of ammonium by the scleractinian coral Stylophora pistillata: effect of feeding, light, and ammonium concentrations. Limnol Oceanogr 47:782–790

Grover R, Maguer JF, Allemand D, Ferrier-Pagès C (2008) Uptake of dissolved free amino acids by the scleractinian coral Stylophora pistillata. J Exp Biol 211:860–865

Hadaidi G, Röthig T, Yum LK et al (2017) Stable mucus-associated bacterial communities in bleached and healthy corals of Porites lobata from the Arabian seas. Sci Rep 7:45362

Hadaidi G, Gegner HM, Ziegler M, Voolstra CR (2018) Carbohydrate composition of mucus from scleractinian corals from the Central Red Sea. Coral Reefs 38:21. https://doi.org/10.1007/s00338-018-01758-5

Hansell DA, Carlson CA (2014) Biogeochemistry of marine dissolved organic matter. Academic

Hardie DG, Scott JW, Pan DA, Hudson ER (2003) Management of cellular energy by the AMP-activated. FEBS Lett 546:113

Haydon TD, Matthews JL, Seymour JR et al (2023) Metabolomic signatures of corals thriving across extreme reef habitats reveal strategies of heat stress tolerance. Proc Biol Sci 290:20221877

Heneghan RF, Everett JD, Blanchard JL et al (2023) Climate-driven zooplankton shifts cause large-scale declines in food quality for fish. Nat Clim Chang 13:470–477

Hinderstein LM, Marr JCA, Martinez FA et al (2010) Theme section on "Mesophotic coral ecosystems: characterization, ecology, and management". Coral Reefs 29:247–251

Hoadley KD, Hamilton M, Poirier CL et al (2021) Selective uptake of pelagic microbial community members by Caribbean reef corals. Appl Environ Microbiol 87:e03175-20. https://doi.org/10.1128/AEM.03175-20

Hoegh-Guldberg O (1999) Climate change, coral bleaching and the future of the world's coral reefs. Mar Freshw Res 50:839–866

Houlbrèque F, Ferrier-Pagès C (2009) Heterotrophy in tropical scleractinian corals. Biol Rev Camb Philos Soc 84:1–17

Houlbrèque F, Tambutté E, Ferrier-Pagès C (2003) Effect of zooplankton availability on the rates of photosynthesis, and tissue and skeletal growth in the scleractinian coral Stylophora pistillata. J Exp Mar Bio Ecol 296:145–166

Houlbrèque F, Tambutté E, Allemand D, Ferrier-Pagès C (2004a) Interactions between zooplankton feeding, photosynthesis and skeletal growth in the scleractinian coral Stylophora pistillata. J Exp Biol 207:1461–1469

Houlbrèque F, Tambutté E, Richard C, Ferrier-Pagès C (2004b) Importance of a micro-diet for scleractinian corals. Mar Ecol Prog Ser 282:151–160

Huettel M, Wild C, Gonelli S (2006) Mucus trap in coral reefs: formation and temporal evolution of particle aggregates caused by coral mucus. Mar Ecol Prog Ser 307:69–84

Huffmyer AS, Johnson CJ, Epps AM et al (2021) Feeding and thermal conditioning enhance coral temperature tolerance in juvenile Pocillopora acuta. R Soc Open Sci 8:210644

Hughes AD, Grottoli AG (2013) Heterotrophic compensation: a possible mechanism for resilience of coral reefs to global warming or a sign of prolonged stress? PLoS One 8:e81172. https://doi.org/10.1371/journal.pone.0081172

Hughes TP, Tanner JE (2000) Recruitment failure, life histories, and long-term decline of Caribbean corals. Ecology 81:2250–2263

Hughes TP, Graham NAJ, Jackson JBC et al (2010) Rising to the challenge of sustaining coral reef resilience. Trends Ecol Evol 25:633–642

Jacquet S, Delesalle B, Torréton JP, Blanchot J (2006) Response of phytoplankton communities to increased anthropogenic influences (southwestern lagoon, New Caledonia). Mar Ecol Prog Ser 320:65–78. https://doi.org/10.3354/meps320065

Jaffe MD, Padilla-Gamiño JL, Nunn BL, Rodrigues LJ (2023) Coral trophic pathways impact the allocation of carbon and nitrogen for egg development after bleaching. Front Ecol Evol 11:1251220

Hughes TP, Kerry JT, Álvarez-Noriega M et al (2017) Global warming and recurrent mass bleaching of corals. Nature 543:373–377

Hughes TP, Anderson KD, Connolly SR et al (2018) Spatial and temporal patterns of mass bleaching of corals in the Anthropocene. Science 359:80–83

Hume BC, Mejia-Restrepo A, Voolstra CR, Berumen ML (2020) Fine-scale delineation of Symbiodiniaceae genotypes on a previously bleached central Red Sea reef system demonstrates a prevalence of coral host-specific associations. Coral Reefs 39(3):583–601

Imbs AB (2013) Fatty acids and other lipids of corals: composition, distribution, and biosynthesis. Russ J Mar Biol 39:153–168

Jackson AE, Miller DJ, Yellowlees D (1989) Phosphorus metabolism in the coral–zooxanthellae symbiosis: characterization and possible roles of two acid phosphatases in the algal symbiontSymbiodiniumsp. Proc R Soc Lond 238:193–202

Jaspers C, Fraune S, Arnold AE et al (2019) Resolving structure and function of metaorganisms through a holistic framework combining reductionist and integrative approaches. Zoology 133:81–87

Johannes RE, Tepley L (1974) Examination of feeding of the reef coral Porites lobata in situ using time lapse photography. Proc 2nd Int Coral Reef Symp 1:127–131

Johnson AS, Sebens KP (1993) Consequences of a flattened morphology: effects of flow on feeding rates of the scleractinian coral Meandrina meandrites. Mar Ecol Prog Ser 99:99–114

Joppien M, Westphal H, Stuhr M, Doo SS (2022) Microplastics alter feeding strategies of a coral reef organism. Limnol Oceanogr Lett 7:131. https://doi.org/10.1002/lol2.10237

Kahng SE, Akkaynak D, Shlesinger T et al (2019) Light, temperature, photosynthesis, heterotrophy, and the lower depth limits of Mesophotic coral ecosystems. In: Loya Y, Puglise KA, Bridge TCL (eds) Mesophotic coral ecosystems. Springer International Publishing, Cham, pp 801–828

Kavousi J, Keppel G (2017) Clarifying the concept of climate change refugia for coral reefs. ICES J Mar Sci 75:43–49

Keighan R, van Woesik R, Yalon A et al (2023) Moderate chlorophyll-a environments reduce coral bleaching during thermal stress in Yap. Micronesia Sci Rep 13:9338

Knowlton N, Rohwer F (2003) Multispecies microbial mutualisms on coral reefs: the host as a habitat. Am Nat 162:S51–S62

Kochman N'a-R, Grover R, Rottier C et al (2021) The reef building coral Stylophora pistillata uses stored carbohydrates to maintain ATP levels under thermal stress. Coral Reefs 40:1473–1485

Kramer N, Tamir R, Eyal G, Loya Y (2020) Coral morphology portrays the spatial distribution and population size-structure along a 5–100 m depth gradient. Front Mar Sci 7:615. https://doi.org/10.3389/fmars.2020.00615

Krueger T, Bodin J, Horwitz N et al (2018) Temperature and feeding induce tissue level changes in autotrophic and heterotrophic nutrient allocation in the coral symbiosis–a NanoSIMS study. Sci Rep 8:12710. https://doi.org/10.1038/s41598-018-31094-1

Kwiatkowski L, Aumont O, Bopp L (2019) Consistent trophic amplification of marine biomass declines under climate change. Glob Chang Biol 25:218–229

Lachs L, Beauchamp E, Bukurou L et al (2023) Breeding corals for heat tolerance. Res Sq Prepr:1–12

Ladd MC, Shantz AA (2020) Trophic interactions in coral reef restoration: a review. Food Webs 24:e00149

LaJeunesse TC, Parkinson JE, Gabrielson PW et al (2018) Systematic revision of Symbiodiniaceae highlights the antiquity and diversity of coral endosymbionts. Curr Biol 28:2570–2580

Leal MC, Nejstgaard JC, Calado R et al (2014) Molecular assessment of heterotrophy and prey digestion in zooxanthellate cnidarians. Mol Ecol 23:3838–3848

Legendre L, Demers S, Delesalle B, Harnois C (1988) Biomass and photosynthetic activity of phototrophic picoplankton in coral reef waters (Moorea Island, French Polynesia). Mar Ecol Prog Ser 47:153–160

Leggat W, Seneca F, Wasmund K et al (2011) Differential responses of the coral host and their algal symbiont to thermal stress. PLoS One 6:e26687

Leggat WP, Camp EF, Suggett DJ et al (2019) Rapid coral decay is associated with marine heatwave mortality events on reefs. Curr Biol 29:2723–2730.e4

Leichter JJ, Shellenbarger G, Genovese SJ, Wing SR (1998) Breaking internal waves on a Florida (USA) coral reef:a plankton pump at work? Mar Ecol Prog Ser 166:83–97

Lenhoff M, Howard HW (1977) Aquatic invertebrates: model systems for study of receptor activation and evolution of receptor proteins. Annu Rev Pharmacol Toxicol 17:243–258

Lesser MP (1997) Oxidative stress causes coral bleaching during exposure to elevated temperatures. Coral Reefs 16:187–192

Lesser MP (2011) Coral bleaching: causes and mechanisms. In: Dubinsky Z, Stambler N (eds) Coral reefs: an ecosystem in transition. Springer, Dordrecht, pp 405–419

Lesser MP, Farrell JH (2004) Exposure to solar radiation increases damage to both host tissues and algal symbionts of corals during thermal stress. Coral Reefs 23:367–377

Lesser MP, Slattery M, Leichter JJ (2009) Ecology of mesophotic coral reefs. J Exp Mar Biol Ecol 375:1–8

Leuzinger S, Anthony KRN, Willis BL (2003) Reproductive energy investment in corals: scaling with module size. Oecologia 136:524–531

Levas S, Grottoli AG, Warner ME et al (2015) Organic carbon fluxes mediated by corals at elevated pCO2 and temperature. Mar Ecol Prog Ser 519:153–164

Levas S, Grottoli AG, Schoepf V et al (2016) Can heterotrophic uptake of dissolved organic carbon and zooplankton mitigate carbon budget deficits in annually bleached corals? Coral Reefs 35:495–506

Lewandowska AM, Boyce DG, Hofmann M et al (2014) Effects of sea surface warming on marine plankton. Ecol Lett 17:614–623

Lewis JB (1974) The importance of light and food upon the early growth of the reef coral favia fragum (esper). J Exp Mar Bio Ecol 15:299–304

Lewis JB, Price WS (1975) Feeding mechanisms and feeding strategies of Atlantic reef corals. J Zool 176:527–544

Liu C-Y, Zhang F, Sun Y-F et al (2020) Effects of nitrate enrichment on respiration, photosynthesis, and fatty acid composition of reef coral Pocillopora damicornis larvae. Front Mar Sci 7:531. https://doi.org/10.3389/fmars.2020.00531

Lopez GR, Levinton JS (1987) Ecology of deposit-feeding animals in marine sediments. Q Rev Biol 62:235–260

Loya Y, Sakai K, Yamazato K et al (2001) Coral bleaching: the winners and the losers. Ecol Lett 4:122–131

Loya Y, Eyal G, Treibitz T et al (2016) Theme section on mesophotic coral ecosystems: advances in knowledge and future perspectives. Coral Reefs 35:1–9

Lyndby NH, Holm JB, Wangpraseurt D et al (2020) Effect of temperature and feeding on carbon budgets and O2 dynamics in Pocillopora damicornis. Mar Ecol Prog Ser 652:49–62

Maggioni F, Bell JJ, Pujo-Pay M, Shaffer M, Cerrano C, Lemonnier H, Letourneur Y, Rodolfo-Metalpa R (2023) Sponge organic matter recycling: Reduced detritus production under extreme environmental conditions. Mar Pollut Bull 190:114869. https://doi.org/10.1016/j.marpolbul.2023.114869. Accessed 5 Apr 2023

Maire J, Tandon K, Collingro A et al (2022) Endozoicomonas-chlamydiae interactions in cell-associated microbial aggregates of the coral Pocillopora acuta. bioRxiv:2022.11.28.517745

Martinez S, Kolodny Y, Shemesh E et al (2020) Energy sources of the depth-generalist Mixotrophic coral Stylophora pistillata. Front Mar Sci 7:988

Martinez S, Grover R, Baker DM, Ferrier-Pagès C (2022) Symbiodiniaceae are the first site of heterotrophic nitrogen assimilation in reef-building corals. MBio 13:e0160122

Martinez S, Grover R, Ferrier-Pagès C (2024) Unveiling the importance of heterotrophy for coral symbiosis under heat stress. mBio 15:e01966–24. https://doi.org/10.1128/mbio.01966-24

McConnaughey T (1989) 13C and 18O isotopic disequilibrium in biological carbonates: I Patterns. Geochim Cosmochim Acta 53:151–162

Mckinnon D, Richardson AJ, Burford MA, Furnas MJ (2007) Chapter 6 Vulnerability of great barrier reef plankton to climate change. The Great Barrier Reef Marine Park Authority

McLachlan RH, Price JT, Muñoz-Garcia A et al (2022) Physiological acclimatization in Hawaiian corals following a 22-month shift in baseline seawater temperature and pH. Sci Rep 12:3712

McNally SP, Parsons RJ, Santoro AE, Apprill A (2017) Multifaceted impacts of the stony coral Porites astreoides on picoplankton abundance and community composition. Limnol Oceanogr 62:217–234

McWilliam M, Hoogenboom MO, Baird AH et al (2018) Biogeographical disparity in the functional diversity and redundancy of corals. Proc Natl Acad Sci 115:3084–3089

Mertz EM, McDonald A (2021) Ecosystem pushing: coral restoration in Refugia as an unexplored answer to climate change adaptation. IOP Conf Ser: Earth Environ Sci 809:012011

Meunier V, Bonnet S, Pernice M et al (2019) Bleaching forces coral's heterotrophy on diazotrophs and Synechococcus. ISME J 13:2882–2886

Meunier V, Bonnet S, Benavides M et al (2021) Diazotroph-derived nitrogen assimilation strategies differ by Scleractinian coral species. Front Mar Sci 8. https://doi.org/10.3389/fmars.2021.692248

Meunier V, Bonnet S, Camps M et al (2022) Ingestion of Diazotrophs makes corals more resistant to heat stress. Biomolecules 12:537

Minagawa M, Wada E (1984) Stepwise enrichment of 15N along food chains: further evidence and the relation between δ15N and animal age. Geochim Cosmochim Acta 48:1135–1140

Muir PR, Wallace CC, Done T, Aguirre JD (2015) Limited scope for latitudinal extension of reef corals. Science 348:1135–1138

Mumby PJ, Elliott IA, Eakin CM et al (2011) Reserve design for uncertain responses of coral reefs to climate change. Ecol Lett 14:132–140

Muscatine L (1990) The role of symbiotic algae in carbon and energy flux in reef corals. Coral Reefs: Ecosyst World 25:75–87

Muscatine L, Porter JW (1977) Reef corals: mutualistic symbioses adapted to nutrient-poor environments. Bioscience 27:454–460

Muscatine L, D'Elia CF (1978) The uptake, retention, and release of ammmonium by reef corals. Limnol Oceanogr 23:725–734

Muscatine L, Falkowski PG, Porter JW, Dubinsky Z (1984) Fate of photosynthetic fixed carbon in light- and shade-adapted colonies of the symbiotic coral Stylophora pistillata. Proc R Soc Lond 222:181–202

Nakabayashi A, Yamakita T, Nakamura T et al (2019) The potential role of temperate Japanese regions as refugia for the coral Acropora hyacinthus in the face of climate change. Sci Rep 9:1892

Naumann MS, Tolosa I, Taviani M et al (2015) Trophic ecology of two cold-water coral species from the Mediterranean Sea revealed by lipid biomarkers and compound-specific isotope analyses. Coral Reefs 34:1165–1175

Orejas C, Gori A, Rad-Menéndez C et al (2016) The effect of flow speed and food size on the capture efficiency and feeding behaviour of the cold-water coral Lophelia pertusa. J Exp Mar Bio Ecol 481:34–40

Osinga R, Schutter M, Griffioen B et al (2011) The biology and economics of coral growth. Mar Biotechnol 13:658–671

Palardy JE, Grottoli AG, Matthews KA (2005) Effects of upwelling, depth, morphology and polyp size on feeding in three species of Panamanian corals. Mar Ecol Prog Ser 300:79–89

Palardy JE, Grottoli AG, Matthews KA (2006) Effect of naturally changing zooplankton concentrations on feeding rates of two coral species in the eastern Pacific. J Exp Mar Bio Ecol 331:99–107

Palardy JE, Rodrigues LJ, Grottoli AG (2008) The importance of zooplankton to the daily metabolic carbon requirements of healthy and bleached corals at two depths. J Exp Mar Bio Ecol 367:180–188

Palenik B, Morel FMM (1990) Amino acid utilization by marine phytoplankton: a novel mechanism. Limnol Oceanogr 35:260–269

Parrish CC (2013) Lipids in marine ecosystems. ISRN Oceano 2013(9)

Peixoto RS, Rosado PM, de Assis Leite DC et al (2017) Beneficial microorganisms for corals (BMC): proposed mechanisms for coral health and resilience. Front Microbiol 8:341

Perry CT, Kench PS, Smithers SG et al (2011) Implications of reef ecosystem change for the stability and maintenance of coral reef islands. Glob Chang Biol 17:3679–3696

Petersen D, Wietheger A, Laterveer M (2008) Influence of different food sources on the initial development of sexual recruits of reef building corals in aquaculture. Aquaculture 277:174–178

Picciano M, Ferrier-Pagès C (2007) Ingestion of pico- and nanoplankton by the Mediterranean red coral Corallium rubrum. Mar Biol 150:773–782

Pittera J, Humily F, Thorel M et al (2014) Connecting thermal physiology and latitudinal niche partitioning in marine Synechococcus. ISME J 8:1221–1236

Pogoreutz C, Oakley CA, Rädecker N et al (2022) Coral holobiont cues prime Endozoicomonas for a symbiotic lifestyle. ISME J 16:1883. https://doi.org/10.1038/s41396-022-01226-7

Price NN, Muko S, Legendre L et al (2019) Global biogeography of coral recruitment: tropical decline and subtropical increase. Mar Ecol Prog Ser 621:1–17

Price JT, McLachlan RH, Jury CP (2021) Isotopic approaches to estimating the contribution of heterotrophic sources to Hawaiian corals. Limnol Oceanogr 66(6):2393–2407

Putnam HM, Edmunds PJ, Fan TY (2010) Effect of a fluctuating thermal regime on adult and larval reef corals. Invertebr Biol 129(3):199–209

Pyle RL, Copus JM (2019) Mesophotic coral ecosystems: introduction and overview. In: Loya Y, Puglise KA, Bridge TCL (eds) Mesophotic coral ecosystems. Springer International Publishing, Cham, pp 3–27

Rädecker N, Pogoreutz C, Voolstra CR et al (2015) Nitrogen cycling in corals: the key to understanding holobiont functioning? Trends Microbiol 23:490–497

Rädecker N, Pogoreutz C, Gegner HM (2021) Heat stress destabilizes symbiotic nutrient cycling in corals. https://doi.org/10.1073/pnas.2022653118/-/DCSupplemental

Raz-Bahat M, Douek J, Moiseeva E et al (2016) The digestive system of the stony coral Stylophora pistillata. Cell Tissue Res 368:311. https://doi.org/10.1007/s00441-016-2555-y

Ritchie KB (2006) Regulation of microbial populations by coral surface mucus and mucus-associated bacteria. Mar Ecol Prog Ser 322:1–14

Rodrigues LJ, Grottoli AG (2007) Energy reserves and metabolism as indicators of coral recovery from bleaching. Limnol Oceanogr 52:1874–1882

Rohwer F, Seguritan V, Azam F, Knowlton N (2002) Diversity and distribution of coral-associated bacteria. Mar Ecol Prog Ser 243:1–10

Rosenberg E, Koren O, Reshef L et al (2007) The role of microorganisms in coral health, disease and evolution. Nat Rev Microbiol 5:355–362

Rosenfeld M, Bresler V, Abelson A (1999) Sediment as a possible source of food for corals. Ecol Lett 2:345

Sangmanee K, Casareto BE, Nguyen TD et al (2020) Influence of thermal stress and bleaching on heterotrophic feeding of two scleractinian corals on pico-nanoplankton. Mar Pollut Bull 158:111405

Santiago-Valentín JD, Zavala-Casas DA, Hernández-Penagos LE et al (2023) Reproductive and physiological plasticity of scleractinian corals in response to changes in environmental conditions. Hydrobiologia 850:577–590

Saper J (2022) Augmented coral health: linking heterotrophic diets to the acquisition of probiotics in symbiotic scleractinia. James Cook University, Rmasters

Saper J, Raina JB, Humphrey C, Høj L, Bourne DG (2025) Microbial processes and nutrient uptake in the coral Holobiont and reef ecosystems. In: Peixoto RS, Voolstra CR (eds) Coral reef microbiome. Coral reefs of the world, vol 20. Springer, Cham. https://doi.org/10.1007/978-3-031-76692-3_9

Sebens KP, Vandersall KS, Savina LA, Graham KR (1996) Zooplankton capture by two scleractinian corals, Madracis mirabilis and Montastrea cavernosa, in a field enclosure. Mar Biol 127:303–317

Sebens KP, Witting J, Helmuth B (1997) Effects of water flow and branch spacing on particle capture by the reef coral Madracis mirabilis (Duchassaing and Michelotti). J Exp Mar Bio Ecol 211:1–28

Séré MG, Massé LM, Perissinotto R, Schleyer MH (2010) Influence of heterotrophic feeding on the sexual reproduction of Pocillopora verrucosa in aquaria. J Exp Mar Bio Ecol 395:63–71

Setälä O, Fleming-Lehtinen V, Lehtiniemi M (2014) Ingestion and transfer of microplastics in the planktonic food web. Environ Pollut 185:77–83

Shlesinger T, van Woesik R (2023) Oceanic differences in coral-bleaching responses to marine heatwaves. Sci Total Environ 871:162113

Skinner C, Cobain MRD, Zhu Y et al (2022) Progress and direction in the use of stable isotopes to understand complex coral reef ecosystems: a review. Oceanogr Mar Biol Annu Rev 60:373–432

Sorokin YI (1991) Parameters of productivity and metabolism of coral reef ecosystems off Central Vietnam. Estuar Coast Shelf Sci 33:259–280

Starko S, Fifer J, Claar DC et al (2023) Marine heatwaves threaten cryptic coral diversity and erode associations amongst coevolving partners. bioRxiv:2023.01.07.522953

Stock CA, Dunne JP, John JG (2014) Global-scale carbon and energy flows through the marine planktonic food web: an analysis with a coupled physical–biological model. Prog Oceanogr 120:1–28

Sturaro N, Hsieh YE, Chen Q et al (2021) Trophic plasticity of mixotrophic corals under contrasting environments. Funct Ecol 35:2841–2855

Suggett DJ, Smith DJ (2011) Interpreting the sign of coral bleaching as friend vs. foe. Glob Chang Biol 17:45–55

Sully S, Burkepile DE, Donovan MK et al (2019) A global analysis of coral bleaching over the past two decades. Nat Commun 10:1264

Sully S, Hodgson G, van Woesik R (2022) Present and future bright and dark spots for coral reefs through climate change. Glob Chang Biol 28:4509–4522

Sutton DC, Hoegh-Guldberg O (1990) Hostzooxanthella interactions in four temperate marine invertebrate symbioses: assessment of effect of host extracts on symbionts. Biol Bull 178(2):175–186

Tada K, Sakai K, Nakano Y et al (2003) Size-fractionated phytoplankton biomass in coral reef waters off Sesoko Island, Okinawa, Japan. J Plankton Res 25:991–997

Tagliafico A, Rangel S, Kelaher B, Christidis L (2018) Optimizing heterotrophic feeding rates of three commercially important scleractinian corals. Aquaculture 483:96–101

Tanvet C, Camp EF, Sutton J et al (2023) Corals adapted to extreme and fluctuating seawater pH increase calcification rates and have unique symbiont communities. Ecol Evol 13:e10099

Thibault M, Lorrain A, Houlbrèque F (2021) Comment on trophic strategy and bleaching resistance in reef-building corals. Sci Adv 7

Thompson DM, van Woesik R (2009) Corals escape bleaching in regions that recently and historically experienced frequent thermal stress. Proc Biol Sci 276:2893–2901

Toh TC, Peh JWK, Chou LM (2013) Heterotrophy in recruits of the scleractinian coral Pocillopora damicornis. Mar Freshw Behav Physiol 46:313–320

Toh TC, Ng CSL, Peh JWK et al (2014) Augmenting the post-transplantation growth and survivorship of juvenile scleractinian corals via nutritional enhancement. PLoS One 9:e98529

Tolosa I, Treignier C, Grover R, Ferrier-Pagès C (2011) Impact of feeding and short-term temperature stress on the content and isotopic signature of fatty acids, sterols, and alcohols in the scleractinian coral Turbinaria reniformis. Coral Reefs 30:763–774

Travaglione N, Evans R, Moustaka M et al (2023) Scleractinian corals rely on heterotrophy in highly turbid environments. Coral Reefs 42:997

Treignier C, Grover R, Ferrier-Pagès C, Tolosa I (2008) Effect of light and feeding on the fatty acid and sterol composition of zooxanthellae and host tissue isolated from the scleractinian coral Turbinaria reniformis. Limnol Oceanogr 53:2702–2710

Tremblay P, Grover R, Jean François Maguer LLA, Ferrier-Pagès C (2012a) Autotrophic caron budget in coral tissue: a new 13C-based model of photosynthate translocation. J Exp Biol 215:1384–1393

Tremblay P, Naumann MS, Sikorski S et al (2012b) Experimental assessment of organic carbon fluxes in the scleractinian coral Stylophora pistillata during a thermal and photo stress event. Mar Ecol Prog Ser 453:63–77

Trench RK (1979) The cell biology of plant-animal symbiosis. Ann Rev Pl Physiol 30:485–531

Turner JA, Babcock RC, Hovey R, Kendrick GA (2017) Deep thinking: a systematic review of mesophotic coral ecosystems. ICES J Mar Sci 74:2309–2320

van Woesik R, Shlesinger T, Grottoli AG et al (2022) Coral-bleaching responses to climate change across biological scales. Glob Chang Biol 28:4229–4250

Viladrich N, Roca A, Gori A, Baena P, Linares C (2024) Heterotrophy in parental coral colonies enhances larval survival independently of heat stress. Coral Reefs. https://doi.org/10.21203/rs.3.rs-5418525/v1

Wall CB, Ricci CA, Foulds GE et al (2018) The effects of environmental history and thermal stress on coral physiology and immunity. Mar Biol 165:56

Wall CB, Kaluhiokalani M, Popp BN et al (2020) Divergent symbiont communities determine the physiology and nutrition of a reef coral across a light-availability gradient. ISME J 14:945–958

Wang Q, Zheng X, Zhou X, Zhang H, Cai L, Leung JY, Huang L (2024a) Symbiont genus determines the trophic strategy of corals: implications for intraspecific competition for energy sources in coral reefs. Ecol Indic 158:111477

Wang Q, Zhou X, Wang J, Zhang H, Fang H, Cai F et al (2024b) Heterotrophy confers corals with resistance but limits their range expansion: a case of marginal coral communities. Ecosyst Health Sustain 10:0246

Ward S (1995) Two patterns of energy allocation for growth, reproduction and lipid storage in the scleractinian coral Pocillopora damicornis. Coral Reefs 14:87–90

Ward S, Harrison P (2000) Changes in gametogenesis and fecundity of acroporid corals that were exposed to elevated nitrogen and phosphorus during the ENCORE experiment. J Exp Mar Bio Ecol 246:179–221

Warner ME, Fitt WK, Schmidt GW (1999) Damage to photosystem II in symbiotic dinoflagellates: a determinant of coral bleaching. PNAS 96(14):8007–8012

Weis VM (2008) Cellular mechanisms of cnidarian bleaching: stress causes the collapse of symbiosis. J Exp Biol 211:3059–3066

Wijgerde T, Diantari R, Lewaru MW et al (2011) Extracoelenteric zooplankton feeding is a key mechanism of nutrient acquisition for the scleractinian coral Galaxea fascicularis. J Exp Biol 214:3351–3357

Wild C, Woyt H, Huettel M (2005) Influence of coral mucus on nutrient fluxes in carbonate sands. Mar Ecol Prog Ser 287:87–98

Wolanski E, Drew E, Abel KM, O'Brien J (1988) Tidal jets, nutrient upwelling and their influence on the productivity of the alga Halimeda in the ribbon reefs, great barrier reef. Estuar Coast Shelf Sci 26:169–201

Yates KK, Rogers CS, Herlan JJ et al (2014) Diverse coral communities in mangrove habitats suggest a novel refuge from climate change. Biogeosciences 11:4321–4337

Yonge CM, Nicholls AG (1931) Studies on the physiology of corals IV. The structure, distribution and physiology of the zooxanthelke. Great barrier reef expedition, 1928–1929. Sci Repts Brit Mus I pp 135–176

Ziegler M, Grupstra CGB, Barreto MM et al (2019) Coral bacterial community structure responds to environmental change in a host-specific manner. Nat Commun 10:3092. https://doi.org/10.1038/s41467-019-10969-5

The Biodiversity of Metazoan Parasites of Coral Reef Fishes of the Central and Eastern Indo-Pacific

9

Christina Pavloudi, Geoffrey A. Boxshall, Rodney A. Bray, Thomas H. Cribb, Scott C. Cutmore, Terrence L. Miller, Nicholas Wee, and Pierre Sasal

Abstract

The Central and Eastern Indo-Pacific is a vast geographical region, ranging from southern Japan (north), French Polynesia (east), Great Barrier Reef (south), and Coral Triangle (west); it contains the world's best-developed and richest coral reefs which support an extraordinary richness of fishes. While there is extensive understanding of the richness and biogeography of these fishes, far less is known of their parasites. Here we review what is currently known for metazoan parasites, focusing on copepods and trematodes, infecting fishes of this region. Most of the research has been highly concentrated geographically, and just a small portion of the parasitic fauna has been characterized. Both copepods and trematodes in this region are strongly specific to their hosts, and most fish species have not yet been surveyed thoroughly, if at all, for their parasites. Ultimately, fish parasite richness in the region will far exceed that of their hosts. Reliable identification of fish parasites faces many problems, and molecular approaches are becoming increasingly important, especially for certain cases (e.g., cryptic species, larval stages). Critical "next steps" for improving our understanding of the parasitic taxonomy and ecology are to examine more fish species, to investigate under-sampled ecoregions, and to expand the molecular characterization.

Keywords

Central and Eastern Indo-Pacific · Coral reef fish · Parasites · Copepoda · Trematoda

C. Pavloudi
UAR 3278 CRIOBE, PSL Research University, EPHE-UPVD-CNRS, Perpignan, France

Laboratoire d'Excellence "CORAIL",
Moorea, French Polynesia

European Marine Biological Resource Centre (EMBRC-ERIC),
Paris, France
e-mail: christina.pavloudi@embrc.eu

G. A. Boxshall · R. A. Bray
Department of Life Sciences, Natural History Museum,
London, UK
e-mail: G.Boxshall@nhm.ac.uk; rod@braymaster.co.uk

T. H. Cribb · N. Wee
Queensland Museum, Biodiversity and Geosciences Program,
South Brisbane, QLD, Australia
e-mail: t.cribb@uq.edu.au; nicholas.wee@qm.qld.gov.au

S. C. Cutmore · T. L. Miller
Queensland Museum, Biodiversity and Geosciences Program,
South Brisbane, QLD, Australia

The University of Queensland, School of the Environment,
St Lucia, QLD, Australia
e-mail: scott.cutmore@qm.qld.gov.au; terry.miller@qm.qld.gov.au

P. Sasal (✉)
UAR 3278 CRIOBE, PSL Research University, EPHE-UPVD-CNRS, Perpignan, France

Laboratoire d'Excellence "CORAIL",
Moorea, French Polynesia
e-mail: sasal@univ-perp.fr

9.1 Introduction

Coral reefs, despite being one of the ecosystems with the highest biodiversity (Justine 2010), are highly sensitive to and threatened by both natural and anthropogenic factors (Riegl et al. 2009). Among the latter, undoubtedly global warming and ocean acidification are the most pronounced and, at the same time, challenging to tackle. However, there are also locally induced threats, such as overfishing, that can severely impact the integrity of the coral reef ecosystem (Riegl et al. 2009) since removing fish species alters the reefs' balance. Another impact of overfishing is the simplification of the food web which affects also fish parasites by reducing the number of potentially available hosts (Lafferty et al. 2008). In addition, the introduction of alien species can also impact negatively the reefs, given that some

S. Planes (ed.), *The Future of Coral Reefs*, Coral Reefs of the World 21, https://doi.org/10.1007/978-3-031-98584-3_9

of them can become invasive but also due to the fact that they might introduce their own parasites to the native fauna (Dunn et al. 2012).

Challenges such as global warming impact not only the coral species but also the associated reef fauna. The strong association between corals and fish has been known for decades; loss of live corals causes decline in reef fish abundances and vice versa, which in turn affects the provision of ecosystem services (Pratchett et al. 2014). Increased temperature causes coral bleaching, which results in habitat loss coral reef fish, thus impacting the latter indirectly; additionally, higher temperature directly impacts fishes due to physiological and behavioral induced changes (Munday et al. 2008). Moreover, reduced abundance of cleaner fish results in increased exposure of common fish species to parasites (Grutter et al. 2018), which suggests that coral loss can cause a potential cascading effect with unknown consequences. In order to break this vicious cycle and maintain coral reef biodiversity stable, a thorough assessment of all the biodiversity components is needed.

It is well-known that parasites constitute a high proportion of global biodiversity (Dobson et al. 2008) and that they have a substantial impact in ecosystem functioning, through direct and indirect effects on host populations and host fitness (Frainer et al. 2018). In many cases, selection induced by parasites forces hosts to adapt in order to reduce the cost of infection, which in turn triggers parasite adaptation, leading to the so-called host-parasite coevolution (Ebert and Fields 2020). However, despite being crucial ecosystem components, parasites are frequently neglected due to their small size and the complexity of laboratory technical tasks required for their accurate identification (Justine 2010). Within the broad field of parasitology, differences in difficulty have, undoubtedly, guided research toward certain groups of parasites, e.g., helminths, while others, such as protists, have received less attention (Frainer et al. 2018).

From the parasite perspective, highly specialized species are more susceptible to extinction than generalists, i.e., having only one or a few hosts inevitably causes them to be more vulnerable to the reduction in abundance or, in extreme cases, local extinction of their host species (Dallas and Cornelius 2015). Given that it is feared that many (host) species will become endangered or even extinct in the near future as a response to climate change and anthropogenic activities, this would lead to the extinction of their parasites due to interspecific dependencies (Bellay et al. 2020).

Efforts to identify and characterize new parasite species, their life cycles, and their respective hosts, are important not only because they contribute to the global knowledge of biodiversity but because they also provide insights into the regulation of the host population, the complex trophic interactions, ecosystem health, and functioning (Poulin et al. 2020). However, there are major gaps in knowledge linked to a poor parasite discovery efficiency, understudy of some biogeographical areas and taxonomic groups, as well as lack of taxonomists (Gaston and May 1992). In addition, the lack of incorporation of molecular tools for parasite identification is also a crucial problem.

Studies on parasites of fish have been conducted both on aquaculture fishes, with the ultimate aim to control, manage, and eradicate the parasites so as to maximize fish production (Timi and Mackenzie 2015), as well as on wild fishes (Wood et al. 2010). Various fish traits, such as body size, type of environment, trophic level, and geographical range have been found to affect parasitic composition and load (Luque and Poulin 2008). Coral-reef fishes are affected by parasites at various ontogenetic stages, although most studies have been conducted on adult, rather than larval, fishes (Peyrusse et al. 2012). Parasites have a valuable ecological role since they are regulating the symbiotic associations in the coral reef ecosystems (Zhokhov and Mikheev 2015), and therefore parasites in (wild) coral reef fishes should be thoroughly studied.

Here, we focus on the parasites of coral reef fishes of the Central and Eastern Indo-Pacific (CEI-P) realms, as categorized by Spalding et al. (2007) (Fig. 9.1). These realms are further divided into 18 provinces (25–42) and 52 ecoregions (20112–25,154). We focus on the CEI-P as a biogeographically cohesive area based on the composition of the fish fauna; observations of coral reef fishes anywhere in the CEI-P will reveal highly similar fishes, many of which are shared over most, or all, of the area. We are interested in determining the nature and richness of the parasites that infect these fishes. Our goal is to summarize the basic biodiversity understanding for two major groups of parasites of fishes of this region and to analyze the nature, composition, and distribution of the parasitic fauna and its hosts, as well as attempt to understand patterns of host specificity.

9.2 Methodological Approach

First, we retrieved fish occurrences from FishBase (Froese and Pauly 2023) and Ocean Biodiversity Information System (OBIS 2023), for each of the Marine Ecoregions of the World (MEOW) of interest, as identified by Spalding et al. (2007). It should be noted that other biogeographical classifications exist, e.g., that by Kulbicki et al. (2013) which is based on reef fishes; however, we chose to base our work on the MEOW classification since its delineations were based on multiple phyla, levels of endemism, and also environmental factors. We also incorporated fish occurrences from the barcode reference library of French Polynesian shore fishes (Delrieu-Trottin et al. 2019) and a cleaner reef fish fauna checklist (Quimbayo et al. 2021), as well as occurrences from locally stored databases on

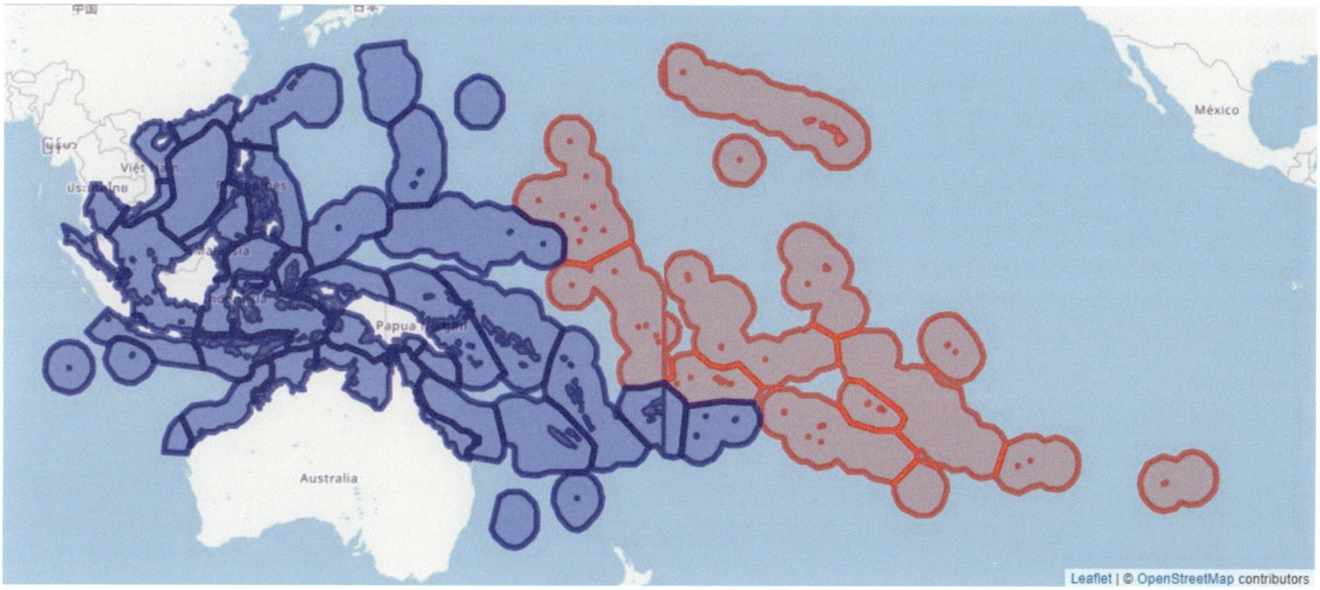

Fig. 9.1 The Central (blue) and Eastern Indo-Pacific (red) realms; ecoregions are presented with darker borders

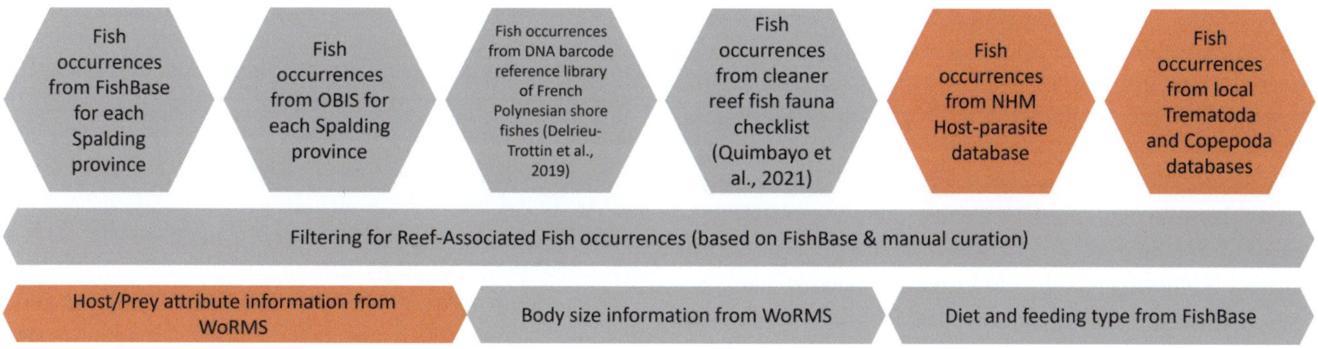

Fig. 9.2 Schematic diagram of the methodological approach. Grey represents the fish-related steps of the approach and orange the steps related both to parasites as well as hosts

Trematoda (information from 536 original publications) and Copepoda parasitizing fish. For Trematoda and for certain areas which we could assign to MEOW provinces, we also retrieved information from the host-parasite database of the Natural History Museum, London. Using this list of unique fish species, we retrieved information on their ecology (enabling restriction of the list to reef-associated species) and trophic level from FishBase, as well as the species that parasitize them and their body size from the World Register of Marine Species (WoRMS) (Ahyong et al. 2023). More methodological information can be found in Pavloudi (2023) (Fig. 9.2).

Our methodological approach led to lists of parasites and hosts which are inevitably incomplete, as a consequence of information missing from FishBase, WoRMS, and OBIS, as well as from the checklists and databases we added. No database is complete, i.e., data collection is not done comprehensively; there are certainly records of hosts and parasites without specific location which makes data retrieval problematic given our focus on specific ecoregions. Also, there are cases where information on location does exist, but it is encoded in a different way and without mentioning the MEOW ecoregions. We also encountered difficulty in identifying which fish species are genuinely in the region, and in particular, which species are in each MEOW province. Finally, it should be noted that determining which of the MEOW provinces to attribute records from the complex pattern of regions seen in Asian waters, was problematic. When standardized terms and ontologies are not used to describe certain species attributes, information is ultimately lost. For the parasites, we chose to focus on the two most abundant parasitic groups, Trematoda and Copepoda, as there are the groups for which the best-organized data are available.

Table 9.1 Most abundant reef fish families represented in the provinces of interest

Family	Number of genera	Number of species	% Total reef fish species
Gobiidae	95	603	12.41
Labridae	52	373	7.68
Pomacentridae	25	328	6.75
Apogonidae	33	309	6.36
Serranidae	49	256	5.27
Blenniidae	36	240	4.94
Syngnathidae	29	128	2.63
Muraenidae	14	118	2.43
Ophichthidae	29	108	2.22
Scorpaenidae	13	108	2.22

9.3 Diversity of Fish

The fauna of fishes of the CEI-P is immense; depending on the source of records, there are well over 7000 species known for the CEI-P in over 400 families, comprising over 20% of total global fish richness. Arguably, the brightly colored and easily seen "reef-associated" species, such as chaetodontids, labrids, pomacentrids, and scarids, are the most immediately characteristic component of this fauna, but other ecological groups are also highly significant. The Indo-Pacific, as well as especially the Indo-Australian Archipelago, is characterized by the highest reef fish richness (Parravicini et al. 2013), so it is not surprising that our analyses identified 4859 unique reef-associated fish species, belonging to 1004 genera and 190 families for the provinces of interest. Notably, the most abundant reef fish families account for ~53% of the total reef fish species (Table 9.1).

As shown in Table 9.2, fish richness varies dramatically between provinces. The richest province, the Western Coral Triangle with 3315 species, contrasts sharply with the 163 species reported from Easter Island (Fig. 9.3).

9.4 Diversity and Host Specificity of Fish Parasites

Although fishes of the CEI-P harbor multiple metazoan parasite groups, here we focus on two groups that are both abundant and relatively well documented: Trematoda (Platyhelminthes) and Copepoda (Crustacea). Other, less studied, groups include Acanthocephala, Cestoda, Hirudinea, Isopoda, Monogenea, Myxozoa, and Nematoda.

Table 9.2 Unique occurrences of reef-associated fish species per MEOW province

Province	Total fish	Reef fish
Eastern Coral Triangle	3429	2775
Java Transitional	1899	1442
Lord Howe and Norfolk Islands	1510	878
Northeast Australian Shelf	3506	2297
Northwest Australian Shelf	3375	2123
Sahul Shelf	3295	2236
South China Sea	3496	1911
South Kuroshio	3821	2375
Sunda Shelf	3426	2644
Tropical Northwestern Pacific	2579	2124
Tropical Southwestern Pacific	4550	2845
Western Coral Triangle	5296	3315
Central Polynesia	1790	1486
Easter Island	182	163
Hawaii	1524	802
Marquesas	808	607
Marshall, Gilbert, and Ellis Islands	1904	1621
Southeast Polynesia	1681	1351

9.4.1 Trematodes

Trematodes are endoparasitic platyhelminths that infect the digestive tract, circulatory system, and many other organs of aquatic and terrestrial vertebrates. They have complex life cycles that vary considerably (Collins 2017). In fish, they can also be found under the scales, in the eyes and on or in the gills (Klimpel et al. 2019). They include nearly 20,000 described species (Collins 2017).

We found records of 1597 fully identified species of adult trematodes from reef fishes of the CEI-P, belonging to 468 genera and 52 families (Table 9.3). The ten richest families comprise ~70% of the species, contrasting with the distribution of richness of the host fishes, for which the ten richest families account for ~53% of total reef fish richness. All the richest trematode families are also important in other parts of the world. No major taxon of marine fish trematodes is entirely lacking from the fauna, although perhaps the Fellodistomidae, at ~2%, seems depauperate relative to its significance in the Atlantic and deeper sea fishes. Nine families (Cladorchiidae, Deropristidae, Gibsonivermidae, Mesometridae, Microphallidae, Ptychogonimidae, Sanguinicolidae, Stenakridae, and Tandanicolidae) are represented by only single species. Some of these are small families globally (e.g., Mesometridae), whereas others are at their richest in freshwater fishes (e.g., Cladorchiidae), and some are principally parasites of tetrapods (e.g., Microphallidae); none of the families occurs only in the region of interest.

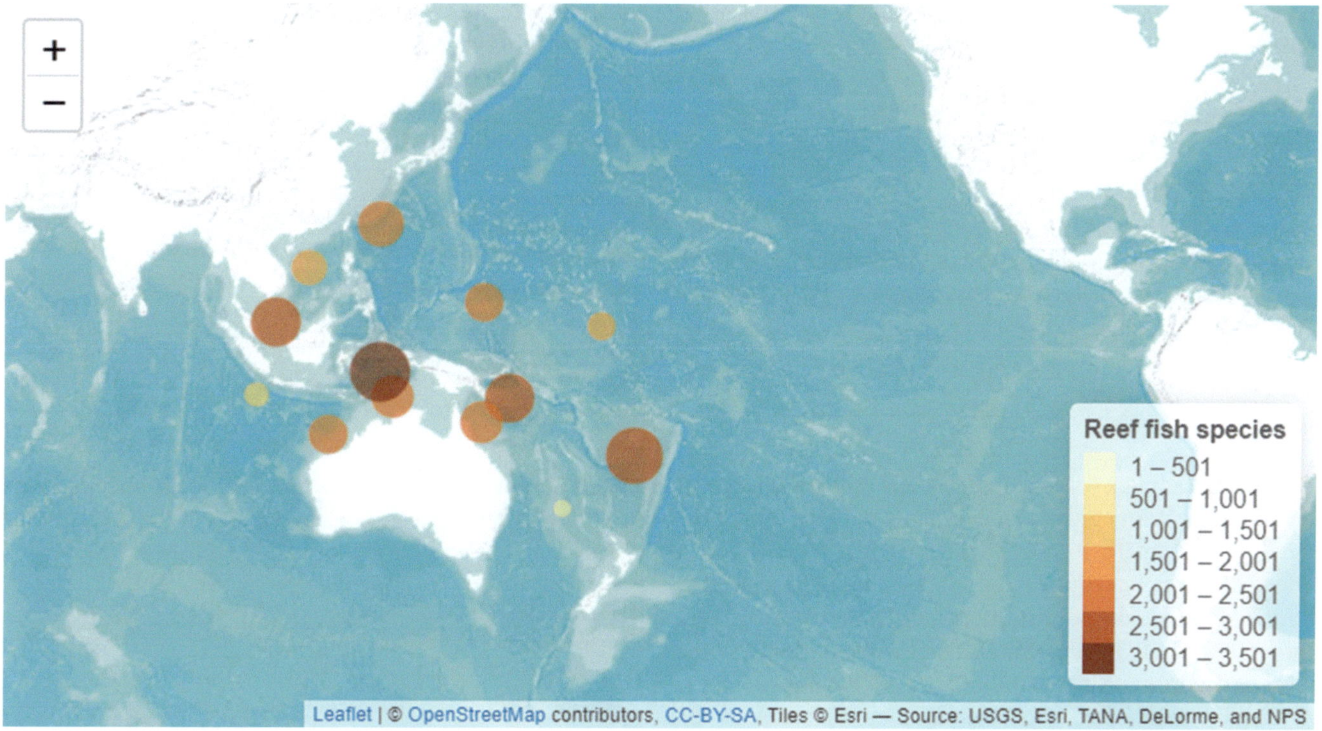

Fig. 9.3 Richness of reef-associated fish species per province in the CEI-P

Table 9.3 Most abundant trematode families represented in the provinces of interest

Family	Number of genera	Number of species	% Total trematode species
Opecoelidae	54	247	15.47
Hemiuridae	33	149	9.33
Bucephalidae	20	146	9.14
Lepocreadiidae	34	119	7.45
Monorchiidae	41	106	6.64
Cryptogonimidae	31	101	6.32
Didymozoidae	37	83	5.20
Aporocotylidae	19	72	4.51
Acanthocolpidae	10	62	3.88
Bunocotylidae	14	44	2.76

The richness of most genera is low; 221 genera are monotypic, and another 79 have only two species. All the richest genera (Table 9.4) occur elsewhere, at least in the Indian Ocean, but notably *Paradiscogaster* (23 species) and *Transversotrema* (30 species) are not known from the Atlantic despite the presence of apparently suitable hosts there.

These 1597 trematode species have been reported from 110 families, 328 genera, and 841 species of fishes, almost exclusively bony fishes (~99%). Trematode richness per host species varies dramatically. The richest hosts (*Euthynnus affinis*, *Mugil cephalus*, *Selar crumenophthalmus*, *Lateolabrax japonicus*) have 30, or more, trematode species (Table 9.5). In contrast, ~29% of the fish species have only a single known trematode species, and ~17% have only two.

Table 9.4 Richest genera of trematodes represented in the provinces of interest

Family	Genus	Number of species
Hemiuridae	*Lecithochirium*	41
Acanthocolpidae	*Stephanostomum*	38
Bucephalidae	*Prosorhynchus*	35
Transversotrematidae	*Transversotrema*	30
Cryptogonimidae	*Siphoderina*	24
Faustulidae	*Paradiscogaster*	23
Monorchiidae	*Hurleytrematoides*	22
Bucephalidae	*Prosorhynchoides*	22
Bucephalidae	*Bucephalus*	19
Opecoelidae	*Plagioporus*	19

Table 9.5 Fish species hosting greatest species richness of trematodes

Reef fish species	Number of trematode species
Euthynnus affinis	45
Mugil cephalus	40
Selar crumenophthalmus	35
Lateolabrax japonicus	30
Epinephelus fasciatus	28
Siganus fuscescens	28
Seriola dumerili	27
Platycephalus indicus	25
Rachycentron canadum	25
Epinephelus coioides	24
Lethrinus nebulosus	24
Lutjanus fulviflamma	24

Table 9.6 Most abundant host families, ranked by the number of unique host/parasite combinations for Trematoda

Family	Number of species	Number of unique host/parasite combinations
Carangidae	57	382
Serranidae	49	323
Chaetodontidae	50	317
Lutjanidae	38	284
Acanthuridae	33	201
Pomacentridae	55	185
Labridae	50	153
Mullidae	23	150
Lethrinidae	22	148
Siganidae	19	144

Table 9.7 Most abundant families of parasitic copepods represented in the provinces of interest

Family	Number of genera	Number of species	% Total copepod species
Caligidae	20	144	33.96
Hatschekiidae	1	62	14.62
Lernanthropidae	7	42	9.91
Taeniacanthidae	8	41	9.67
Bomolochidae	13	31	7.31
Philichthyidae	3	21	4.95
Pennellidae	7	15	3.54
Lernaeopodidae	7	14	3.30
Ergasilidae	3	12	2.83
Pandaridae	6	10	2.36

The fish families with the most unique host-parasite combinations are Carangidae and Serranidae (Table 9.6); both families account for ~13% of the reef fish species found infected with trematodes. Of the 110 fish families reported as hosts for trematodes, the 10 most heavily reported account for ~60% of all host/parasite combinations. Notably, the level of reporting of fish families as hosts for trematodes in the region of interest is only weakly correlated with the actual richness of the families in the area. Six of the ten richest families (Apogonidae, Blenniidae, Gobiidae, Scorpaenidae, Syngnathidae, and Ophichthidae) do not appear on this list.

Trematodes of this fauna are overwhelmingly strongly specific to their hosts; of the 841 reported host species, 241 (~19%) are so far reported with only a single trematode species and a further 146 (~17%) with only two. Typically, where species have multiple hosts, the fishes are closely related. It is important to note, however, that high specificity is not universal. Certain hemiurids appear especially broad in their specificity, and molecular studies have shown that isolated species in some other families also infect a surprisingly wide range of fishes (Miller et al. 2011).

9.4.2 Copepods

Copepoda is one of the largest classes of crustaceans, consisting of more than 11,500 marine species, of which more than 5000 are parasitic to hosts ranging from sponges to chordates. Most parasitic copepods are host specific and have direct life cycles, although a few utilize two hosts to complete their life cycle (Boxshall and Hayes 2019).

We found records of 424 species of copepods from reef fishes of the CEI-P, belonging to 92 genera and 20 families (Table 9.7). All the major families of fish parasitic copepods are represented, with only six families absent. These six are small families globally, comprising a total of only 11 species in 9 genera. Four families (Caligidae, Hatschekiidae, Lernanthropidae, and Taeniacanthidae) comprise ~68% of the copepod species.

Table 9.8 Richest genera of parasitic copepods represented in the provinces of interest

Family	Genus	Number of species
Caligidae	*Caligus*	89
Hatschekiidae	*Hatschekia*	62
Lernanthropidae	*Lernanthropus*	20
Taeniacanthidae	*Taeniacanthus*	20
Philichthyidae	*Colobomatus*	19
Caligidae	*Lepeophtheirus*	16

The richest genera of copepods, which represent ~53% of the copepod species are shown in Table 9.8. *Caligus* is globally the most species-rich genus in the entire Copepoda, with over 270 valid species (Walter and Boxshall 2025). A total of 46 genera are monotypic and another 30 only have 2 species.

These 424 copepod species have been reported from 70 families, 195 genera, and 389 species of reef fishes, mainly (~95%) bony fishes (Teleostei), as for trematodes. All the richest genera are ectoparasitic, with the exception of *Colobomatus* which inhabits the subsurface canals of the lateral line system of its fish hosts.

The richest fish hosts are *Acanthopagrus australis*, *Acanthopagrus latus* (13 copepod species each), and *Caranx ignobilis* (11 copepod species) (Table 9.9); about ~48% of the fish species were reported with only a single copepod species and a further ~25% with only two.

The host families with the highest number of unique host-parasite combinations are, as for the Trematoda, Carangidae, and Serranidae (Table 9.10); both families account for ~15% of the reef fish species found infected with copepods. Again, many of the most species-rich fish families are missing from this table.

Table 9.9 Fish species hosting greatest species richness of copepods

Reef fish species	Number of copepod species
Acanthopagrus australis	13
Acanthopagrus latus	13
Caranx ignobilis	11
Acanthopagrus berda	10
Mugil cephalus	10
Sillago sihama	10
Ablennes hians	9
Euthynnus affinis	9
Epinephelus malabaricus	8
Scomberomorus commerson	8
Tylosurus crocodilus	8

Table 9.10 Most abundant host families, ranked by the number of unique host/parasite combinations for Copepoda

Family	Number of species	Number of unique host/parasite combinations
Carangidae	28	83
Serranidae	30	71
Belonidae	10	46
Lutjanidae	20	45
Scombridae	12	44
Sparidae	5	40
Acanthuridae	16	35
Balistidae	17	34
Carcharhinidae	11	29
Monacanthidae	19	27

9.5 Biogeography

Known copepod and trematode richness in the provinces of the CEI-P is summarized in Table 9.11. The key observation to emerge from these data is the unevenness of sampling across the region as a whole and its weakness in most provinces. The ratio of known fish species to parasite species to combined copepod and trematode richness ranges from 439 (Lord Howe and Norfolk Islands) to just 2.2 for Hawaii. For host/parasite combinations, the Eastern Coral Triangle has the highest ratio (47.8) and Hawaii again the lowest (0.8). Although far from the richest province in either fish or parasites, Hawaii is clearly the best-known province. Overall, the data suggest that four provinces (Hawaii, Northeast Australian Shelf, South China Sea, and tropical southwestern Pacific) are moderately well-known. Data for the rest of the provinces are, unfortunately, inadequate. The unevenness of the coverage of the fauna is shown in Fig. 9.4.

9.6 Reliability of Molecular Identification of Fish Parasites

Historically, the overwhelmingly dominant approach to the description and identification of parasite species (of all groups) in this region has been using morphological data, often supported by host and geographical distinctions. The gold standard was morphometrics based on a robust set of well-fixed samples. For taxonomic groups rich in sclerotized morphological features, e.g., copepods and monogeneans, such identifications are generally still considered scientifically convincing. However, for groups lacking diagnosable hard structures, e.g., myxosporeans and trematodes, identifications based on morphology alone are now increasingly appearing to be unreliable. Probably for all parasite groups, levels of interspecific variation, host specificity, host-related morphological variation, and geographical distribution vary considerably, so incautious inference may readily mislead. The rise of the use of molecular data has made it apparent that issues of morphological plasticity and cryptic speciation (Nadler and León 2011) are indeed significant for marine parasites.

Molecular data began to be incorporated in fish parasite species identifications in the region about 30 years ago (e.g., Anderson 1998; Anderson and Barker 1993; Cribb et al. 1998; Lo et al. 2001). Early data was mainly for two rRNA gene regions (the second internal transcribed spacer region (ITS2 rRNA) and the large (28S) ribosomal subunit RNA coding region). The ITS2 region has become the most widely used region in trematode identification, with hundreds of tropical Indo-West Pacific trematodes characterized by these data. Although ITS2 has proven effective for many trematode groups, recently variability in this marker has been demonstrated to be too low for accurate identification of some combinations of closely related trematode species in multiple families (e.g., Bray et al. 2022, 2023; Cribb et al. 2022; Cutmore et al. 2023). Several other markers have been trialed for trematode identifications (e.g., COI, Cytb, ND1, ND4), but only the mitochondrial cytochrome oxidase subunit I (COI) gene has been widely adopted across multiple research groups. COI data have proven highly effective in distinguishing closely related species and thus disentangling cryptic complexes over wide geographical ranges (e.g., Bray et al. 2022; Cutmore and Cribb 2022; Huston et al. 2021; Wee et al. 2022).

Copepods (and other taxa with informative sclerotized parts) have received far less sequencing attention than their soft-bodied counterparts. However, there is now some molecular evidence of host-related morphological variation and taxonomically unreliable morphology in some parasitic groups (Hua et al. 2019; Suyama et al. 2021; Montes et al. 2022).

Overall, we consider the incorporation of molecular data essential for reliable trematode descriptions and identifications and worthy of at least periodic testing for parasitic copepods. Not only do such data provide for improved reliability of identification, but they are also critical for the development of accurate phylogenetic systematics (Caira et al. 2014) and matching morphologically distinct life cycle stages. Given that only a relatively small proportion of even

Table 9.11 Unique species occurrences and host/parasite combinations per MEOW province

Province	Reef fish species	Copepoda Species	Copepoda Unique host/parasite combinations	Trematoda Species	Trematoda Unique host/parasite combinations	Hosts/parasites	Hosts/combinations
Eastern Coral Triangle	2775	17	23	5	35	126.1	47.8
Java Transitional	1442	12	12	24	96	40.1	13.4
Lord Howe and Norfolk Islands	878	0	0	2	25	439.0	35.1
Northeast Australian Shelf	2297	141	217	498	1586	3.6	1.3
Northwest Australian Shelf	2123	29	43	67	282	22.1	6.5
Sahul Shelf	2236	38	45	9	98	47.6	15.6
South China Sea	1911	123	208	522	1345	3.0	1.2
South Kuroshio	2375	68	91	229	649	8.0	3.2
Sunda Shelf	2644	73	82	63	255	19.4	7.8
Tropical Northwestern Pacific	2124	16	20	64	147	26.6	12.7
Tropical Southwestern Pacific	2845	70	100	520	1785	4.8	1.5
Western Coral Triangle	3315	89	136	197	684	11.6	4.0
Central Polynesia	1486	25	42	1	1	57.2	34.6
Easter Island	163	2	5	0	0	81.5	32.6
Hawaii	802	44	94	325	860	2.2	0.8
Marquesas	607	0	0	5	38	121.4	16.0
Marshall, Gilbert and Ellis Islands	1621	22	30	1	13	70.5	37.7
Southeast Polynesia	1351	9	9	43	206	26.0	6.3

the known parasitic fauna of fishes of region have been sequenced, we can expect much change in the classification (both synonymy and species splitting) into the future.

9.7 Host Size and Parasite Richness

Meta-analyses of the drivers of parasite richness (e.g., Kamiya et al. 2014) suggest that host size, host geographical distribution, and host abundance are all positively associated with parasite richness. Of these three potential correlates, only host size (measured by maximum species length as reported by FishBase and WoRMS) can be easily analyzed for our data. It seems that there is a weak positive relationship between maximum fish length and number of known parasite species (Fig. 9.5). However, there are two obvious weaknesses in this analysis. First, objectively, length is an imperfect correlate of overall size, and, additionally, body size data (i.e., length data) are not available for all the host species in our dataset. Second, sampling of the fish species represented here is highly uneven; presumably for some fish species, the list of parasites is almost complete, whereas for others it has barely started.

Despite these problems, we suspect that the available data are best interpreted as suggesting that large fish size tends to facilitate high parasite richness (but may not guarantee it) but

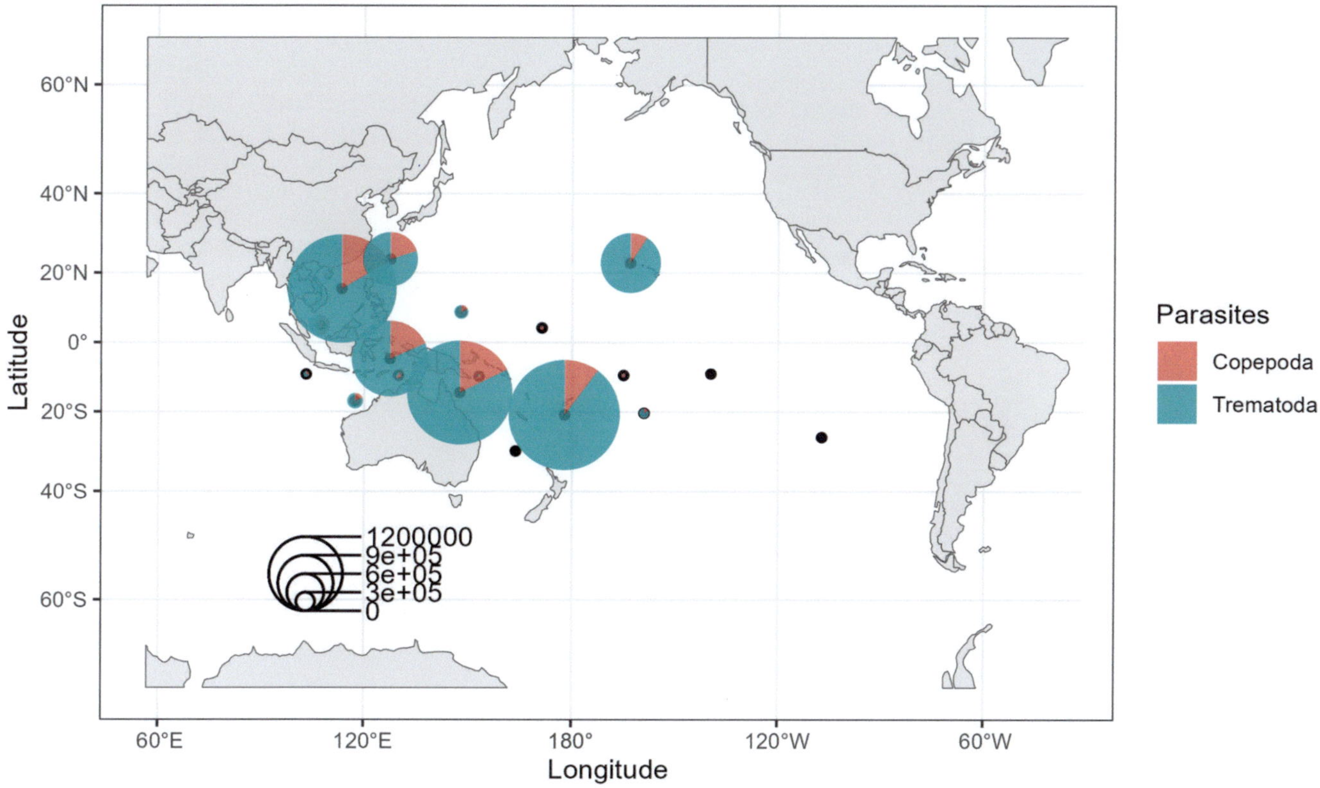

Fig. 9.4 Geographical distribution of parasites. Size of pie charts is proportional to the number of species for each group (not at the original scale, to aid visualization)

that "large" can be considered to set in well below a length of 50 cm. A small size broadly correlates with low richness (but the relationship is poorly understood), and for the bulk of fish between the extremes, a wide range of factors determines richness and composition.

The smallest fish species infected with adult trematodes are *Cirrhitichthys falco* and *Chromis nitida* (7 cm each) (Dyer et al. 1988). Other fishes of similar length (7.5 cm) are *Neopomacentrus azysron*, *Cirripectes filamentosus*, *Chrysiptera rollandi*, and *Pomacentrus melanochir*, each infected by one trematode species, *Rudarius ercodes* infected by two, and *Pomacentrus tripunctatus*, infected by three trematode species. For copepods, infections have been found in even smaller species; the smallest (2.5 cm) were *Eviota sebreei* and *Pleurosicya micheli*, both infected by the same copepod species, *Cardiodectes shini*.

In our dataset of reef fish species, and for those that body size data are available, ~38% have a maximum length ≤10 cm, with tiny fish (here arbitrarily identified as those with a maximum length of ≤6 cm) comprising ~20% of the fish reef species. However, just 33 of 841 (3.9%) fish species infected by trematodes and 19 of 389 (4.9%) of those infected by copepods have a maximum length of ≤10 cm; of those, only 5 can be characterized as tiny fish and are mostly found infected by copepods. Tiny fish are problematic to study on several bases; first, their likely depauperate parasite fauna makes them uninviting for study (perhaps generating a self-fulfilling prophecy). Second, the species are poorly dealt with by most of the available fish guides and, even with good guides, their identification is challenging. Third, it is probable that tiny size (and corresponding short life and limited diet) is correlated with low parasite prevalence so that large fish numbers must be collected to effectively sample the parasitic fauna.

Collecting large numbers of single species is logistically challenging and rarely encouraged by permitting bodies. The nature and richness of the parasite fauna of the truly small and tiny fishes is a major gap in our understanding and has important implications for understanding total richness of the fauna. We note that recently Zhokhov et al. (2020) reported on parasites of 61 individuals of 8 species of "small cryptic coral reef fish" and found no adult copepods or trematodes.

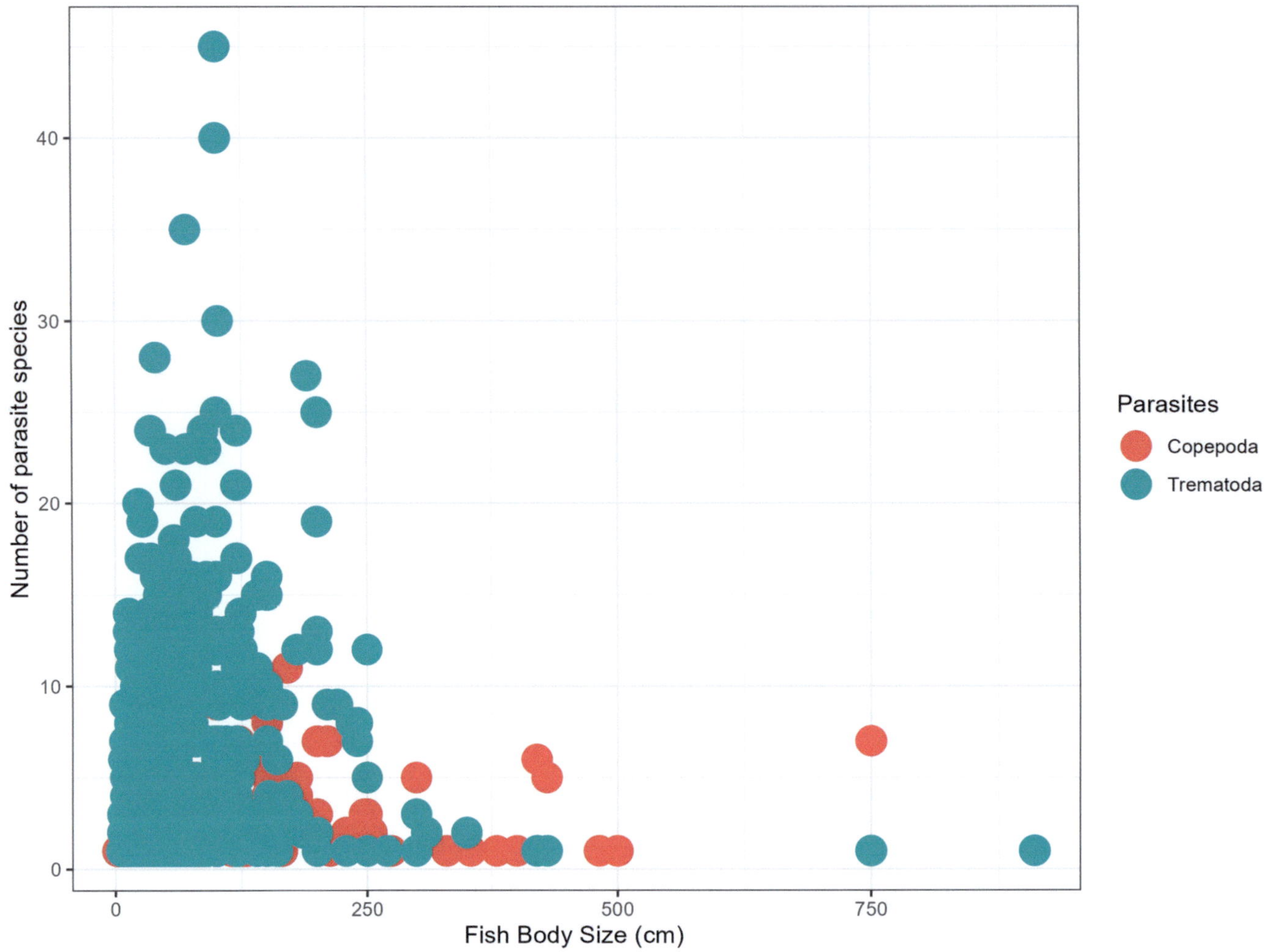

Fig. 9.5 Relationship between maximum fish length in cm and known richness of parasite species

9.8 Host Diet and Trematode Richness

Beyond considerations of host size and the intuitively likely but largely unexplored correlates of host geographical distribution and abundance, host diet is the key factor that drives the structure of parasite communities in the CEI-P and elsewhere. However, certain groups of parasites, e.g., trematodes, rely on host diet-dependent modes of transmission much more than others, e.g., copepods.

The most abundant families of parasitic copepods infect the gills or gill filaments of their host, either exclusively (Hatschekiidae, Lernanthropidae) or among other surfaces such as body, oral, and branchial cavities (Caligidae, Taeniacanthidae, Pennellidae, Lernaeopodidae). Parasitic copepods typically have free-swimming nauplius larvae which are nonfeeding and serve as the main dispersal phase in the life cycle. Following the nauplius phase is the first copepod stage which is the primary infective stage in virtually all fish parasitic copepods. This stage actively seeks out potential hosts. Therefore, it is the copepod behavior that takes them into the precise zone of the water column where the probability of encountering a suitable host is maximized.

For the above reasons, this section is mainly focusing on trematodes. Apart from the Aporocotylidae and Transversotrematidae, all trematodes are transmitted in the diet of their hosts. There are strong patterns, generally family specific, for where the cercariae or metacercariae (the stage that must be ingested) occur.

Cercariae may be ingested directly from the plankton or substrates. This mode of transmission is generally considered primitive and probably applies only to some species of Bivesiculidae, Fellodistomidae, and Tandanicolidae. Unsurprisingly, the key hosts are mainly planktivores (Apogonidae, Blenniidae, Hemiramphidae, Megalopidae, Mugilidae, Pomacentridae). Some life cycles for these families are also evidently extended to three-host cycles leading to the infection of serranids, holocentrids, and muraenids. This could explain the highest average number of trematode species in hosts that are browsing on substrates or filtering plankton (Fig. 9.6).

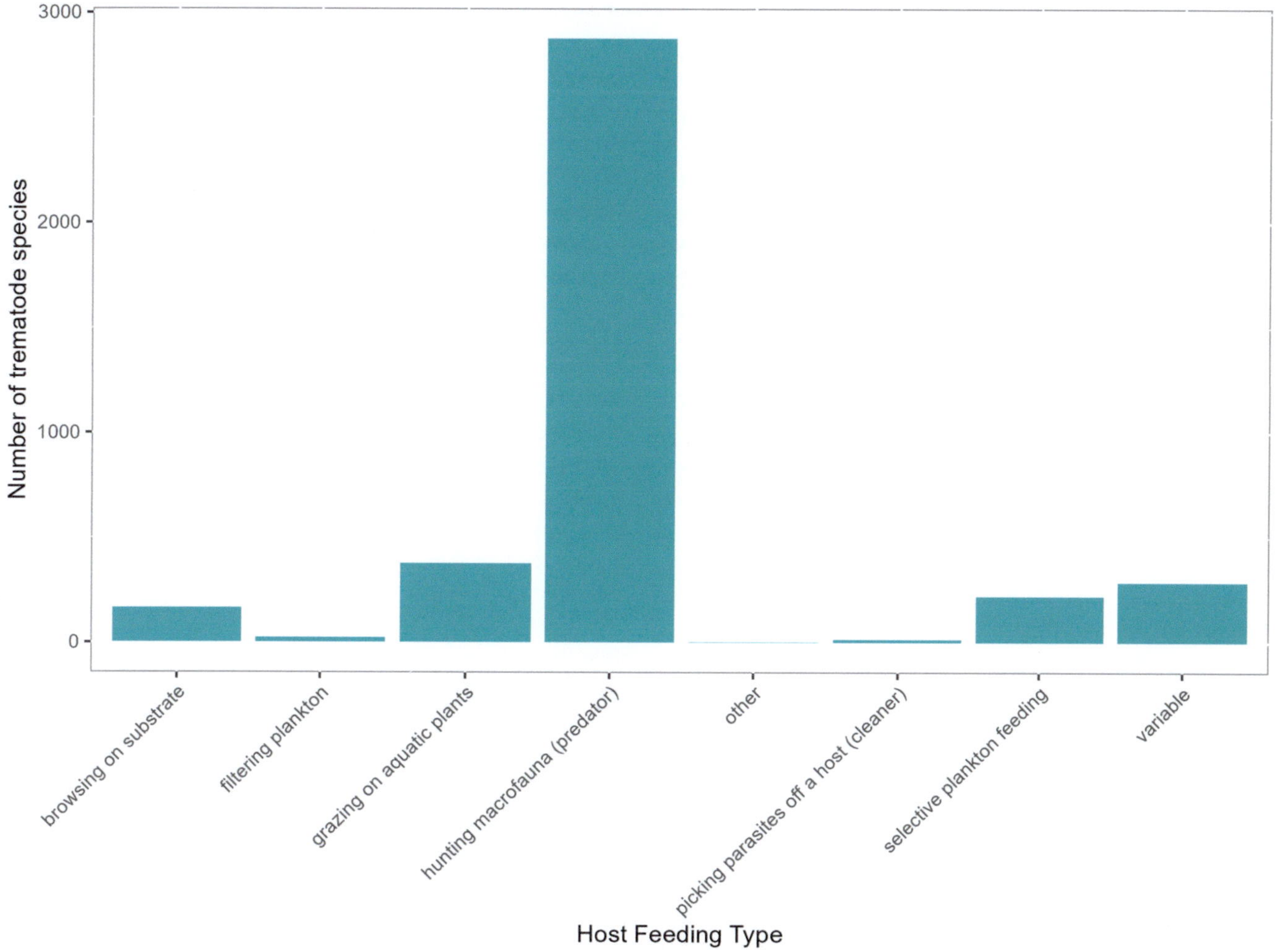

Fig. 9.6 Total number of parasite species relative to host feeding type

Cercariae may form metacercariae in the open. The most important representation of this form of transmission is where the cercaria encysts on algae or seagrass to be ingested by herbivores. This form of transmission has either been demonstrated or inferred for the Cladorchiidae, Emprostiotrematidae, Enenteridae, Gorgocephalidae, Gyliauchenidae, Haploporidae, Haplosplanchnidae, Mesometridae, and Microscaphidiidae. Infections of these families occur overwhelmingly in the browser and herbivore families Acanthuridae, Hemiramphidae, Kyphosidae, Mugilidae, Pomacanthidae, Scaridae, Siganidae, Zanclidae, and some Chaetodontidae and Pomacentridae. This pattern is also observed in our data, as the average number of trematodes in hosts that graze on aquatic plants is quite high (Fig. 9.6).

Cercariae may encyst in an intermediate host in which, typically, further growth occurs. Almost all groups of marine animals act as second intermediate hosts for some trematodes. Typically, families have distinct patterns of specificity for second intermediate hosts: the Acanthocolpidae, Bucephalidae, and Cryptogonimidae overwhelmingly infect fishes; the Faustulidae, Lecithasteridae, Lepidapedidae, Lepocreadiidae, Megaperidae, Monorchiidae, and Zoogonidae overwhelmingly infect invertebrates; the Didymozoidae, Hemiuridae, and Opecoelidae infect both invertebrates and fishes. This flexibility perhaps explains partly why these are the three richest families in the CEI-P. Trematode families with this life cycle dominate the fauna of most of the major families of predatory fishes—Carangidae, Chaetodontidae, Lethrinidae, Lutjanidae, Nemipteridae, Scombridae, Serranidae, and many others. This pattern is represented by the host feeding type "hunting macrofauna (predator)" in Fig. 9.6.

Two trematode families in the CEI-P are transmitted to marine fishes independent of the diet of the host: Aporocotylidae and Transversotrematidae. Aporocotylidae, the fish blood flukes, are transmitted by the direct penetration of their hosts by the cercaria. As adults, aporocotylids are probably most commonly found in the lumen of the heart but also occur in the blood vessels of the gills, the head, the mesenteries, the liver, the kidney, and even the musculature. Despite this flexibility, their host distribution is exceptionally patchy. Transversotrematidae are remarkable parasites that live beneath the scales of their hosts. Infection occurs by the direct attachment of the large and highly developed cercariae. This habitat would appear to render a very large range of fishes susceptible to infection, but so far, they have been found on only 29 families of fishes with the greatest richness associated with Lutjanidae (10 species) and Mullidae and Haemulidae (6 species).

Finally, we suspect that a handful of dietary specializations can be inferred to have enabled "escape" from some trematode infections. Scarine labrids (parrotfishes) have specialized diets of "scraping" and "excavating" (Bellwood and Choat 1990), enabled by their prominently fused teeth; they eat few if any whole animals that can be recognized in their gut contents. These fish are relatively large; *Bolbometopon muricatum* reaches 120 cm, and probably no species attains less than 25 cm. Despite these impressive sizes, from the 67 species known in the region, only 16 have been found infected with trematodes (and 5 with copepods, infection by which is not host diet dependent). Further, they are infected by both diet-dependent (Emprostiotrematidae, Gyliauchenidae, Megaperidae, Haplosplanchnidae, Opecoelidae, Bunocotylidae) and diet-independent (Aporocotylidae and Transversotrematidae) trematode families. This modest fauna infecting scarines contrasts with that infecting the non-scarine Labridae; their derived diet correlated with dramatically reduced trematode parasitism for scarines. There is uncertainty associated with these conclusions because of the lack of evidence of how heavily the Scaridae has been studied. Negative data are rarely reported and almost never aggregated into effective evidence. However, there is no reason to think that these abundant fishes should have received significantly less attention than other comparable taxa.

Overall, we interpret the composition of the trematode fauna of fishes in the CEI-P as one that dynamically tracks evolution of host diet. No attempt has yet been made to correlate the evolution of trematodes with the specifics of that feeding habits of the fish lineages (e.g., Bellwood et al. 2010; Lobato et al. 2014; Streelman et al. 2002; Tebbett et al. 2022), but the field is tantalizing.

9.9 Predictions of Parasite Richness

A core goal for biodiversity studies is to find and describe all the species. For a handful of taxa (e.g., birds and mammals), this goal is close (Costello et al. 2012); however, for groups of invertebrates, this goal is variably distant (Dobson et al. 2008). Predictions of total richness for any fauna are useful for framing approaches to the task of completing the work. It is well-understood, however, that estimates of total richness tend to be become reliable only as the final goal is approached; the enormous variation in predictions of terrestrial arthropod richness makes the point very clear (Ødegaard 2000).

Approaches to estimates of parasite richness have, among their difficulties, one clear advantage. Parasites depend on the presence of suitable hosts so, if that number is known, we have one hard parameter that can be used to constrain the estimate. Our studies found 4859 reef-associated fish species for the CEI-P. The number grows annually with new descriptions and extensions of geographical range so that a figure of 5000 is perhaps both neat and plausible for consideration of the total size of the parasite fauna. A simple extrapolation of the current figures for copepods and trematodes (389 copepod species in 424 fish species; 1597 trematode species in 841 fish species) for 5000 fishes would suggest 5449 and 9494 species, respectively. However, it seems highly improbable that rates of richness identified in the fish surveyed so far will be maintained for the thousands of under-surveyed parasite species.

Parasite richness may be overestimated due to various reasons. First, parasitologists will naturally seek to study the most rewarding (i.e., the most parasite-rich) species. Furthermore, richness in small, rare, and deep-water fishes is likely to be lower than that reported so far. Moreover, since many species of parasites are shared by multiple hosts, the examination of previously under-surveyed fish species may not add much (if any) new richness because the parasites may already be known from previously surveyed relatives.

Similarly, parasite richness can be underestimated; at least some parasite species have limited distributions in the region and are thus not identified by surveys in just a few provinces. Also, surveys of those fishes for which there are some parasite records have certainly not been finalized.

In our view, the combination of these complexities renders the reliable prediction of total parasite richness for copepods and trematodes of the reef-associated fishes of the CEI-P effectively impossible. We think the issues are illustrated in the accumulation curves (Fig. 9.7) of trematodes (shown in blue) and copepods (shown in red) with the increasing numbers of host fish species in the CEI-P. Both curves rise steeply initially but then flatten progressively as the (randomized) fishes become (on average) more and more closely related; more than twice as many CEI-P fishes have been reported as hosts for trematodes as for copepods. Our view is that we cannot presently predict the size of the parasitic fauna with useful reliability; however, we think it unlikely that final parasite richness could be less than double

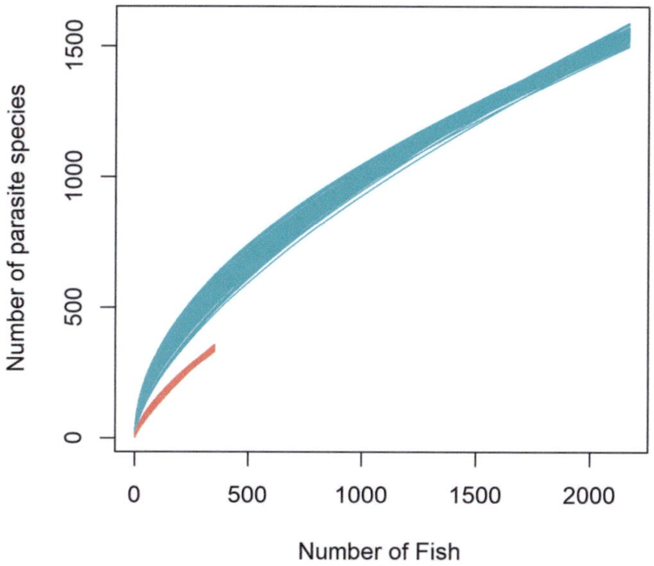

Fig. 9.7 Species accumulation curve relative to the number of fish species examined for the CEI-P (after 100 randomizations and fitted Arrhenius models to all random accumulations). Blue: trematodes; red: copepods

of what is known now. In turn, this lack of knowledge has tremendous implications for our understanding of the coral reef ecosystem functioning.

9.10 Knowledge Gaps and Future Directions

It is well established that some taxonomic groups are far better known than others and that these discrepancies are a major issue in the study of biodiversity (Troudet et al. 2017). Unfortunately, this taxonomic bias has changed little since the 1950s; while many species remain to be discovered and described, the greatest effort is still focused on the same, better known, taxa (Gaston and May 1992). This is true generally, but it is most pronounced for groups that are inherently difficult to study, such as parasites. In addition, the lack of funding for (so-called) fundamental research should not be taken lightly as, undoubtedly, funding guides researchers toward more applied studies (Löbl et al. 2023). From the meta-analysis we have conducted, the group for which most information was retrieved was Trematoda, followed by Copepoda, Monogenea, Nematoda, and Cestoda. Future research should aim to collect more data and bridge the gaps between the taxonomic groups, since the current level of information does not allow for safe conclusions to be drawn.

Arguably, the biogeography of parasites in the region is one of the two greatest gaps in our understanding of the fauna; the depth of sampling of individual species over range so far allows only superficial inferences. Another gap is our understanding of the biological importance of parasitism involved. Trematodes typically castrate their molluscan first intermediate hosts. Similarly, some pennellid copepods are highly pathogenic to their fish hosts. However, the impact of infections of most copepods and trematodes is much lower and subtle (in the definitive host) and almost entirely unstudied. We expect a range of effects from the undetectable (probably common) to the profound (awaiting discovery).

The evidence suggests that most fish species in the region reaching above a length threshold of perhaps 10 cm are infected by parasites. There are many hundreds of such species not yet reported as hosts for trematodes or copepods. The best-known component of the fauna is that of the easily accessible shallow-water species; deeper-water species of all habitats are poorly known. The smallest fishes probably contribute little to the overall richness of the parasitic fauna, but this subject is conspicuously poorly characterized. Most parasites in the region are highly host-specific with few infecting more than a single family of fishes; it has been suggested that the greater host specificity of parasites in tropical regions may be a result of more stable host populations at low latitudes due to limited climatic variability (Poulin 2021), which could also be the case for CEI-P.

For several provinces in the CEI-P, there is negligible information on parasites. Although this is to be expected, since some locations might be difficult to reach and sample, more effort should be given to study those under-represented regions. Such study is necessary to enable a more complete knowledge on the connectivity patterns of parasites and their transmission pathways.

Species discovery efforts should be conducted with the appropriate techniques for each taxonomic group (Cutmore et al. 2025), and they should also be combined with molecular studies. Reference databases should be created with vouchered specimens, and future sampling campaigns should be more interdisciplinary; a joint collaboration of taxonomists and molecular biologists would result in a vast increase in available curated and high-quality information for parasites.

As for all biological data, those for parasites should be findable, accessible, interoperable, and reusable (FAIR) (Wilkinson et al. 2016), so that scientific conclusions can be derived with greater confidence and the data can be used in multiple studies. In addition, species data already included in public databases should be enhanced to include metadata information, such as the ecoregion and province.

Coral reefs are at risk due to a variety of pressures, with the most profound being climate change and ocean acidification (Pandolfi et al. 2011). Reduction of the coral reefs causes a subsequent decrease of the fish fauna that is associated with them (Wilson et al. 2006), which, in turn, adversely affects the species that parasitize them. While such loss will be more evident for specialized parasites, it may likely affect the generalized ones too, in ways we cannot predict since we lack a comprehensive understanding of their life cycles. Most of the parasites in our study were found to be host-specific; therefore, conservation measures should be taken to prevent further loss of coral reefs, not only to protect the coral but also their associated fish and their associated parasites; only through a holistic biodiversity approach, we can bridge all the aforementioned gaps, identify and implement climate change mitigation measures, and break this vicious cycle.

References

Ahyong S, Boyko CB, Bailly N et al (2023) World Register of Marine Species (WoRMS). Available from https://www.marinespecies.org at VLIZ https://doi.org/10.14284/170

Anderson GR (1998) Inference of phylogeny and taxonomy within the Didymozoidae (Digenea) from the second internal transcribed spacer (ITS2) of ribosomal DNA. Syst Parasitol 41:87–94. https://doi.org/10.1023/A:1006024128098

Anderson GR, Barker SC (1993) Species differentiation in the didymozoidae (Digenea): restriction fragment length differences in internal transcribed spacer and 5.8S ribosomal DNA. Int J Parasitol 23:133–136. https://doi.org/10.1016/0020-7519(93)90106-9

Bellay S, de Oliveira EF, Almeida-Neto M, Takemoto RM (2020) Ectoparasites are more vulnerable to host extinction than co-occurring endoparasites: evidence from metazoan parasites of freshwater and marine fishes. Hydrobiologia 847:2873–2882. https://doi.org/10.1007/s10750-020-04279-x

Bellwood DR, Choat JH (1990) A functional analysis of grazing in parrotfishes (family Scaridae): the ecological implications. In: Bruton MN (ed) Alternative life-history styles of fishes. Springer, Dordrecht, pp 189–214

Bellwood DR, Klanten S, Cowman PF et al (2010) Evolutionary history of the butterflyfishes (f: Chaetodontidae) and the rise of coral feeding fishes. J Evol Biol 23:335–349. https://doi.org/10.1111/j.1420-9101.2009.01904.x

Boxshall G, Hayes P (2019) Biodiversity and taxonomy of the parasitic Crustacea. In: Smit NJ, Bruce NL, Hadfield KA (eds) Parasitic Crustacea: state of knowledge and future trends. Springer International Publishing, Cham, pp 73–134

Bray RA, Cutmore SC, Cribb TH (2022) A paradigm for the recognition of cryptic trematode species in tropical Indo-west Pacific fishes: the problematic genus Preptetos (Trematoda: Lepocreadiidae). Int J Parasitol 52:169–203. https://doi.org/10.1016/j.ijpara.2021.08.004

Bray RA, Cutmore SC, Cribb TH (2023) Proposal of a new genus, Doorochen (Digenea: Lepocreadioidea), for reef-inhabiting members of the genus Postlepidapedon Zdzitowiecki, 1993. Parasitol Int 93:102710. https://doi.org/10.1016/j.parint.2022.102710

Caira JN, Jensen K, Waeschenbach A et al (2014) Orders out of chaos – molecular phylogenetics reveals the complexity of shark and stingray tapeworm relationships. Int J Parasitol 44:55–73. https://doi.org/10.1016/j.ijpara.2013.10.004

Collins JJ (2017) Platyhelminthes. Curr Biol 27:R252–R256. https://doi.org/10.1016/j.cub.2017.02.016

Costello MJ, Wilson S, Houlding B (2012) Predicting total global species richness using rates of species description and estimates of taxonomic effort. Syst Biol 61:871. https://doi.org/10.1093/sysbio/syr080

Cribb TH, Anderson GR, Adlard RD, Bray RA (1998) A DNA-based demonstration of a three-host life-cycle for the Bivesiculidae (Platyhelminthes: Digenea). Int J Parasitol 28:1791–1795. https://doi.org/10.1016/S0020-7519(98)00127-1

Cribb TH, Bray RA, Justine J-L et al (2022) A world of taxonomic pain: cryptic species, inexplicable host-specificity, and host-induced morphological variation among species of Bivesicula Yamaguti, 1934 (Trematoda: Bivesiculidae) from Indo-Pacific Holocentridae, Muraenidae and Serranidae. Parasitology 149:831–853. https://doi.org/10.1017/S0031182022000282

Cutmore SC, Cribb TH (2022) New collections of blood flukes (Aporocotylidae) from fishes of the tropical Indo-west Pacific, including a new genus, two new species and molecular evidence that Elaphrobates chaetodontis (Yamaguti, 1970) is widespread in the region. Parasitol Int 88:102565. https://doi.org/10.1016/j.parint.2022.102565

Cutmore SC, Corner RD, Cribb TH (2023) Morphological constraint obscures richness: a mitochondrial exploration of cryptic richness in Transversotrema (Trematoda: Transversotrematidae). Int J Parasitol 53:595–635. https://doi.org/10.1016/j.ijpara.2023.06.006

Cutmore SC, Bray RA, Huston DC et al (2025) Twenty thousand fishes under the seas: insights into the collection and storage of trematodes from the examination of 20,000 fishes in the tropical Indo west-Pacific. J Helminthol 99:e45. https://doi.org/10.1017/S0022149X24000968

Dallas T, Cornelius E (2015) Co-extinction in a host-parasite network: identifying key hosts for network stability. Sci Rep 5:13185. https://doi.org/10.1038/srep13185

Delrieu-Trottin E, Williams JT, Pitassy D et al (2019) A DNA barcode reference library of French Polynesian shore fishes. Sci Data 6:114. https://doi.org/10.1038/s41597-019-0123-5

Dobson A, Lafferty KD, Kuris AM et al (2008) Homage to Linnaeus: how many parasites? How many hosts? Proc Natl Acad Sci 105:11482–11489. https://doi.org/10.1073/pnas.0803232105

Dunn AM, Torchin ME, Hatcher MJ et al (2012) Indirect effects of parasites in invasions. Funct Ecol 26:1262–1274. https://doi.org/10.1111/j.1365-2435.2012.02041.x

Dyer WG, Williams EH, Williams LB (1988) Digenetic trematodes of marine fishes of Okinawa, Japan. J Parasitol 74:638–645. https://doi.org/10.2307/3282183

Ebert D, Fields PD (2020) Host–parasite co-evolution and its genomic signature. Nat Rev Genet 21:754–768. https://doi.org/10.1038/s41576-020-0269-1

Frainer A, McKie BG, Amundsen P-A et al (2018) Parasitism and the biodiversity-functioning relationship. Trends Ecol Evol 33:260–268. https://doi.org/10.1016/j.tree.2018.01.011

Froese R, Pauly D (2023) FishBase. World Wide Web electronic publication. www.fishbase.org

Gaston KJ, May RM (1992) Taxonomy of taxonomists. Nature 356:281–282. https://doi.org/10.1038/356281a0

Grutter AS, De Brauwer M, Bshary R et al (2018) Parasite infestation increases on coral reefs without cleaner fish. Coral Reefs 37:15–24. https://doi.org/10.1007/s00338-017-1628-z

Hua CJ, Zhang D, Zou H et al (2019) Morphology is not a reliable taxonomic tool for the genus Lernaea: molecular data and experimental infection reveal that L. cyprinacea and L. cruciata are conspecific. Parasit Vectors 12:579. https://doi.org/10.1186/s13071-019-3831-y

Huston DC, Cutmore SC, Miller TL et al (2021) Gorgocephalidae (Digenea: Lepocreadioidea) in the Indo-West Pacific: new species, life-cycle data and perspectives on species delineation over geographic range. Zool J Linnean Soc 193:1416–1455. https://doi.org/10.1093/zoolinnean/zlab002

Justine J-L (2010) Parasites of coral reef fish: how much do we know? With a bibliography of fish parasites in New Caledonia. Belg J Zool 140:155–190

Kamiya T, O'Dwyer K, Nakagawa S, Poulin R (2014) What determines species richness of parasitic organisms? A meta-analysis across animal, plant and fungal hosts. Biol Rev 89:123–134. https://doi.org/10.1111/brv.12046

Klimpel S, Kuhn T, Münster J et al (2019) Parasitic groups. In: Klimpel S, Kuhn T, Münster J et al (eds) Parasites of marine fish and cephalopods: a practical guide. Springer International Publishing, Cham, pp 29–76

Kulbicki M, Parravicini V, Bellwood DR et al (2013) Global biogeography of reef fishes: a hierarchical quantitative delineation of regions. PLoS One 8:e81847. https://doi.org/10.1371/journal.pone.0081847

Lafferty KD, Shaw JC, Kuris AM (2008) Reef fishes have higher parasite richness at unfished Palmyra Atoll compared to fished Kiritimati Island. EcoHealth 5:338–345. https://doi.org/10.1007/s10393-008-0196-7

Lo CM, Morgan JAT, Galzin R, Cribb TH (2001) Identical digeneans in coral reef fishes from French Polynesia and the Great Barrier Reef (Australia) demonstrated by morphology and molecules. Int J Parasitol 31:1573–1578. https://doi.org/10.1016/S0020-7519(01)00306-X

Lobato FL, Barneche DR, Siqueira AC et al (2014) Diet and diversification in the evolution of coral reef fishes. PLoS One 9:e102094. https://doi.org/10.1371/journal.pone.0102094

Löbl I, Klausnitzer B, Hartmann M, Krell F-T (2023) The silent extinction of species and taxonomists—an appeal to science policymakers and legislators. Diversity 15:1053. https://doi.org/10.3390/d15101053

Luque JL, Poulin R (2008) Linking ecology with parasite diversity in Neotropical fishes. J Fish Biol 72:189–204. https://doi.org/10.1111/j.1095-8649.2007.01695.x

Miller TL, Bray RA, Cribb TH (2011) Taxonomic approaches to and interpretation of host specificity of trematodes of fishes: lessons from the Great Barrier Reef. Parasitology 138:1710–1722. https://doi.org/10.1017/S0031182011000576

Montes MM, Castro-Romero R, Öktener A et al (2022) Molecular evidence supports the validity of three Parabrachiella species infecting mugilids. Acta Trop 225:106211. https://doi.org/10.1016/j.actatropica.2021.106211

Munday PL, Jones GP, Pratchett MS, Williams AJ (2008) Climate change and the future for coral reef fishes. Fish Fish 9:261–285. https://doi.org/10.1111/j.1467-2979.2008.00281.x

Nadler SA, León GP-PD (2011) Integrating molecular and morphological approaches for characterizing parasite cryptic species: implications for parasitology. Parasitology 138:1688–1709. https://doi.org/10.1017/S003118201000168X

OBIS (2023) Ocean biodiversity information system. Intergovernmental Oceanographic Commission of UNESCO

Ødegaard F (2000) How many species of arthropods? Erwin's estimate revised. Biol J Linn Soc 71:583–597. https://doi.org/10.1111/j.1095-8312.2000.tb01279.x

Pandolfi JM, Connolly SR, Marshall DJ, Cohen AL (2011) Projecting coral reef futures under global warming and ocean acidification. Science 333:418–422. https://doi.org/10.1126/science.1204794

Parravicini V, Kulbicki M, Bellwood DR et al (2013) Global patterns and predictors of tropical reef fish species richness. Ecography 36:1254–1262. https://doi.org/10.1111/j.1600-0587.2013.00291.x

Pavloudi C (2023) Repository containing scripts and input files for the analysis of the LabEx CORAIL book chapter on fish parasites. Available at https://github.com/cpavloud/LabEx_CORAIL

Peyrusse K, Sasal P, Lecchini D (2012) Parasite acquisition by larval coral-reef fishes. J Fish Biol 81:2040–2043. https://doi.org/10.1111/j.1095-8649.2012.03431.x

Poulin R (2021) Functional biogeography of parasite traits: hypotheses and evidence. Philos Trans R Soc B Biol Sci 376:20200365. https://doi.org/10.1098/rstb.2020.0365

Poulin R, Presswell B, Jorge F (2020) The state of fish parasite discovery and taxonomy: a critical assessment and a look forward. Int J Parasitol 50:733–742. https://doi.org/10.1016/j.ijpara.2019.12.009

Pratchett MS, Hoey AS, Wilson SK (2014) Reef degradation and the loss of critical ecosystem goods and services provided by coral reef fishes. Curr Opin Environ Sustain 7:37–43. https://doi.org/10.1016/j.cosust.2013.11.022

Quimbayo JP, Mendes TC, Barneche DR et al (2021) Patterns of taxonomic and functional diversity in the global cleaner reef fish fauna. J Biogeogr 48:2469–2485. https://doi.org/10.1111/jbi.14214

Riegl B, Bruckner A, Coles SL et al (2009) Coral reefs. Ann N Y Acad Sci 1162:136–186. https://doi.org/10.1111/j.1749-6632.2009.04493.x

Spalding MD, Fox HE, Allen GR et al (2007) Marine ecoregions of the world: a bioregionalization of coastal and shelf areas. Bioscience 57:573–583. https://doi.org/10.1641/B570707

Streelman JT, Alfaro M, Westneat MW et al (2002) Evolutionary history of the parrotfishes: biogeography, ecomorphology, and comparative diversity. Evolution 56:961–971. https://doi.org/10.1111/j.0014-3820.2002.tb01408.x

Suyama S, Yanagimoto T, Nakai K et al (2021) A taxonomic revision of Pennella Oken, 1815 based on morphology and genetics (Copepoda: Siphonostomatoida: Pennellidae). J Crustac Biol 41:ruab040. https://doi.org/10.1093/jcbiol/ruab040

Tebbett SB, Siqueira AC, Bellwood DR (2022) The functional roles of surgeonfishes on coral reefs: past, present and future. Rev Fish Biol Fish 32:387–439. https://doi.org/10.1007/s11160-021-09692-6

Timi JT, Mackenzie K (2015) Parasites in fisheries and mariculture. Parasitology 142:1–4. https://doi.org/10.1017/S0031182014001188

Troudet J, Grandcolas P, Blin A et al (2017) Taxonomic bias in biodiversity data and societal preferences. Sci Rep 7:9132. https://doi.org/10.1038/s41598-017-09084-6

Walter TC, Boxshall G (2025) World of copepods database. Caligus Müller OF, 1785. Accessed through: World Register of Marine Species at: https://www.marinespecies.org/aphia.php?p=taxdetails&id=135566 on 13 Mar 2025

Wee NQ-X, Cribb TH, Shirakashi S, Cutmore SC (2022) Three new species of Helicometroides Yamaguti, 1934 from Japan and Australia, with new molecular evidence of a widespread species. Parasitology 149:622–639. https://doi.org/10.1017/S0031182022000051

Wilkinson MD, Dumontier M, IjJ A et al (2016) The FAIR guiding principles for scientific data management and stewardship. Sci Data 3:160018. https://doi.org/10.1038/sdata.2016.18

Wilson SK, Graham NA, Pratchett MS et al (2006) Multiple disturbances and the global degradation of coral reefs: are reef fishes at risk or resilient? Glob Chang Biol 12:2220–2234. https://doi.org/10.1111/j.1365-2486.2006.01252.x

Wood CL, Lafferty KD, Micheli F (2010) Fishing out marine parasites? Impacts of fishing on rates of parasitism in the ocean. Ecol Lett 13:761–775. https://doi.org/10.1111/j.1461-0248.2010.01467.x

Zhokhov AE, Mikheev VN (2015) Symbiotic relationships of coral fish influence their infection by macroparasites. Dokl Biol Sci 462:134–137. https://doi.org/10.1134/S0012496615030072

Zhokhov AE, Pugacheva MN, Thi HV, Mikheev VN (2020) Parasites of small cryptic coral reef fish from the South China Sea. Russ J Mar Biol 46:88–96. https://doi.org/10.1134/S1063074020020121

Snapshots of Coral Reef Biodiversity

10

Félix Pellerin, Diane Espel, Jane Ballard, Jessica Salaün, Ehsan Kayal, Mayalen Zubia, Grégory Lasne, Michel Pichon, Pauline Bosserelle, Julien Wickel, Claire Garrigue, Yves Letourneur, Mehdi Adjeroud, François Guilhaumon, Véronique Berteaux-Lecellier, Gaël Lecellier, Mireille M. M. Guillaume, Malika René-Trouillefou, Clémence Gatti Howell, Thierry Jauffrais, and Mohsen Kayal

Abstract

Coral reefs are sanctuaries for a large portion of our planet's biodiversity, but their rapid decline begs critical questions about what exactly is being lost, what driving forces determine decline or resilience, and what implications for life on Earth. To help address these questions, this chapter synthesizes the state of knowledge on the diversity of coral reef life forms at different scales of biological and ecological organization, encompassing variability in genes, biomolecules, cells, organs, organisms, populations, communities, ecosystems, socio-ecosystems, and biogeoclimatic systems, with a particular focus on French tropical overseas. We discuss research pathways to address knowledge gaps in various scientific disciplines, covering genetics (genomics, transcriptomics, proteomics, metabolomics), morphology, ethology, bioacoustics, demography, functional ecology, macro-ecology, landscape ecology, human and social sciences, bioclimatology, and their cross-disciplinary interactions. Based on contributions from coral reef experts working in different coral reef regions around the world, our synthesis supports a better understanding of the various dimensions of coral reef biodiversity and proposes future research orientations for improving knowledge and conservation. Given the rapid disappearance of coral reefs and how much is still unknown, this chapter underscores

Synopsis

This chapter provides a holistic vision of coral reef biodiversity across vast scales of biological and ecological organization, from genes and molecules to ecosystems and their geoclimatic and social environments. We compare biodiversity metrics among key ecological communities and regions as we synthesize the current state of knowledge on coral reef biodiversity, pinpoint important knowledge gaps, and identify prospective research.

F. Pellerin · M. Adjeroud
Laboratoire d'Excellence CORAIL, Paris, France

USR 3278 CRIOBE, PSL Université Paris, EPHE-UPVD-CNRS, Perpignan, France

UMR 250 ENTROPIE, IRD-IFREMER-CNRS-University of Reunion-University of New Caledonia, Nouméa, New Caledonia

D. Espel
UMR 250 ENTROPIE, IRD-IFREMER-CNRS-University of Reunion-University of New Caledonia, Nouméa, New Caledonia

CNRS, ECOBIO, UMR 6553, University of Rennes, Rennes, France

J. Ballard
Tour du Valat Research Institute for the Conservation of Mediterranean Wetlands, Arles, France

J. Salaün
Laboratoire d'Excellence CORAIL, Paris, France

USR 3278 CRIOBE, PSL Université Paris, EPHE-UPVD-CNRS, Perpignan, France

E. Kayal
Department of Ecology, Evolution, and Organismal Biology, Iowa State University, Ames, IA, USA

M. Zubia
Laboratoire d'Excellence CORAIL, Paris, France

University of French Polynesia, UMR-241-SECOPOL (ILM, Ifremer, IRD), Campus Outumaoro, Tahiti, French Polynesia

G. Lasne
Biocénose marine SARL, BP 8229, 98807 Nouméa cedex, Nouméa, New Caledonia

M. Pichon
Biodiversity and Geosciences Program, Queensland Museum, Townsville, QLD, Australia

S. Planes (ed.), *The Future of Coral Reefs*, Coral Reefs of the World 21, https://doi.org/10.1007/978-3-031-98584-3_10

the urgency of cross-scale biodiversity assessments for a systemic understanding of the distribution and dynamics of biodiversity.

Keywords

Systems science · Biodiversity distribution · Collaborative cross-system research · Coral reef management · Sustainability

10.1 Introduction

10.1.1 Coral Reefs in the Twenty-First Century

Coral reefs are arguably the most biodiverse and productive ecosystems on Earth (Fig. 10.1). They host an estimated 38% of all marine species (Fisher et al. 2015) and contribute fundamental benefits to humanity and the natural world Moberg and Folke 1999; Speers et al. 2016 Woodhead et al. 2019; Carlot et al. 2023) despite making up a mere 0.08% of the ocean floor (Smith 1978), or 0.17% including associated lagoonal habitats (Spalding et al. 2001), and lying in tropical waters that are poor in nutrients (de Goeij et al. 2013). However, coral reefs are declining due to the high vulnerability of several of their ecologically key functional groups to human-driven and climatic stressors (Brandl et al. 2019a; IPBES 2019; Calvin et al. 2023). In particular, unsustainable rapid growth in coastal development, pollution, fishing, and climate change are increasingly reducing the regeneration of reef habitats by their slow-growing foundational species, the calcifying corals, and altering the capacity of the food webs to preserve their functional equilibrium (Fig. 10.2; Vajed Samiei et al. 2015; Hoegh-Guldberg et al. 2017; Hughes

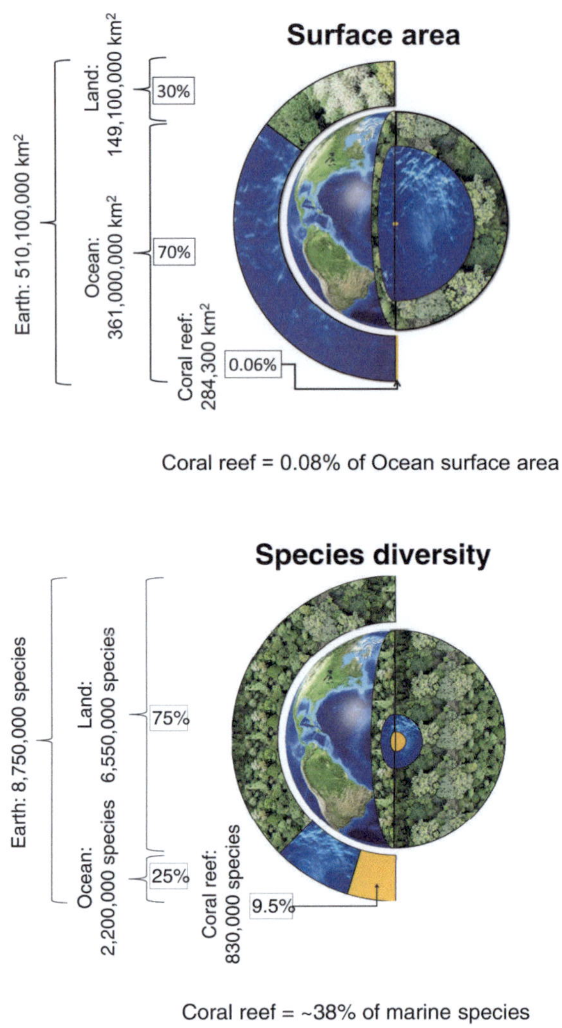

Fig. 10.1 Coral reef contributions in terms of surface area and species diversity, based on total species number estimates according to Mora et al. (2011) and Fisher et al. (2015)

P. Bosserelle
Consultante indépendante en écologie récifale, Papeete, French Polynesia

J. Wickel
GIE MAREX – Marine Expertise Company, Piton Saint Leu, Reunion

C. Garrigue · Y. Letourneur · F. Guilhaumon · T. Jauffrais
Laboratoire d'Excellence CORAIL, Paris, France

UMR 250 ENTROPIE, IRD-IFREMER-CNRS-University of Reunion-University of New Caledonia, Nouméa, New Caledonia

V. Berteaux-Lecellier
Laboratoire d'Excellence CORAIL, Paris, France

EMR 9001 SantEco – UMR 250 ENTROPIE, IRD-CNRS-University of Reunion-University of New Caledonia, Nouméa, New Caledonia

G. Lecellier
Laboratoire d'Excellence CORAIL, Paris, France

ISEA EA 7484, University of New Caledonia, Noumea, New Caledonia

M. M. M. Guillaume · M. René-Trouillefou
Laboratoire d'Excellence CORAIL, Paris, France

UMR 8UMR 8067 BOREA, Muséum National d'Histoire Naturelle, SU, CNRS, IRD, UA, Aviv, Paris, France

C. G. Howell
Institut Louis Malardé (ILM), UMR 241-SECOPOL (IFREMER, ILM, IRD, UPF), Laboratory of Marine Biotoxins, Papeete, Tahiti, French Polynesia

M. Kayal (✉)
Laboratoire d'Excellence CORAIL, Paris, France

UMR 250 ENTROPIE, IRD-IFREMER-CNRS-University of Reunion-University of New Caledonia, Nouméa, New Caledonia

Station Marine d'Endoume, Marseille, France
e-mail: mohsen.kayal@ird.fr

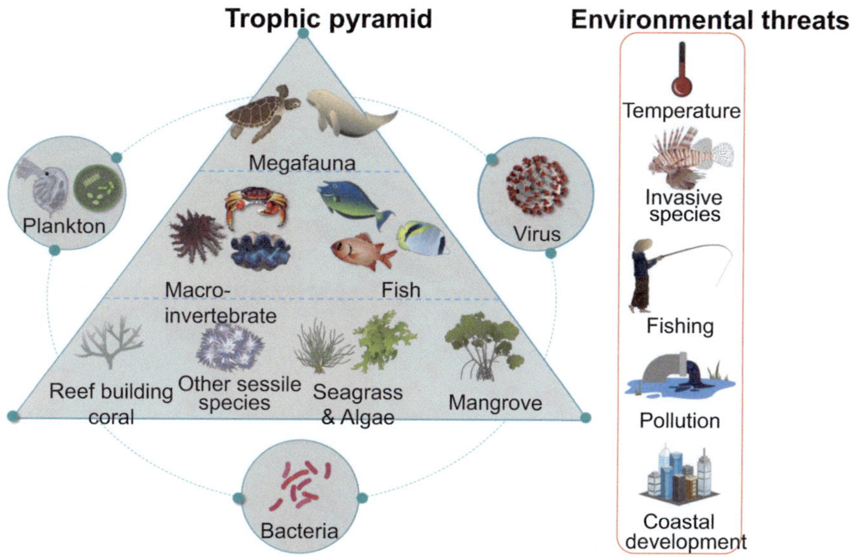

Fig. 10.2 Key ecological communities of coral reef ecosystems and associated seagrass and algae beds, mangrove forests, and sandy lagoonal plains. The communities are organized in a trophic pyramid, with species occupying distinct ecological niches and contributing a set of ecological functions that enable ecosystem functionality in a state of relative stability. Environmental stressors alter species abundances, threatening the equilibrium and functionality of the system

et al. 2018; Quattrini et al. 2020; Ouédraogo et al. 2021, 2023; Andrello et al. 2022). In many places, eutrophication and overfishing have reduced the capacity of herbivores to control algae communities, leading to major ecosystem shifts (Hughes et al. 2017; Schmitt et al. 2019). The widespread decline of coral reefs and consequently of the benefits they provide to society has made coral reef conservation a societal priority in many regions and a major target for humanity globally (IPBES 2019; Maxwell et al. 2020; Eddy et al. 2021). Within these efforts and in the larger scope of knowledge and conservation of the Earth's biodiversity, it is fundamental to comprehend the extent to which coral reefs contribute to biodiversity, biodiversity distribution in relation to coral reef conditions and human influences, and the consequences of coral reef decline for marine life and humanity. To begin, what exactly is coral reef biodiversity?

10.1.2 What Is Biodiversity?

Biodiversity is defined as the variability of life forms. Despite the simplicity of the concept, biodiversity remains a major scientific mystery, as variability of life is found at a multitude of biological and ecological scales, from the very small to the very large (Fig. 10.3). In longer terms, as defined by Article 2 of the Convention on Biological Diversity (www.cbd.int): "biological diversity means the variability among living organisms from all sources including, inter alia, terrestrial, marine and other aquatic ecosystems and the ecological complexes of which they are part; this includes diversity within species, between species and of ecosystems." In fact, variability of life is characterized in as many dimensions as there are metrics to assess, resulting in a plethora of biodiversity measurements and disciplinary sub-definitions. While this fractal nature of biodiversity measurements has complexified interdisciplinary interactions, biodiversity assessments remain largely incomplete, as most existing metrics have not yet reached a final estimate and biodiversity metrics continue to increase with the development of new scientific tools and disciplines (Caley et al. 2014; Guerra et al. 2020; Galand et al. 2023. Overall, a global evaluation of Earth's biodiversity is necessary for a better understanding of the organization of life on our planet and the consequences of the biological losses it suffers (Costello et al. 2013; IPBES 2019; Bradshaw et al. 2021). As sanctuaries for a large portion of marine biodiversity, coral reefs are a good place to begin this endeavor.

10.1.3 Objectives of the Chapter

Using contributions from a team of coral reef experts working in different coral reef ecoregions around the world, we compile information from various scientific disciplines of biology and environmental sciences to synthesize knowledge on major metrics of coral reef biodiversity at different scales of biological (i.e., relating to organisms) and ecological organization (i.e., relating to interactions among organisms and with their environments), from genes to ecosystems and their natural and human-associated environments (Fig. 10.3). We evaluate differences among key coral reef biological communities and geographical regions, identify major knowledge gaps, and discuss prospective research. Our synthesis provides an understanding of the various dimensions in which coral reef biodiversity takes form and highlights future research for improving our knowledge of nature and its preservation.

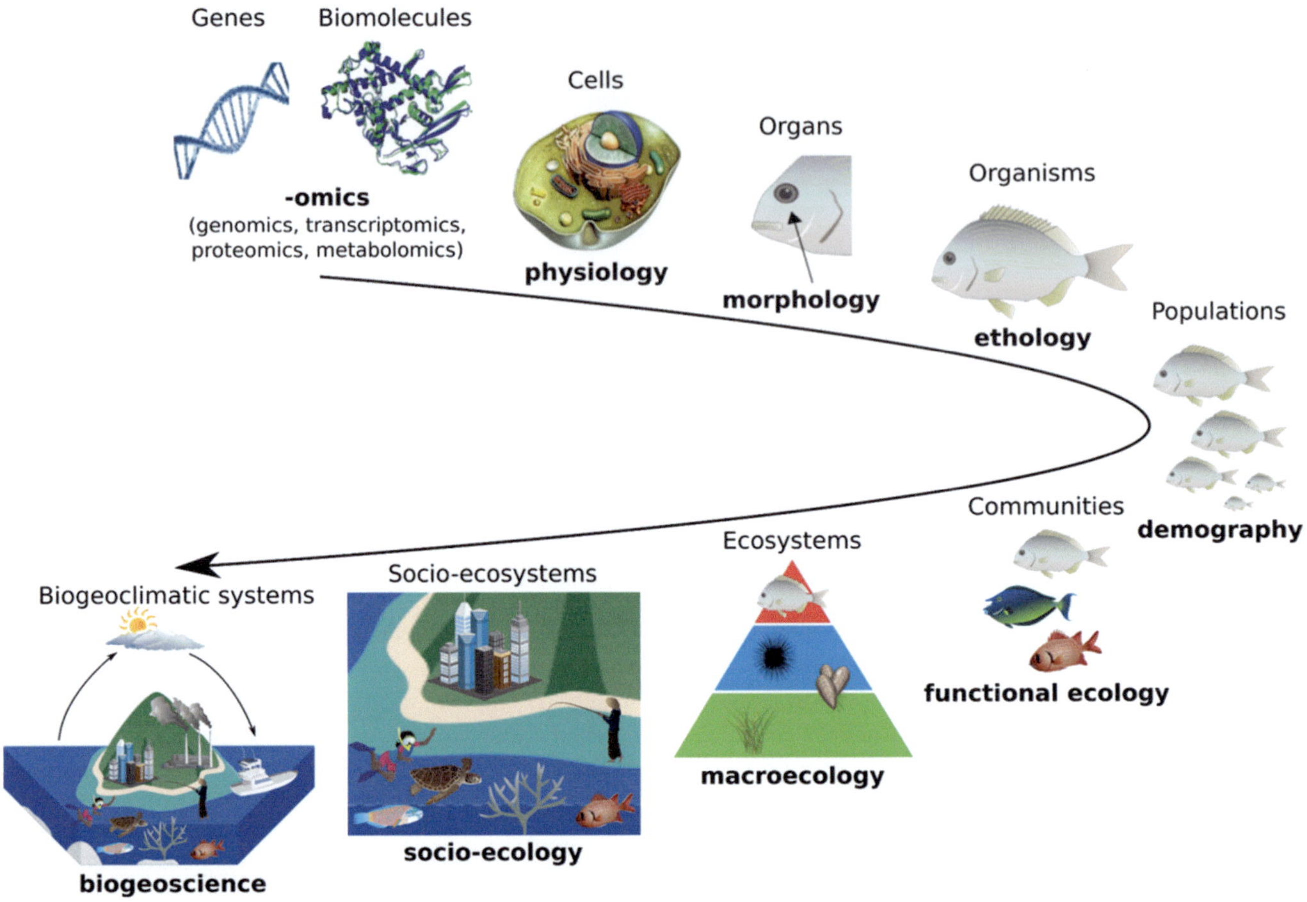

Fig. 10.3 Hierarchical organization of the living, from genes to biogeoclimatic systems, from which biodiversity is the result of variabilities at the different scales. Text indicates the different scales of biological and ecological organization (top) and the corresponding scientific disciplines (bottom)

10.2 Methods

We have leveraged the coral reef expertise of the French Laboratory of Excellence LabEx CORAIL network across 12 French tropical overseas localities situated in the Caribbean, Indian, and Pacific regions (Fig. 10.4). This represents ~5% of the world's constructed hard-substrate reefs, ~9% of all coral reef surface area when including associated sandy lagoonal habitats, and encompasses 46° of latitude (18°N–28°S) and all major coral reef basins (Smith 1978; Spalding et al. 2001; Andréfouët et al. 2008).

We characterize coral reef biodiversity at the different scales of biological and ecological organization, from genes to biogeoclimatic systems, by compiling quantitative metrics from the various scientific disciplines (Fig. 10.3) for different coral reef biological communities and regions (Fig. 10.4). Our focus is on major biological groups considered as key ecological compartments of coral reef ecosystems, including reef-building corals and other sessile animals (soft-corals, sponges, ascidians, bryozoans), micro- and macroalgae, vagile invertebrates (crustaceans, mollusks, echinoderms, annelids, platyhelminthes, sipunculid worms, hemichordates), fish (actinopterygians, elasmobranchii), megafauna (sea turtles, dugongs, dolphins, whales, sharks and rays, sea snakes), bacteria, and viruses (Fig. 10.2). Data on seagrasses, mangroves, and sandy lagoonal plains are also integrated as adjacent ecosystems with species distributions that often overlap with coral reef habitats (Moberg and Folke 1999; Andréfouët et al. 2008; Guannel et al. 2016). Environmental data consists of metrics characterizing the geologic, climatic, and societal aspects of the coral reef systems.

Biodiversity data is gathered thanks to the contribution of experts from various disciplines and regions, complemented with data obtained from literature and global databases. Particular attention has been given to the most documented and commonly used biodiversity metrics, and those viewed as essential biodiversity variables (https://geobon.org/ebvs/; Pereira et al. 2013). Experts have also been asked to identify major knowledge gaps and prospective research to address them.

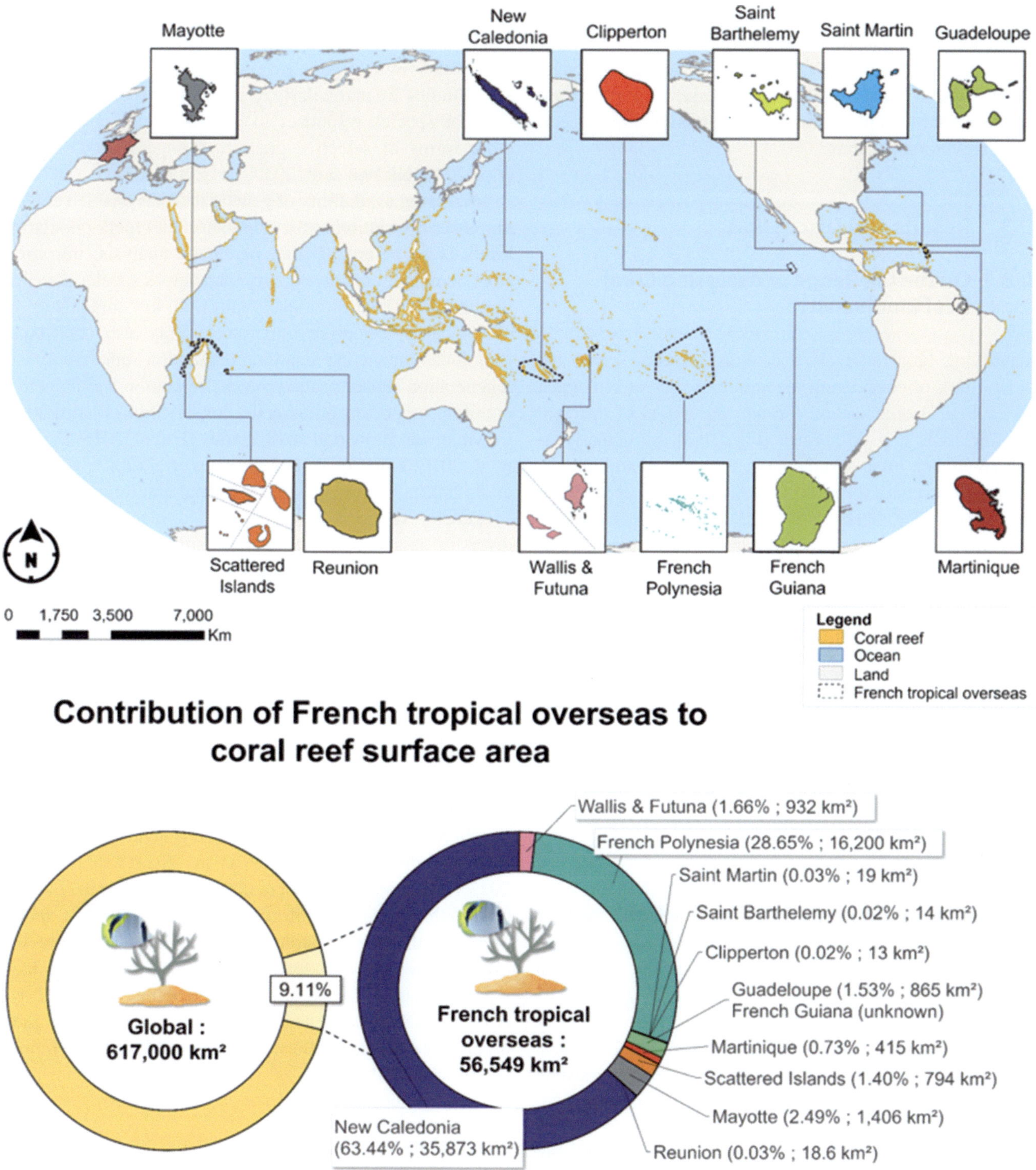

Fig. 10.4 Map of the world's coral reefs and contributions of French tropical overseas. Coral reef surface areas account for constructed hard reef and sandy lagoonal plain habitats (cf. Spalding et al. 2001)

All maps and graphics have been produced using the R software environment (R v4.2.3, R Core Team 2023) complemented with the "ade4," "stringr," "ggplot2," "png," "RcolorBrewer," and "Cowplot" packages. Graphics are complemented with iconography from https://ian.umces.edu and https://biorender.com.

10.3 Results and Discussion

10.3.1 On the Challenge of Assessing Coral Reef Biodiversity

Quantifying biodiversity is a daunting exercise given the colossal task of considering the various scales of biological and ecological organization through the lenses of different scientific tools and disciplines (Fig. 10.3), particularly as most are in their infancy and none have achieved a finite estimate of biodiversity (Costello et al. 2013; Caley et al. 2014; Guerra et al. 2020; Galand et al. 2023). Nevertheless, estimating biodiversity is a necessity for humanity when rapid biodiversity loss is a global reality with irrevocable consequences for life on Earth (Kayal et al. 2019; IPBES 2019; Calvin et al. 2023). Due to their central role in hosting a large portion of the world's biodiversity and their vulnerability to collapse, coral reefs are certainly a good focus point for this endeavor. The following sections synthesize coral reef biodiversity based on a multitude of metrics that characterize variability of coral reef systems at various biological and ecological scales, providing a snapshot of the work in progress.

10.3.2 Genes

Unknown until recent times, the birth of molecular biology in the 1950s opened paths to a fascinating new vision of how biological features are controlled by inheritable biological units, otherwise known as genes, organized in long sequences of deoxyribonucleic acid, or DNA (Gayon 2016). This revelation confirmed early work by Mendel (1865) on biological inheritance. Since, genetic investigation has been in constant acceleration, with the development of new tools and techniques providing means to deepen our exploration of biological diversity at the level of genomes. Despite significant increases in the number of studies in recent years, the genetic exploration of coral reef ecosystems remains in its infancy. Genomic resources have been developed for some model species within different ecological communities (Fig. 10.5), providing various metrics for measuring genetic diversity within and among taxa, including length, composition, and organization of coding and noncoding sequences (Fig. 10.6) and their degree of similarity between different evolutionary paths (Fig. 10.7). These metrics can be accessed at the DNA and RNA levels, the latter also providing functional information in the form of gene expression data (transcriptomics). Such studies are particularly relevant to ecological investigations of species responses to environmental changes and other forms of selective pressure (Gingeras 2007; Fuller et al. 2020; Nielsen et al. 2020; Levy et al. 2021).

Increasing availability of genetic data has enabled exploring genetic variation at the population and meta-population levels (e.g., heterozygosity, parentage analysis, introgression, demographic connectivity) and across species distribution ranges (i.e., phylogeography) in key organisms of ecological or scientific importance. This also contributes ecological knowledge regarding population connectivity and regeneration capacity, and species origination and dispersal, therefore supporting biodiversity conservation (Baums 2008; Cowman and Bellwood 2013; Edmunds et al. 2016; Siqueira et al. 2019; Dubé et al. 2020; Swain et al. 2021; Cantalice et al. 2022). Exploring both the spatial and temporal nature of genetic variation is paramount to identifying genetic diversity hotspots and the mechanisms underlying adaptation to changing environments, a field that necessitates further investigation (Fig. 10.5). New approaches combining genetic exploration with descriptors of environmental condition and biodiversity at larger scales of bioecological organization will help in this endeavor (Nielsen et al. 2020; Selmoni et al. 2020; Manel et al. 2020; Bongaerts et al. 2021).

By enabling species identification based on short signature sequences, barcoding data represents the prevalent and accessible form of genetic resource, particularly useful for accessing hidden communities such as bacteria, viruses, symbionts, and cryptic organisms using environmental-DNA (eDNA) approaches (Robbins et al. 2021; Galand et al. 2023). GenBank (www.ncbi.nlm.nih.gov/genbank/) is the main public repository of genetic data in the public sphere (Leray et al. 2019), though project-specific datasets are also available. Recent decreases in the cost of sequencing have led to a sharp increase in shallow genomic surveys. Low-coverage DNA sequencing provides access to complete organellar genomes as these small molecules are often present in several thousands of copies compared to the larger nuclear genomes present as a single copy per cell. Mitochondrial DNA (mtDNA) data is a favorite molecular tool for the study of species diversity in animals (e.g., Shinzato et al. 2021), whereas chloroplast DNA (cpDNA) data is preferably used in plants and algae, mainly owing to differing rates of sequence evolution in these organisms. In addition to sequence data, complete organelle genomes provide morphological characters in the form of genome organization and molecular conformation, providing insights into genetic expression (Fig. 10.7; Kayal et al. 2012a, b, 2015a, b; Lavrov and Pett 2016).

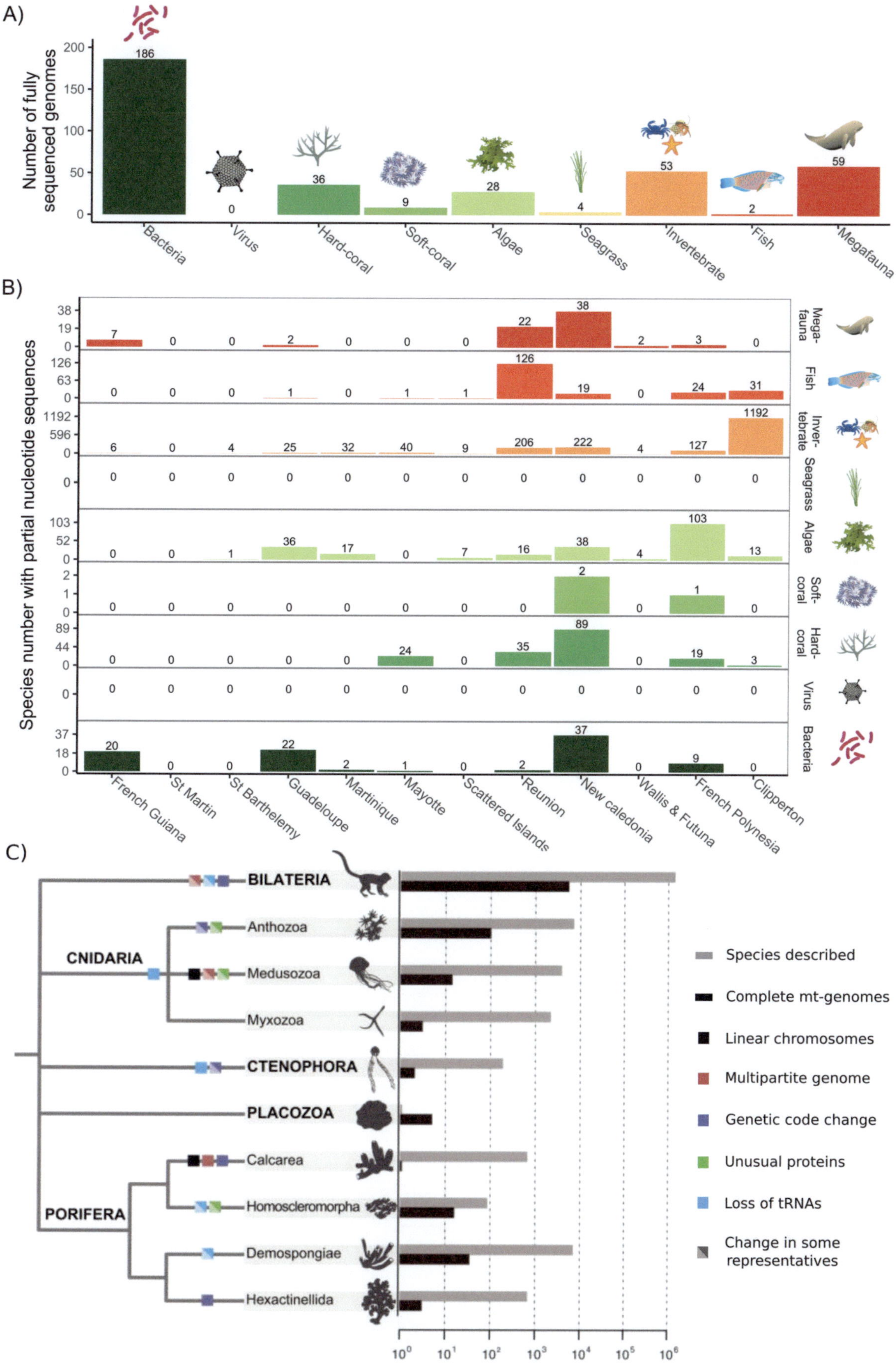

Fig. 10.5 Number of coral reef-associated species with characterized whole-genome sequences (**a**), number of species with available partial DNA sequences across French tropical overseas (**b**), and number of described species and complete mitochondrial genomes per group (**c**). Data from GenBank (**a**, **b**) and from Lavrov and Pett 2016 (**c**)

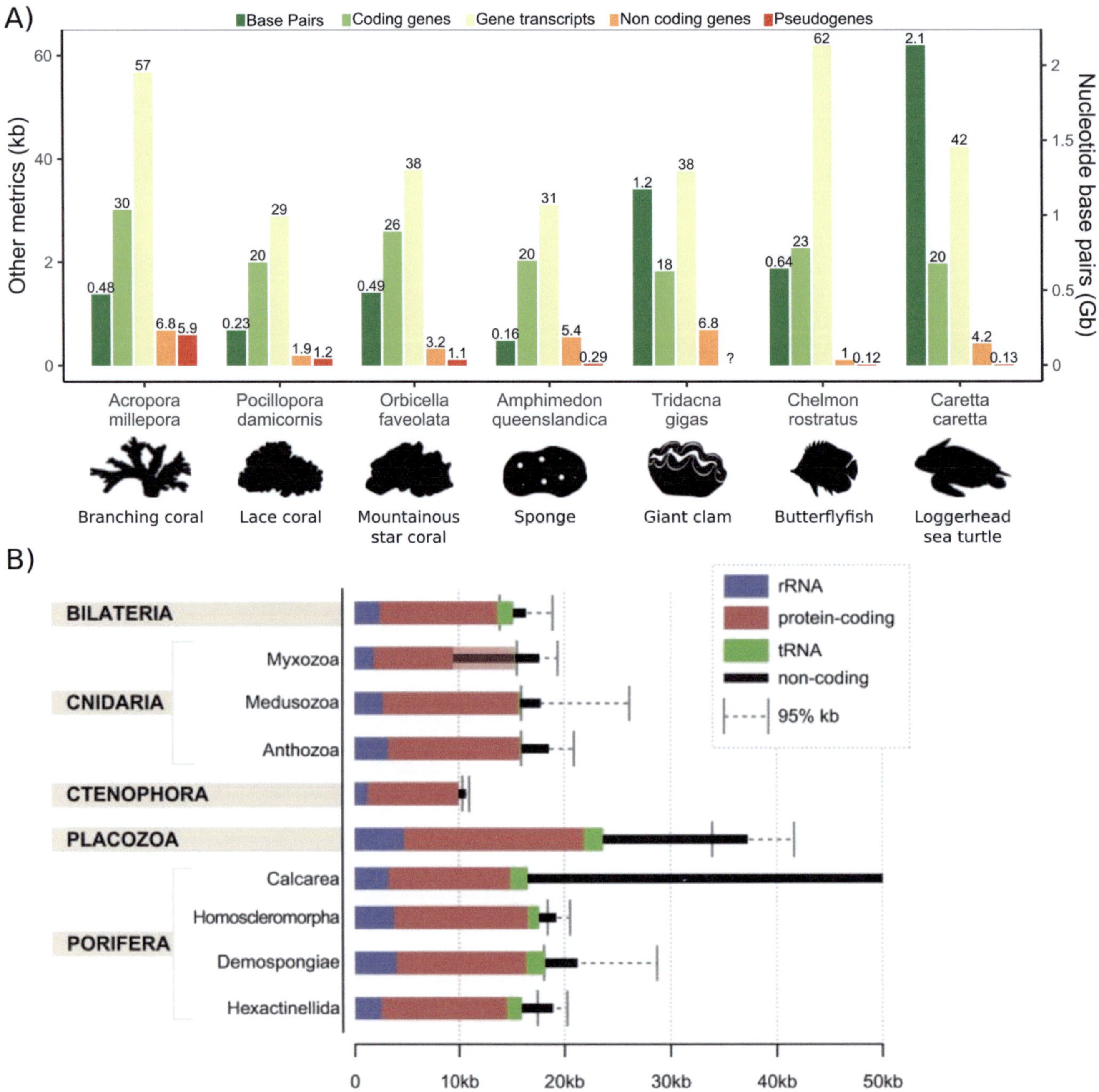

Fig. 10.6 Differences in nuclear (**a**, data from https://rapid.ensembl.org) and mitochondrial (**b**, from Lavrov and Pett 2016) genome size among key coral reef species groups

The taxonomic diversity of coral reefs makes them ideal laboratories for exploring evolutionary patterns in living organisms. Taxonomic comparisons indicate a high level of diversity in genome size, composition, and structural organization, with no apparent relationship between genome complexities and species evolutionary positions (Fig. 10.6), although similarity between genetic sequences decreases with increasing evolutionary distance (Ciccarelli et al. 2006). However, genetic characterization of several branches of the tree of life, including major animal clades such as Porifera, Bryozoa, and Platyhelminthes, remains scarce even in barcode libraries (Mugnai et al. 2021). Such paucity of genetic information is typical of marine, compared to land, ecosystems, impeding large-scale comparisons of genetic and genomic diversity among coral reefs (Fig. 10.5). Cross-regional initiatives such as the recent Tara Pacific campaigns contribute greatly to remedy such shortcomings (Planes and Allemand 2023).

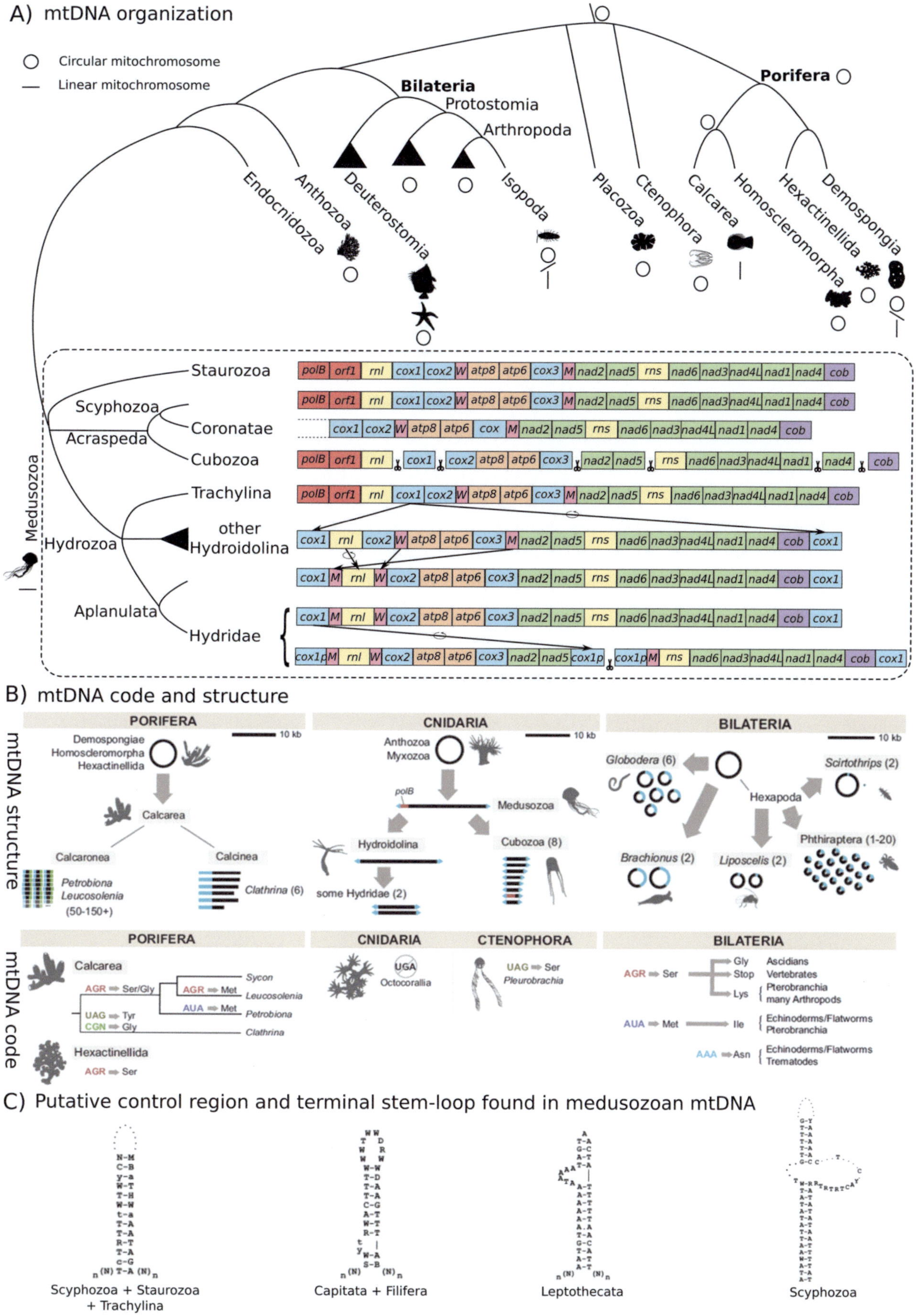

Fig. 10.7 Evolution of mitochondrial DNA (mtDNA) in reef-associated animals with zoom on Medusozoa. Differences in mtDNA result from linearization, fragmentation, gene rearrangement, and changes in coding sequences, which affect DNA molecular structure and the level of genetic expression. (Adapted from Kayal et al. (2012a, b) and Lavrov and Pett (2016))

10.3.3 Biomolecules

Coral reefs are host to a high diversity of molecular compounds, most of which are still uncharacterized, released by their myriad of local inhabitants. MarinLit (https://marinlit.rsc.org) is a major database that documents the diversity of biomolecules of marine origin with medicinal and engineering potentials. Among the ~25,000 marine natural products registered in this database in 2019, a large proportion originates from reef-associated organisms, particularly sponges (33%) and cnidarians (28%) (Fig. 10.8; Carroll et al. 2019). These numbers increase rapidly, with ~1400 new marine biomolecules being described each year, a majority (888 in 2021) originating from tropical regions and sponges and cnidarians (37% in 2021; Carroll et al. 2023). This natural wealth is of particular interest to biopharmaceutical and biotech industries using the bioactive chemicals as cytotoxic, anticancer, pain relief, and antifouling products (Ghareeb et al. 2020; Gomez-Banderas 2022; Wegley Kelly et al. 2022). Research in this field benefits from the attention of active industries and international groups and is likely to uncover many more potential uses as screening of coral reef compounds for natural products continues, notably on less emblematic and cryptic organisms that remain understudied.

By investigating proteins and other small molecules present in organisms and their environments, proteomic and metabolomic approaches provide an additional layer of complexity to molecular biodiversity exploration (Jiang et al. 2021). While regulation of macroalgae is failing on many reefs due to eutrophication and declines in herbivory (Hughes et al. 2017), exploitation of macroalgae can provide a large source for biomaterials and constitute a win-win for humans and ecosystems alike in a time when the sustainable exploitation of ocean resources (i.e., blue economy) is in full expansion (Urban Jr and Ittekkot 2022). Indeed, coral reef macroalgae offer a huge and attractive potential for developing new drugs and biomaterials, an area that is expected to increase with the development of macroalgal farms, especially in French tropical overseas (Stiger-Pouvreau and Zubia 2020).

10.3.4 Cells

Variability in cell organization is a major feature of evolutionary biology, with the acquisition of cellular membranes and organelles being determinant in the development and diversification of life López-García and Moreira 2023). The large diversity in cell structure and organization found in coral reefs reflects that of taxonomic groups from various portions of the phylogenetic tree, ranging from the most basic virus and simpler prokaryotic cells to the most composite cells endowed with complex membranes and organelles acquired through symbiotic associations (Fig. 10.9a). The evolution of multicellularity has enabled cell specialization and the development of organs, leading to a diversification of cell features and functions within organisms (Fig. 10.9b). For instance, one species of reef-building coral harbors 40 different cell types throughout its life cycle (Levy et al. 2021). Such cell diversification confers niche extension and competitive advantage to organisms, influencing biodiversity patterns at higher levels of bioecological organization (population, community).

Cell diversity can reach its paroxysm in symbiotic (paired living beings) and holobiontic (multiple living beings functioning as a unit) organisms, such as reef-building corals finding mutual benefits between heterotrophy in animals and autotrophy in plants to colonize oligotrophic waters (Fig. 10.9b; Jiang et al. 2021). Physiological studies on coral reef species have specifically focused on key metabolic functions, such as respiration, photosynthesis, and calcification, to assess species performances under various environmental conditions, providing insights into their capacity to survive future reef conditions (Vajed Samiei et al. 2015; Hoegh-Guldberg et al. 2017; Andrello et al. 2022). Recent development in genomics, transcriptomics, and proteomics methods enables the exploration of new cell types and functions, providing further insights into the physiological functioning of cells and organisms (Konopka and Wilkins 2012; Levy et al. 2021). New approaches in meta-“omics” (e.g., metagenomics, meta-transcriptomics, meta-proteomics) applied through the lens of the holobiont concept shed light on the hidden functional and cellular diversity of complex marine ecosystems. In fact, marine holobionts are an emerging field of research that shows promising outcomes for our understanding of coral reefs and other ecosystems (Dittami et al. 2021). For instance, monitoring of marine holobionts can provide clues into environmental stressors while providing explanatory mechanisms for adaptation to changing climate (González-Pech et al. 2023). Such knowledge can in turn be used in hierarchical, multi-scale quantitative approaches to bridge the gap to higher bioecological levels such as populations, communities, and ecosystems (Edmunds et al. 2014; Smallegange et al. 2017; Condie et al. 2018; Dubois et al. 2019; Carturan et al. 2020; Morais and Bellwood 2020).

10.3.5 Organs

The diversity of organ shapes and functions was a foundation for Charles Darwin’s work on species evolution (Darwin 1859). Work in this realm has paid particular attention to the diversity of key functional organs and transmission of their characters throughout evolutionary pathways, including locomotion, sensory, communication, and reproductive

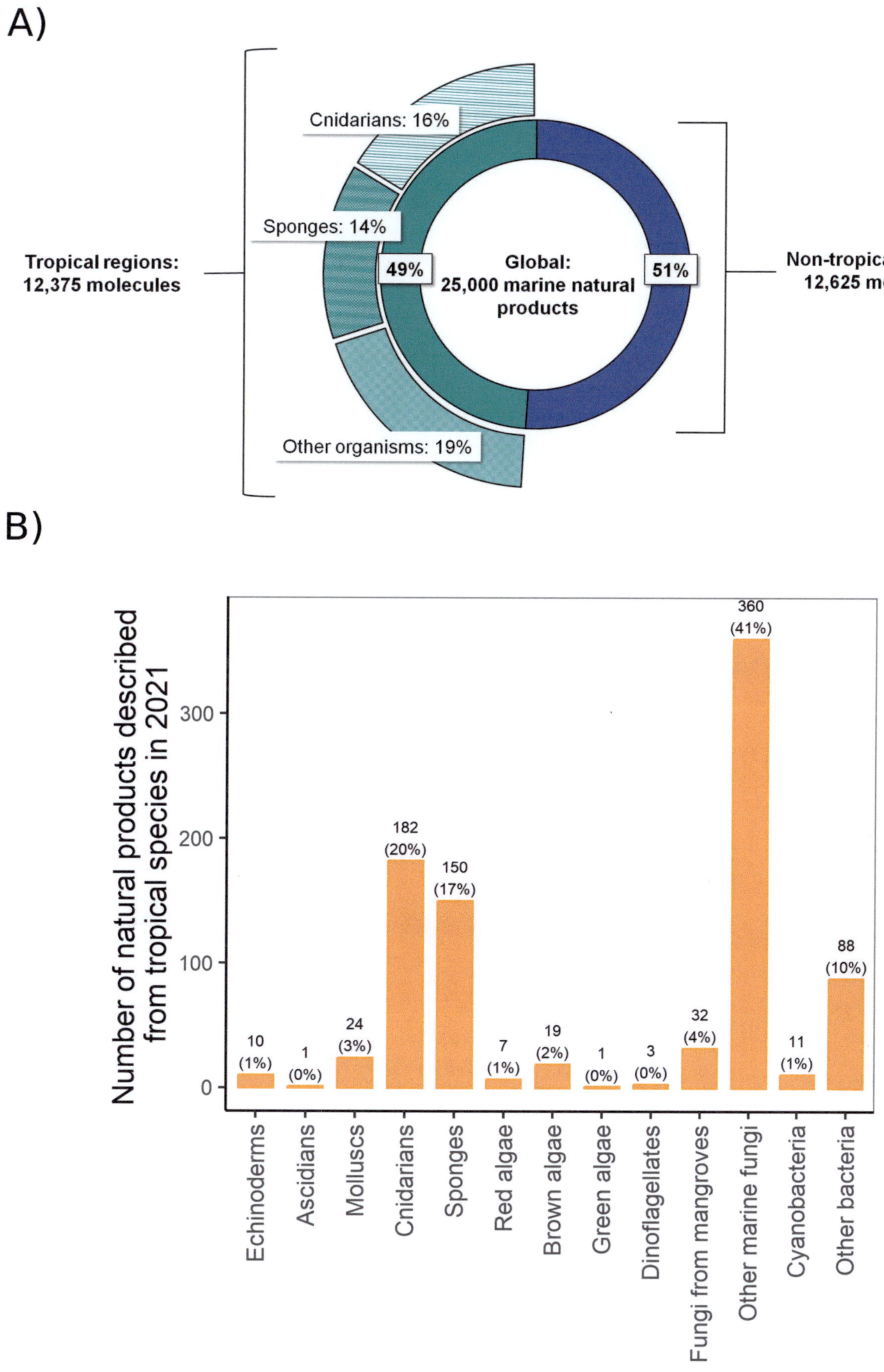

Fig. 10.8 Diversity of biomolecules associated with coral reef species. (**a**) Proportion of marine natural products associated with tropical reef organisms recorded in the MarinLit database in 2019 and (**b**) number of natural marine products described in 2021 extracted from tropical species. (Data from Carroll et al. (2019, 2023))

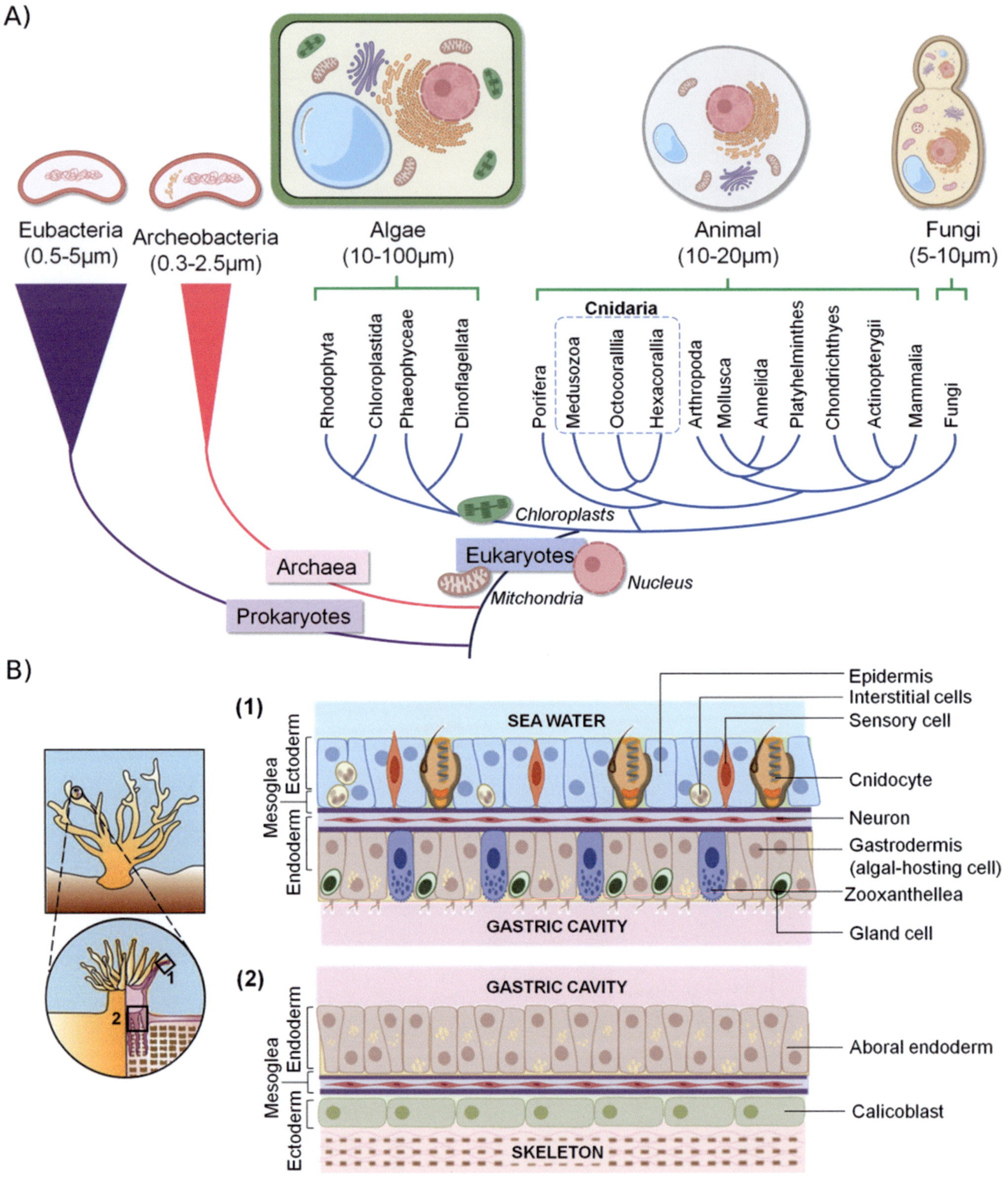

Fig. 10.9 Cell diversity across the tree of life and coral reef organisms. (**a**) Diversity in cell structures and organelles composition. (**b**) Cell specialization in reef-building corals. (Adapted from Boilard et al. 2020; Levy et al. 2021)

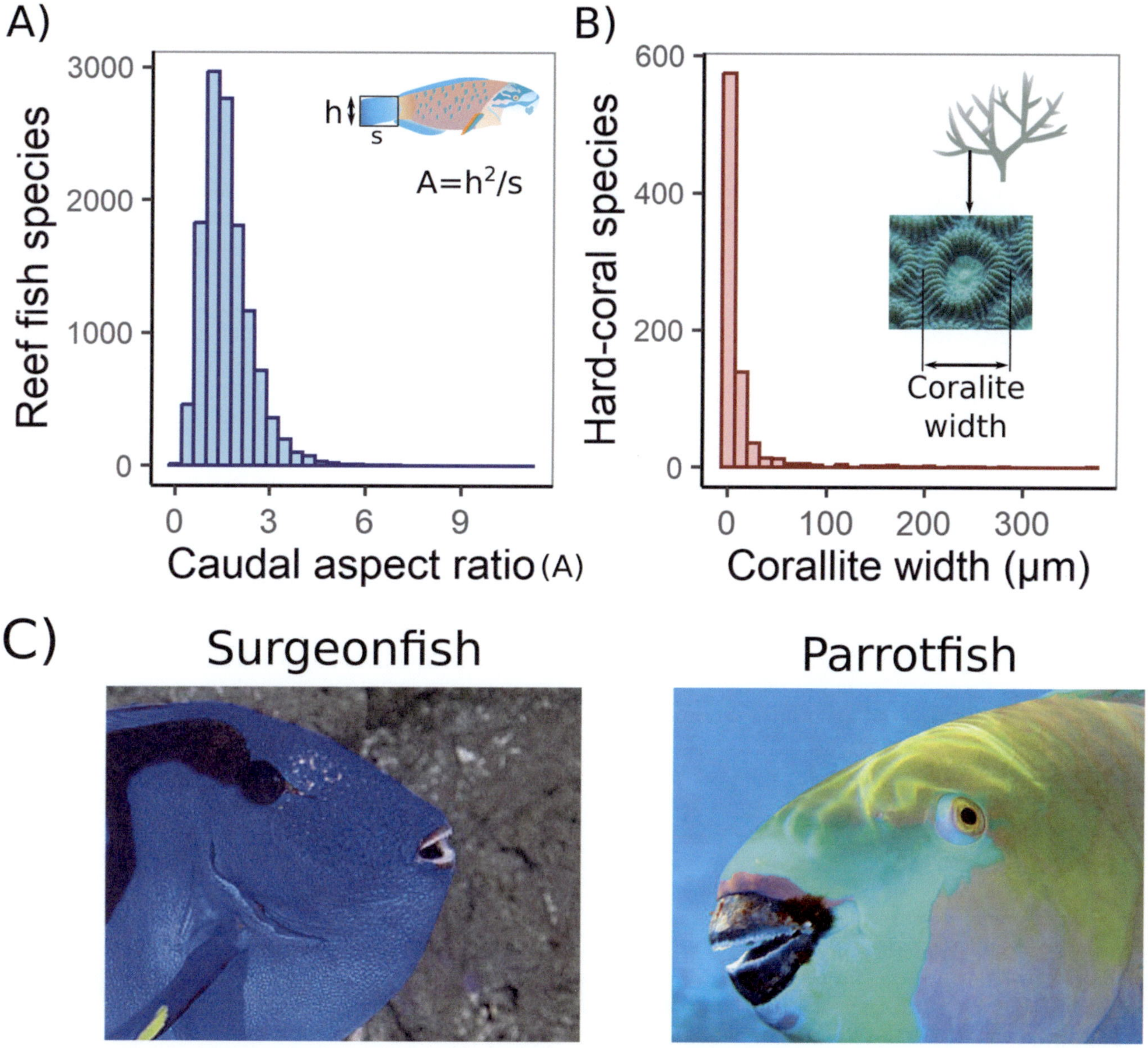

Fig. 10.10 Examples of organ diversity in coral reef species. (**a**) Variability in the shape of reef fish tails (caudal aspect ratio *A*) as an indicator of species levels of activity and swimming speeds (Pauly 1989). (**b**) Variability in the size of reef-building coral corallites. (**c**) Mouth specialization in herbivorous surgeonfish and microphage parrotfish. (Data from www.fishbase.se and McWilliam et al. (2018))

organs, although often without a specific focus on coral reef species Cernuda-Cernuda and García-Fernández 1996 Maruska 2001 Gregory 2008, Flammang 2014Fine and Parmentier 2015 Majoris et al. 2021. More focused are studies on species biological traits, which typically rely on organism- and organ-level characteristics for species identification and ecological comparisons (Fig. 10.10). Such information is recorded in reference databases such as https://coraltraits.org and www.fishbase.se. While analyses of organ variabilities are scarce, the most common metrics are found for organs used for ecological investigation such as fish otoliths and scales, mineral bodies of the inner ear and surface layer that are used for sclerochronologic estimations of organism age, and ontologic investigations of diet and life-cycle dynamics (Heather et al. 2018; Morat et al. 2020). Recent advancements in this realm show a similar use of the vertebrate eye lens to explore dietary and spatial variations throughout organisms' lifetime (Quaeck-Davies et al. 2018), an area of exploration also investigated in human dental cementum (Bertrand et al. 2022). Similar investigations are performed on marine plants and invertebrates, although using a different set of organs (Duarte et al. 1994; Rypel et al. 2008; Ouréns et al. 2013). In coral reefs, special attention has been given to the diversity of organs associated with key ecological functions for ecosystem health, such as the role of parrotfishes whose beaks specialize in particular in scraping algae with cyanobacteria, microorganisms, and invertebrate endosymbionts off substrates as compared to the mouths of surgeonfishes more adapted to grazing (Fig. 10.10c; Clements et al. 2017). Physiological and phylogenetic studies investigate the efficiency of such functions and their ecological role in coral reefs, with particular emphasis on the importance of diversity and redundancy of the functions associated with such organs for ecosystem resilience (Guillemot et al. 2011; Siqueira et al. 2019; Pimiento et al. 2020; Robinson et al. 2020).

10.3.6 Organisms

Organisms living in coral reefs vary in many aspects, including in their morphometric sizes and shapes (Fig. 10.11), and biological traits of colors and behaviors, resulting in a level of diversity and curiosity of life-forms that is unique to these ecosystems. Global databases describing these characteristics for reef-building corals and fish include https://coraltraits.org and www.fishbase.se.

Variability in the size of organisms relates to differences in species biological traits and also to characteristics intrinsic to individuals (e.g., age) and extrinsic features of their environments (e.g., temperature, food availability). Variability in size is often used as an ecological indicator of species health in given environments, given that body size is easier to measure than age, and is the result of a variety of past demographic processes, such as survival and growth, that are sensitive to environmental conditions, including natural disturbance and human stressors. Body size is also a good descriptor of current species performance capacities in demographic processes that determine the ecological success of populations into the future (Darling et al. 2012; Kayal et al. 2015b; Morais and Bellwood 2020), although other ecological descriptors such as age also play an important role (Rapuano et al. 2023). Typically, the largest individuals are found in the most suitable and less disturbed environments, reflecting species optimal environmental niches and success in ecosystem management (McCook et al. 2010; Cinner et al. 2018; Dietzel et al. 2020).

Body shape is often used for species identification and taxonomic studies on species diversification (Veron 2000; Ramírez-Portilla et al. 2022; Tea et al. 2022), sometimes in relation to species performances in key ecological processes such as fish swimming capacity or wave resistance and light absorption in corals, and the associated energetics (Fig. 10.11a, b; Walker et al. 2013; Madin et al. 2014; Rossi et al. 2018; Cresswell et al. 2020; Kramer et al. 2022; Carlot et al. 2023). A large body of work is particularly dedicated to relating the weight of individual organisms to their body length for estimations of population biomass, species stock assessments, and ecosystem productivity (Wolfe et al. 2020; Pacey et al. 2023; Tekwa et al. 2023). Using a large sample size of 53,800 specimens from 788 reef species from New Caledonia, Kulbicki et al. (2005) estimated variability in reef fish length-weight relationships to be 9.4% at the species level, 11.4% at the genus level, 13.5% at the family level, and 13.2% within morphological groups (Fig. 10.12b). Based on their analysis of reef fish morphologies, the authors concluded that diversity in fish shapes is constrained by physical limits. There are a dozen major morphological groups in corals and 30 of reef fishes (Fig. 10.11; Veron 2000; Kulbicki et al. 2005). Another major research path focuses on species ecological functions, notably for habitat-forming organisms like corals whose morphologies determine the abundance and diversity of species hosted in their interstices (Darling et al. 2017; Richardson et al. 2017b; Aston et al. 2022).

Diversity in colors is one of the most striking and unique features of (healthy) coral reefs, investigated predominantly in studies on ecosystem health and aesthetics values (Haas et al. 2015; Langlois et al. 2022). Other studies explore the evolutionary and molecular determinants leading to the diversity of colors found on coral reefs, sometimes in relation to jewelry industries, and how some species use colors for communication and camouflage (Marshall 2000; Lemer et al. 2015; Tea et al. 2022). Sometimes, striking colors are indicative of stress, as observed during coral bleaching events (Bollati et al. 2020). However, the mechanisms behind, and consequences of, diversity in colors remain largely to be explored (Hemingson et al. 2019; Hodge et al. 2020).

Relatively little knowledge is available about the behaviors of individual organisms, how they adapt to local environments, and change with human influences. Most studies focus on species of interest for their ecological or societal role. A relatively large portion of studies on marine organism behavior focus on large and charismatic animals which are easier to track and are often a center of touristic attraction and, in particular sharks which can also cause human casualties (Juhel et al. 2019; Taglioni et al. 2019; Lassauce et al. 2020; Maillaud et al. 2022; Riley et al. 2022; Séguigne et al. 2023). The studies also explore habitat use, feeding and mating behaviors, and social interactions of the organisms (Papastamatiou et al. 2006; Mourier et al. 2012; Mourier et al. 2013; Mourier et al. 2016; Guillaume and Séret 2021; Lassauce et al. 2022; Lassauce et al. 2023), as well as behavioral sensitivity to human stressors (Holles et al. 2013; Gil et al. 2020; Clark et al. 2020; Udyawer et al. 2021). Another center of interest has been the coral predatory seastar, crown-of-thorns starfish (*Acanthaster* spp.), whose population outbreaks are major causes of coral mortality throughout the Indo-Pacific (Pratchett et al. 2014). The studies have particularly focused on movement and feeding behaviors of the sea-

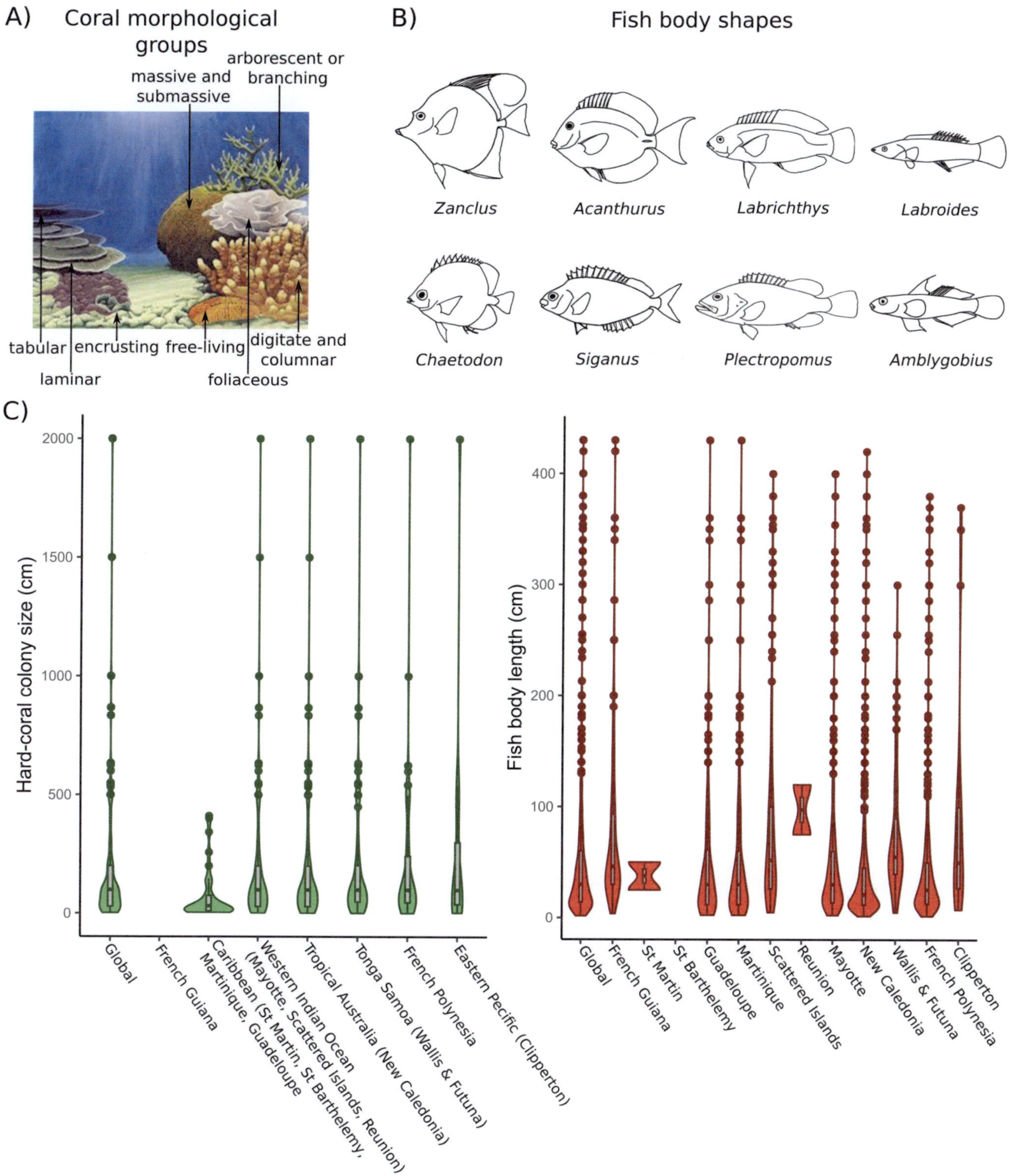

Fig. 10.11 Variability in coral and fish shapes and sizes. Major morphological groups in corals (**a**, adapted from Veron (2000) and fish (**b**, adapted from Walker et al. (2013). Examples of size distribution (whiskers plots presenting median, first and third quartiles, and outliers; violin plots presenting data densities) in corals and fish among French tropical overseas (**c**, data extracted from McWilliam et al. (2018) for corals and from www.fishbase.se for fish)

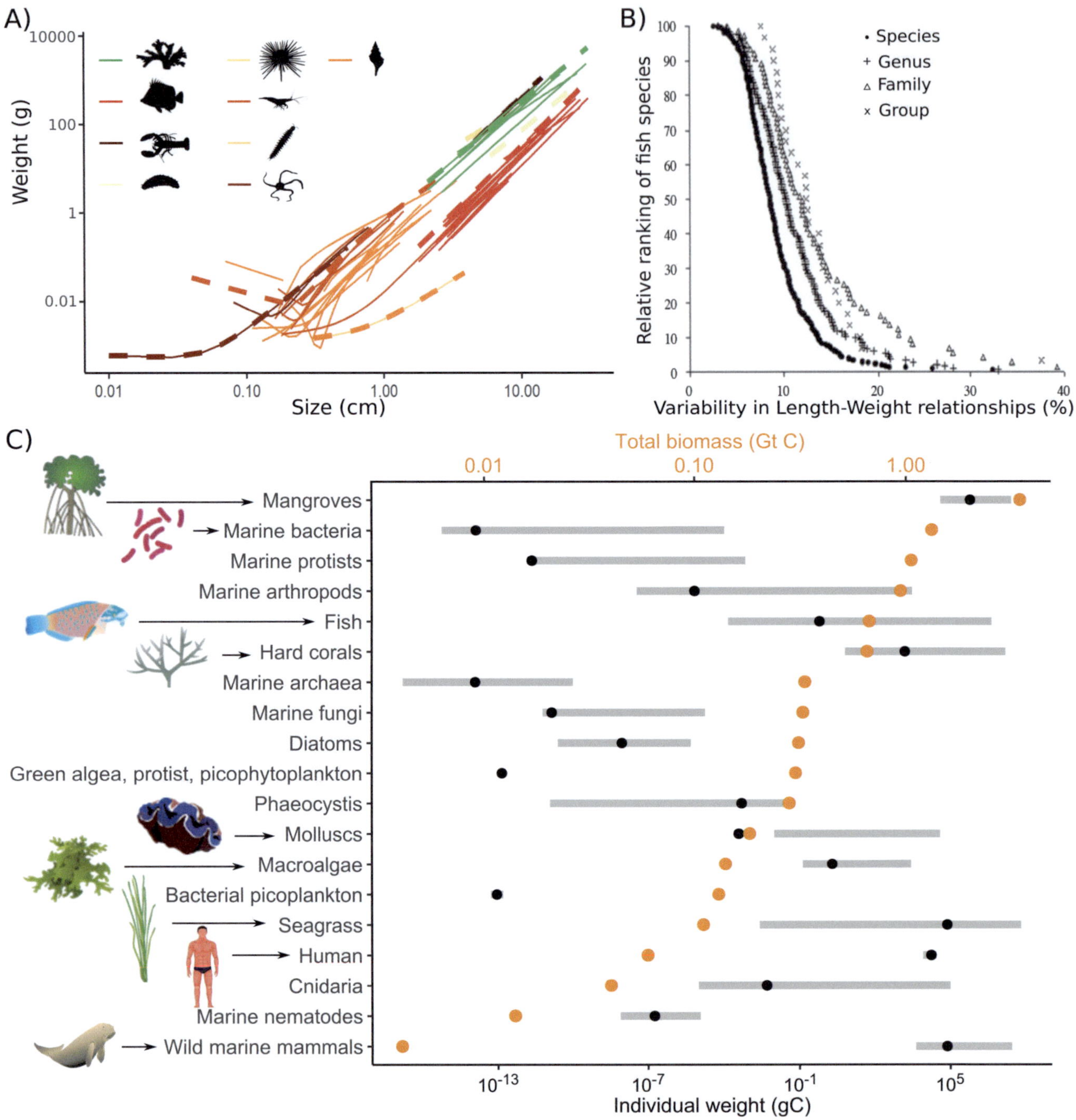

Fig. 10.12 Length-weight relationships (**a**) of various coral reef organisms by species (solid lines) and species groups (dashed lines), variability in length-weight relationships among species, genera, families, and morphological groups of coral reef fish in New Caledonia (**b**, adapted from Kulbicki et al. 2005), and individual body weight (range and median) and global biomass of various marine species as compared to humans (**c**). Length-weight relationship data in (A) from Akiona et al. (2022) for fish, Pacey et al. (2023) for corals; Wolfe et al. (2020) for Annelida, Mollusca, Arthropoda, and Echinodermata; Lee et al. (2018) for Holothuria; Rahman et al. (2012) for urchins; Aisyah and Triharyuni (2017) for lobsters. Individual body weight and global biomass data in (**c**) from West et al. (2009), Bar-On et al. (2018), Tekwa et al. (2023)

star as key drivers of coral decline De'ath and Moran 1998; Pratchett 2001; Kayal et al. 2011, 2012b, 2017a, b; Petie et al. 2016; Pratchett et al. 2020) and the identification of its natural predators, which can help mitigate the outbreaks (Bos et al. 2008; Cowan et al. 2017; Kayal and Lenihan 2025), recently with the help of DNA technology (Kroon et al. 2020). Other studies investigated predator-prey, competitive, and mutualistic behavior in different reef organisms (Pratchett 2005; Mills and Côté 2010; Leray et al. 2013; Kayal et al. 2017a; Kayal and Adjeroud 2022). Facing widespread coral reef decline, a large body of work on species behaviors is dedicated to finding solutions for mitigating ecosystem vulnerability to disturbance or boosting recovery through active restoration (Kayal et al. 2011; McKeon and Moore 2014; Kayal and Kayal 2017; Gordon et al. 2019). Sometimes, original studies emerge from unplanned, opportunistic observations of species unusual behaviors, such as bleached corals compensating energy loss by feeding on neighboring macroalgae, and more recently on internal symbiotic microalgae, potentially leading to new investigation channels (Marhaver 2011; Wiedenmann et al. 2023). While the level of plasticity of species behaviors in the face of environmental changes remains largely unquantified, the development and democratization of imagery tools such as underwater cameras and drones have greatly increased research capacity in this realm by enabling increased observations of organisms in their natural environments (Cleguer et al. 2021; Heudier et al. 2023). Further developments in artificial intelligence enable more effective analysis of ecosystem imagery and acoustic recordings for characterizing species attributes and behaviors under various natural and human conditions (González-Rivero et al. 2020; Jamil et al. 2021; Ditria et al. 2022; Ouassine et al. 2024).

10.3.7 Populations

Population abundance estimates are commonly used for stock assessments and evaluations of ecosystem health. Common metrics include population size and distribution range, density, biomass, percent-cover, and size-structure (Fig. 10.13; Adjeroud et al. 2018a, b; Dietzel et al. 2020, 2021; Pacey et al. 2023; Tekwa et al. 2023). In coral reefs as in other ecosystems, many species are often observed at population abundances and size-structures that are near their lower bound (Figs. 10.11c and 10.13), while knowledge on ultimate demographic states achievable by populations and the maximum carrying capacity of ecosystems remains limited (Levin 2000). Nevertheless, comparisons between healthy and altered ecosystem states provide insights on near optimal population abundance, distribution, and size-structure, and the degree of reduction imposed by various stressors (McCook et al. 2010; Cinner et al. 2018; Darling et al. 2019; Morais et al. 2021). Amidst rapid environmental change, key research questions focus on population resilience and the capacity of species to maintain their abundance through the demographic processes of survival, growth, reproduction, and migration (Darling et al. 2012; Mumby et al. 2014; Kayal et al. 2015b; Morais and Bellwood 2020). Global databases with information on these processes include https://coraltraits.org and www.fishbase.se. Demographic investigations use empirical measurements of species performances in these processes (Fig. 10.14) to evaluate the diversity of life-history strategies and identify, through quantitative models, the mechanisms leading to population bottlenecks and species ecological success and coexistence in various environments (Madin et al. 2014; Bozec et al. 2016; Kayal et al. 2018; Riegl et al. 2018; Darling et al. 2019; Carturan et al. 2020). While such investigations depend on fine-scale empirical data on species dynamics, data hitherto limited to few highly studied coral reef sites, current developments in imagery approaches are opening paths to their democratization (González-Rivero et al. 2020; Collin et al. 2021; Jamil et al. 2021; Heudier et al. 2023; Kayal et al. 2023). Increased data availability along with amplified computer power has opened paths to pluridisciplinary quantitative investigations for a better understanding of ecosystem processes across biological, ecological, and biogeoclimatic scales (Jones et al. 2015; Condie et al. 2021; Van Woesik et al. 2022; Carlot et al. 2023; Donovan et al. 2023).

As an alternative to comprehensive approaches designed to integrating a multitude of ecological processes driving species dynamics, other investigation routes focus on single indicator processes serving as ecological metrics of species performance and resiliency such as recruitment rates, competitive wars, or recovery capacities (Done et al. 2010; Penin et al. 2010; Osborne et al. 2017; Vercelloni et al. 2019; Kayal and Adjeroud 2022). In particular, coral recruitment is a commonly used, integrated indicator of coral reef resilience capacity, as reflecting both adult coral abundance and reproductive success, and larval dispersal, settlement, and metamorphosis into reef habitats (Adjeroud et al. 2017). It is monitored in different regions using a standardized method quantifying annual settlement rates of coral larvae on artificial substrates (Mundy 2000). Regional comparisons of coral recruitment rates indicate differences in magnitude, taxonomic composition, and seasonality, with maxima observed at mid-latitudinal range in the southwest Pacific, central Great Barrier Reef and southern New Caledonia, where coral communities are dominated by spawning acroporids, with a shift to brooding pocilloporids and spawning poritids in other places (Fig. 10.14e; Hughes et al. 2002a, b; Adjeroud et al. 2017). Quantitative understanding of the coral recruitment process is still nascent but increasing, thanks to dedicated scientific attention (Hughes et al. 2000; Kayal et al.

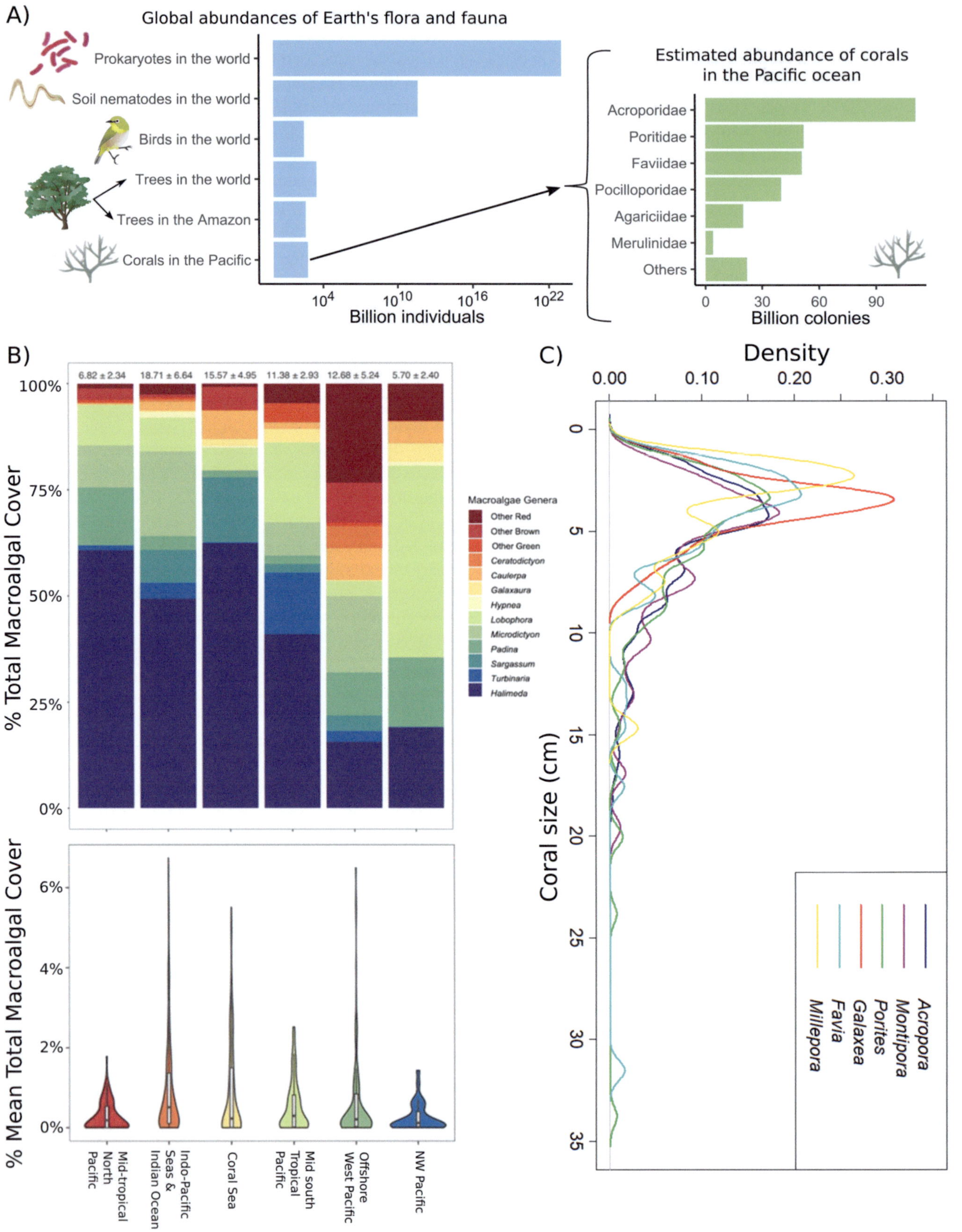

Fig. 10.13 Examples of population state metrics for key coral reef species. Estimated abundance of Pacific reef-building corals as compared to other widespread species (**a**, data from Dietzel et al. 2021). Macroalgae cover in the Indo-Pacific (**b**, adapted from Cannon et al. 2023). Coral size distribution (**c**, adapted from Kayal et al. 2023)

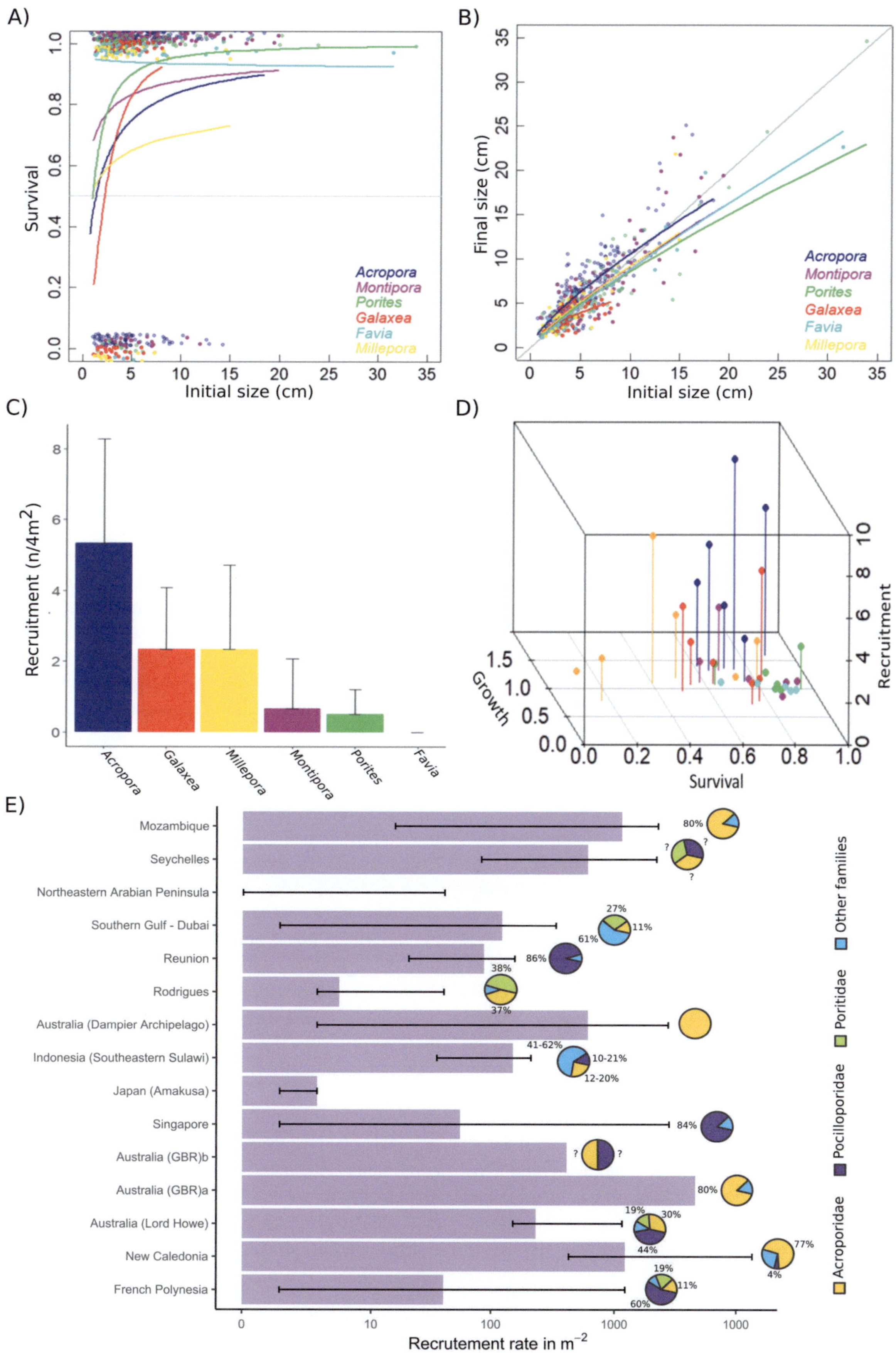

Fig. 10.14 Empirical metrics of coral demographic rates in annual survival (**a**), growth (**b**), and recruitment (**c**), with a visualization of the three processes combined (**d**, adapted from Kayal et al. 2023), and coral settlement rates on artificial substrates in different regions (**e**, data from Adjeroud et al. (2022))

2015b; Bramanti and Edmunds 2016; Doropoulos et al. 2016; Leinbach et al. 2021; Doropoulos et al. 2022). The rise of genetic tools and oceanographic modeling of currents enable estimations of population connectivity in various coral reef species, which exhibit various larval durations and dispersal capabilities (Fig. 10.15; Lester and Ruttenberg 2005; Shanks 2009; Cuif et al. 2014; Cravatte et al. 2015; Edmunds et al. 2016; Hock et al. 2017; Dubé et al. 2020; Lesturgie et al. 2023). Overall, a variety of approaches are deployed to investigate the importance of larval recruitment in the replenishment of coral reef species populations and ecosystem functioning, with direct benefits to biodiversity management and spatial conservation planning (Kayal et al. 2018; Riegl et al. 2018; Brandl et al. 2019b; Manel et al. 2019).

10.3.8 Communities

The number of different species composing a biological community is the most historical and common metric of biodiversity (historically "biological diversity"; Harris 1916). Studies at this scale of observation can be distinguished into two major groups, those focused on taxonomic diversity with the end goal of characterizing community composition and those with an ecological endeavor evaluating changes in community composition and their consequences for ecosystems.

Coral reefs are estimated to host over 2 billion species, including ~830,000 (confidence interval 550,000–1,300,000) multicellular taxa, that is, ~38% of all described marine species (Fig. 10.16; Fisher et al. 2015). Taxonomic work is active in describing new species. Yet large uncertainties surround existing taxonomic delineations in many groups including corals, and species classifications evolve rapidly, particularly through the use of genomics, and many species names are expected to change over time (e.g., Benzoni et al. 2010; Huang et al. 2014a, b; Nakajima et al. 2017; Tea et al. 2022). Up-to-date taxonomic reference databases, such as TAXREF from the Inventaire National du Patrimoine Naturel (INPN, https://inpn.mnhn.fr, Gargominy et al. 2022) and the World Register of Marine Species (WoRMS, www.marinespecies.org) are essential for ensuring coherence in species designations. Current descriptions distinguish 1681 species of hard corals including 995 associated with coral reefs, 19,772 fish, 55,238 crustaceans, 50,877 mollusks, and 3760 hydrozoans (see Fig. 10.16a, b for all groups). Overall, species descriptions and therefore biodiversity estimates are more complete for large species in shallow habitats such as megafauna which are easier to observe and tend to decrease for smaller and more cryptic habitats and species, including sponges, although genetic meta-barcoding and bioacoustic approaches increasingly enable detection (Elise et al. 2019; Nichols et al. 2022; Wolfe et al. 2023; Galand et al. 2023). Biogeographic studies on species distributions assess spatial variations in species diversity and the mechanisms behind speciation and extinctions (Cowman and Bellwood 2013; Clements et al. 2017; Tea et al. 2022; Cantalice et al. 2022). Regional comparisons identify a pronounced gradient in species diversity with the spatial extent of coral reef systems (Bellwood et al. 2005) as well as their geographic distance to the center of marine biodiversity situated in Southeast Asia (Hughes et al. 2002a; Dietzel et al. 2021; Lutzenkirchen et al. 2023) and to the equator (Jones et al. 2019), features that are reflected across French overseas (Figs. 10.17, 10.18 and 10.19). These patterns concord with our general understanding that the level of biodiversity of a system is a function of the extent and diversity of ecological niches available, which relates to spatial as well as temporal heterogeneity (e.g., increased seasonality at higher latitudes promoting species coexistence through competitive reversals; Logan 1984; Kayal and Adjeroud 2022), and the variety of organisms able to occupy these niches. Nevertheless, current estimates of species richness should be considered with caution for certain species groups, since estimation methods and sampling efforts may differ among regions and result in potential bias (e.g., differences in coral diversity estimates in Fig. 10.16c). For instance, both molecular and morphological approaches are used to assess species diversity, which often results in contrasting outcomes, and a recent extrapolation-based study accounting for sampling effort suggests that coral diversity may in fact be higher in the western Indian ocean than in the coral triangle (Kusumoto et al. 2020). Overall, how species richness and other biodiversity metrics vary with spatial scales or other units of sampling effort remains poorly understood (Nash et al. 2014; Chase et al. 2018; Donovan et al. 2023).

A large body of work exists on using the dynamics of key communities of ecological or social importance, such as corals and fish, to assess coral reef ecosystem health and associated management strategies (Lamy et al. 2016; Kayal et al. 2018; Adjeroud et al. 2018a; Vercelloni et al. 2019; Carturan et al. 2020; Graham et al. 2020). Some studies investigate more specifically the ecological functions associated with different communities, such as primary production, herbivory, nutrient transport and recycling, and habitat formation or alteration (Brandl et al. 2019a; Pimiento et al. 2020), and the ecological processes through which changes in one community spread to other ecosystem components (Leray et al. 2012; Kayal et al. 2012b; Condie et al. 2018; Dubois et al. 2019). Nevertheless, the complexity of biotic interactions, including predator-prey, competition versus mutualism, symbiosis versus parasitism, and how they are modulated by environmental conditions remains largely unexplored, particularly in biodiverse and ecologically oscillating ecosystems such as coral reefs (Connell 1978; Schmitt et al.

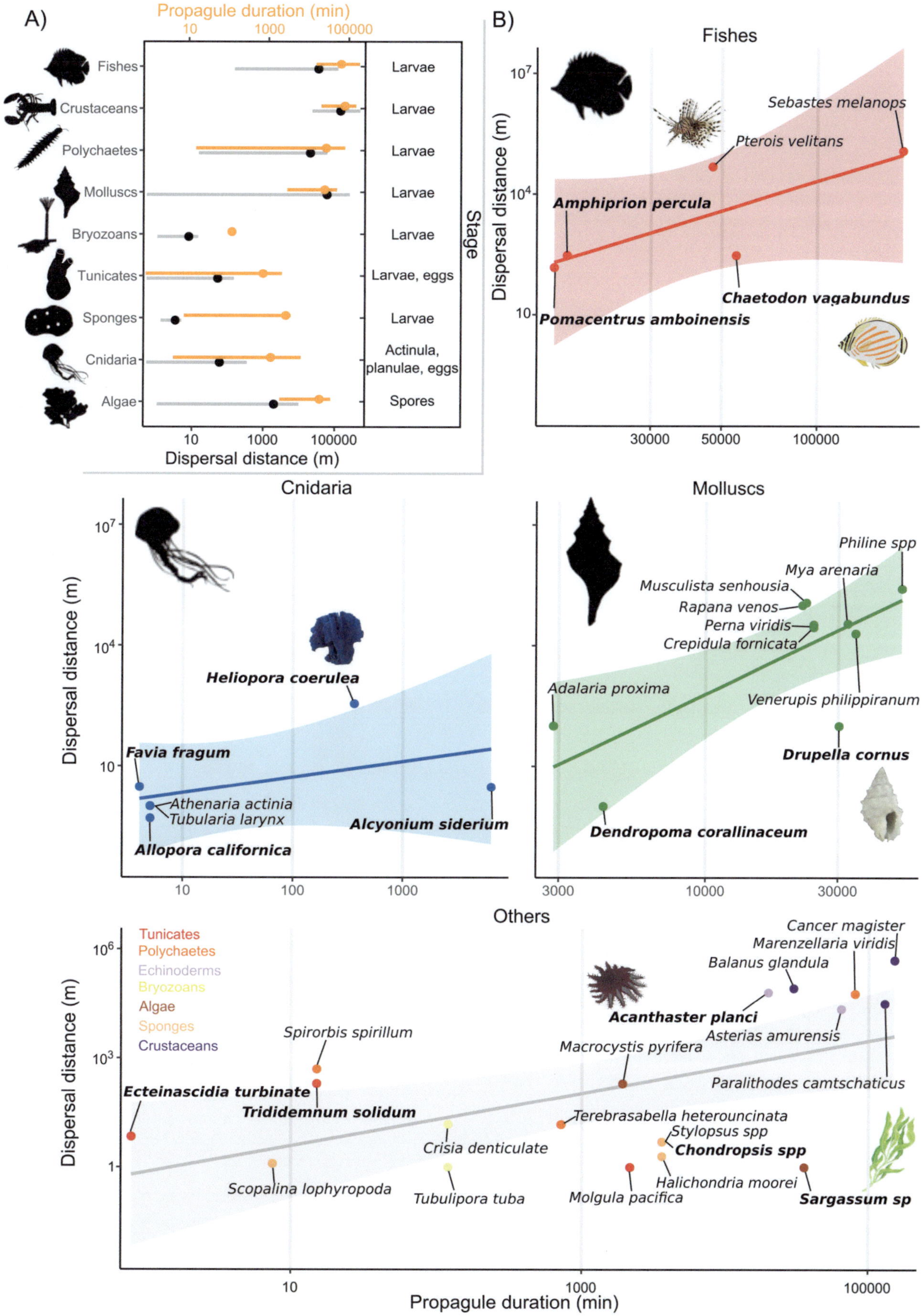

Fig. 10.15 Dispersal capabilities of various marine organisms with characterized propagule duration and dispersal range (not restricted to coral reef species, data from Shanks 2009). (**a**) Dispersal stage, distance, and duration (mean ± range) for different taxonomic groups and (**b**) relationships between propagule dispersal and duration. Coral reef species are identified in bold

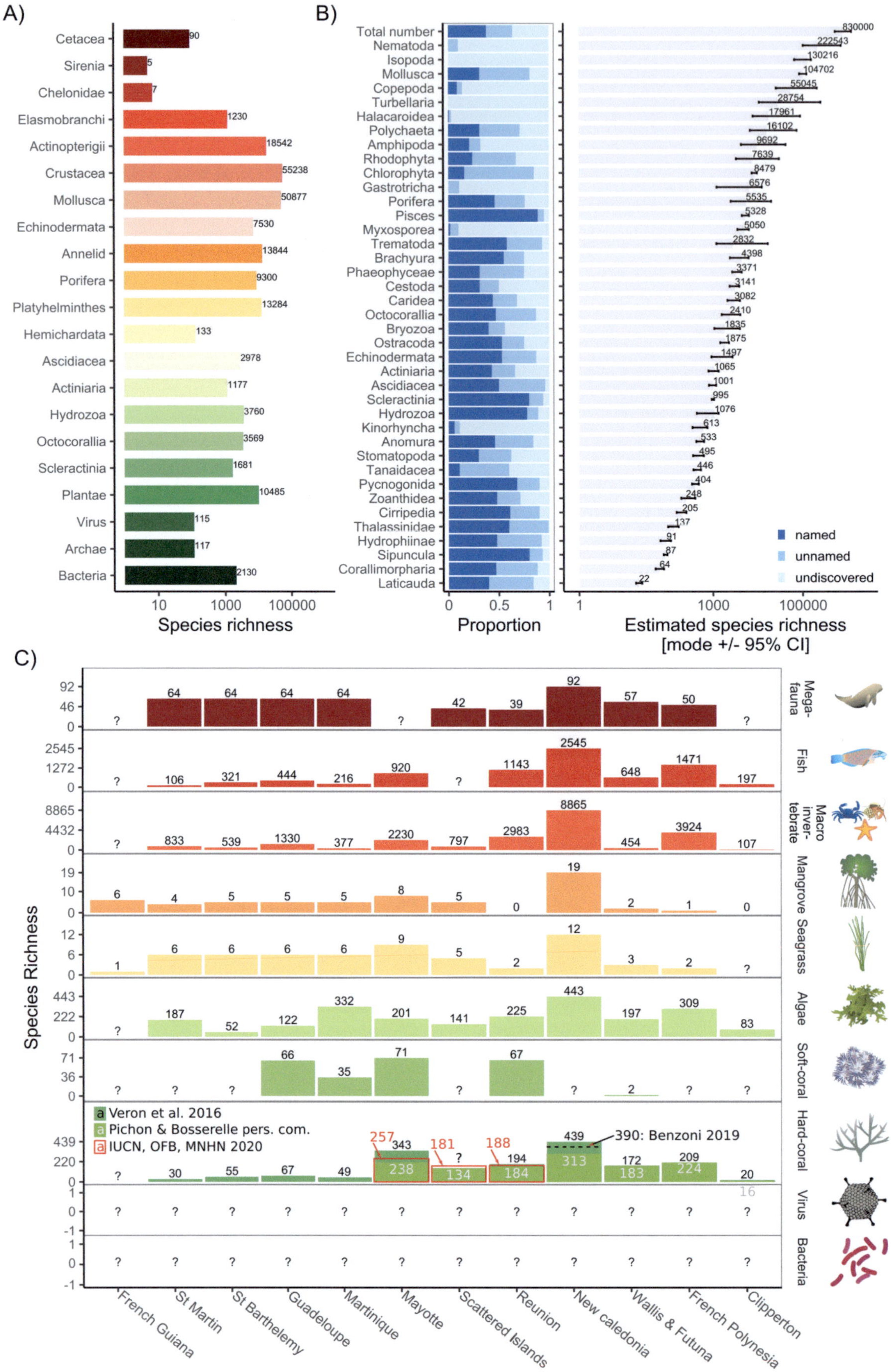

Fig. 10.16 Species diversity of major coral reef communities and associated lagoonal habitats. (**a**) Number of all named marine species (not restricted to coral reef-associated species) per taxonomic group (data from the World Register of Marine Species). (**b**) Estimated species richness

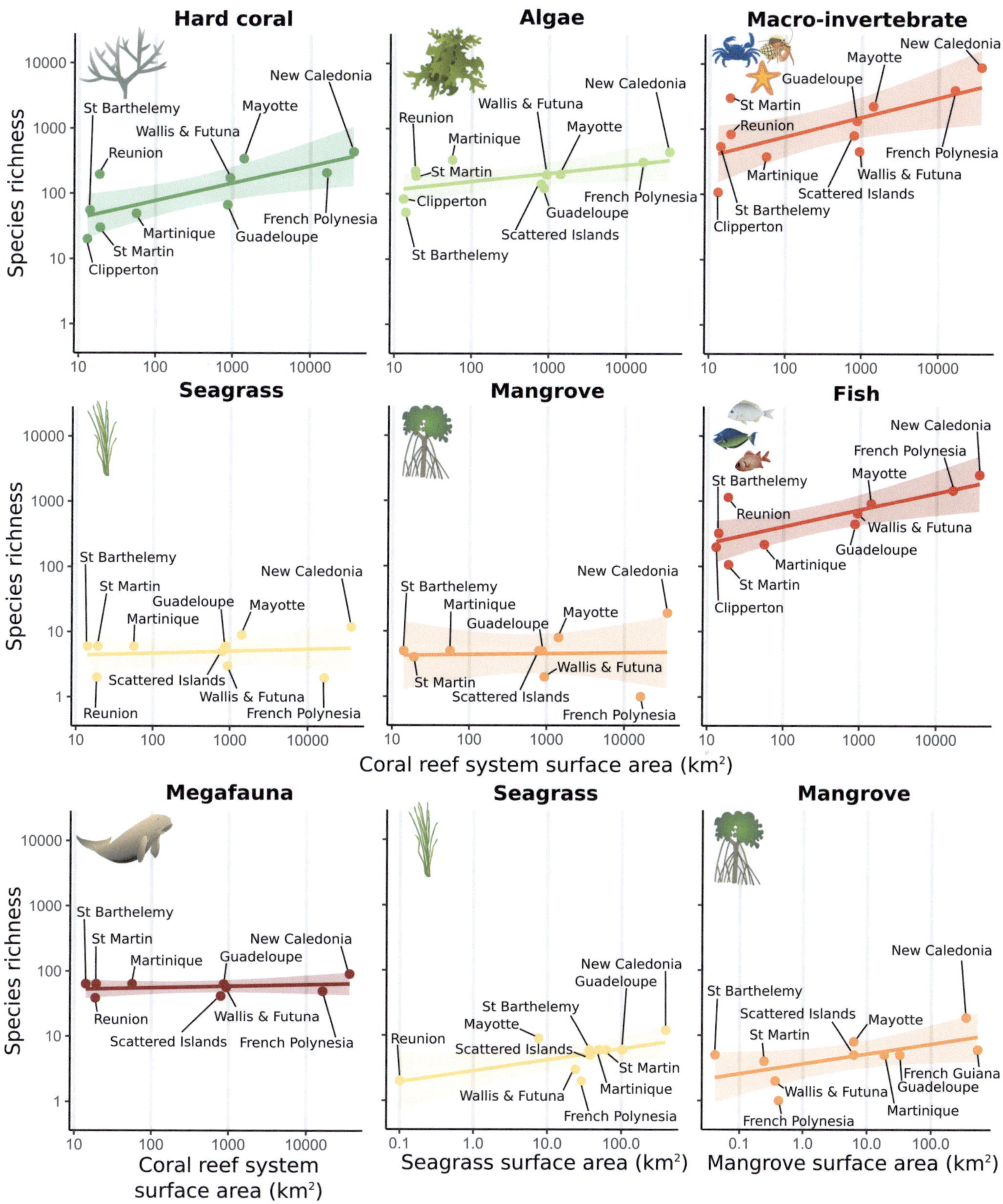

Fig. 10.17 Species diversity of major coral reef ecological communities as a function of ecosystem surface area (coral reef system surface area includes constructed hard-substrate reef and associated lagoonal habitats, c.f. Spalding et al. 2001)

Fig. 10.16 (continued) of coral reef-associated multicellular communities. (Adapted from Fisher et al. 2015). (**c**) Species richness for key coral reef communities in French overseas. For reef-building corals, estimates from different sources are provided. Data from Pichon and Bosserelle (pers. com.), Benzoni (2019) and IUCN comité français et al. (2020) result from a more in-depth analysis of sample specimens from the area. Megafauna comprises large marine species groups: sea turtles, dugongs, dolphins, whales, sharks and rays, and sea snakes

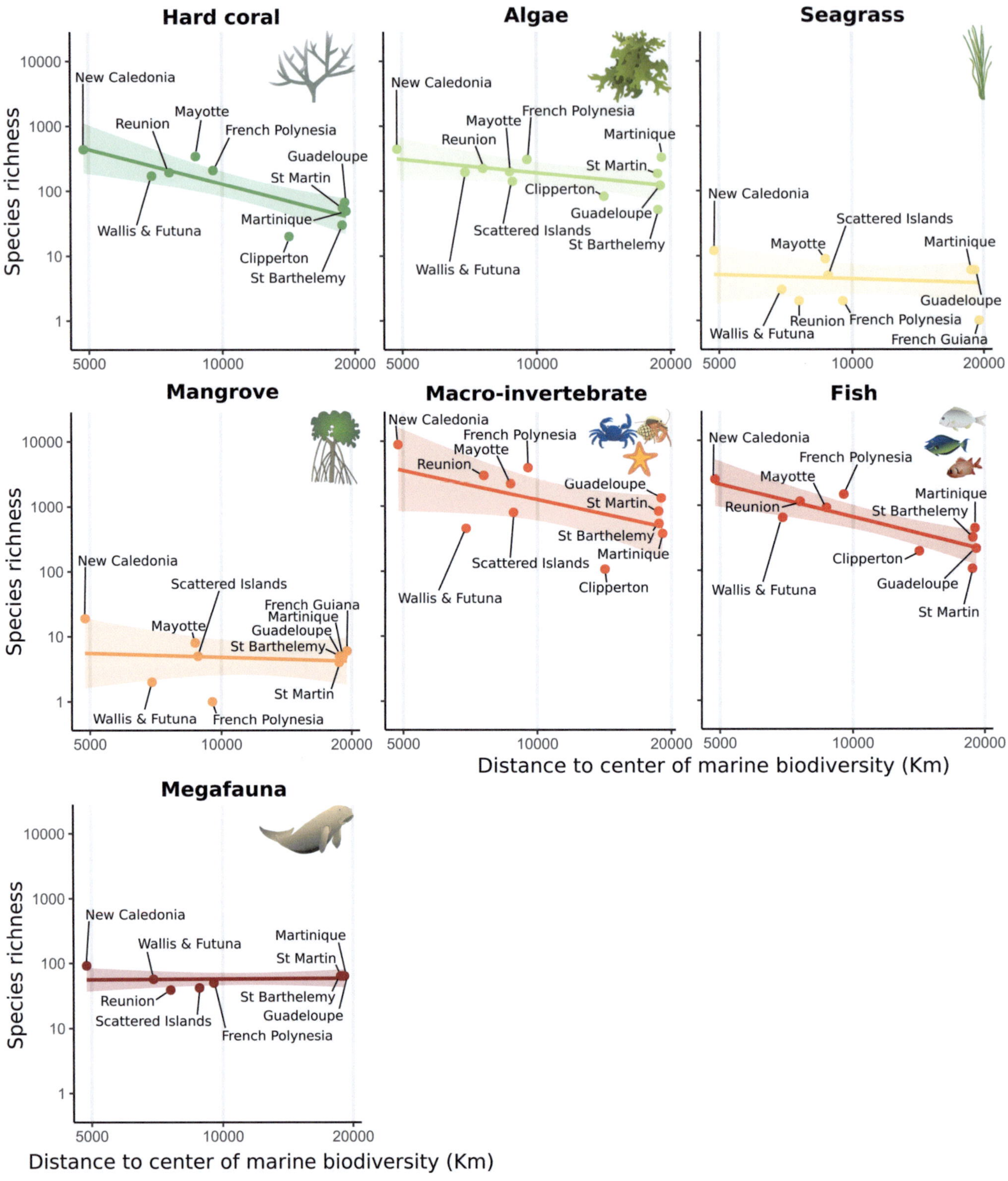

Fig. 10.18 Species richness of major coral reef ecological communities as a function of distance to the center of marine biodiversity (reference point set as western Banda Sea, latitude –5.461885, longitude 122.936220)

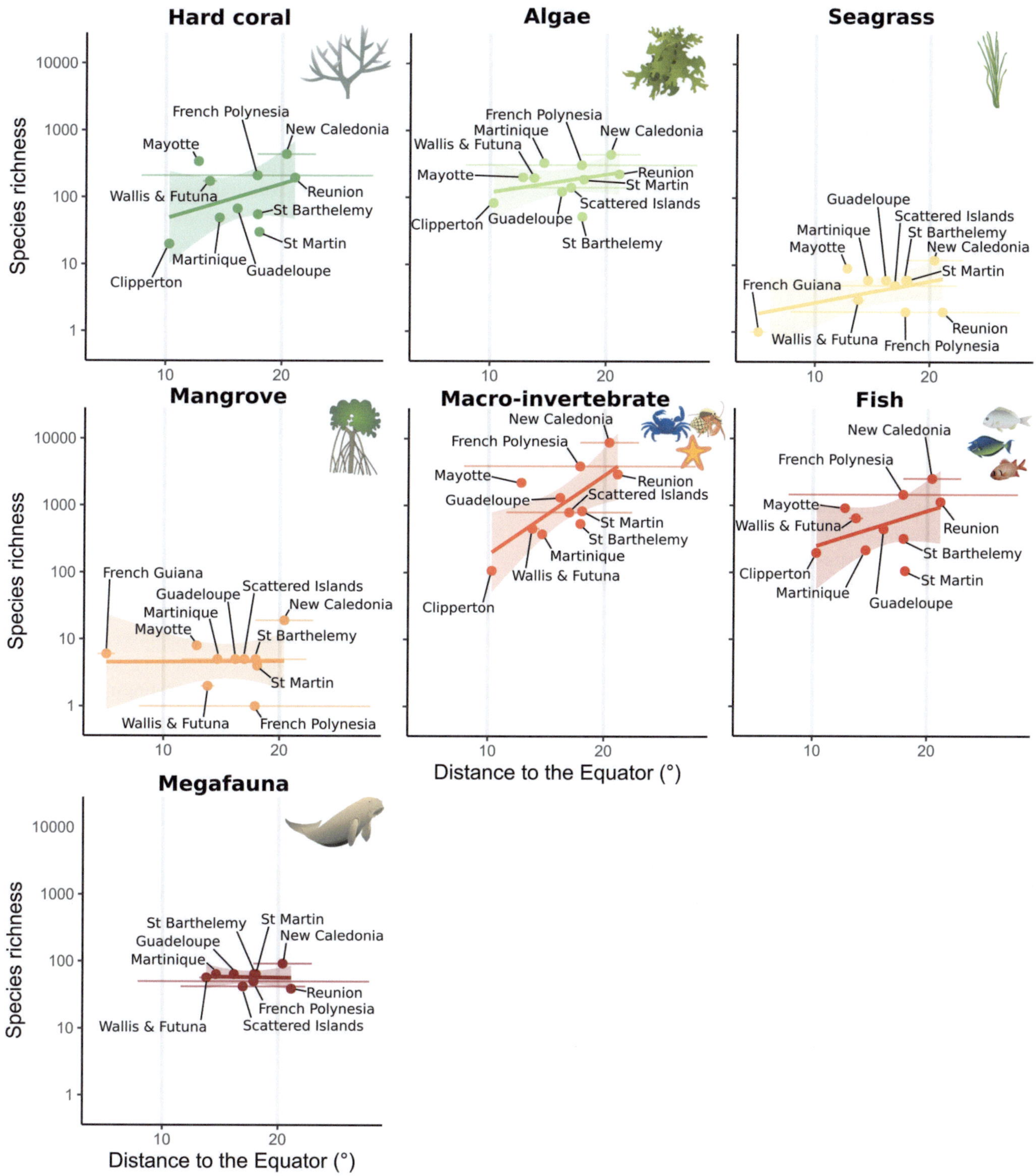

Fig. 10.19 Species richness of major coral reef ecological communities as a function of distance to the Equator (absolute average latitude ±range, in degree)

2009; Mills and Côté 2010; Kayal et al. 2011; McKeon and Moore 2014; Kayal et al. 2017a; Kayal and Adjeroud 2022). It is only recently that fine-scale ecosystem time-series data have enabled empirical investigations on major ecological processes driving ecosystem resilience (Mumby et al. 2007; Riegl et al. 2018; Cresswell et al. 2023). To strengthen coral reef conservation actions, more knowledge is needed on the ecological processes driving community dynamics, including on the less emblematic and cryptic communities such as microorganisms, algae, and sponges that are understudied (de Goeij et al. 2013; Brandl et al. 2019b; Cannon et al. 2023; Galand et al. 2023).

10.3.9 Ecosystems

Coral reef ecosystems around the world differ in size, composition, health, and dynamics, resulting in a myriad of ecosystem states with unique ecological features. However, some key ecosystem characteristics enable differentiating coral reef systems into broad functional categories along ecological gradients. At a broad consideration, the size and degree of isolation have strong consequences on the structure and fate of coral reef ecosystems. On one side of the spectrum, large continuous systems in the southwestern Pacific, such as the Australian Great Barrier Reef with 348,700 km^2 of reef habitats and New Caledonia's reef system with 35,873 km^2 contrast with small and isolated systems on the other side of the spectrum, such as the atoll of Clipperton with its 13 km^2 of reefs located approximately 950 km from the closest coral reef system (Romero-Torres et al. 2018; Figs. 10.4 and 10.20). Within this spectrum, many systems of various sizes are found with differing degrees of geographic fragmentation and ecological isolation, sometimes due to the presence of natural barriers to dispersal (Raitsos et al. 2017; Hock et al. 2017; Romero-Torres et al. 2018). For example, the 19 km^2 of coral reefs surrounding the single island of Reunion in the Indian Ocean are situated roughly 200 km from the closest reef system of Mauritius and 700 km from Madagascar, whereas the 16,200 km^2 of reefs in French Polynesia are scattered among 5 archipelagoes, 118 islands, and atolls (1576 land units including islets, Fig. 10.20b) across 2,500,000 km^2 of ocean in the Central South Pacific. These broad geomorphological characteristics of reef systems are relatively well-known for shallow habitats in clear waters, thanks in big part to satellite imagery, but knowledge decreases rapidly with water depth and turbidity as evidenced by the Great Amazon Reef that was only recently discovered under the Amazonian river plume and extends ~1000 km, including into French Guiana's coastal waters (Moura et al. 2016; Francini-Filho et al. 2018). The size and degree of isolation of the reefs determine the level of chemical and biological connectivity between ecosystems and therefore the productivity, regeneration capacity, and resilience of the diverse species populations they host (Edmunds et al. 2016; Gounand et al. 2018). In the face of disturbances, small and isolated coral reef systems are therefore generally more vulnerable to full-system collapse and exhibit limited recovery capacities, while for larger interconnected coral reef systems, disturbance impacts are often restricted to small portions, resulting in higher rates of demographic replenishment from adjacent unaffected populations (Wolff et al. 2016; Riegl et al. 2018; Vercelloni et al. 2019; Adjeroud et al. 2022). Larger reef systems also host higher proportions of biodiversity, thanks to a wider geographical expansion and the amount and diversity of ecological habitats and niches available (Eddy et al. 2021), as illustrated with New Caledonia's large ecosystem surface areas and associated biodiversity metrics (Fig. 10.17). Meta-ecosystem approaches characterizing chemical and demographic flow between ecosystems are still nascent, and their development should deepen our understanding of ecological drivers of ecosystem productivity and resilience (Gounand et al. 2018; Condie et al. 2021).

Ecosystem productivity and trophic organization are other key attributes characterizing different states in coral reef ecosystems. However, comprehensive ecosystem-level evaluations remain scarce, as most ecological investigations are conducted at the population or community scale, seldom integrating the various communities constituting coral reef ecosystems (Fig. 10.2; Rogers et al. 2015; Condie et al. 2018; Dubois et al. 2019). Nevertheless, simplified evaluations based on the representation of a limited number of ecological components or functions capture key species interactions such as habitat provision, predation, and competition, therefore characterizing major ecosystem dynamics (Kayal et al. 2012b; Mourier et al. 2016; Brandl et al. 2019a; Pimiento et al. 2020; Adam et al. 2021; Wolfe et al. 2023). Overall, the studies indicate that the magnitude of ecosystem biomass and its distribution among different ecological compartments vary in space and time as a function of the carrying capacity of the systems, the length and complexity of the food webs, the species richness of the different communities, and their dynamics in relation to the natural and human environments (Trebilco et al. 2013; Jacquet et al. 2020; Nagelkerken et al. 2020; Mehner et al. 2022). As such, healthy coral reef ecosystems situated in protected areas near the center of marine biodiversity tend to be characterized by more developed food webs and higher overall productivity, thanks to a more effective exploitation of ecological niches (McCook et al. 2010; Mellin et al. 2016; Morais et al. 2021; Heudier et al. 2023). In contrast, disturbed, degraded, and less diverse coral reefs tend to show shorter food webs and lower ecosystem productivity due to a limited exploitation of ecological niches, mortality events, and eventually the collapse of most vulnerable species populations (D'agata et al.

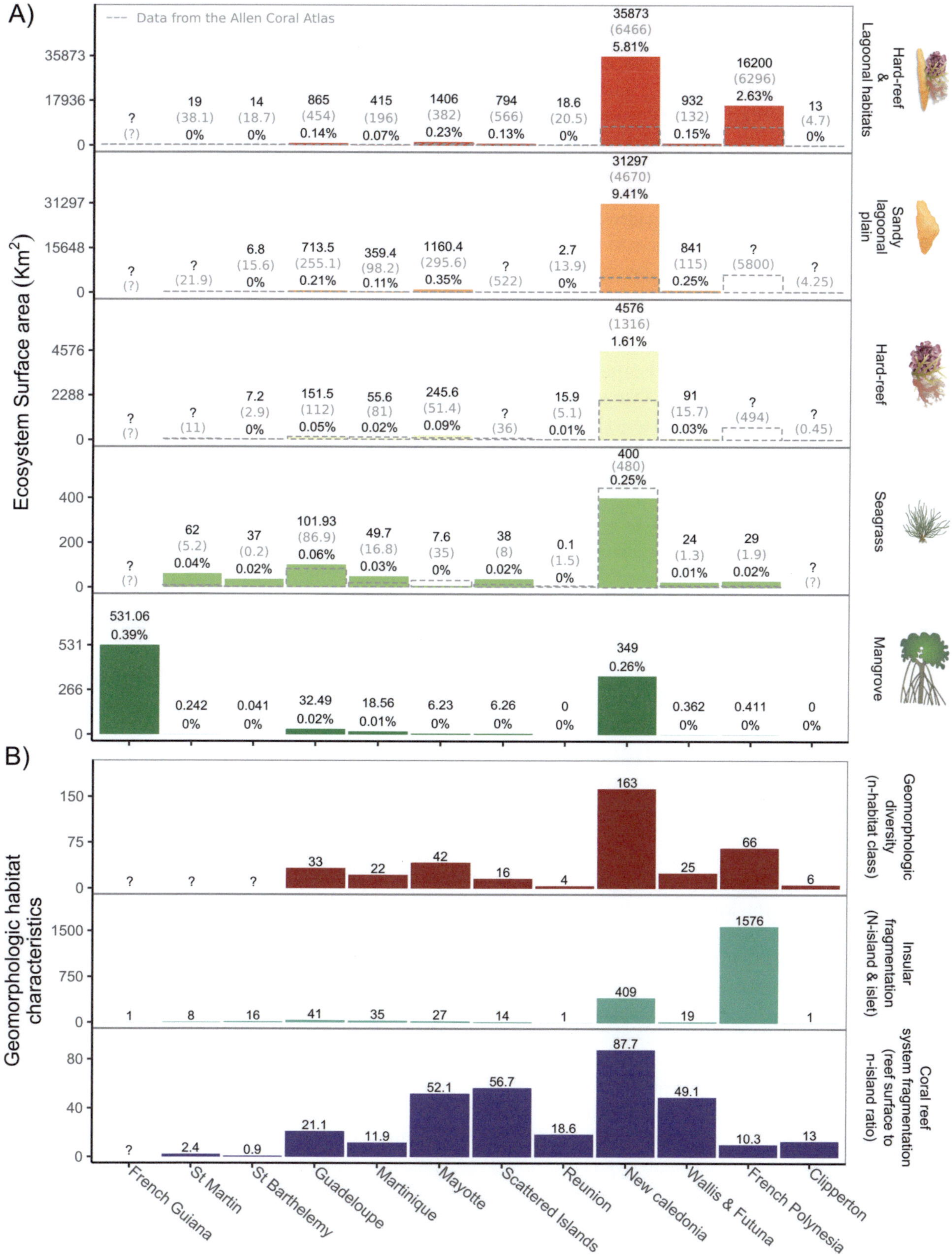

Fig. 10.20 Characteristics of coral reefs and associated ecosystems across French tropical overseas. Ecosystem surface areas and the percentage of the global values they represent are illustrated (data from and Spalding et al. 2001 IFRECOR 2021, as well as https://allencoralatlas.org for comparison), along with the diversity in reef geomorphologic habitats (number of level 5 teledetection classification out of a total of 800 coral reef habitat geomorphologies found on Earth, data from Andréfouët et al. (2008) and system fragmentation indices

2016; Cinner et al. 2018; Darling et al. 2019; Duprey et al. 2020; Eddy et al. 2021; Zhang et al. 2023). Coral decline, nutrient discharge, and the collapse of herbivore populations have led to drastic ecosystem upheavals, including shifts in dominance from corals to algae in many places (McManus and Polsenberg 2004; Schmitt et al. 2019; Morais et al. 2020; Adam et al. 2021; Reverter et al. 2022). While the ecological mechanisms controlling ecosystem resistance and resilience to such abrupt transition are generally understood, ecosystem vulnerability to various forms of stressors acting in isolation or in concert remains largely to be quantified. The development of new methods for monitoring coral reef ecosystems such as bioacoustics (Elise et al. 2019; Dimoff et al. 2021), remote-sensing (Hedley et al. 2016), and eDNA (Alexander et al. 2020; West et al. 2020), boosted by machine learning algorithms (Burns et al. 2022; Ditria et al. 2022), can help accelerate data acquisitions and processing, including in cryptic and mesophotic coral reef habitats for which relatively little is known (Brandl et al. 2019b; Pérez-Rosales et al. 2022; Wolfe et al. 2023).

Among the major ecological compartments of coral reef ecosystems, species diversity generally culminates in invertebrates, followed by fishes, corals, and algae, although the relative contributions of the different communities to species diversity vary among regions (Figs. 10.16 and 10.21a). While differences in data availability among regions blurs to some degree the level of certainty on these patterns, furthering these trans-community evaluations will bring heightened understanding of the magnitude and organization of ecosystem biodiversity among regions and taxonomic groups. Inter-regional comparisons indicate a marked contrast in coral reef ecological states between oceanic basins, with corals prevailing as primary constituents of reef habitats in the Pacific Ocean and macroalgae dominating the substrate on Caribbean reefs, with an intermediate state observed in the Indian Ocean (Fig. 10.21b; Reverter et al. 2022). These regional differences in dominance between corals and macroalgae concord with global gradients in coral species diversity and robustness of herbivory, which culminate in the Pacific Ocean and are weakest in the Caribbean (Hughes et al. 2002b; Bellwood et al. 2005; Siqueira et al. 2019; Cramer et al. 2021). Similarly, the abundance of various species groups changes with the latitudinal gradient, including sponges which are more prevalent near the equator and soft-corals at higher latitudes (Reverter et al. 2022). A positive relationship is observed between community diversity and biomass across regions for corals and fish, but not for algae in which the species diversity to biomass relationship is negative (Fig. 10.22). These patterns concord with our general understanding that species diversity supports higher ecological productivity, particularly at higher trophic levels, in structurally complex ecosystems dominated by corals, and declines when reefs are dominated by algae (Mumby and Steneck 2008; Graham and Nash 2013; Nash et al. 2014; Rogers et al. 2015; Clements et al. 2017). They indicate biomass maxima in coral and fish communities are achieved by species additions and therefore a diversified occupation of ecological niches, whereas high algal productivity tends to be driven by few proliferating species. These community-level differences have cascading effects that spread to other ecosystem compartments through species ecological functions and trophic and other forms of biotic interactions. Diversified coral communities contribute structural habitats for a myriad of organisms, including for species necessitating relatively large and long-lasting structures, therefore amplifying biodiversity and trophic food chains beyond the capacity of algal structures (Richardson et al. 2017a; Urbina-Barreto et al. 2021). This topic necessitates deeper investigation to comprehend the links between biodiversity, ecological productivity, habitat, and spatial scale, and the consequences of the ongoing changes of coral reef ecosystems.

A synthesis of the major environmental disturbances impacting coral reefs across French tropical overseas over the last decades indicates coral reef systems differ in their exposure and vulnerability to different types of disturbance (Fig. 10.23). While some disturbances, such as cyclones and increasingly coral bleaching events, occur globally in the tropics (Hughes et al. 2018; Puotinen et al. 2020), others are restricted to specific regions. This is the case with major diseases causing coral and sea-urchin *Diadema antillarum* die-offs, outbreaks of the holopelagic macroalgae *Sargassum fluitans* and *S. natans*, and rapid population expansions of invasive seagrass *Halophila stipulacea* and lionfish *Pterois* spp. in the Caribbean Sea (Côté and Smith 2018; Winters et al. 2020; Marsh et al. 2021; Hewson et al. 2023). Similarly, outbreaks of the coral predator crown-of-thorns starfish *Acanthaster* spp. are restricted to the Indo-Pacific, whereas tsunami impacts to coral reefs are mostly reported in the southwest Pacific (Løvholt et al. 2012; Pratchett et al. 2014). These large-scale spatial patterns emerging from inter-regional comparisons are also modulated by smaller, intra-system differences, particularly in broad and fragmented coral reef systems such as French Polynesia, New Caledonia, Scattered Islands, and Great Barrier Reef where disturbance impacts are spatially contained (Kayal et al. 2012b; Wolff et al. 2016; Adjeroud et al. 2018a; Mellin et al. 2019; Vercelloni et al. 2019; Pérez-Rosales et al. 2021a).

Despite significant differences in the level of knowledge available on environmental disturbances among regions, it appears that some coral reef systems like the ones in French Polynesia, New Caledonia, and the Caribbean islands are situated in areas prone to high frequency of disturbance, while others in Wallis and Futuna and Scattered Islands (though Tromelin is in a cyclone route) seem to lie in relatively stable environmental sanctuaries that suffer fewer

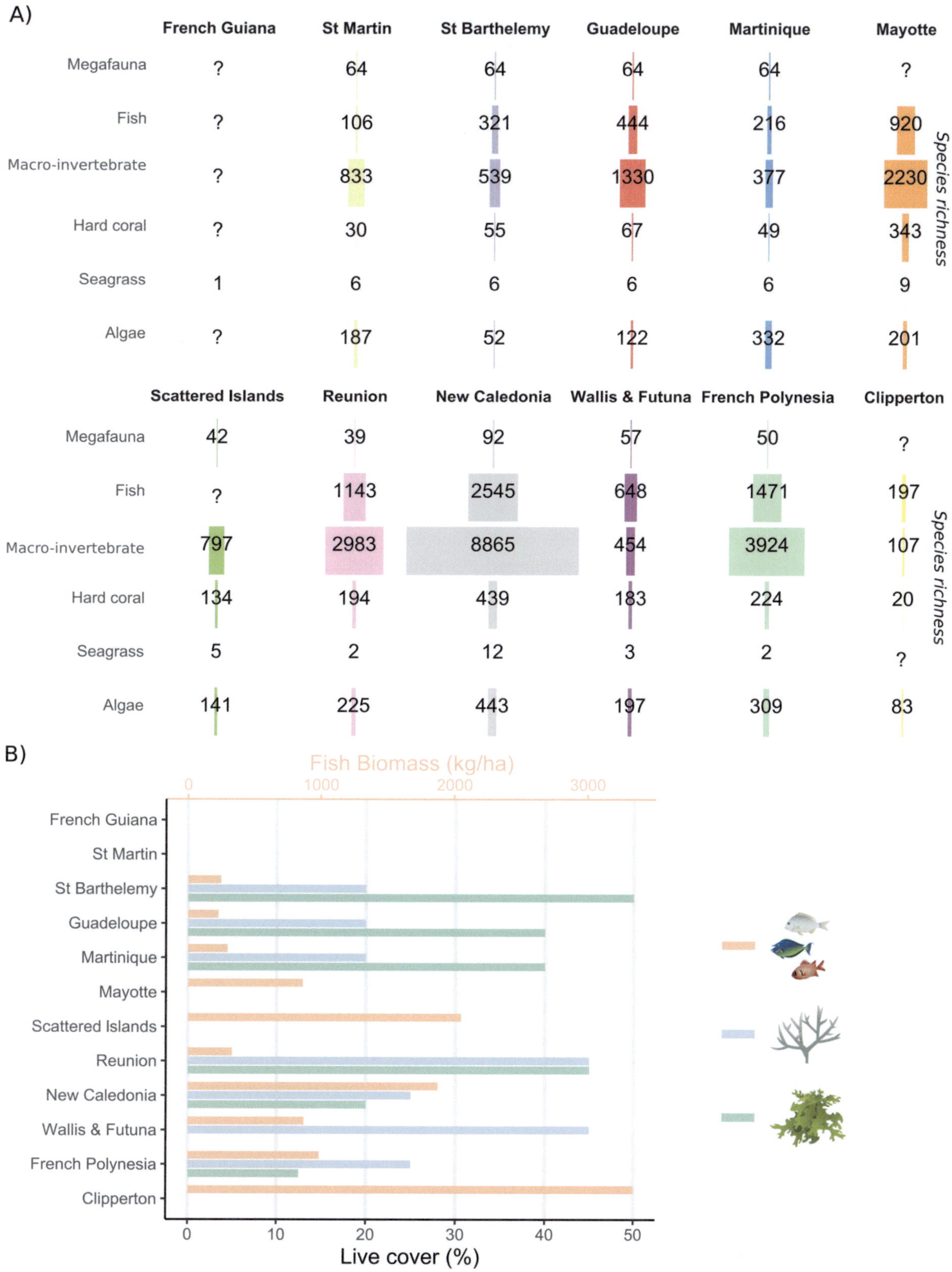

Fig. 10.21 Coral reef ecosystem diversity and productivity across French tropical overseas. Species richness of key ecological communities constituting the coral reef trophic pyramid (**a**, see Fig. 10.16 for data sources) and percent cover of corals and algae and fish biomass (**b**, data from McClanahan et al. 2019, 2021; Souter et al. 2021)

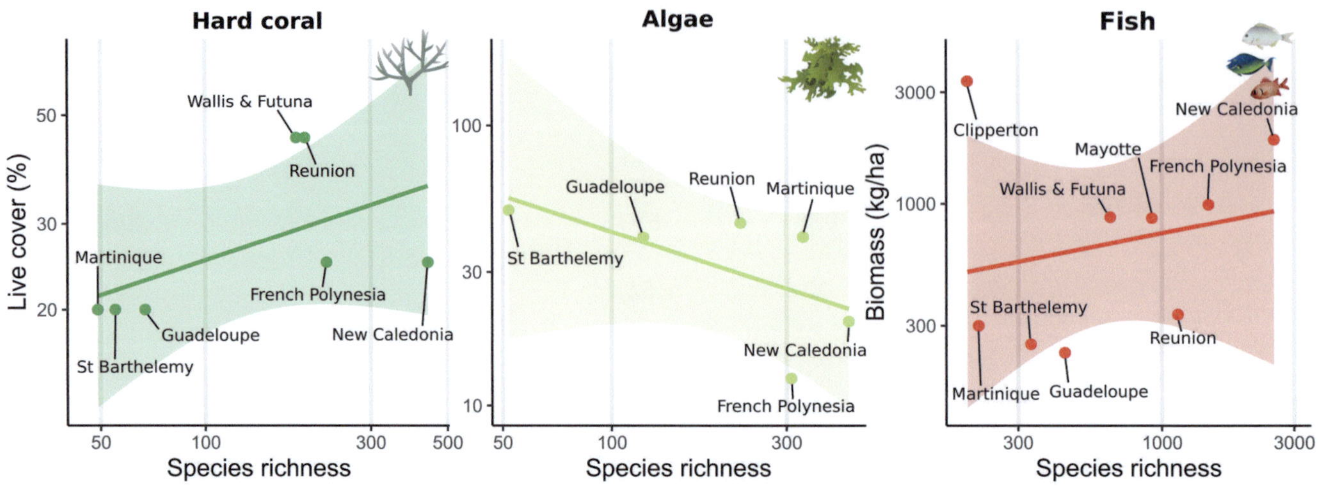

Fig. 10.22 Relationships between species diversity (richness) and biomass (reflected by percent cover in sessile communities) in corals, algae, and fish. Percent cover and biomass data from McClanahan et al. (2019, 2021) and Souter et al. (2021); see Fig. 10.16 for species richness data sources

disturbance events (Fig. 10.23b; Løvholt et al. 2012; Puotinen et al. 2020; Kuempel et al. 2022). While coral reefs have long evolved with environmental disturbances whose nonequilibrium effects on ecosystems are seen as a major source of their exceptional biodiversity (Connell 1978; Quattrini et al. 2020), the consequences of recent changes in disturbance types and regimes for coral reef resilience and functionality remain insufficiently understood to support adequate management, especially in regions of high exposure and vulnerability Adjeroud et al. 2018a; Vercelloni et al. 2019; Leinbach et al. 2021; Pérez-Rosales et al. 2021a; Carlot et al. 2023; Zhang et al. 2023). Eutrophication and fishing are suspected to generate a higher occurrence and intensity of *Acanthaster* outbreaks and associated mass coral mortality, whereas the frequency and magnitude of temperature anomaly-induced coral bleaching events and cyclonic storms have risen with ongoing climate changes (Sweatman 2008; McCook et al. 2010; Mellin et al. 2016; Vanhatalo et al. 2017; Puotinen et al. 2020; Cresswell et al. 2023). Time-series analyses of coral cover indicate *Acanthaster* outbreaks have resulted in strongest impacts on reefs through mass coral mortality and cascading effects on various reef communities, followed by cyclones and then coral bleaching events, although coral mortality associated with regional and global bleaching is becoming more prominent (Osborne et al. 2011; De'Ath et al. 2012; Kayal et al. 2012b; Adjeroud et al. 2018a; Vercelloni et al. 2019; Pérez-Rosales et al. 2021a; Gilmour et al. 2022). Coral bleaching events have become a common part of coral reef dynamics in most tropical regions, with variable inter-annual occurrences depending on climatic oscillations (El Niño Southern Oscillation, North Atlantic Oscillation) and the ecological state (abundance, composition) of coral communities (Sully et al. 2019). Higher sea water temperatures have also been related to increases in coral disease (Burke et al. 2023). The National Oceanic and Atmospheric Administration (NOAA)'s (https://coralreefwatch.noaa.gov) tool has been a major information source for marine heatwave and coral bleaching alerts and forecasts (and https://www.nhc.noaa.gov for tropical cyclones), although its continued availability is uncertain due to current government instability and public service meltdown in the USA.

Over recent decades, much research attention has been dedicated to elucidating the mystery behind *Acanthaster* outbreaks, whose drivers have remained elusive (Pratchett et al. 2014, 2017), although pieces of knowledge add up with new studies performed during each new outbreak cycle. Accumulated evidence points toward the role of nutrient limitations on larval stages (Fabricius et al. 2010; Wooldridge and Brodie 2015), predatory control particularly on juveniles (Sweatman 2008; Cowan et al. 2017; Westcott et al. 2020), and, more recently, the capacity of the juvenile seastars to remain in a latent (ecologically "dormant") herbivorous stage where cryptic populations hidden in dead-coral rubble can build up over years before shifting to corallivory and full-grown mature adults once live coral is abundant (Deaker et al. 2020; Wilmes et al. 2020; Deaker and Byrne 2022). Recurrent observations of outbreak episodes have enabled identification of initiation sites and propagation patterns of outbreaks in several regions, along with a good knowledge of their impacts to coral communities, repeatedly observed patterns that therefore appear predictable (Kayal et al. 2012b; Adjeroud et al. 2018b; Babcock et al. 2020; Wilmes et al. 2020; Condie et al. 2021). Among French tropical overseas, *Acanthaster* outbreaks are recorded in French Polynesia, New Caledonia, and Mayotte with a periodicity of

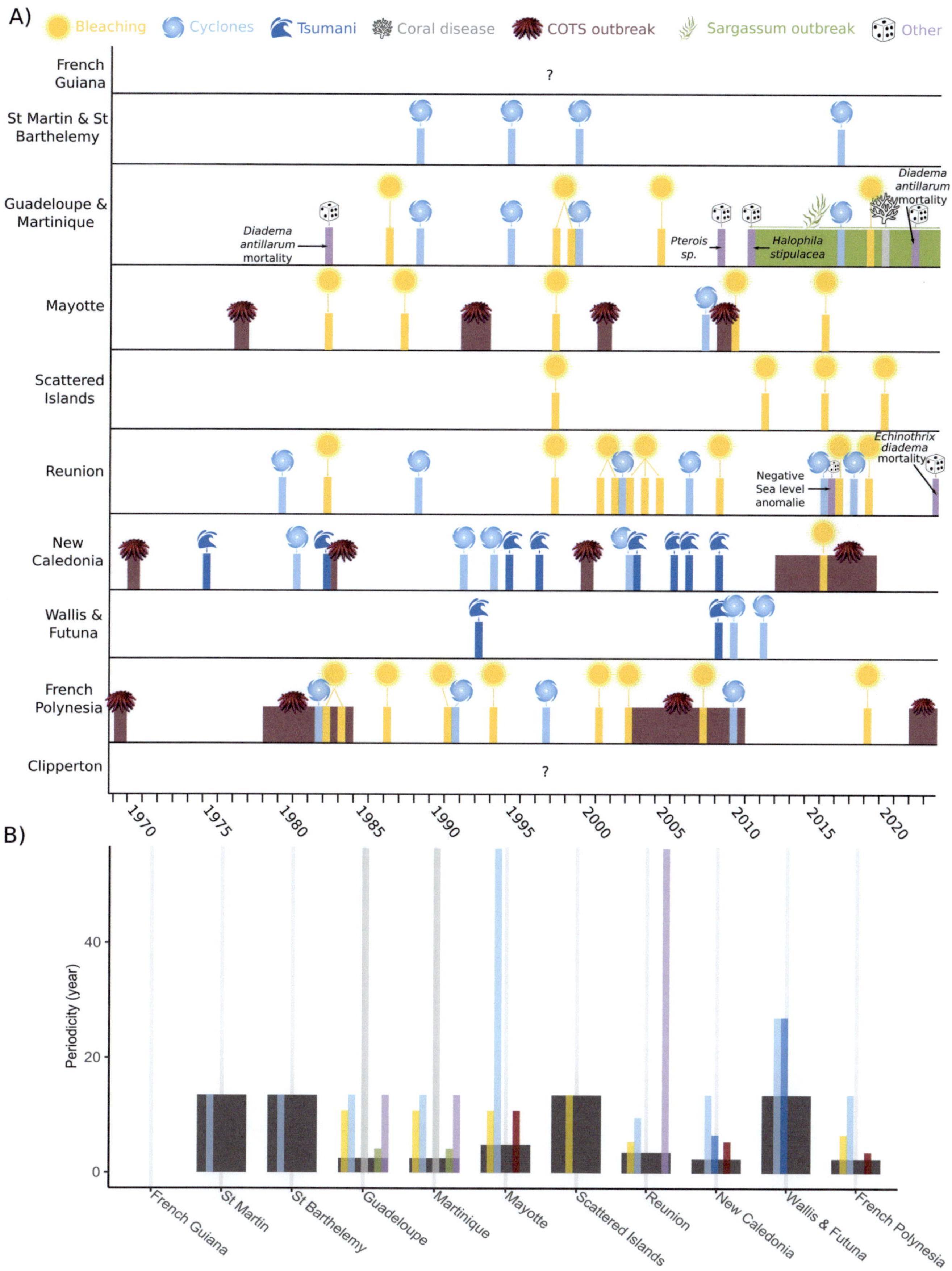

Fig. 10.23 Major environmental disturbances to coral reefs recorded in French tropical overseas over years (**a**) and mean periodicity of the events (**b**). In **b**, occurrence periodicities are shown per disturbance type (color bars) and considering all types of disturbances (black bars)

~15 years (Fig. 10.23), which is consistent with the frequency of outbreaks reported in other regions such as Australia's Great Barrier Reef (Kayal et al. 2012b; Pratchett et al. 2017; Adjeroud et al. 2018a; Babcock et al. 2020; Richards et al. 2021). Occurrences of the outbreaks seem correlated with ocean productivity in relation to regional weather oscillations influencing patterns of temperature, rainfall, and wind (Fabricius et al. 2010; Houk and Raubani 2010; Wooldridge and Brodie 2015; Matthews et al. 2020), with hitherto no direct evidence of recent increases in frequencies or intensities of the outbreaks. Yet, comparative coral reef health assessments highlight the importance of preserving ecosystem integrity for higher chances of resilience in the face of *Acanthaster* outbreaks (Sweatman 2008; McCook et al. 2010; Mellin et al. 2016; Westcott et al. 2020; Kroon et al. 2021). Recently, much effort has been dedicated to identifying reef predators able to keep *Acanthaster* populations in check as a natural solution to outbreaks (Cowan et al. 2017; Desbiens et al. 2023; Kayal and Lenihan 2025). Within-system studies indicate the seastar outbreaks are restricted to reefs subject to coastal influence and propagate from distinct initiation sites that are exposed to high nutrient discharge, including mid-shelf reefs near the towns of Cairns in Australia's Great Barrier Reef and Noumea in New Caledonia and the north shore of the island of Moorea in French Polynesia (Fabricius et al. 2010; Kayal et al. 2012b; Wooldridge and Brodie 2015; Adjeroud et al. 2018b; Adam et al. 2021). As a new cycle of *Acanthaster* outbreaks spreads in the Pacific, interregional connectivity between *Acanthaster* populations among adjacent systems and the degree of synchronicity of outbreaks in response to regional weather patterns remain to be explored. Among major environmental disturbances impacting coral reefs, *Acanthaster* outbreaks are the easiest to try to mitigate, and control programs are performed in various regions (Condie et al. 2021; Castro-Sanguino et al. 2023). Yet, culling of the seastar populations is costly, has shown relatively limited efficiency over large spatial scales, and addresses a symptom rather than the root causes of the outbreaks. Nevertheless, despite its notorious reputation as a harmful species, a pest (Pratchett et al. 2014; Kayal et al. 2017b), *Acanthaster* is also recognized for its role as a keystone predator benefiting biodiversity by feeding preferentially upon competitive acroporids (particularly genera *Acropora* and *Montipora*) that otherwise have the capacity to outcompete other corals (Kayal et al. 2011; Darling et al. 2012; Pratchett et al. 2017; Kayal and Adjeroud 2022). The degree to which *Acanthaster* outbreaks are triggered by large abundance of their preferred prey species remains to be investigated, though long-term observations in Moorea, French Polynesia, indicate swarms of seastars emerge when competition for space among corals is particularly prevalent (Fig. 10.24).

10.3.10 Socio-ecosystems

Coral reef socio-ecosystems around the world come in a variety of forms, combining differing levels of natural ecosystem and human society features (Figs. 10.20, 10.21, 10.22, 10.23 and 10.25). The prevalence and impacts of human populations, as well as their ties to ecosystems for coastal protection, food, and economic resources, are key to ecosystem health and the resulting level of sustainability and benefits to nature and society (McCook et al. 2010; Cinner et al. 2018; Comte and Pendleton 2018; Williams et al. 2019; Darling et al. 2019; Castro-Cadenas et al. 2022). Small natural systems supporting large human populations tend to suffer from stronger impacts from coastal development, overexploitation, and pollution, resulting in reduced ecosystem health, management deficiencies, and limited socio-ecosystem resilience (Duprey et al. 2020; Boyce et al. 2020; Ouédraogo et al. 2021; Andrello et al. 2022; Ouédraogo et al. 2023). This begs the long-standing question of the carrying capacity of natural systems, not only in terms of the size of the human population they can support but most importantly the overall level of stress they can sustain without declining. Other socio-ecosystem aspects, including the ecological and geomorphological contexts as well as the economic and cultural ties to the reef, further contribute to the vulnerabilities of the systems (Mumby et al. 2007; Anthony et al. 2015; Siqueira et al. 2019; Cramer et al. 2021). When management strategies fail to preserve coral reef ecosystem health, declines in biodiversity, biological productivity, and ecological functionality reduce the level of ecosystem services provided to society, often with a rise of negative interactions and ecosystem disservices (Speers et al. 2016; Woodhead et al. 2019; Carlot et al. 2023). Harmful algal blooms such as outbreaks of benthic ciguatoxin-producing dinoflagellates occur more frequently in degraded environments where dead coral skeletons and macroalgae surfaces provide substrate for the microorganism, with strong societal consequences for human health and economy (Fig. 10.26; Alves de Souza et al. 2022). Described for centuries in French tropical overseas, ciguatera poisonings are on the rise under the possible influence of global changes (marine heat waves, ocean acidification, etc.), and the distribution range of ciguatoxin-producing species and their accumulation in trophic food webs is expanding to subtropical/temperate regions, though the statistics of human poisonings remains highly underestimated as most cases go undocumented (Fig. 10.27; Rongo et al. 2009; Chinain et al. 2021). The CiguaWatch Initiative, led by French Polynesia,

Fig. 10.24 Photographs illustrating the co-occurrence of outbreaks of the coral predatory seastar crown-of-thorns starfish (*Acanthaster*) with periods of high abundance of their favorite prey, the acroporid corals (particularly genera *Acropora* and *Montipora*). Photographs were taken in the lagoon of Moorea, French Polynesia, at the onsets of two distinct episodes of *Acanthaster* outbreaks occurring ~15 years apart and causing widespread coral mortality (Fig. 10.23). The pictures show the capacity of encrusting *Montipora* to overgrow other corals, particularly massive *Porites* which is foundational to forming large habitat structures in the lagoon. Because *Acanthaster* preferentially preys upon acroporids, their episodic outbreaks prevent *Montipora* to fully outcompete *Porites*, promoting species coexistence. Bottom-right picture shows *Acropora* and *Montipora* colonies (underscored names) bearing white feeding scars characteristic of recent predation by the seastar. See text for references on the topic

provides a regional platform for training and data sharing on Ciguatera poisoning surveillance in the South Pacific (https://ciguawatch.ilm.pf). Ongoing research particularly seeks a better understanding of ciguatera-related algal bloom ecology for anticipating and monitoring outbreaks, evaluating their consequences on societies, and improving ciguatoxin detection tests, patient diagnostic tools, and medical care (Rongo et al. 2009; Chinain et al. 2020; Chinain et al. 2021). Other examples of negative feedback from system alterations include recent concerns about increasing outbreaks of the coral predatory seastar *Acanthaster* throughout the Indo-Pacific and *Sargassum*

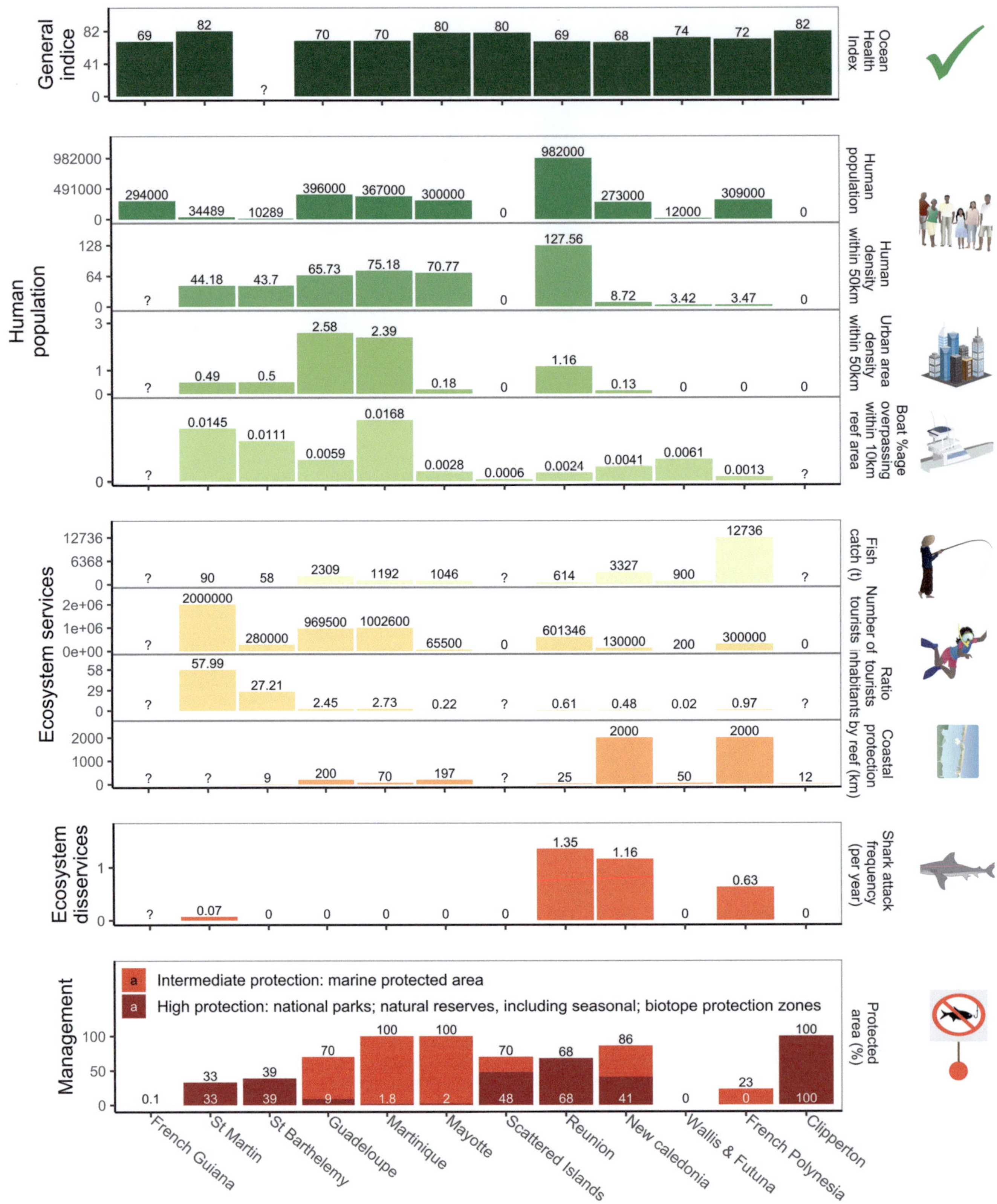

Fig. 10.25 Differences in coral reef socio-ecosystems among French tropical overseas. The Ocean health index summarizes ecosystem state and resilience on a scale of 0–100 (Halpern et al. 2012). Shark attack frequency was calculated over the 1980–2023 period. (Data from https://oceanhealthindex.org, https://recifs.epfl.ch, https://www.iedom.fr, https://public.opendatasoft.com and IFRECOR (2021))

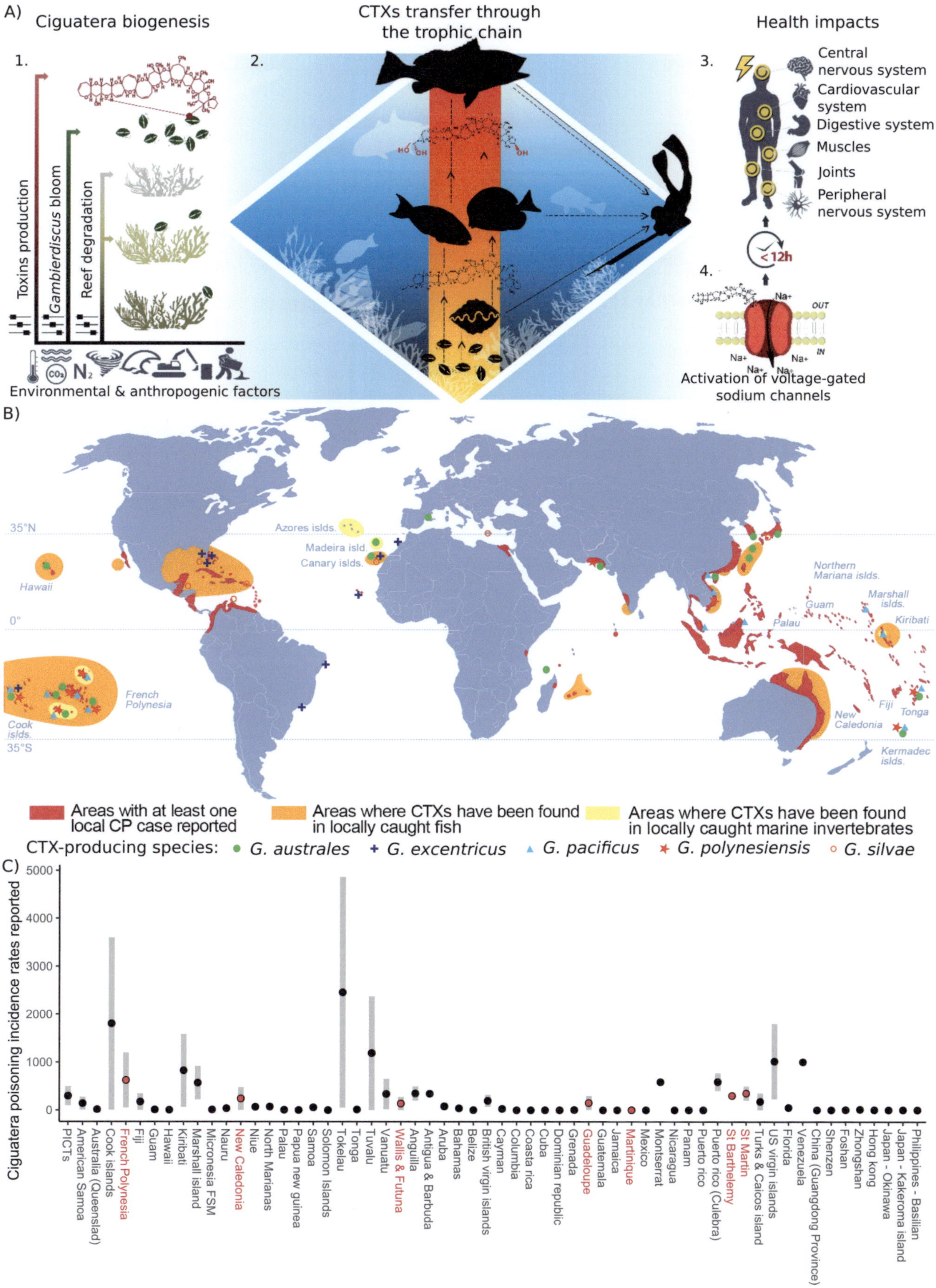

Fig. 10.26 Ciguatera poisoning biogenesis and transmission (**a**, adapted from Chinain et al. 2023), distribution of ciguatoxin (CTX) producing species (*Gambierdiscus* spp.) Ciguatera poisoning prevalence areas (**b**, adapted from Chinain et al. 2021), and Ciguatera poisoning (CP) incidence rates among regions (**c**, data adapted from Chinain et al. 2021 and references therein). PICTs for Pacific Island Countries and Territories

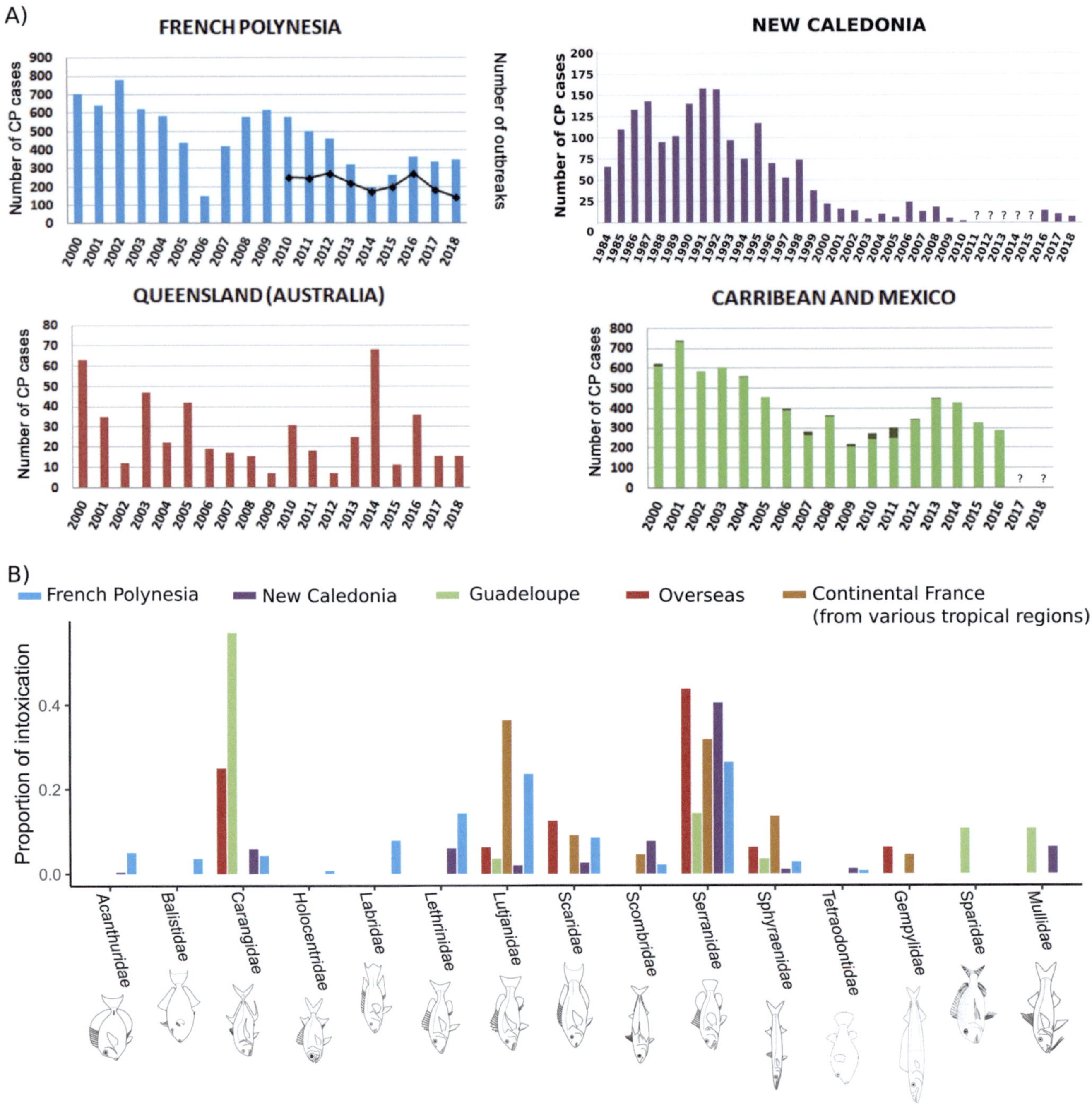

Fig. 10.27 Documented number of human ciguatera poisoning (CP) cases through time per region (**a**, adapted from DASS 2018; Chinain et al. 2021) and vectors implicated in both autochthonous and imported poisonings (**b**, data from Boucaud-Maitre et al. 2017; DASS 2018; Gatti and Chinain 2023)

strandings in the Caribbean (Fig. 10.23; Pratchett et al. 2014; Matthews et al. 2020; Marsh et al. 2021) and critical shark attacks on humans in Australia, Reunion, and New Caledonia (Fig. 10.25; Taglioni et al. 2019; Maillaud et al. 2022; Riley et al. 2022), although there is still limited understanding of underlying drivers. In the face of uncertainties and lack of incentives to address root causes, *Sargassum* removals and culling of *Acanthaster* and shark populations are common practice in different coral reef socio-ecosystems, practices that are often hotly debated (Papastamatiou et al. 2006; Vanhatalo et al. 2017; Pratchett and Cumming 2019; Babcock et al. 2020; Condie et al. 2021; Gray et al. 2021; Maillaud et al. 2022; Borsa et al. 2023). Fortunately, positive examples of human-wildlife interactions with potentially deadly animals also exist, although the mechanisms determining sustainability and their thresholds remain in big part to be understood (Udyawer et al. 2021).

Coral reef biodiversity and associated ecosystem productivity and services have long supported the development of human societies in the tropics. Yet in various places, rapid and ecologically unsustainable growth has caused coastal alterations that are beyond the tolerance capacity of ecosystems, threatening the sustainability of the coral reef socio-ecosystems (Woodhead et al. 2019; IPBES 2019; Duprey et al. 2020). Achieving sustainable coral reef socio-ecosystem development is however not easy, as local human activities often impact marine ecosystems over large spatial scales and are increasingly combined with global environmental changes (Donovan et al. 2023), as many human concentrations have already caused significant and long-lasting damage to their surrounding ecosystems and as many societies still prioritize unsustainable development paths (Brodie and Waterhouse 2012; Cinner et al. 2018; Daskin and Pringle 2018; Kayal et al. 2019; Darling et al. 2019; OECD 2020; Cramer et al. 2021; Ouédraogo et al. 2021; Ouédraogo et al. 2023; Calvin et al. 2023). The most extreme disturbances in French tropical overseas took form as nuclear testing in 1966–1996 on the islands of Mururoa and Fangataufa in French Polynesia, of which 41 were conducted above ground and 140 in volcanic rock under the coral reef (Philippe et al. 2022), and broad use of the highly toxic and long-lasting organochlorine pesticide chlordecone in Guadeloupe and Martinique in 1973–1993 (Multigner et al. 2016), with, apart from cancers in local human populations, largely unknown consequences on the coral reef socio-ecosystems and life more broadly.

Among the >500 million people living in tropical coastal areas protected from erosion and storm waves by coral reefs, ~100 million live below 10 m elevation within 10 km of a coral reef (Ferrario et al. 2014; Beck et al. 2018; Carlot et al. 2023). Among these, ~three million people live in French tropical overseas, mostly in Reunion, Martinique, and Guadeloupe where coastal urbanization and human density are the highest, followed by French Polynesia, Mayotte, New Caledonia, and French Guiana, although the latest is largely dominated by mangroves (Figs. 10.20 and 10.25). Coastal protection by coral reefs is particularly prominent in New Caledonia due to the high exposure to tsunamis, particularly on the eastern coasts, as well as in French Polynesia where relatively young barrier reefs protect the small Central Pacific island community from storm waves (Sahal et al. 2011; Løvholt et al. 2012; Burke and Spalding 2022). Fishing rates are the highest in French Polynesia where seafood constitutes a highest portion of the diet and fish exportation a significant part of the economy (Morin et al. 2016; FAO 2022), whereas tourism is particularly developed in the Caribbean, followed by Reunion and French Polynesia. The degree of reliance of tropical coastal societies on coral reefs varies from one system to another, though decline in coral reef health has strong consequences for exposure to flooding, food security, and economic opportunities (Pendleton 2010; Comte and Pendleton 2018; Alves de Souza et al. 2022; Carlot et al. 2023).

Coral reef socio-ecosystem vulnerability depends in big part on the ability of societies to durably manage ecosystems by preventing degradation and overexploitation through implementation of effective environmental regulations, exploitation quotas, and protected areas (McCook et al. 2010; Anthony et al. 2015; Rogers et al. 2015; D'agata et al. 2016). Overall, there are four major management strategies that target social and ecological vulnerabilities: environmental mitigation, ecosystem protection, restoration, and adaptation, though trade-offs and synergies between the different strategies and prerequisite conditions for success are largely unevaluated (Comte and Pendleton 2018; Thiault et al. 2020a). Comparisons across socio-ecosystems indicate that key characteristics of success include strong regulation enforcement and community engagement in management strategies (Cabral et al. 2018; Thiault et al. 2020a; Thiault et al. 2020b; Maxwell et al. 2020). Coral reef management across French overseas currently includes different levels of regulatory-designated marine protected areas, varying from 0% of ecosystem surface area in Wallis and Futuna to 100% in Clipperton (Fig. 10.25), though this does not account for other forms of locally recognized management. Limitations to design, implement, and enforce sustainable environmental management are often economic in nature, related to the level of inequality, dependency on natural resources and global market, and capacity of protection from local and international poaching (McCook et al. 2010; Cabral et al. 2018; Januchowski-Hartley et al. 2020; Thiault et al. 2020b; Maxwell et al. 2020; FAO 2022). The governance of socio-ecosystems and how different societal actors take part in management decisions are also important factors in socio-ecosystem trajectories (Morrison et al. 2020; Thiault et al. 2020a).

Coral reef socio-ecosystems exhibit a multitude of governance systems, as observed in French tropical overseas characterized by different degrees of decision-making centralization and a hierarchy of bodies performing policy-making, public safety, and territorial planning roles with different jurisdictions (Fig. 10.28). Given increasing environmental changes, and that many coral reefs may already be approaching their maximum tolerance limits, governance systems face the challenge of adapting rapidly while engaging a variety of societal actors at different scales (Donovan et al. 2023). Success in achieving healthy and resilient coral reef systems may depend on the capacity of socio-ecosystems to anticipate and adapt to future challenges (Rogers et al. 2015). Socio-ecosystemic diversity is a key attribute of French tropical overseas that are characterized by

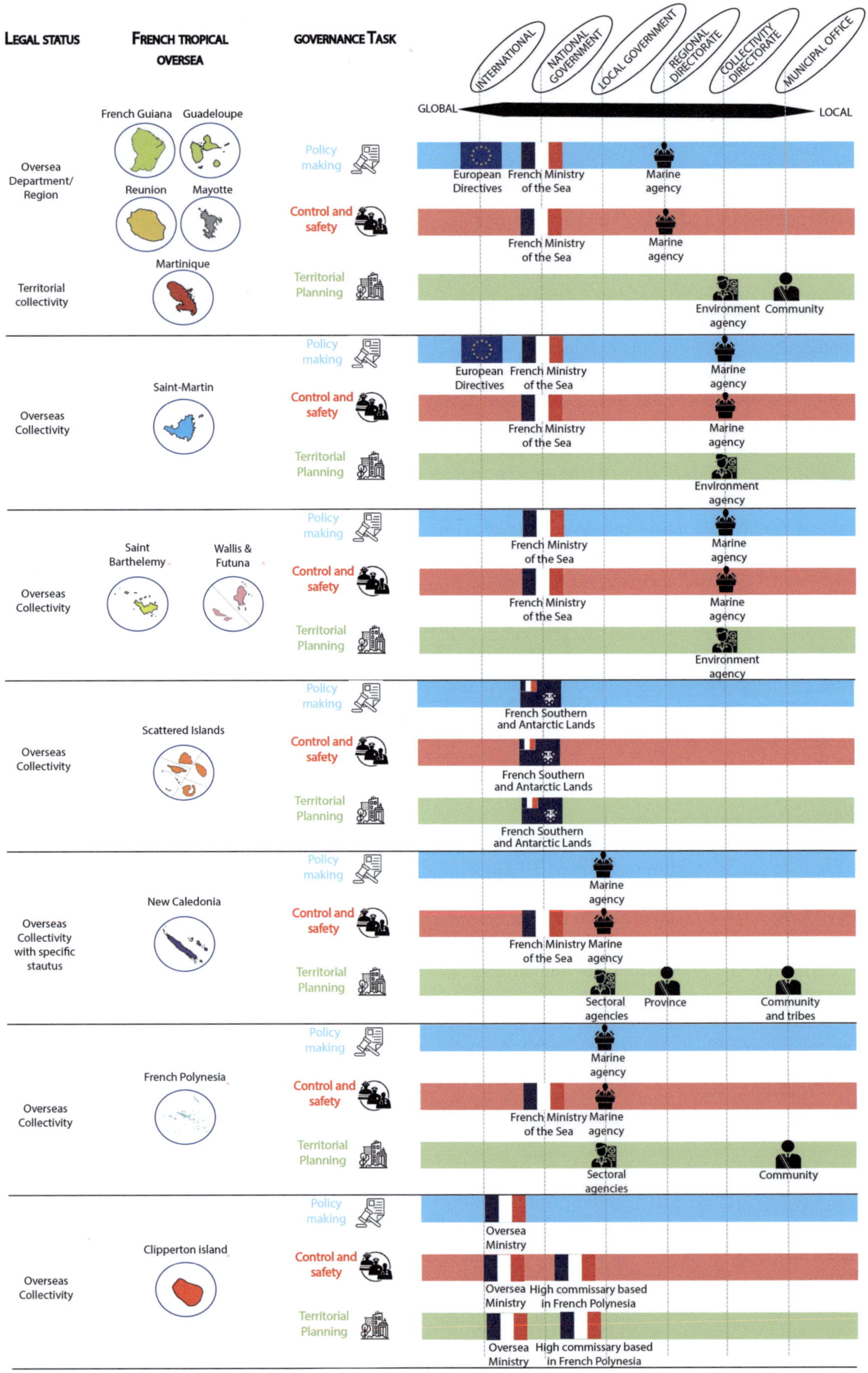

Fig. 10.28 Governance systems of coral reef socio-ecosystems in French tropical overseas. Diagrams represent the hierarchy and jurisdictions of different governing bodies in relation to the three tasks of policymaking, public safety, and territorial planning. Note that the uninhabited atoll of Clipperton is governed by France through a delegation based in French Polynesia

contrasting human histories, social and political organizations, and cultural and economic ties to natural resources. Societal challenges in these dispersed insular systems include finding a balance between local and national identities in a postcolonial era where the search for democracy, local governance, and social development is reliant on ecosystem preservation for coastal protection, food security, and economic development but challenged by pressures for overexploitation from the globalized market, a context increasingly exacerbated by environmental degradation and coral reef decline.

10.3.11 Biogeoclimatic Systems

Coral reefs interact with the Earth's geologic and climatic environments through multiple ocean-land-atmosphere interchanges whose prevalence varies with the size and characteristics of the systems (Fig. 10.29). By acreting carbonated reefs along the coasts, corals sequestrate oceanic carbon, modify ocean alkalinity, and produce shallow water substrate at slow yet potentially large scales (Smith 1978; Spalding et al. 2001; Jones et al. 2015; Carlot et al. 2021). The reefs reduce wave energy protecting coastlines from erosion and submersion and produce large amounts of carbonated sand that accumulates to create new habitats for many marine and terrestrial species, therefore amplifying biodiversity and associated benefits (Figs. 10.20 and 10.29c). Indeed, coral reef accretion and erosion generate conditions favorable to the establishment of diverse biological communities including seagrass and algae beds, mangroves, and the vegetation covering lagoonal islets, all of which contribute additional sets of ecological functions and ecosystem services such as habitat creation, carbon sequestration, and attenuation of wave and wind energy (Moberg and Folke 1999; McIvor et al. 2012; Guannel et al. 2016; Andréfouet et al. 2021; Carlot et al. 2023; Heudier et al. 2023). Among French tropical overseas, the contribution of coral reef accretion to the coastal systems' emerged land and lagoonal systems is most prominent in Scatter Islands, Clipperton, French Polynesia, Wallis and Futuna, Mayotte, Guadeloupe, and New Caledonia (Fig. 10.29c).

Coral reefs also interact with the local climate. By producing large stretches of white sand beaches and clear and shallow underwater habitats, coral reefs modify the local albedo, or how much solar radiation is sent back to the atmosphere, with a direct impact on surface temperature (Hays et al. 2001; McGowan et al. 2019). Coral reefs and associated ecosystems may influence local atmospheric conditions through heat and moisture exchanges and affect local meteorology such as cloud cover and rainfall (McGowan et al. 2022). Coral reefs are a source of dimethyl sulfide, an important aerosol involved in cloud formation, though its contribution to the local and regional aerosol burdens might be low compared to human sources (Yuan et al. 2022; Fiddes et al. 2022). Human societies living off the reef also contribute to local changes in albedo and wind through deforestation, urbanization, and various heat sources (Jin et al. 2019; Ouyang et al. 2022). The relative importance of coral reefs to the local climate might therefore vary from one system to another, depending on local characteristics of land-to-reef ratio, climatic environment, and human emissions of heat, gas, and particles. This includes direct emissions of heat for energy production and use in diverse sectors such as transport and industries, changes in albedo through land-use change including deforestation and urbanization, and emissions of greenhouse gas such as CO_2 and methane, and aerosols (Yuan et al. 2022). This diversity of coral reef biogeoclimatic systems is well illustrated across French tropical overseas (Fig. 10.29), providing a rich study system to investigate how coral reef systems interact with local and regional climate nowadays and into the future. This research avenue may help deepen understanding of mechanisms determining coral reef vulnerability to climate change, identification of potential adaptation and mitigation solutions, and predictions of thermal stress induced coral bleaching under future climatic conditions.

In the objective of socio-environmental sustainability, decarbonation of island emissions remains a real challenge, as many tropical overseas rely heavily on importations for food, material, energy, and economic activities (e.g., tourism), all of which come at a high cost in terms of carbon emissions (Kuo and Chen 2009). Among French tropical overseas, New Caledonia, whose economy is based on the mining industry, has the highest energy consumption and carbon emissions (Fig. 10.29a), with >75% of energy consumption associated with mineral extraction, purification, and export, mostly of nickel and cobalt (Losfeld et al. 2015). It is also the most distant tropical overseas from metropolitan France (Fig. 10.4), amplifying import-export-associated emissions. Despite ample access to solar, aeolian, and hydraulic energy, New Caledonia has a low proportion of renewable energy, though transition to green energy is a local priority and more supporting infrastructure has started to emerge around the archipelago over the recent years. In descending order of emissions, Reunion, Guadeloupe, Martinique, French Polynesia, and Guiana follow human population abundances (Fig. 10.30). Per capita emissions are lowest in Mayotte, which recently transitioned into a French overseas department and exhibits relatively low rates of urbanization (Figs. 10.25 and 10.28; Ninon 2007), and highest in New Caledonia with its mining industry, the two places standing out from the CO_2 emission to human population relationship (Fig. 10.30). Guadeloupe, Reunion, and Martinique are leading the transition to sustainable energy (Fig. 10.29a), with rapid changes over recent years. While

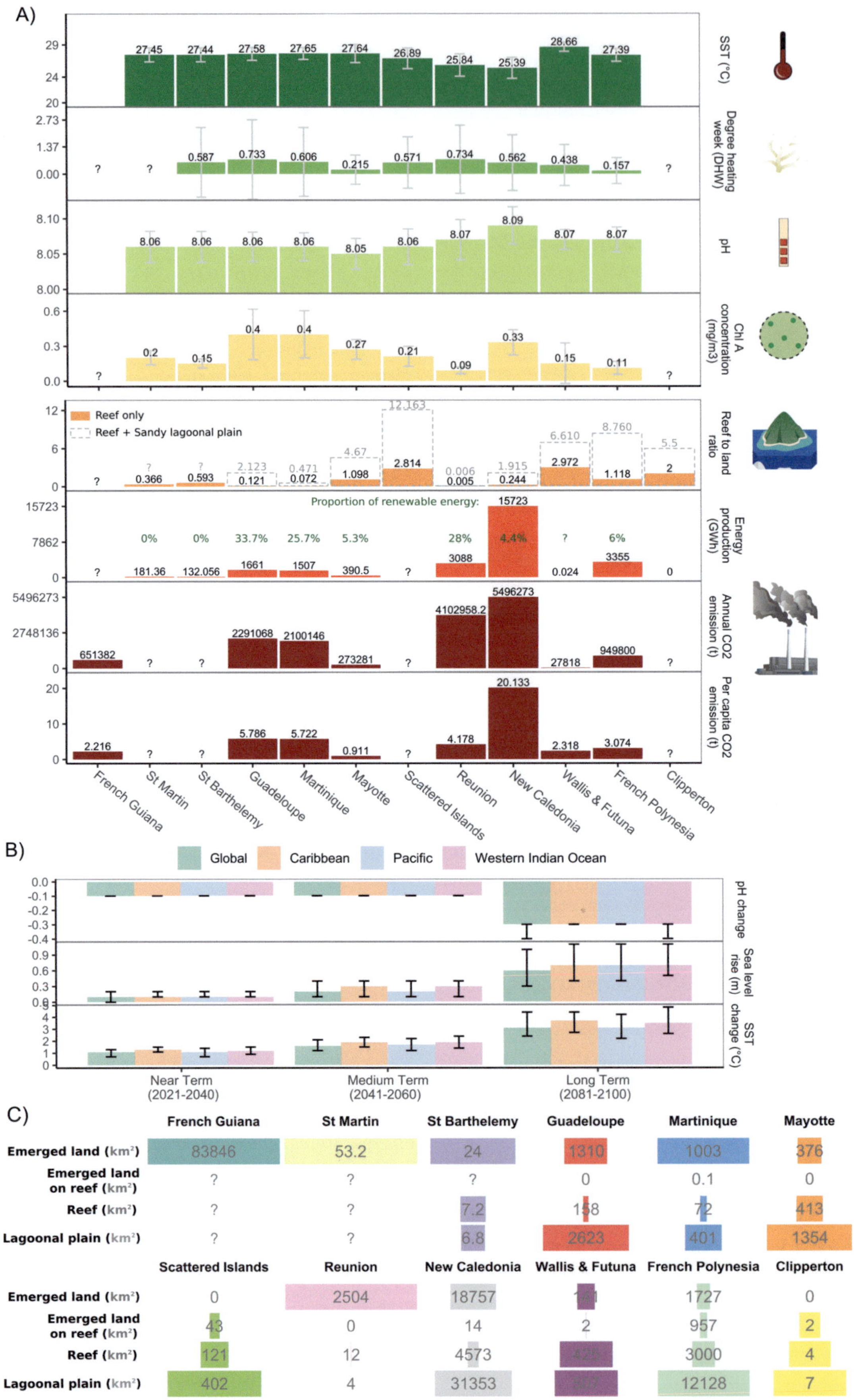

Fig. 10.29 Variability in coral reef biogeoclimatic features among French tropical overseas. (**a**) current characteristics (2021 data from Recif.epfl.ch, Andrefouet et al. 2008, www.worldometers.info, https://www.iedom.fr). (**b**) Predicted changes according to IPCC following the SSP5–8.5 (business-as-usual) scenario (data from https://interactive-atlas.ipcc.ch). (**c**) Surface areas of reef, lagoonal plain, and emerged land per region. (Data from Andréfouët et al. 2008)

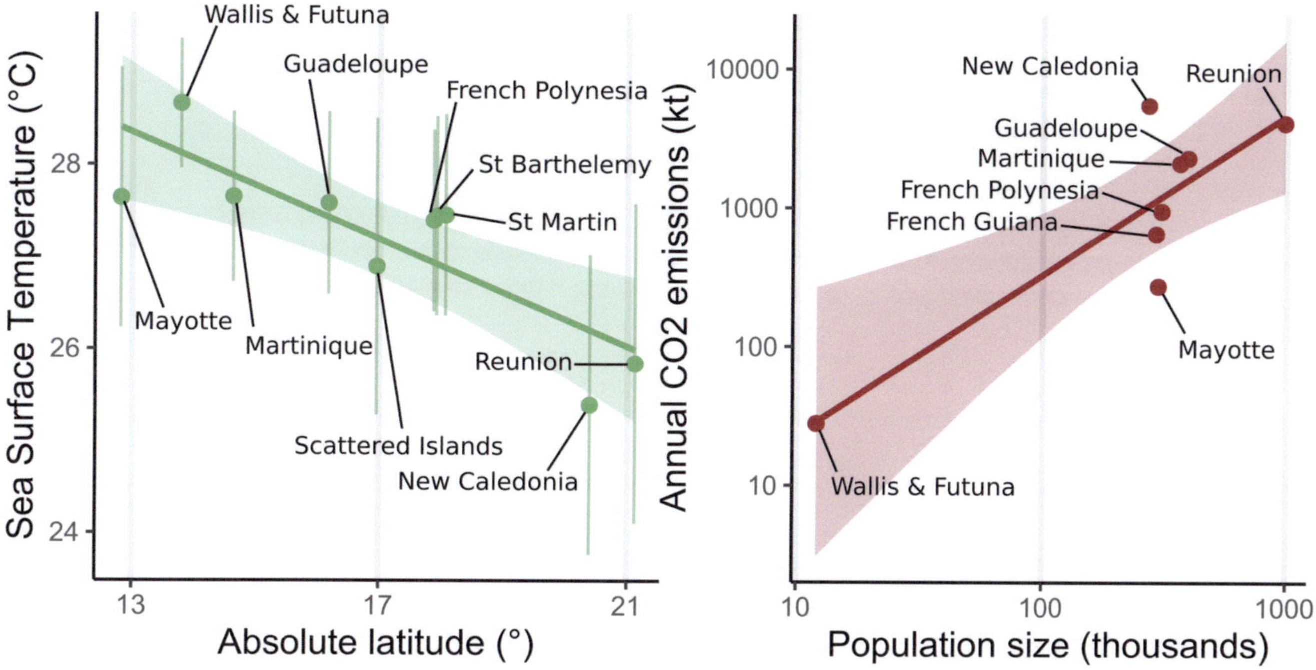

Fig. 10.30 Relationships between sea temperature (mean ± range) and latitude and CO_2 emissions and human demography across French tropical overseas

coral reef systems around the globe are exposed to different temperature regimes in relation to their latitudinal distributions (Fig. 10.30), the Intergovernmental Panel on Climate Change (IPCC) projections predict higher temperatures, sea-level rise, and lower pH, with a higher exposure for Caribbean reefs (Fig. 10.29b) that adds to their already higher vulnerability to coral decline and ecological shifts (Figs. 10.21b and 10.23; Hoegh-Guldberg et al. 2017; Kuempel et al. 2022). In the current battle for ecological and climate sustainability, tropical islands are key protagonists as both major contributors and victims of environmental decline and have a decisive role to play (Hernández-Delgado 2015).

10.3.12 Overview of Coral Reef Biodiversity Metrics Across French Tropical Overseas

Coral reef systems across French tropical overseas differ in many aspects, as expressed by the disciplinary metrics used to assess biodiversity at the multiple scales of biological and ecological organization covered in this chapter, though data remains incomplete and unbalanced (Figs. 10.5–10.30). A multidimensional representation along the various ecological gradients with available data enables visualizing the degrees of similarity and magnitudes of variation between the different systems (Fig. 10.31). This representation distinguishes along a primary axis New Caledonia, with the largest coral reef system and highest rates of species richness and habitat diversity, from the other French overseas that fall into three distinct groups along a secondary axis. On one side of this axis lies the densely populated and touristic systems with a status of overseas department of Martinique, Guadeloupe, and Reunion, contrasting with the other side of less densely populated systems of French Polynesia and Wallis and Futuna in the South Pacific and the uninhabited systems of Clipperton and Scattered Islands, and between the two extremes is the grouping of Saint Martin, Saint Barthelemy, Mayotte, and French Guiana. Along this secondary axis, the less densely populated systems are also characterized by stronger ties to the reef through higher rates of coastal protection and fish catch and higher ecosystem health as reflected by ocean health and coral bleaching indices than the more densely populated localities. While every coral reef system is unique, this multidimensional representation underscores how much New Caledonia differs from other systems, standing as a singular hotspot for coral reef biodiversity. A deeper exploration of coral reef biodiversity is necessary to further our understanding of biodiversity hotspots and their preservation, in French tropical overseas and globally.

10.3.13 Knowledge Gaps

Our synthesis identifies a generally positive correlation between the bioecological scale of organization of coral reef life forms and the level of knowledge available (Figs. 10.32

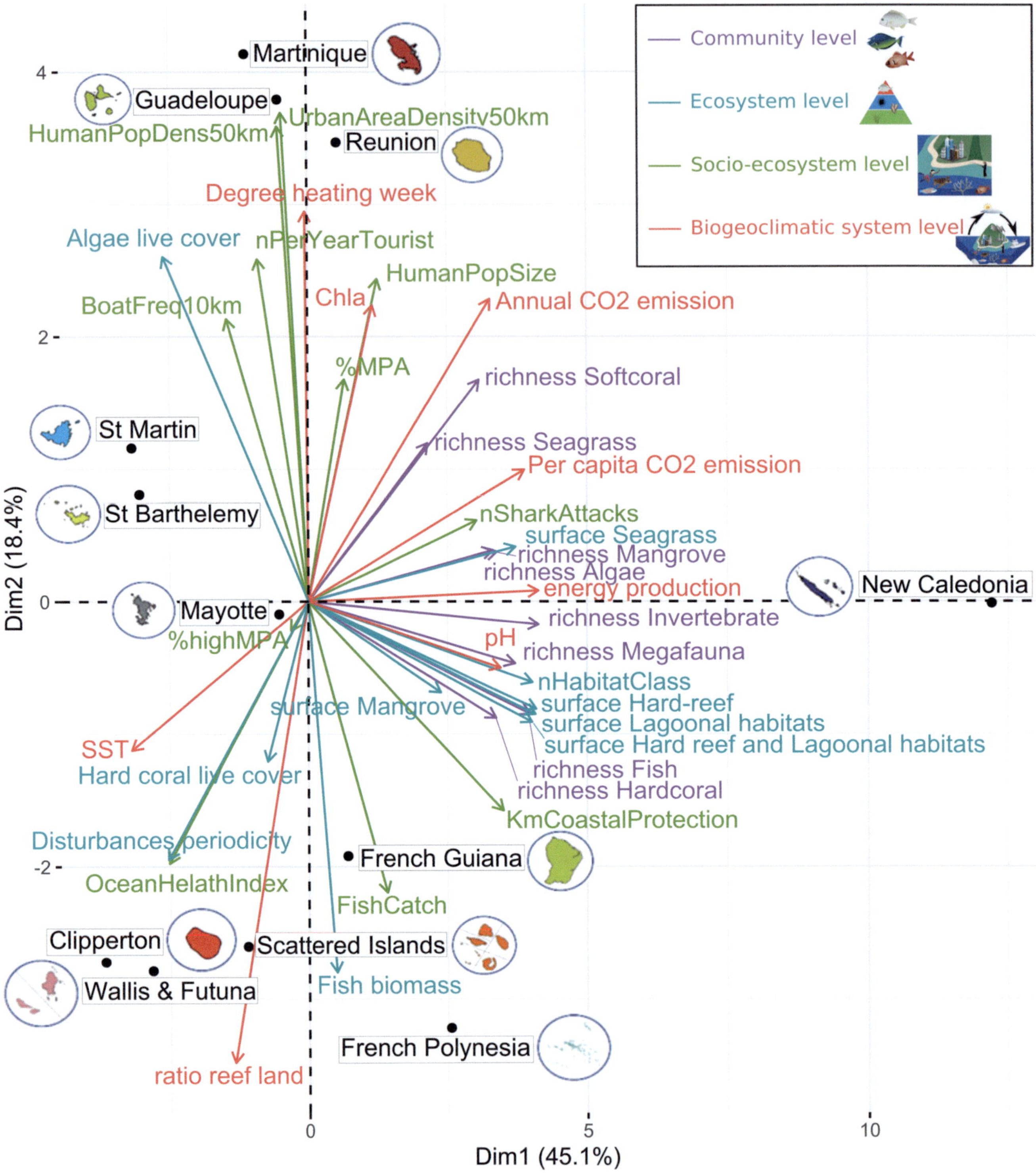

Fig. 10.31 Comparisons of coral reef systems across French tropical overseas. Principal component analysis including all biodiversity metrics considered in this study at the community, ecosystem, socio-ecosystem, and biogeoclimatic system levels of organization. Note that the distributions of the overseas systems and biodiversity variables are calculated in different spaces. The juxtaposition of the two provides information on the relative position of the overseas along the variables rather than an absolute position. Missing values were estimated using R's missMDA package tools (Husson and Josse 2020). The lower scales of genes, molecules, cells, organs, organisms, and populations are not included due to limitations in data availability

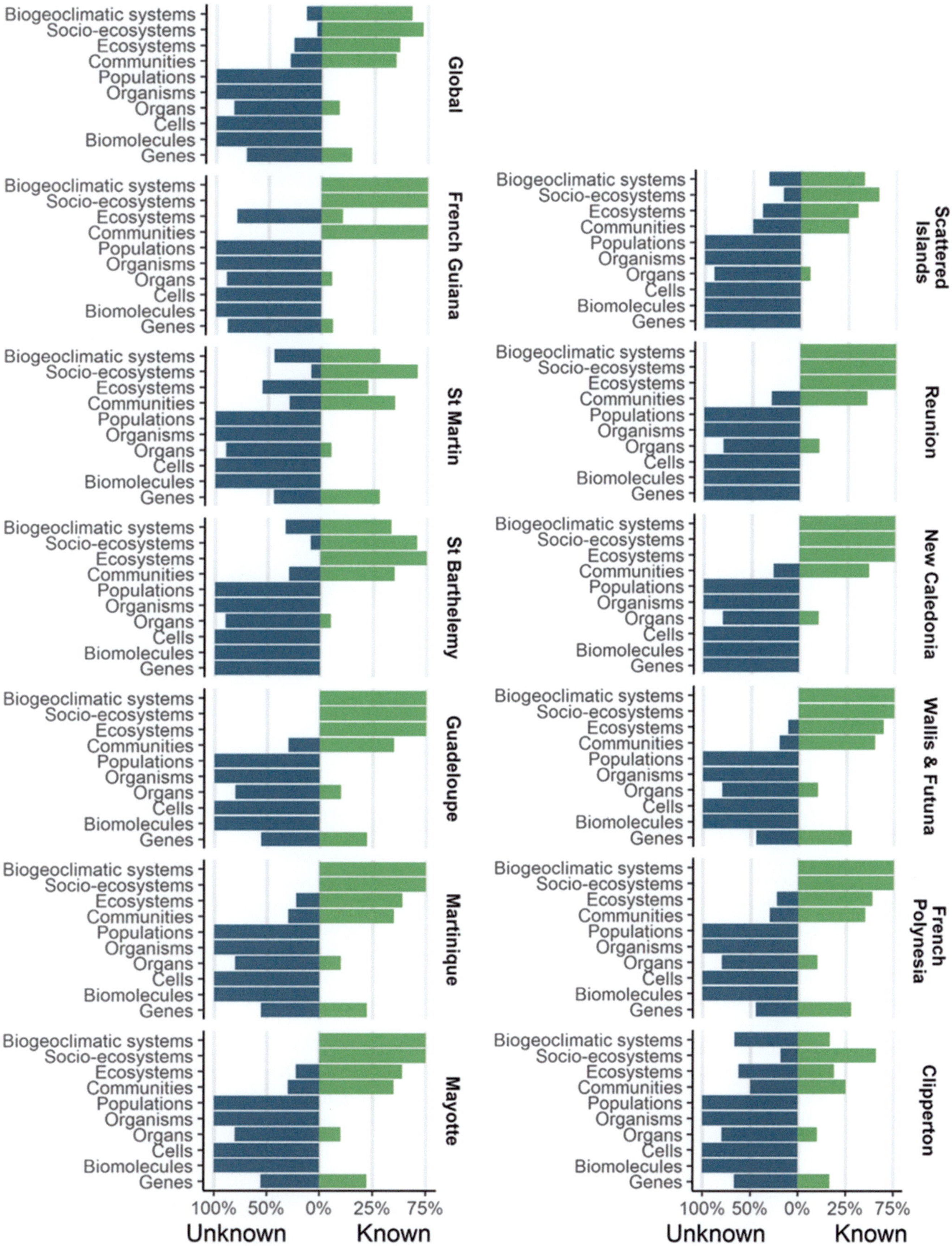

Fig. 10.32 Summarized state of knowledge on coral reef biodiversity by scale of biological and ecological organization and across French tropical overseas. Proportions of knowledge and knowledge gaps are estimated based on data completeness for the different metrics presented in Figs. 10.5–10.30, which is in part influenced by data accessibility for the purpose of the present study

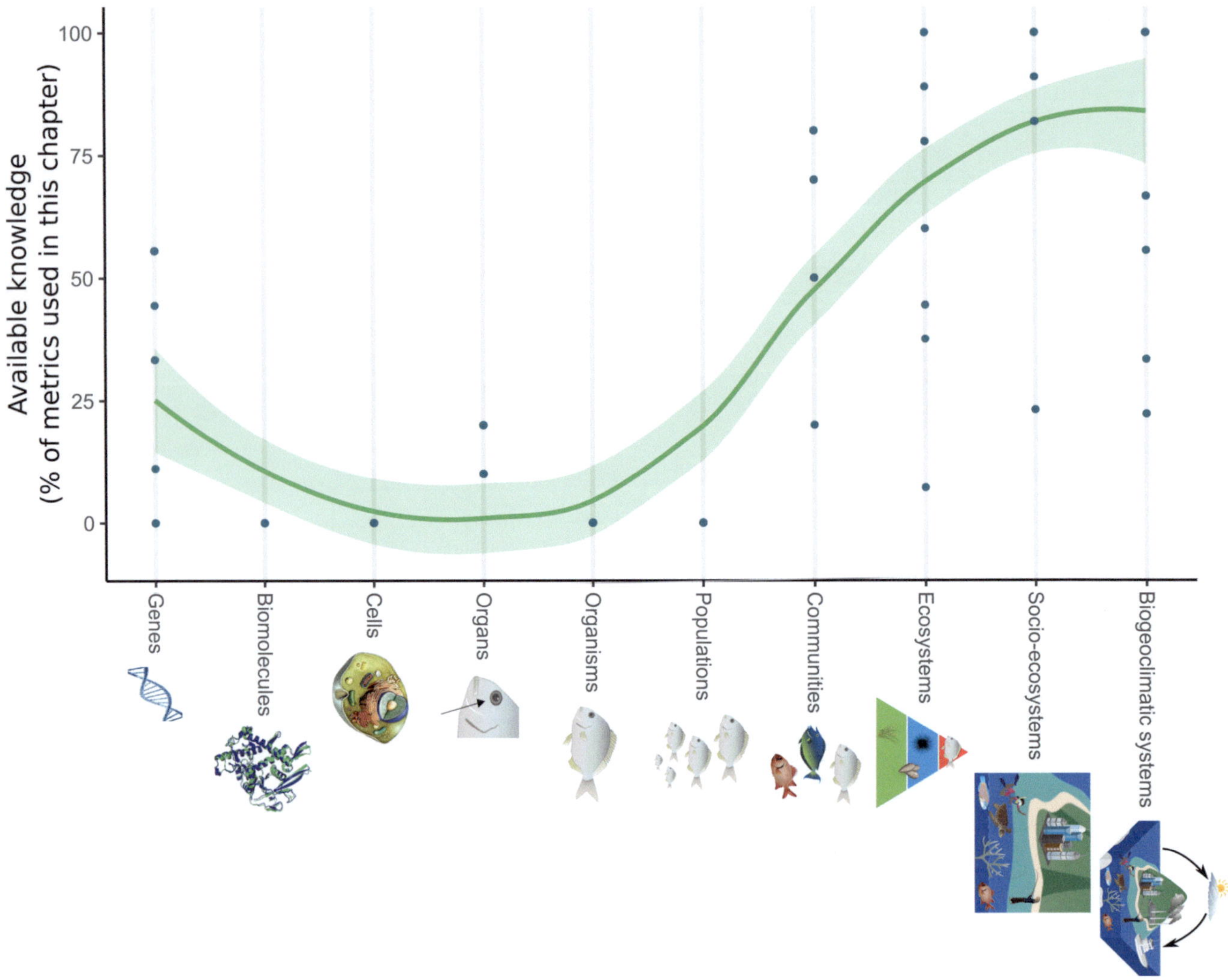

Fig. 10.33 Relationship between the state of knowledge (% of acquired data) and scale of biological and ecological organization, based on Fig. 10.32. Percentage of knowledge is estimated based on data completeness for the different metrics presented in Figs. 10.5–10.30, which is in part influenced by data accessibility for the purpose of the present study

and 10.33), which accords with the young age of scientific disciplines and limitations in exploring biodiversity at smaller spatial scales, although genetic and molecular approaches are becoming more accessible. Similarly, little data is available for small and cryptic organisms such as bacteria, viruses, and fungi, while ecologically key and emblematic macroorganisms such as reef-building corals and fishes have been more studied. Global databases play a central role in synthesizing and making such knowledge available, though are still lacking for many biological groups and disciplines. Coral reef biodiversity metrics are also unbalanced in space as reflected among French tropical overseas, although such shortcomings are also observed in other types of ecosystems where knowledge and conservation are highly skewed (Guerra et al. 2020; Maxwell et al. 2020). Little is known on cryptic and mesophotic habitats, which are considered potential refuges for corals in a warmer climate and whose exploration is only starting, including the recently discovered Great Amazonian Reef, parts of which lie in French Guinea waters (Moura et al. 2016; Francini-Filho et al. 2018; Pérez-Rosales et al. 2021b; Pérez-Rosales et al. 2022). There are still major knowledge gaps in most if not all metrics of biodiversity, and some estimations come with significant uncertainties, including recent questioning on species delineation and the speed of acclimation versus adaptation processes (Galtier 2019; Gibert et al. 2019). Overall, most of our knowledge relates to the state of coral reef biodiversity and its distribution as measured at different scales, whereas our understanding of driving mechanisms controlling biodiversity levels across scales and their dynamics remains yet too limited to enable predictions and therefore more proactive management strategies for biodiversity resiliency in the face of environmental changes (Rogers et al. 2015; Donovan et al. 2023). Nevertheless, some coral reef sites stand out as world references for exploring coral reef biodiversity and its drivers Thirukanthan et al. 2023. This includes portions of the Great Barrier Reef in Australia, such as Lizard Island (https://australian.museum/), the island of Moorea in French Polynesia, which is arguably the most studied reef system on the planet, thanks to two international field research stations acting as hubs for coral reef studies (www.criobe.pf, www.moorea.berkeley.edu) and coral reefs near Noumea in New Caledonia that have benefited from locally based scientific attention since the 1940s (www.ird.fr/nouvelle-caledonie).

10.3.14 Conclusions

Coral reef biodiversity expands in a multitude of dimensions associated with the various scientific disciplines and metrics used to characterize the variability of life forms across scales of biological and ecological organization. While biodiversity metrics are known to differ among regions, declining with distance from the center of marine biodiversity in Southeast Asia, our synthesis highlights how they also vary with the size of coral reef systems and the extent of habitats and diversity of ecological niches they provide. The exploration of coral reef biodiversity is a challenging endeavor that concentrates the efforts of an active global research community, as many knowledge gaps remain with significant unbalance between scientific disciplines and regions, alongside the decline in biodiversity levels, with life forms disappearing before they are characterized. In the face of the current biodiversity crisis, our knowledge of coral reef biodiversity appears quite superficial compared to the in-depth understanding needed to anticipate, detect, and counter ecological declines. This entails new research avenues, particularly transdisciplinary approaches to bridge our understanding of biological, ecological, and biogeoclimatic processes across scales of bioecological organization, endeavors that remain largely insufficient to comprehend the mechanisms behind biodiversity production and maintenance and safeguard their future (Jones et al. 2015; Smallegange et al. 2017; Carturan et al. 2020; Condie et al. 2021; Ouédraogo et al. 2021; Van Woesik et al. 2022; Carlot et al. 2023; Ouédraogo et al. 2023). Unanswered questions critical to human survival throughout the twenty-first century pertain to the carrying capacity of ecological systems and the effectiveness of societal paths toward sustainability. Due to their unique biodiversity and associated level of scientific attention, coral reef systems can be key to addressing these questions. In the global landscape of coral reef biodiversity, French tropical overseas are major players in this realm, thanks to their unique distribution, coverage, and a robust management and research network supported by the public system (Cocquempot et al. 2019). This unique combination provides a large set of data-rich systems that allow exploration of how biodiversity takes form at different scales of bioecological organization and interacts with human and climatic environments. While coral reef biodiversity has received increased national and international attention, restrictions in scientific research funding remain a primary impediment to advance understanding and hence conservation of biodiversity in coral reefs and globally (Huang and Huang 2018; Thirukanthan et al. 2023).

References

Adam TC, Burkepile DE, Holbrook SJ, Carpenter RC, Claudet J, Loiseau C, Thiault L, Brooks AJ, Washburn L, Schmitt RJ (2021) Running head: nutrient enrichment on a coral reef landscape-scale patterns of nutrient enrichment in a coral reef ecosystem: implications for coral to algae phase shifts. Ecol Appl 31:e2227

Adjeroud M, Kayal M, Penin L (2017) Importance of recruitment processes in the dynamics and resilience of coral reef assemblages. Mar Anim For 549:569

Adjeroud M, Kayal M, Iborra-Cantonnet C, Vercelloni J, Bosserelle P, Liao V, Chancerelle Y, Claudet J, Penin L (2018a) Recovery of coral assemblages despite acute and recurrent disturbances on a South Central Pacific reef. Sci Rep 8:9680. https://doi.org/10.1038/s41598-018-27891-3

Adjeroud M, Kayal M, Peignon C, Juncker M, Mills SC, Beldade R, Dumas P (2018b) Ephemeral and localized outbreaks of the coral predator Acanthaster cf. solaris in the southwestern lagoon of New Caledonia. Zool Stud 57

Adjeroud M, Peignon C, Gauliard C, Penin L, Kayal M (2022) Extremely high but localized pulses of coral recruitment in the southwestern lagoon of New Caledonia and implications for conservation. Mar Ecol Prog Ser 692:67–79. https://doi.org/10.3354/meps14073

Aisyah A, Triharyuni S (2017) Production, size distribution, and length weight relationship of lobster landed in the south coast of Yogyakarta, Indonesia. Indones Fish Res J 16:15–24

Akiona AK, Zgliczynski BJ, Sandin SA (2022) Length–weight relationships for 18 coral reef fish species from the Central Pacific. J Appl Ichthyol 38:118–122

Alexander JB, Bunce M, White N, Wilkinson SP, Adam AA, Berry T, Stat M, Thomas L, Newman SJ, Dugal L (2020) Development of a multi-assay approach for monitoring coral diversity using eDNA metabarcoding. Coral Reefs 39:159–171

Alves de Souza C, Mardones JI, Yñiguez AT, Le Bihan V, Guillotreau P, Gatti CMI, Richlen ML, Larsen J, Berdalet E (2022) Harmful Algae. In: Urban ER Jr, Ittekkot V (eds) Blue economy: an ocean science perspective. Springer Nature, Singapore, pp 287–317

Andréfouët S, Chagnaud N, Kranenburg C, Chauvin C (2008) Atlas des récifs coralliens de France d'outre-mer

Andréfouet S, Derville S, Buttin J, Dirberg G, Wabnitz CC, Garrigue C, Payri CE (2021) Nation-wide hierarchical and spatially-explicit framework to characterize seagrass meadows in new-Caledonia, and its potential application to the indo-Pacific. Mar Pollut Bull 173:113036

Andrello M, Darling ES, Wenger A, Suárez-Castro AF, Gelfand S, Ahmadia GN (2022) A global map of human pressures on tropical coral reefs. Conserv Lett 15:e12858

Anthony KRN, Marshall PA, Abdulla A, Beeden R, Bergh C, Black R, Eakin CM, Game ET, Gooch M, Graham NAJ, Green A, Heron SF, van Hooidonk R, Knowland C, Mangubhai S, Marshall N, Maynard JA, McGinnity P, McLeod E, PeterJ M, Nyström M, Obura D, Oliver J, Possingham HP, Pressey RL, Rowlands GP, Tamelander J, Wachenfeld D, Wear S (2015) Operationalizing resilience for

adaptive coral reef management under global environmental change. Glob Change Biol 21:48–61. https://doi.org/10.1111/gcb.12700
Aston EA, Duce S, Hoey AS, Ferrari R (2022) A protocol for extracting structural metrics from 3D reconstructions of corals. Front Mar Sci 9
Babcock RC, Plagányi ÉE, Condie SA, Westcott DA, Fletcher CS, Bonin MC, Cameron D (2020) Suppressing the next crown-of-thorns outbreak on the great barrier reef. Coral Reefs 39:1233–1244
Bar-On YM, Phillips R, Milo R (2018) The biomass distribution on earth. Proc Natl Acad Sci 115:6506–6511
Baums IB (2008) A restoration genetics guide for coral reef conservation. Mol Ecol 17:2796–2811. https://doi.org/10.1111/j.1365-294X.2008.03787.x
Beck MW, Losada IJ, Menéndez P, Reguero BG, Díaz-Simal P, Fernández F (2018) The global flood protection savings provided by coral reefs. Nat Commun 9:2186
Bellwood DR, Hughes TP, Connolly SR, Tanner J (2005) Environmental and geometric constraints on indo-Pacific coral reef biodiversity. Ecol Lett 8:643–651
Benzoni F (2019) Chapitre 5. Les coraux de récifs de Nouvelle-Calédonie, un patrimoine diversifié et précieux. In: Payri CE (ed) Nouvelle-Calédonie: Archipel de corail. IRD Éditions, Marseille, pp 49–59
Benzoni F, Stefani F, Pichon M, Galli P (2010) The name game: morpho-molecular species boundaries in the genus Psammocora (Cnidaria, Scleractinia). Zool J Linnean Soc 160:421–456. https://doi.org/10.1111/j.1096-3642.2010.00622.x
Bertrand B, Vercauteren M, Cunha E, Bécart A, Gosset D, Hédouin V (2022) Automated age-at-death estimation by cementochronology: essential application or additional complication? Am J Biol Anthropol 179:314–326
Boilard A, Dubé CE, Gruet C, Mercière A, Hernandez-Agreda A, Derome N (2020) Defining coral bleaching as a microbial dysbiosis within the coral holobiont. Microorganisms 8:1682
Bollati E, D'Angelo C, Alderdice R, Pratchett M, Ziegler M, Wiedenmann J (2020) Optical feedback loop involving dinoflagellate symbiont and scleractinian host drives colorful coral bleaching. Curr Biol 30:2433–2445
Bongaerts P, Dubé CE, Prata KE, Gijsbers JC, Achlatis M, Hernandez-Agreda A (2021) Reefscape genomics: leveraging advances in 3D imaging to assess Fine-scale patterns of genomic variation on coral reefs. Front Mar Sci 8
Borsa P, Cornaille M, Richer de Forges B (2023) Shark culling at a world heritage site. Nature 620:950–950
Bos AR, Gumanao GS, Salac FN (2008) A newly discovered predator of the crown-of-thorns starfish. Coral Reefs 27:581–581
Boucaud-Maitre D, Ferracci S, Pelczar S (2017) Quels sont les facteurs pronostiques de recours des urgentistes aux centres antipoison et de toxicovigilance? Étude rétrospective de cohorte en Guadeloupe entre 2013 et 2015. Toxicol Anal Clin 29:251–256
Boyce DG, Lotze HK, Tittensor DP, Carozza DA, Worm B (2020) Future Ocean biomass losses may widen socioeconomic equity gaps. Nat Commun 11:2235. https://doi.org/10.1038/s41467-020-15708-9
Bozec Y-M, O'Farrell S, Bruggemann JH, Luckhurst BE, Mumby PJ (2016) Tradeoffs between fisheries harvest and the resilience of coral reefs. Proc Natl Acad Sci 113:4536–4541
Bradshaw CJ, Ehrlich PR, Beattie A, Ceballos G, Crist E, Diamond J, Dirzo R, Ehrlich AH, Harte J, Harte ME (2021) Underestimating the challenges of avoiding a ghastly future. Front Conserv Sci 1:9
Bramanti L, Edmunds PJ (2016) Density-associated recruitment mediates coral population dynamics on a coral reef. Coral Reefs 35:543–553
Brandl SJ, Rasher DB, Côté IM, Casey JM, Darling ES, Lefcheck JS, Duffy JE (2019a) Coral reef ecosystem functioning: eight core processes and the role of biodiversity. Front Ecol Environ 17:445–454
Brandl SJ, Tornabene L, Goatley CH, Casey JM, Morais RA, Côté IM, Baldwin CC, Parravicini V, Schiettekatte NM, Bellwood DR (2019b) Demographic dynamics of the smallest marine vertebrates fuel coral reef ecosystem functioning. Science 364:1189–1192
Brodie J, Waterhouse J (2012) A critical review of environmental management of the 'not so Great'Barrier reef. Estuar Coast Shelf Sci 104:1–22
Burke L, Spalding M (2022) Shoreline protection by the world's coral reefs: mapping the benefits to people, assets, and infrastructure. Mar Policy 146:105311
Burke S, Pottier P, Lagisz M, Macartney EL, Ainsworth T, Drobniak SM, Nakagawa S (2023) The impact of rising temperatures on the prevalence of coral diseases and its predictability: a global meta-analysis. Ecol Lett 26:1466–1481
Burns C, Bollard B, Narayanan A (2022) Machine-learning for mapping and monitoring shallow coral reef habitats. Remote Sens 14:2666
Cabral RB, Mayorga J, Clemence M, Lynham J, Koeshendrajana S, Muawanah U, Nugroho D, Anna Z, Mira GA, Zulbainarni N, Gaines SD, Costello C (2018) Rapid and lasting gains from solving illegal fishing. Nat Ecol Evol 2:650–658. https://doi.org/10.1038/s41559-018-0499-1
Caley MJ, Fisher R, Mengersen K (2014) Global species richness estimates have not converged. Trends Ecol Evol 29:187–188. https://doi.org/10.1016/j.tree.2014.02.002
Calvin K, Dasgupta D, Krinner G, Mukherji A, Thorne PW, Trisos C, Romero J, Aldunce P, Barrett K, Blanco G, Cheung WWL, Connors S, Denton F, Diongue-Niang A, Dodman D, Garschagen M, Geden O, Hayward B, Jones C, Jotzo F, Krug T, Lasco R, Lee Y-Y, Masson-Delmotte V, Meinshausen M, Mintenbeck K, Mokssit A, Otto FEL, Pathak M, Pirani A, Poloczanska E, Pörtner H-O, Revi A, Roberts DC, Roy J, Ruane AC, Skea J, Shukla PR, Slade R, Slangen A, Sokona Y, Sörensson AA, Tignor M, Van Vuuren D, Wei Y-M, Winkler H, Zhai P, Zommers Z, Hourcade J-C, Johnson FX, Pachauri S, Simpson NP, Singh C, Thomas A, Totin E, Arias P, Bustamante M, Elgizouli I, Flato G, Howden M, Méndez-Vallejo C, Pereira JJ, Pichs-Madruga R, Rose SK, Saheb Y, Sánchez Rodríguez R, Ürge-Vorsatz D, Xiao C, Yassaa N, Alegría A, Armour K, Bednar-Friedl B, Blok K, Cissé G, Dentener F, Eriksen S, Fischer E, Garner G, Guivarch C, Haasnoot M, Hansen G, Hauser M, Hawkins E, Hermans T, Kopp R, Leprince-Ringuet N, Lewis J, Ley D, Ludden C, Niamir L, Nicholls Z, Some S, Szopa S, Trewin B, Van Der Wijst K-I, Winter G, Witting M, Birt A, Ha M, Romero J, Kim J, Haites EF, Jung Y, Stavins R, Birt A, Ha M, Orendain DJA, Ignon L, Park S, Park Y, Reisinger A, Cammaramo D, Fischlin A, Fuglestvedt JS, Hansen G, Ludden C, Masson-Delmotte V, Matthews JBR, Mintenbeck K, Pirani A, Poloczanska E, Leprince-Ringuet N, Péan C (2023) IPCC, 2023: Climate Change 2023: Synthesis Report. In: Core Writing Team, Lee H, Romero J (eds) Contribution of Working Groups I, II and III to the Sixth Assessment Report of the Intergovernmental Panel on Climate Change. Intergovernmental Panel on Climate Change (IPCC), Geneva, Switzerland
Cannon SE, Donner SD, Liu A, González Espinosa PC, Baird AH, Baum JK, Bauman AG, Beger M, Benkwitt CE, Birt MJ (2023) Macroalgae exhibit diverse responses to human disturbances on coral reefs, vol 29. Glob Change Biol, p 3318
Cantalice KM, Alvarado-Ortega J, Bellwood DR, Siqueira AC (2022) Rising from the ashes: the biogeographic origins of modern coral reef fishes. Bioscience 72:769–777. https://doi.org/10.1093/biosci/biac045
Carlot J, Kayal M, Lenihan HS, Brandl SJ, Casey JM, Adjeroud M, Cardini U, Merciere A, Espiau B, Barneche DR (2021) Juvenile corals underpin coral reef carbonate production after disturbance. Glob Change Biol 27:2623–2632
Carlot J, Vousdoukas M, Rovere A, Karambas T, Lenihan HS, Kayal M, Adjeroud M, Pérez-Rosales G, Hedouin L, Parravicini V (2023)

Coral reef structural complexity loss exposes coastlines to waves. Sci Rep 13:1683
Carroll RA, Copp RB, Davis AR, Keyzers R, Prinsep RM (2019) Marine natural products. Nat Prod Rep 36:122–173. https://doi.org/10.1039/C8NP00092A
Carroll AR, Copp BR, Davis RA, Keyzers RA, Prinsep MR (2023) Marine natural products. Nat Prod Rep 40:275
Carturan BS, Pither J, Maréchal J-P, Bradshaw CJ, Parrott L (2020) Combining agent-based, trait-based and demographic approaches to model coral-community dynamics. elife 9:e55993
Castro-Cadenas MD, Loiseau C, Reimer JM, Claudet J (2022) Tracking changes in social-ecological systems along environmental disturbances with the ocean health index. Sci Total Environ 841:156423
Castro-Sanguino C, Bozec Y-M, Condie SA, Fletcher CS, Hock K, Roelfsema C, Westcott DA, Mumby PJ (2023) Control efforts of crown-of-thorns starfish outbreaks to limit future coral decline across the Great Barrier Reef. Ecosphere 14:e4580
Cernuda-Cernuda R, García-Fernández JM (1996) Structural diversity of the ordinary and specialized lateral line organs. Microsc Res Tech 34:302–312
Chase JM, McGill BJ, McGlinn DJ, May F, Blowes SA, Xiao X, Knight TM, Purschke O, Gotelli NJ (2018) Embracing scale-dependence to achieve a deeper understanding of biodiversity and its change across communities. Ecol Lett 21:1737–1751
Chinain M, Gatti CM, Martin-Yken H, Mélanie R, Darius HT (2020) Ciguatera poisoning: An increasing burden for Pacific Island communities in light of climate change. Clim Change Mar Freshw Toxins 2nd Ed Bot LM Louzao MC Vilariño N Eds 369–428
Chinain M, Gatti CMI, Darius HT, Quod J-P, Tester PA (2021) Ciguatera poisonings: a global review of occurrences and trends. Harmful Algae 102:101873
Chinain M, Gatti Howell C, Roué M, Ung A, Henry K, Revel T, Cruchet P, Viallon J, Darius HT (2023) Ciguatera poisoning in French Polynesia: a review of the distribution and toxicity of *Gambierdiscus* spp., and related impacts on food web components and human health. Harmful Algae 129:102525. https://doi.org/10.1016/j.hal.2023.102525
Ciccarelli FD, Doerks T, Von Mering C, Creevey CJ, Snel B, Bork P (2006) Toward automatic reconstruction of a highly resolved tree of life. Science 311:1283–1287
Cinner JE, Maire E, Huchery C, MacNeil MA, Graham NA, Mora C, McClanahan TR, Barnes ML, Kittinger JN, Hicks CC (2018) Gravity of human impacts mediates coral reef conservation gains. Proc Natl Acad Sci 115:E6116–E6125
Clark TD, Raby GD, Roche DG, Binning SA, Speers-Roesch B, Jutfelt F, Sundin J (2020) Ocean acidification does not impair the behaviour of coral reef fishes. Nature 577:370–375. https://doi.org/10.1038/s41586-019-1903-y
Cleguer C, Kelly N, Tyne J, Wieser M, Peel D, Hodgson A (2021) A novel method for using small unoccupied aerial vehicles to survey wildlife species and model their density distribution. Front Mar Sci 8:640338
Clements KD, German DP, Piché J, Tribollet A, Choat JH (2017) Integrating ecological roles and trophic diversification on coral reefs: multiple lines of evidence identify parrotfishes as microphages. Biol J Linn Soc 120:729–751
Cocquempot L, Delacourt C, Paillet J, Riou P, Aucan J, Castelle B, Charria G, Claudet J, Conan P, Coppola L (2019) Coastal Ocean and nearshore observation: a French case study. Front Mar Sci 6:324
Collin A, Andel M, Lecchini D, Claudet J (2021) Mapping sub-metre 3D Land-Sea coral Reefscapes using Superspectral WorldView-3 satellite Stereoimagery. In: Oceans. MDPI, pp 315–329
Comte A, Pendleton LH (2018) Management strategies for coral reefs and people under global environmental change: 25 years of scientific research. J Environ Manag 209:462–474. https://doi.org/10.1016/j.jenvman.2017.12.051
Condie SA, Plagányi ÉE, Morello EB, Hock K, Beeden R (2018) Great barrier reef recovery through multiple interventions. Conserv Biol 32:1356–1367. https://doi.org/10.1111/cobi.13161
Condie SA, Anthony KR, Babcock RC, Baird ME, Beeden R, Fletcher CS, Gorton R, Harrison D, Hobday AJ, Plagányi ÉE (2021) Large-scale interventions may delay decline of the Great Barrier Reef. R Soc Open Sci 8:201296
Connell JH (1978) Diversity in tropical rain forests and coral reefs: high diversity of trees and corals is maintained only in a nonequilibrium state. Science 199:1302–1310
Costello MJ, May RM, Stork NE (2013) Can we name earth's species before they go extinct? Science 339:413–416. https://doi.org/10.1126/science.1230318
Côté IM, Smith NS (2018) The lionfish Pterois sp. invasion: has the worst-case scenario come to pass? J Fish Biol 92:660–689
Cowan Z-L, Pratchett M, Messmer V, Ling S (2017) Known predators of crown-of-thorns starfish (Acanthaster spp.) and their role in mitigating, if not preventing, population outbreaks. Diversity 9:7
Cowman PF, Bellwood DR (2013) The historical biogeography of coral reef fishes: global patterns of origination and dispersal. J Biogeogr 40:209–224
Cramer KL, Donovan MK, Jackson JB, Greenstein BJ, Korpanty CA, Cook GM, Pandolfi JM (2021) The transformation of Caribbean coral communities since humans. Ecol Evol 11:10098–10118
Cravatte S, Kestenare E, Eldin G, Ganachaud A, Lefèvre J, Marin F, Menkes C, Aucan J (2015) Regional circulation around New Caledonia from two decades of observations. J Mar Syst 148:249–271
Cresswell AK, Thomson DP, Haywood MDE, Renton M (2020) Frequent hydrodynamic disturbances decrease the morphological diversity and structural complexity of 3D simulated coral communities. Coral Reefs 39:1147–1161. https://doi.org/10.1007/s00338-020-01947-1
Cresswell AK, Renton M, Langlois TJ, Thomson DP, Lynn J, Claudet J (2023) Coral reef state influences resilience to acute climate-mediated disturbances. Glob Ecol Biogeogr geb.13771. https://doi.org/10.1111/geb.13771
Cuif M, Kaplan DM, Lefèvre J, Faure VM, Caillaud M, Verley P, Vigliola L, Lett C (2014) Wind-induced variability in larval retention in a coral reef system: a biophysical modelling study in the south-West lagoon of New Caledonia. Prog Oceanogr 122:105–115. https://doi.org/10.1016/j.pocean.2013.12.006
D'agata S, Mouillot D, Wantiez L, Friedlander AM, Kulbicki M, Vigliola L (2016) Marine reserves lag behind wilderness in the conservation of key functional roles. Nat Commun 7:12000
Darling ES, Alvarez-Filip L, Oliver TA, McClanahan TR, Côté IM (2012) Evaluating life-history strategies of reef corals from species traits. Ecol Lett 15:1378–1386. https://doi.org/10.1111/j.1461-0248.2012.01861.x
Darling ES, Graham NA, Januchowski-Hartley FA, Nash KL, Pratchett MS, Wilson SK (2017) Relationships between structural complexity, coral traits, and reef fish assemblages. Coral Reefs 36:561–575
Darling ES, McClanahan TR, Maina J, Gurney GG, Graham NAJ, Januchowski-Hartley F, Cinner JE, Mora C, Hicks CC, Maire E, Puotinen M, Skirving WJ, Adjeroud M, Ahmadia G, Arthur R, Bauman AG, Beger M, Berumen ML, Bigot L, Bouwmeester J, Brenier A, Bridge TCL, Brown E, Campbell SJ, Cannon S, Cauvin B, Chen CA, Claudet J, Denis V, Donner S, Estradivari FN, Feary DA, Fenner D, Fox H, Franklin EC, Friedlander A, Gilmour J, Goiran C, Guest J, Hobbs J-PA, Hoey AS, Houk P, Johnson S, Jupiter SD, Kayal M, Kuo C, Lamb J, Lee MAC, Low J, Muthiga N, Muttaqin E, Nand Y, Nash KL, Nedlic O, Pandolfi JM, Pardede S, Patankar V, Penin L, Ribas-Deulofeu L, Richards Z, Roberts TE, Rodgers KS, Safuan CDM, Sala E, Shedrawi G, Sin TM, Smallhorn-West P, Smith JE, Sommer B, Steinberg PD, Sutthacheep M, Tan CHJ, Williams GJ, Wilson S, Yeemin T, Bruno JF, Fortin M-J, Krkosek

M, Mouillot D (2019) Social–environmental drivers inform strategic management of coral reefs in the Anthropocene. Nat Ecol Evol 3:1341–1350. https://doi.org/10.1038/s41559-019-0953-8

Darwin C (1859) On the origin of species by means of natural selection, or the preservation of favoured races in the struggle for life, 1st edn. Murray, London

Daskin JH, Pringle RM (2018) Warfare and wildlife declines in Africa's protected areas. Nature 553:328–332

DASS (2018) Situation sanitaire en Nouvelle-Calédonie – La ciguatera

de Goeij JM, van Oevelen D, Vermeij MJA, Osinga R, Middelburg JJ, de Goeij AFPM, Admiraal W (2013) Surviving in a Marine Desert: the sponge loop retains resources within coral reefs. Science 342:108–110. https://doi.org/10.1126/science.1241981

De'ath G, Moran PJ (1998) Factors affecting the behaviour of crown-of-thorns starfish (Acanthaster planci L.) on the Great Barrier Reef:: 1: Patterns of activity. J Exp Mar Biol Ecol 220:83–106. https://doi.org/10.1016/S0022-0981(97)00085-3

De'Ath G, Fabricius KE, Sweatman H, Puotinen M (2012) The 27–year decline of coral cover on the Great Barrier Reef and its causes. Proc Natl Acad Sci 109:17995–17999

Deaker DJ, Byrne M (2022) Crown of thorns starfish life-history traits contribute to outbreaks, a continuing concern for coral reefs. Emerg Top Life Sci 6:67–79

Deaker DJ, Agüera A, Lin H-A, Lawson C, Budden C, Dworjanyn SA, Mos B, Byrne M (2020) The hidden army: corallivorous crown-of-thorns seastars can spend years as herbivorous juveniles. Biol Lett 16:20190849

Desbiens AA, Mumby PJ, Dworjanyn S, Plagányi ÉE, Uthicke S, Wolfe K (2023) Novel rubble-dwelling predators of herbivorous juvenile crown-of-thorns starfish (Acanthaster sp.). Coral Reefs 42:579–591. https://doi.org/10.1007/s00338-023-02364-w

Dietzel A, Bode M, Connolly SR, Hughes TP (2020) Long-term shifts in the colony size structure of coral populations along the Great Barrier Reef. Proc R Soc B Biol Sci 287:20201432. https://doi.org/10.1098/rspb.2020.1432

Dietzel A, Bode M, Connolly SR, Hughes TP (2021) The population sizes and global extinction risk of reef-building coral species at biogeographic scales. Nat Ecol Evol 5:663–669. https://doi.org/10.1038/s41559-021-01393-4

Dimoff SA, Halliday WD, Pine MK, Tietjen KL, Juanes F, Baum JK (2021) The utility of different acoustic indicators to describe biological sounds of a coral reef soundscape. Ecol Indic 124:107435. https://doi.org/10.1016/j.ecolind.2021.107435

Ditria EM, Buelow CA, Gonzalez-Rivero M, Connolly RM (2022) Artificial intelligence and automated monitoring for assisting conservation of marine ecosystems: a perspective. Front Mar Sci 9:918104

Dittami SM, Arboleda E, Auguet J-C, Bigalke A, Briand E, Cárdenas P, Cardini U, Decelle J, Engelen AH, Eveillard D (2021) A community perspective on the concept of marine holobionts: current status, challenges, and future directions. PeerJ 9:e10911

Done TJ, DeVantier LM, Turak E, Fisk DA, Wakeford M, Van Woesik R (2010) Coral growth on three reefs: development of recovery benchmarks using a space for time approach. Coral Reefs 29:815–833

Donovan MK, Alves C, Burns J, Drury C, Meier OW, Ritson-Williams R, Cunning R, Dunn RP, Goodbody-Gringley G, Henderson LM, Knapp ISS, Levy J, Logan CA, Mudge L, Sullivan C, Gates RD, Asner GP (2023) From polyps to pixels: understanding coral reef resilience to local and global change across scales. Landsc Ecol 38:737–752. https://doi.org/10.1007/s10980-022-01463-3

Doropoulos C, Roff G, Bozec Y-M, Zupan M, Werminghausen J, Mumby PJ (2016) Characterizing the ecological trade-offs throughout the early ontogeny of coral recruitment. Ecol Monogr 86:20–44

Doropoulos C, Bozec Y-M, Gouezo M, Priest MA, Thomson DP, Mumby PJ, Roff G (2022) Cryptic coral recruits as dormant 'seed banks': an unrecognised mechanism of rapid reef recovery. Ecology 103:1–14

Duarte C, Marbá N, Agawin N, Cebrián J, Enriquez S, Fortes M, Gallegos M, Merino M, Olesen B, Sand-Jensen K, Uri J, Vermaat J (1994) Reconstruction of seagrass dynamics: age determinations and associated tools for the seagrass ecologist. Mar Ecol Prog Ser 107:195–209. https://doi.org/10.3354/meps107195

Dubé CE, Boissin E, Mercière A, Planes S (2020) Parentage analyses identify local dispersal events and sibling aggregations in a natural population of Millepora hydrocorals, a free-spawning marine invertebrate. Mol Ecol 29:1508–1522. https://doi.org/10.1111/mec.15418

Dubois M, Gascuel D, Coll M, Claudet J (2019) Recovery debts can be revealed by ecosystem network-based approaches. Ecosystems 22:658–676. https://doi.org/10.1007/s10021-018-0294-5

Duprey NN, Wang TX, Kim T, Cybulski JD, Vonhof HB, Crutzen PJ, Haug GH, Sigman DM, Martínez-García A, Baker DM (2020) Megacity development and the demise of coastal coral communities: evidence from coral skeleton δ15N records in the Pearl River estuary. Glob Change Biol 26:1338–1353. https://doi.org/10.1111/gcb.14923

Eddy T, Lam V, Reygondeau G, Cisneros-Montemayor A, Greer K, Palomares M, Bruno J, Ota Y, Cheung W (2021) Global decline in capacity of coral reefs to provide ecosystem services. One Earth 4:1278. https://doi.org/10.1016/j.oneear.2021.08.016

Edmunds PJ, Burgess SC, Putnam HM, Baskett ML, Bramanti L, Fabina NS, Han X, Lesser MP, Madin JS, Wall CB (2014) Evaluating the causal basis of ecological success within the scleractinia: an integral projection model approach. Mar Biol 161:2719–2734

Edmunds PJ, Leichter JJ, Johnston EC, Tong EJ, Toonen RJ (2016) Ecological and genetic variation in reef-building corals on four S ociety I slands. Limnol Oceanogr 61:543–557

Elise S, Urbina-Barreto I, Pinel R, Mahamadaly V, Bureau S, Penin L, Adjeroud M, Kulbicki M, Bruggemann JH (2019) Assessing key ecosystem functions through soundscapes: a new perspective from coral reefs. Ecol Indic 107:105623. https://doi.org/10.1016/j.ecolind.2019.105623

Fabricius KE, Okaji K, De'Ath G (2010) Three lines of evidence to link outbreaks of the crown-of-thorns seastar Acanthaster planci to the release of larval food limitation. Coral Reefs 29:593–605

FAO (2022) The state of world fisheries and aquaculture 2022: towards blue transformation. FAO, Rome

Ferrario F, Beck MW, Storlazzi CD, Micheli F, Shepard CC, Airoldi L (2014) The effectiveness of coral reefs for coastal hazard risk reduction and adaptation. Nat Commun 5:3794

Fiddes SL, Woodhouse MT, Utembe S, Schofield R, Alexander SP, Alroe J, Chambers SD, Chen Z, Cravigan L, Dunne E, Humphries RS, Johnson G, Keywood MD, Lane TP, Miljevic B, Omori Y, Protat A, Ristovski Z, Selleck P, Swan HB, Tanimoto H, Ward JP, Williams AG (2022) The contribution of coral-reef-derived dimethyl sulfide to aerosol burden over the great barrier reef: a modelling study. Atmospheric Chem Phys 22:2419–2445. https://doi.org/10.5194/acp-22-2419-2022

Fine ML, Parmentier E (2015) Mechanisms of fish sound production. In: Ladich F (ed) Sound communication in fishes. Springer, Vienna, pp 77–126

Fisher R, O'Leary RA, Low-Choy S, Mengersen K, Knowlton N, Brainard RE, Caley MJ (2015) Species richness on coral reefs and the pursuit of convergent global estimates. Curr Biol 25:500–505

Flammang BE (2014) The fish tail as a derivation from axial musculoskeletal anatomy: an integrative analysis of functional morphology. Zoology 117:86–92. https://doi.org/10.1016/j.zool.2013.10.001

Francini-Filho RB, Asp NE, Siegle E, Hocevar J, Lowyck K, D'Avila N, Vasconcelos AA, Baitelo R, Rezende CE, Omachi CY (2018) Perspectives on the Great Amazon Reef: extension, biodiversity, and threats. Front Mar Sci 142

Fuller ZL, Mocellin VJ, Morris LA, Cantin N, Shepherd J, Sarre L, Peng J, Liao Y, Pickrell J, Andolfatto P (2020) Population genetics of the coral Acropora millepora: toward genomic prediction of bleaching. Science 369:eaba4674

Galand PE, Ruscheweyh H-J, Salazar G, Hochart C, Henry N, Hume BCC, Oliveira PH, Perdereau A, Labadie K, Belser C, Boissin E, Romac S, Poulain J, Bourdin G, Iwankow G, Moulin C, Armstrong EJ, Paz-García DA, Ziegler M, Agostini S, Banaigs B, Boss E, Bowler C, de Vargas C, Douville E, Flores M, Forcioli D, Furla P, Gilson E, Lombard F, Pesant S, Reynaud S, Thomas OP, Troublé R, Zoccola D, Voolstra CR, Thurber RV, Sunagawa S, Wincker P, Allemand D, Planes S (2023) Diversity of the Pacific Ocean coral reef microbiome. Nat Commun 14:3039. https://doi.org/10.1038/s41467-023-38500-x

Galtier N (2019) Delineating species in the speciation continuum: a proposal. Evol Appl 12:657–663

Gargominy O, Tercerie S, Régnier C, Ramage T, Dupont P, Daszkiewicz P, Poncet L (2022) TAXREF, référentiel taxonomique pour la France: méthodologie, mise en øeuvre et diffusion. PhD Thesis, PatriNat (OFB-CNRS-MNHN)

Gatti CMI, Chinain M (2023) Bilan 2022 de la ciguatéra en Polynésie française. In: Inst Louis Malardé https://www.ilm.pf/actualites-ilm/bilan-2022-de-la-ciguatera-en-polynesie-francaise/. Accessed 3 Sep 2023

Gayon J (2016) From Mendel to epigenetics: history of genetics. C R Biol 339:225–230. https://doi.org/10.1016/j.crvi.2016.05.009

Ghareeb MA, Tammam MA, El-Demerdash A, Atanasov AG (2020) Insights about clinically approved and Preclinically investigated marine natural products. Curr Res Biotechnol 2:88–102. https://doi.org/10.1016/j.crbiot.2020.09.001

Gibert P, Debat V, Ghalambor CK (2019) Phenotypic plasticity, global change, and the speed of adaptive evolution. Curr Opin Insect Sci 35:34–40

Gil MA, Baskett ML, Munch SB, Hein AM (2020) Fast behavioral feedbacks make ecosystems sensitive to pace and not just magnitude of anthropogenic environmental change. Proc Natl Acad Sci 117:25580–25589

Gilmour JP, Cook KL, Ryan NM, Puotinen ML, Green RH, Heyward AJ (2022) A tale of two reef systems: local conditions, disturbances, coral life histories, and the climate catastrophe. Ecol Appl 32:e2509

Gingeras TR (2007) Origin of phenotypes: genes and transcripts. Genome Res 17:682–690. https://doi.org/10.1101/gr.6525007

Gomez-Banderas J (2022) Marine natural products: a promising source of environmentally friendly antifouling agents for the maritime industries. Front Mar Sci 218

González-Pech RA, Li VY, Garcia V, Boville E, Mammone M, Kitano H, Ritchie KB, Medina M (2023) The evolution, assembly, and dynamics of marine Holobionts. Annu Rev Mar Sci 16

González-Rivero M, Beijbom O, Rodriguez-Ramirez A, Bryant DEP, Ganase A, Gonzalez-Marrero Y, Herrera-Reveles A, Kennedy EV, Kim CJS, Lopez-Marcano S, Markey K, Neal BP, Osborne K, Reyes-Nivia C, Sampayo EM, Stolberg K, Taylor A, Vercelloni J, Wyatt M, Hoegh-Guldberg O (2020) Monitoring of coral reefs using artificial intelligence: a feasible and cost-effective approach. Remote Sens 12:489. https://doi.org/10.3390/rs12030489

Gordon TA, Radford AN, Davidson IK, Barnes K, McCloskey K, Nedelec SL, Meekan MG, McCormick MI, Simpson SD (2019) Acoustic enrichment can enhance fish community development on degraded coral reef habitat. Nat Commun 10:5414

Gounand I, Little CJ, Harvey E, Altermatt F (2018) Cross-ecosystem carbon flows connecting ecosystems worldwide. Nat Commun 9:4825

Graham NA, Nash KL (2013) The importance of structural complexity in coral reef ecosystems. Coral Reefs 32:315–326

Graham NAJ, Robinson JPW, Smith SE, Govinden R, Gendron G, Wilson SK (2020) Changing role of coral reef marine reserves in a warming climate. Nat Commun 11:2000. https://doi.org/10.1038/s41467-020-15863-z

Gray LA, Bisonó León AG, Rojas FE, Veroneau SS, Slocum AH (2021) Caribbean-wide, negative emissions solution to Sargassum spp. low-cost collection device and sustainable disposal method. Phycology 1:49–75

Gregory TR (2008) The evolution of complex organs. Evol Educ Outreach 1:358–389

Guannel G, Arkema K, Ruggiero P, Verutes G (2016) The power of three: coral reefs, seagrasses and mangroves protect coastal regions and increase their resilience. PLoS One 11:e0158094

Guerra CA, Heintz-Buschart A, Sikorski J, Chatzinotas A, Guerrero-Ramírez N, Cesarz S, Beaumelle L, Rillig MC, Maestre FT, Delgado-Baquerizo M (2020) Blind spots in global soil biodiversity and ecosystem function research. Nat Commun 11:3870

Guillaume MM, Séret B (2021) Observations of sharks (Elasmobranchii) at Europa Island, a remote marine protected area important for shark conservation in the southern Mozambique Channel. PLoS One 16:e0253867

Guillemot N, Kulbicki M, Chabanet P, Vigliola L (2011) Functional redundancy patterns reveal non-random assembly rules in a species-rich marine assemblage. PLoS One 6:e26735

Haas AF, Guibert M, Foerschner A, Calhoun S, George E, Hatay M, Dinsdale E, Sandin SA, Smith JE, Vermeij MJ (2015) Can we measure beauty? Computational evaluation of coral reef aesthetics. PeerJ 3:e1390

Halpern BS, Longo C, Hardy D, McLeod KL, Samhouri JF, Katona SK, Kleisner K, Lester SE, O'Leary J, Ranelletti M, Rosenberg AA, Scarborough C, Selig ER, Best BD, Brumbaugh DR, Chapin FS, Crowder LB, Daly KL, Doney SC, Elfes C, Fogarty MJ, Gaines SD, Jacobsen KI, Karrer LB, Leslie HM, Neeley E, Pauly D, Polasky S, Ris B, St Martin K, Stone GS, Sumaila UR, Zeller D (2012) An index to assess the health and benefits of the global ocean. Nature 488:615–620. https://doi.org/10.1038/nature11397

Harris JA (1916) The variable desert. Sci Mon 3:41–50

Hays GC, Ashworth JS, Barnsley MJ, Broderick AC, Emery DR, Godley BJ, Henwood A, Jones EL (2001) The importance of sand albedo for the thermal conditions on sea turtle nesting beaches. Oikos 93:87–94

Heather FJ, Childs DZ, Darnaude AM, Blanchard JL (2018) Using an integral projection model to assess the effect of temperature on the growth of gilthead seabream Sparus aurata. PLoS One 13:e0196092

Hedley JD, Roelfsema CM, Chollett I, Harborne AR, Heron SF, Weeks S, Skirving WJ, Strong AE, Eakin CM, Christensen TR (2016) Remote sensing of coral reefs for monitoring and management: a review. Remote Sens 8:118

Hemingson CR, Cowman PF, Hodge JR, Bellwood DR (2019) Colour pattern divergence in reef fish species is rapid and driven by both range overlap and symmetry. Ecol Lett 22:190–199

Hernández-Delgado EA (2015) The emerging threats of climate change on tropical coastal ecosystem services, public health, local economies and livelihood sustainability of small islands: cumulative impacts and synergies. Mar Pollut Bull 101:5–28

Heudier M, Mouillot D, Mannocci L (2023) Assessing the effects of coral reef habitat and marine protected areas on threatened megafauna using aerial surveys. Aquat Conserv Mar Freshw Ecosyst 33:286–297

Hewson I, Ritchie IT, Evans JS, Altera A, Behringer D, Bowman E, Brandt M, Budd KA, Camacho RA, Cornwell TO (2023) A scuticociliate causes mass mortality of Diadema antillarum in the Caribbean Sea. Sci Adv 9:eadg3200

Hock K, Wolff NH, Ortiz JC, Condie SA, Anthony KRN, Blackwell PG, Mumby PJ (2017) Connectivity and systemic resilience of the Great Barrier Reef. PLoS Biol 15:e2003355. https://doi.org/10.1371/journal.pbio.2003355

Hodge JR, Santini F, Wainwright PC (2020) Colour dimorphism in labrid fishes as an adaptation to life on coral reefs. Proc R Soc B 287:20200167

Hoegh-Guldberg O, Poloczanska ES, Skirving W, Dove S (2017) Coral reef ecosystems under climate change and ocean acidification. Front Mar Sci 4:158

Holles S, Simpson S, Radford A, Berten L, Lecchini D (2013) Boat noise disrupts orientation behaviour in a coral reef fish. Mar Ecol Prog Ser 485:295–300. https://doi.org/10.3354/meps10346

Houk P, Raubani J (2010) Acanthaster planci outbreaks in Vanuatu coincide with ocean productivity, furthering trends throughout the Pacific Ocean. J Oceanogr 66:435–438

Huang M-H, Huang M-J (2018) An analysis of global research funding from subject field and funding agencies perspectives in the G9 countries. Scientometrics 115:833–847

Huang D, Benzoni F, Arrigoni R, Baird AH, Berumen ML, Bouwmeester J, Chou LM, Fukami H, Licuanan WY, Lovell ER, Meier R, Todd PA, Budd AF (2014a) Towards a phylogenetic classification of reef corals: the indo-Pacific genera Merulina, Goniastrea and Scapophyllia (Scleractinia, Merulinidae). Zool Scr 43:531–548. https://doi.org/10.1111/zsc.12061

Huang D, Benzoni F, Fukami H, Knowlton N, Smith ND, Budd AF (2014b) Taxonomic classification of the reef coral families Merulinidae, Montastraeidae, and Diploastraeidae (Cnidaria: Anthozoa: Scleractinia). Zool J Linnean Soc 171:277–355. https://doi.org/10.1111/zoj.12140

Hughes TP, Baird AH, Dinsdale EA, Moltschaniwskyj NA, Pratchett MS, Tanner JE, Willis BL (2000) Supply-side ecology works both ways: the link between benthic adults, fecundity, and larval recruits. Ecology 81:2241–2249. https://doi.org/10.1890/0012-9658(2000)081[2241:SSEWBW]2.0.CO;2

Hughes TP, Baird AH, Dinsdale EA, Harriott VJ, Moltschaniwskyj NA, Pratchett MS, Tanner JE, Willis BL (2002a) Detecting regional variation using meta-analysis and large-scale sampling: latitudinal patterns in recruitment. Ecology 83:436–451. https://doi.org/10.1890/0012-9658(2002)083[0436:DRVUMA]2.0.CO;2

Hughes TP, Bellwood DR, Connolly SR (2002b) Biodiversity hotspots, centres of endemicity, and the conservation of coral reefs. Ecol Lett 5:775–784. https://doi.org/10.1046/j.1461-0248.2002.00383.x

Hughes TP, Barnes ML, Bellwood DR, Cinner JE, Cumming GS, Jackson JB, Kleypas J, Van De Leemput IA, Lough JM, Morrison TH (2017) Coral reefs in the Anthropocene. Nature 546:82–90

Hughes TP, Anderson KD, Connolly SR, Heron SF, Kerry JT, Lough JM, Baird AH, Baum JK, Berumen ML, Bridge TC (2018) Spatial and temporal patterns of mass bleaching of corals in the Anthropocene. Science 359:80–83

Husson F, Josse J (2020) missMDA: handling missing values with multivariate data analysis

IPBES (2019) Summary for policymakers of the global assessment report on biodiversity and ecosystem services. Zenodo

IUCN comité français, OFB, MNHN (2020) La Liste rouge des espèces menacées en France – Chapitre des coraux constructeurs de récifs de La Réunion. de Mayotte et des îles Éparses, Paris

Jacquet C, Gounand I, Altermatt F (2020) How pulse disturbances shape size-abundance pyramids. Ecol Lett 23:1014–1023

Jamil S, Rahman M, Haider A (2021) Bag of features (BoF) based deep learning framework for bleached corals detection. Big Data Cogn Comput 5:53

Januchowski-Hartley FA, Vigliola L, Maire E, Kulbicki M, Mouillot D (2020) Low fuel cost and rising fish price threaten coral reef wilderness. Conserv Lett 13:e12706

Jiang J, Wang A, Deng X, Zhou W, Gan Q, Lu Y (2021) How Symbiodiniaceae meets the challenges of life during coral bleaching. Coral Reefs 40:1339–1353

Jin K, Wang F, Chen D, Liu H, Ding W, Shi S (2019) A new global gridded anthropogenic heat flux dataset with high spatial resolution and long-term time series. Sci Data 6:139

Jones NS, Ridgwell A, Hendy EJ (2015) Evaluation of coral reef carbonate production models at a global scale. Biogeosciences 12:1339–1356

Jones LA, Mannion PD, Farnsworth A, Valdes PJ, Kelland S-J, Allison PA (2019) Coupling of palaeontological and neontological reef coral data improves forecasts of biodiversity responses under global climatic change. R Soc Open Sci 6:182111

Juhel J-B, Vigliola L, Wantiez L, Letessier TB, Meeuwig JJ, Mouillot D (2019) Isolation and no-entry marine reserves mitigate anthropogenic impacts on grey reef shark behavior. Sci Rep 9:2897. https://doi.org/10.1038/s41598-018-37145-x

Kayal M, Kayal E (2017) Colonies of the fire coral Millepora platyphylla constitute scleractinian survival oases during Acanthaster outbreaks in French Polynesia. Mar Biodivers 47:255–258. https://doi.org/10.1007/s12526-016-0465-6

Kayal M, Adjeroud M (2022) The war of corals: patterns, drivers and implications of changing coral competitive performances across reef environments. R Soc Open Sci 9:220003. https://doi.org/10.1098/rsos.220003

Kayal M, Lenihan HS (2025) Vacuums of the sea: ecological function of large coral reef benthic scavengers in suppressing crown-of-thorns starfish (COTS) outbreaks. Ideas Ecol Evol 18:1–31. https://doi.org/10.24908/iee.2025.18.1.n

Kayal M, Lenihan HS, Pau C, Penin L, Adjeroud M (2011) Associational refuges among corals mediate impacts of a crown-of-thorns starfish Acanthaster planci outbreak: indirect positive interactions in communities. Coral Reefs 30:827–837. https://doi.org/10.1007/s00338-011-0763-1

Kayal E, Bentlage B, Collins AG, Kayal M, Pirro S, Lavrov DV (2012a) Evolution of linear mitochondrial genomes in medusozoan cnidarians. Genome Biol Evol 4:1–12

Kayal M, Vercelloni J, de Loma TL, Bosserelle P, Chancerelle Y, Geoffroy S, Stievenart C, Michonneau F, Penin L, Planes S, Adjeroud M (2012b) Predator crown-of-thorns starfish (Acanthaster planci) outbreak, Mass mortality of corals, and cascading effects on reef fish and benthic communities. PLoS One 7:e47363. https://doi.org/10.1371/journal.pone.0047363

Kayal E, Bentlage B, Cartwright P, Yanagihara AA, Lindsay DJ, Hopcroft RR, Collins AG (2015a) Phylogenetic analysis of higher-level relationships within Hydroidolina (Cnidaria: Hydrozoa) using mitochondrial genome data and insight into their mitochondrial transcription. PeerJ 3:e1403. https://doi.org/10.7717/peerj.1403

Kayal M, Vercelloni J, Wand MP, Adjeroud M (2015b) Searching for the best bet in life-strategy: a quantitative approach to individual performance and population dynamics in reef-building corals. Ecol Complex 23:73–84. https://doi.org/10.1016/j.ecocom.2015.07.003

Kayal M, Ballard J, Adjeroud M (2017a) Multi-species consumer jams and the fall of guarded corals to crown-of-thorns seastar outbreaks. F1000Res 6:1991. https://doi.org/10.12688/f1000research.13118.2

Kayal M, Bosserelle P, Adjeroud M (2017b) Bias associated with the detectability of the coral-eating pest crown-of-thorns seastar and implications for reef management. R Soc Open Sci 4:170396. https://doi.org/10.1098/rsos.170396

Kayal M, Lenihan HS, Brooks AJ, Holbrook SJ, Schmitt RJ, Kendall BE (2018) Predicting coral community recovery using multi-species population dynamics models. Ecol Lett 21:1790–1799. https://doi.org/10.1111/ele.13153

Kayal M, Lewis H, Ballard J, Kayal E (2019) Humanity and the 21 st century's resource gauntlet: a commentary on Ripple et al. 's article "World scientists' warning to humanity: a second notice". Rethink Ecol 4:21–30. https://doi.org/10.3897/rethinkingecology.4.32116

Kayal M, Mevrel E, Ballard J (2023) Video transect-based coral demographic investigation. Coasts 3(4):370–382. https://doi.org/10.3390/coasts3040022
Konopka A, Wilkins M (2012) Application of meta-transcriptomics and –proteomics to analysis of in situ physiological state. Front Microbiol 3
Kramer N, Guan J, Chen S, Wangpraseurt D, Loya Y (2022) Morpho-functional traits of the coral Stylophora pistillata enhance light capture for photosynthesis at mesophotic depths. Commun Biol 5:1–11. https://doi.org/10.1038/s42003-022-03829-4
Kroon FJ, Lefèvre CD, Doyle JR, Patel F, Milton G, Severati A, Kenway M, Johansson CL, Schnebert S, Thomas-Hall P (2020) DNA-based identification of predators of the corallivorous Crown-of-Thorns Starfish (Acanthaster cf. solaris) from fish faeces and gut contents. Sci Rep 10:8184
Kroon FJ, Barneche DR, Emslie MJ (2021) Fish predators control outbreaks of Crown-of-Thorns Starfish. Nat Commun 12:6986
Kuempel CD, Tulloch VJ, Giffin AL, Simmons BA, Hagger V, Phua C, Hoegh-Guldberg O (2022) Identifying management opportunities to combat climate, land, and marine threats across less climate exposed coral reefs. Conserv Biol 36:e13856
Kulbicki M, Guillemot N, Amand M (2005) A general approach to length-weight relationships for new Caledonian lagoon fishes. Cybium 29:235–252
Kuo N-W, Chen P-H (2009) Quantifying energy use, carbon dioxide emission, and other environmental loads from Island tourism based on a life cycle assessment approach. J Clean Prod 17:1324–1330
Kusumoto B, Costello MJ, Kubota Y, Shiono T, Wei C-L, Yasuhara M, Chao A (2020) Global distribution of coral diversity: biodiversity knowledge gradients related to spatial resolution. Ecol Res 35:315–326
Lamy T, Galzin R, Kulbicki M, Lison de Loma T, Claudet J (2016) Three decades of recurrent declines and recoveries in corals belie ongoing change in fish assemblages. Coral Reefs 35:293–302
Langlois J, Guilhaumon F, Baletaud F, Casajus N, Braga CDA, Fleuré V, Kulbicki M, Loiseau N, Mouillot D, Renoult JP (2022) The aesthetic value of reef fishes is globally mismatched to their conservation priorities. PLoS Biol 20:e3001640
Lassauce H, Chateau O, Erdmann MV, Wantiez L (2020) Diving behavior of the reef manta ray (Mobula alfredi) in New Caledonia: more frequent and deeper night-time diving to 672 meters. PLoS One 15:e0228815
Lassauce H, Dudgeon CL, Armstrong AJ, Wantiez L, Carroll EL (2022) Evidence of fine-scale genetic structure for reef manta rays Mobula alfredi in New Caledonia. Endanger Species Res 47:249–264
Lassauce H, Chateau O, Wantiez L (2023) Spatial ecology of the population of reef manta rays, Mobula alfredi (Krefft, 1868), in New Caledonia using satellite telemetry 1–horizontal behaviour. Fishes 8:328
Lavrov DV, Pett W (2016) Animal mitochondrial DNA as we do not know it: mt-genome organization and evolution in nonbilaterian lineages. Genome Biol Evol 8:2896–2913
Lee S, Ford AK, Mangubhai S, Wild C, Ferse SC (2018) Length-weight relationship, movement rates, and in situ spawning observations of Holothuria scabra (sandfish) in Fiji
Leinbach SE, Speare KE, Rossin AM, Holstein DM, Strader ME (2021) Energetic and reproductive costs of coral recovery in divergent bleaching responses. Sci Rep 11:23546
Lemer S, Saulnier D, Gueguen Y, Planes S (2015) Identification of genes associated with shell color in the black-lipped pearl oyster, Pinctada margaritifera. BMC Genomics 16:1–14
Leray M, Béraud M, Anker A, Chancerelle Y, Mills SC (2012) Acanthaster planci outbreak: decline in coral health, coral size structure modification and consequences for obligate decapod assemblages. PLoS One 7:e35456
Leray M, Yang JY, Meyer CP, Mills SC, Agudelo N, Ranwez V, Boehm JT, Machida RJ (2013) A new versatile primer set targeting a short fragment of the mitochondrial COI region for metabarcoding metazoan diversity: application for characterizing coral reef fish gut contents. Front Zool 10:34. https://doi.org/10.1186/1742-9994-10-34
Leray M, Knowlton N, Ho S-L, Nguyen BN, Machida RJ (2019) GenBank is a reliable resource for 21st century biodiversity research. Proc Natl Acad Sci 116:22651–22656. https://doi.org/10.1073/pnas.1911714116
Lester SE, Ruttenberg BI (2005) The relationship between pelagic larval duration and range size in tropical reef fishes: a synthetic analysis. Proc R Soc B Biol Sci 272:585–591
Lesturgie P, Braun CD, Clua E, Mourier J, Thorrold SR, Vignaud T, Planes S, Mona S (2023) Like a rolling stone: colonization and migration dynamics of the gray reef shark (Carcharhinus amblyrhynchos). Ecol Evol 13:e9746
Levin SA (2000) Encyclopedia of biodiversity. Elsevier Science
Levy S, Elek A, Grau-Bové X, Menéndez-Bravo S, Iglesias M, Tanay A, Mass T, Sebé-Pedrós A (2021) A stony coral cell atlas illuminates the molecular and cellular basis of coral symbiosis, calcification, and immunity. Cell 184:2973–2987
Logan A (1984) Interspecific aggression in hermatypic corals from Bermuda. Coral Reefs 3:131–138
López-García P, Moreira D (2023) The symbiotic origin of the eukaryotic cell. C R Biol 346:55–73. https://doi.org/10.5802/crbiol.118
Losfeld G, L'huillier L, Fogliani B, Jaffré T, Grison C (2015) Mining in New Caledonia: environmental stakes and restoration opportunities. Environ Sci Pollut Res 22:5592–5607
Løvholt F, Glimsdal S, Harbitz CB, Zamora N, Nadim F, Peduzzi P, Dao H, Smebye H (2012) Tsunami hazard and exposure on the global scale. Earth-Sci Rev 110:58–73
Lutzenkirchen LL, Duce SJ, Bellwood DR (2023) The global biogeography of reef morphology. Glob Ecol Biogeogr 32:1353
Madin JS, Baird AH, Dornelas M, Connolly SR (2014) Mechanical vulnerability explains size-dependent mortality of reef corals. Ecol Lett 17:1008–1015
Maillaud C, Tirard P, Borsa P, Guittonneau A-L, Fournier J, Nour M (2022) Shark attacks in New Caledonia from 1958 to 2020: a review of cases. Med Trop Sante Int 2:mtsi-v2i1
Majoris JE, Foretich MA, Hu Y, Nickles KR, Di Persia CL, Chaput R, Schlatter E, Webb JF, Paris CB, Buston PM (2021) An integrative investigation of sensory organ development and orientation behavior throughout the larval phase of a coral reef fish. Sci Rep 11:12377. https://doi.org/10.1038/s41598-021-91640-2
Manel S, Loiseau N, Andrello M, Fietz K, Goñi R, Forcada A, Lenfant P, Kininmonth S, Marcos C, Marques V (2019) Long-distance benefits of marine reserves: myth or reality? Trends Ecol Evol 34:342–354
Manel S, Guerin P-E, Mouillot D, Blanchet S, Velez L, Albouy C, Pellissier L (2020) Global determinants of freshwater and marine fish genetic diversity. Nat Commun 11:692. https://doi.org/10.1038/s41467-020-14409-7
Marhaver KL (2011) Bleaching corals of two species appear to feed from neighboring algal turfs. Coral Reefs 30:651–651
Marsh R, Addo KA, Jayson-Quashigah P-N, Oxenford HA, Maxam A, Anderson R, Skliris N, Dash J, Tompkins EL (2021) Seasonal predictions of holopelagic Sargassum across the tropical Atlantic accounting for uncertainty in drivers and processes: the SARTRAC ensemble forecast system. Front Mar Sci 8:722524
Marshall NJ (2000) Communication and camouflage with the same 'bright'colours in reef fishes. Philos Trans R Soc Lond Ser B Biol Sci 355:1243–1248
Maruska KP (2001) Morphology of the Mechanosensory lateral line system in elasmobranch fishes: ecological and behavioral considerations. Environ Biol Fish 60:47–75. https://doi.org/10.1023/A:1007647924559

Matthews SA, Mellin C, Pratchett MS (2020) Larval connectivity and water quality explain spatial distribution of crown-of-thorns starfish outbreaks across the Great Barrier Reef. In: Advances in marine biology. Elsevier, pp 223–258

Maxwell SL, Cazalis V, Dudley N, Hoffmann M, Rodrigues ASL, Stolton S, Visconti P, Woodley S, Kingston N, Lewis E, Maron M, Strassburg BBN, Wenger A, Jonas HD, Venter O, Watson JEM (2020) Area-based conservation in the twenty-first century. Nature 586:217–227. https://doi.org/10.1038/s41586-020-2773-z

McClanahan TR, Schroeder RE, Friedlander AM, Vigliola L, Wantiez L, Caselle JE, Graham NA, Wilson S, Edgar GJ, Stuart-Smith RD (2019) Global baselines and benchmarks for fish biomass: comparing remote reefs and fisheries closures. Mar Ecol Prog Ser 612:167–192

McClanahan TR, Friedlander AM, Graham NA, Chabanet P, Bruggemann JH (2021) Variability in coral reef fish baseline and benchmark biomass in the central and western Indian Ocean provinces. Aquat Conserv Mar Freshw Ecosyst 31:28–42

McCook LJ, Ayling T, Cappo M, Choat JH, Evans RD, De Freitas DM, Heupel M, Hughes TP, Jones GP, Mapstone B (2010) Adaptive management of the great barrier reef: a globally significant demonstration of the benefits of networks of marine reserves. Proc Natl Acad Sci 107:18278–18285

McGowan H, Sturman A, Saunders M, Theobald A, Wiebe A (2019) Insights from a decade of research on coral reef—atmosphere energetics. J Geophys Res Atmospheres 124:4269–4282

McGowan H, Lensky NG, Abir S, Saunders M (2022) Coral reef coupling to the atmospheric boundary layer through exchanges of heat, moisture, and momentum: case studies from tropical and desert fringing coral reefs. Front Mar Sci 9

McIvor AL, Möller I, Spencer T, Spalding M (2012) Reduction of wind and swell waves by mangroves. The Nature Conservancy and Wetlands International

McKeon CS, Moore JM (2014) Species and size diversity in protective services offered by coral guard-crabs. PeerJ 2:e574

McManus JW, Polsenberg JF (2004) Coral–algal phase shifts on coral reefs: ecological and environmental aspects. Prog Oceanogr 60:263–279

McWilliam M, Hoogenboom MO, Baird AH, Kuo C-Y, Madin JS, Hughes TP (2018) Biogeographical disparity in the functional diversity and redundancy of corals. Proc Natl Acad Sci 115:3084–3089

Mehner T, Attermeyer K, Brauns M, Brothers S, Hilt S, Scharnweber K, van Dorst RM, Vanni MJ, Gaedke U (2022) Trophic transfer efficiency in lakes. Ecosystems 25:1628–1652

Mellin C, Aaron MacNeil M, Cheal AJ, Emslie MJ, Julian Caley M (2016) Marine protected areas increase resilience among coral reef communities. Ecol Lett 19:629–637

Mellin C, Thompson A, Jonker MJ, Emslie MJ (2019) Cross-shelf variation in coral community response to disturbance on the great barrier reef. Diversity 11:38

Mendel G (1865) Versuche uber pflanzen-hybriden. Vorgelegt Den Sitzungen

Mills SC, Côté IM (2010) Crime and punishment in a roaming cleanerfish. Proc R Soc B Biol Sci 277:3617–3622

Moberg F, Folke C (1999) Ecological goods and services of coral reef ecosystems. Ecol Econ 29:215–233

Mora C, Tittensor DP, Adl S, Simpson AG, Worm B (2011) How many species are there on Earth and in the ocean? PLoS Biol 9:e1001127

Morais RA, Bellwood DR (2020) Principles for estimating fish productivity on coral reefs. Coral Reefs 39:1221–1231

Morais RA, Connolly SR, Bellwood DR (2020) Human exploitation shapes productivity–biomass relationships on coral reefs. Glob Change Biol 26:1295–1305. https://doi.org/10.1111/gcb.14941

Morais RA, Siqueira AC, Smallhorn-West PF, Bellwood DR (2021) Spatial subsidies drive sweet spots of tropical marine biomass production. PLoS Biol 19:e3001435

Morat F, Wicquart J, Schiettekatte NMD, de Sinéty G, Bienvenu J, Casey JM, Brandl SJ, Vii J, Carlot J, Degregori S, Mercière A, Fey P, Galzin R, Letourneur Y, Sasal P, Parravicini V (2020) Individual back-calculated size-at-age based on otoliths from Pacific coral reef fish species. Sci Data 7:370. https://doi.org/10.1038/s41597-020-00711-y

Morin E, Gatti C, Bambridge T, Chinain M (2016) Ciguatera fish poisoning: incidence, health costs and risk perception on Moorea Island (society archipelago, French Polynesia). Harmful Algae 60:1–10

Morrison TH, Adger N, Barnett J, Brown K, Possingham H, Hughes T (2020) Advancing coral reef governance into the Anthropocene. One Earth 2:64–74

Moura RL, Amado-Filho GM, Moraes FC, Brasileiro PS, Salomon PS, Mahiques MM, Bastos AC, Almeida MG, Silva JM Jr, Araujo BF (2016) An extensive reef system at the Amazon River mouth. Sci Adv 2:e1501252

Mourier J, Vercelloni J, Planes S (2012) Evidence of social communities in a spatially structured network of a free-ranging shark species. Anim Behav 83:389–401

Mourier J, Buray N, Schultz JK, Clua E, Planes S (2013) Genetic network and breeding patterns of a sicklefin lemon shark (Negaprion acutidens) population in the Society Islands. French Polynesia PLoS One 8:e73899

Mourier J, Maynard J, Parravicini V, Ballesta L, Clua E, Domeier ML, Planes S (2016) Extreme inverted trophic pyramid of reef sharks supported by spawning groupers. Curr Biol 26:2011–2016. https://doi.org/10.1016/j.cub.2016.05.058

Mugnai F, Meglécz E, Abbiati M, Bavestrello G, Bertasi F, Bo M, Capa M, Chenuil A, Colangelo MA, De Clerck O, Gutiérrez JM, Lattanzi L, Leduc M, Martin D, Matterson KO, Mikac B, Plaisance L, Ponti M, Riesgo A, Rossi V, Turicchia E, Waeschenbach A, Wangensteen OS, Costantini F (2021) Are well-studied marine biodiversity hotspots still blackspots for animal barcoding? Glob Ecol Conserv 32:e01909. https://doi.org/10.1016/j.gecco.2021.e01909

Multigner L, Kadhel P, Rouget F, Blanchet P, Cordier S (2016) Chlordecone exposure and adverse effects in French West Indies populations. Environ Sci Pollut Res 23:3–8

Mumby PJ, Steneck RS (2008) Coral reef management and conservation in light of rapidly evolving ecological paradigms. Trends Ecol Evol 23:555–563

Mumby PJ, Hastings A, Edwards HJ (2007) Thresholds and the resilience of Caribbean coral reefs. Nature 450:98–101

Mumby PJ, Chollett I, Bozec Y-M, Wolff NH (2014) Ecological resilience, robustness and vulnerability: how do these concepts benefit ecosystem management? Curr Opin Environ Sustain 7:22–27

Mundy CN (2000) An appraisal of methods used in coral recruitment studies. Coral Reefs 19:124–131

Nagelkerken I, Goldenberg SU, Ferreira CM, Ullah H, Connell SD (2020) Trophic pyramids reorganize when food web architecture fails to adjust to ocean change. Science 369:829–832

Nakajima Y, Wepfer PH, Suzuki S, Zayasu Y, Shinzato C, Satoh N, Mitarai S (2017) Microsatellite markers for multiple Pocillopora genetic lineages offer new insights about coral populations. Sci Rep 7:6729

Nash KL, Allen CR, Angeler DG, Barichievy C, Eason T, Garmestani AS, Graham NA, Granholm D, Knutson M, Nelson RJ (2014) Discontinuities, cross-scale patterns, and the organization of ecosystems. Ecology 95:654–667

Nichols PK, Timmers M, Marko PB (2022) Hide 'n seq: direct versus indirect metabarcoding of coral reef cryptic communities. Environ DNA 4:93–107. https://doi.org/10.1002/edn3.203

Nielsen ES, Henriques R, Beger M, Toonen RJ, Von der Heyden S (2020) Multi-model seascape genomics identifies distinct environmental drivers of selection among sympatric marine species. BMC Evol Biol 20:1–17

Ninon J (2007) La dynamique urbaine à Mayotte: l'étalement de Mamoudzou et la «périphérisation» des centres petits-terriens. Arch Sci Soc Relig 240:305–318

OECD (2020) A comprehensive overview of global biodiversity finance – recherche Google. https://www.google.com/search?client=firefox-b-d&q=A+Comprehensive+Overview+of+Global+Biodiversity+Finance. Accessed 11 Oct 2023

Osborne K, Dolman AM, Burgess SC, Johns KA (2011) Disturbance and the dynamics of coral cover on the Great Barrier Reef (1995–2009). PLoS One 6:e17516

Osborne K, Thompson AA, Cheal AJ, Emslie MJ, Johns KA, Jonker MJ, Logan M, Miller IR, Sweatman HP (2017) Delayed coral recovery in a warming ocean. Glob Change Biol 23:3869–3881

Ouassine Y, Zahir J, Conruyt N, Kayal M, Martin PA, Chenin E, Bigot L, Vignes Lebbe R (2024) Automatic coral Morphotypes detection with YOLO: a deep learning approach for efficient and accurate coral reef monitoring. In: Mercier-Laurent E, Kayakutlu G, Owoc ML, Wahid A, Mason K (eds) Artificial intelligence for knowledge management, energy and sustainability. Springer Nature Switzerland, Cham, pp 177–188

Ouédraogo D-Y, Delaunay M, Sordello R, Hédouin L, Castelin M, Perceval O, Domart-Coulon I, Burga K, Ferrier-Pagès C, Multon R (2021) Evidence on the impacts of chemicals arising from human activity on tropical reef-building corals; a systematic map. Environ Evid 10:1–18

Ouédraogo D-Y, Mell H, Perceval O, Burga K, Domart-Coulon I, Hédouin L, Delaunay M, Guillaume MM, Castelin M, Calvayrac C (2023) What are the toxicity thresholds of chemical pollutants for tropical reef-building corals? A systematic review. Environ Evid 12:1–38

Ouréns R, Flores L, Fernández L, Freire J (2013) Habitat and density-dependent growth of the sea urchin Paracentrotus lividus in Galicia (NW Spain). J Sea Res 76:50–60

Ouyang Z, Sciusco P, Jiao T, Feron S, Lei C, Li F, John R, Fan P, Li X, Williams CA (2022) Albedo changes caused by future urbanization contribute to global warming. Nat Commun 13:3800

Pacey KI, Caballes CF, Pratchett MS (2023) Using size-weight relationships to estimate biomass of heavily targeted aquarium corals by Australia's coral harvest fisheries. Sci Rep 13:1448. https://doi.org/10.1038/s41598-023-28447-w

Papastamatiou YP, Wetherbee BM, Lowe CG, Crow GL (2006) Distribution and diet of four species of carcharhinid shark in the Hawaiian Islands: evidence for resource partitioning and competitive exclusion. Mar Ecol Prog Ser 320:239–251

Pauly D (1989) A simple index of metabolic level in fishes

Pendleton LH (2010) The economic and market value of coasts and estuaries: what's at stake? Econ Mark Value Coasts Estuaries Whats Stake

Penin L, Michonneau F, Baird AH, Connolly SR, Pratchett MS, Kayal M, Adjeroud M (2010) Early post-settlement mortality and the structure of coral assemblages. Mar Ecol Prog Ser 408:55–64

Pereira HM, Ferrier S, Walters M, Geller GN, Jongman RH, Scholes RJ, Bruford MW, Brummitt N, Butchart SH, Cardoso AC (2013) Essential biodiversity variables. Science 339:277–278

Pérez-Rosales G, Brandl SJ, Chancerelle Y, Siu G, Martinez E, Parravicini V, Hédouin L (2021a) Documenting decadal disturbance dynamics reveals archipelago-specific recovery and compositional change on Polynesian reefs. Mar Pollut Bull 170:112659

Pérez-Rosales G, Rouzé H, Torda G, Bongaerts P, Pichon M, Consortium UTP, Parravicini V, Hédouin L (2021b) Mesophotic coral communities escape thermal coral bleaching in French Polynesia. R Soc Open Sci 8:210139

Pérez-Rosales G, Pichon M, Rouzé H, Villéger S, Torda G, Bongaerts P, Carlot J, Consortium UTP, Parravicini V, Hédouin L (2022) Mesophotic coral ecosystems of French Polynesia are hotspots of alpha and beta generic diversity for scleractinian assemblages. Divers Distrib 28:1391–1403

Petie R, Hall MR, Hyldahl M, Garm A (2016) Visual orientation by the crown-of-thorns starfish (Acanthaster planci). Coral Reefs 35:1139–1150. https://doi.org/10.1007/s00338-016-1478-0

Philippe S, Schoenberger S, Ahmed N (2022) Radiation exposures and compensation of victims of French atmospheric nuclear tests in Polynesia. Sci Glob Secur 30:62–94

Pimiento C, Leprieur F, Silvestro D, Lefcheck JS, Albouy C, Rasher DB, Davis M, Svenning J-C, Griffin JN (2020) Functional diversity of marine megafauna in the Anthropocene. Sci Adv 6:eaay7650

Planes S, Allemand D (2023) Insights and achievements from the Tara Pacific expedition. Nat Commun 14:3131. https://doi.org/10.1038/s41467-023-38896-6

Pratchett MS (2001) Influence of coral symbionts on feeding preferences of crown-of-thorns starfish Acanthaster planci in the western Pacific. Mar Ecol Prog Ser 214:111–119

Pratchett MS (2005) Dietary overlap among coral-feeding butterflyfishes (Chaetodontidae) at Lizard Island, northern great barrier reef. Mar Biol 148:373–382

Pratchett MS, Cumming GS (2019) Managing cross-scale dynamics in marine conservation: pest irruptions and lessons from culling of crown-of-thorns starfish (Acanthaster spp.). Biol Conserv 238:108211

Pratchett M, Caballes C, Rivera-Posada J, Sweatman H (2014) Limits to understanding and managing outbreaks of crown-of- thorns starfish (ACANTHASTER Spp.). Oceanogr Mar Biol 52:133–200. https://doi.org/10.1201/b17143-4

Pratchett MS, Caballes CF, Wilmes JC, Matthews S, Mellin C, Sweatman HPA, Nadler LE, Brodie J, Thompson CA, Hoey J, Bos AR, Byrne M, Messmer V, Fortunato SAV, Chen CCM, Buck ACE, Babcock RC, Uthicke S (2017) Thirty years of research on crown-of-thorns starfish (1986–2016): scientific advances and emerging opportunities. Diversity 9:41. https://doi.org/10.3390/d9040041

Pratchett MS, Caballes CF, Messmer V, Fletcher CS, Westcott DA (2020) Movement patterns of Pacific crown-of-thorns starfish (Acanthaster cf. solaris) linked to habitat structure and prey availability

Puotinen M, Drost E, Lowe R, Depczynski M, Radford B, Heyward A, Gilmour J (2020) Towards modelling the future risk of cyclone wave damage to the world's coral reefs. Glob Change Biol 26:4302–4315

Quaeck-Davies K, Bendall VA, MacKenzie KM, Hetherington S, Newton J, Trueman CN (2018) Teleost and elasmobranch eye lenses as a target for life-history stable isotope analyses. PeerJ 6:e4883

Quattrini AM, Rodríguez E, Faircloth BC, Cowman PF, Brugler MR, Farfan GA, Hellberg ME, Kitahara MV, Morrison CL, Paz-García DA (2020) Palaeoclimate Ocean conditions shaped the evolution of corals and their skeletons through deep time. Nat Ecol Evol 4:1531–1538

R Core Team (2023) R: a language and environment for statistical computing

Rahman MA, Amin SMN, Yusoff FM, Arshad A, Kuppan P, Nor Shamsudin M (2012) Length weight relationships and fecundity estimates of long-spined sea urchin, Diadema setosum, from the Pulau Pangkor, peninsular Malaysia. Aquat Ecosyst Health Manag 15:311–315

Raitsos DE, Brewin RJ, Zhan P, Dreano D, Pradhan Y, Nanninga GB, Hoteit I (2017) Sensing coral reef connectivity pathways from space. Sci Rep 7:9338

Ramírez-Portilla C, Bieger IM, Belleman RG, Wilke T, Flot J-F, Baird AH, Harii S, Sinniger F, Kaandorp JA (2022) Quantitative three-dimensional morphological analysis supports species discrimination in complex-shaped and taxonomically challenging corals. Front Mar Sci 9

Rapuano H, Shlesinger T, Roth L, Bronstein O, Loya Y (2023) Coming of age: annual onset of coral reproduction is determined by age rather than size. Iscience 26:106533

Reverter M, Helber SB, Rohde S, De Goeij JM, Schupp PJ (2022) Coral reef benthic community changes in the Anthropocene: biogeographic heterogeneity, overlooked configurations, and methodology. Glob Change Biol 28:1956–1971. https://doi.org/10.1111/gcb.16034

Richards ZT, Juszkiewicz DJ, Hoggett A (2021) Spatio-temporal persistence of scleractinian coral species at Lizard Island, Great Barrier Reef. Coral Reefs 40:1369–1378

Richardson LE, Graham NA, Hoey AS (2017a) Cross-scale habitat structure driven by coral species composition on tropical reefs. Sci Rep 7:7557

Richardson LE, Graham NA, Pratchett MS, Hoey AS (2017b) Structural complexity mediates functional structure of reef fish assemblages among coral habitats. Environ Biol Fish 100:193–207

Riegl B, Johnston M, Purkis S, Howells E, Burt J, Steiner SCC, Sheppard CRC, Bauman A (2018) Population collapse dynamics in Acropora downingi, an Arabian/Persian gulf ecosystem-engineering coral, linked to rising temperature. Glob Change Biol 24:2447–2462. https://doi.org/10.1111/gcb.14114

Riley M, Meagher P, Huveneers C, Leto J, Peddemors VM, Slip D, West J, Bradshaw CJA (2022) The Australian Shark-Incident Database for quantifying temporal and spatial patterns of shark-human conflict. Sci Data 9:378. https://doi.org/10.1038/s41597-022-01453-9

Robbins SJ, Song W, Engelberts JP, Glasl B, Slaby BM, Boyd J, Marangon E, Botté ES, Laffy P, Thomas T, Webster NS (2021) A genomic view of the microbiome of coral reef demosponges. ISME J 15:1641–1654. https://doi.org/10.1038/s41396-020-00876-9

Robinson JP, McDevitt-Irwin JM, Dajka J-C, Hadj-Hammou J, Howlett S, Graba-Landry A, Hoey AS, Nash KL, Wilson SK, Graham NA (2020) Habitat and fishing control grazing potential on coral reefs. Funct Ecol 34:240–251

Rogers A, Harborne AR, Brown CJ, Bozec Y-M, Castro C, Chollett I, Hock K, Knowland CA, Marshell A, Ortiz JC, Razak T, Roff G, Samper-Villarreal J, Saunders MI, Wolff NH, Mumby PJ (2015) Anticipative management for coral reef ecosystem services in the 21st century. Glob Change Biol 21:504–514. https://doi.org/10.1111/gcb.12725

Romero-Torres M, Treml EA, Acosta A, Paz-García DA (2018) The Eastern Tropical Pacific coral population connectivity and the role of the Eastern Pacific Barrier. Sci Rep 8:9354. https://doi.org/10.1038/s41598-018-27644-2

Rongo T, Bush M, Van Woesik R (2009) Did ciguatera prompt the late Holocene Polynesian voyages of discovery? J Biogeogr 36:1423–1432

Rossi S, Schubert N, Brown D, Soares M de O, Grosso V, Rangel-Huerta E, Maldonado E (2018) Linking host morphology and symbiont performance in octocorals. Sci Rep 8:12823

Rypel AL, Haag WR, Findlay RH (2008) Validation of annual growth rings in freshwater mussel shells using cross dating. Can J Fish Aquat Sci 65:2224–2232

Sahal A, Morin J, Schindelé F, Lavigne F (2011) A catalog of tsunamis in La Réunion Island from august 27th 1883 to october 26th 2010. Sci Tsunami Hazards 30

Schmitt RJ, Holbrook SJ, Brooks AJ, Lape JC (2009) Intraguild predation in a structured habitat: distinguishing multiple-predator effects from competitor effects. Ecology 90:2434–2443

Schmitt RJ, Holbrook SJ, Davis SL, Brooks AJ, Adam TC (2019) Experimental support for alternative attractors on coral reefs. Proc Natl Acad Sci 116:4372–4381

Séguigne C, Bègue M, Meyer C, Mourier J, Clua É (2023) Provisioning ecotourism does not increase tiger shark site fidelity. Sci Rep 13:7785. https://doi.org/10.1038/s41598-023-34446-8

Selmoni O, Rochat E, Lecellier G, Berteaux-Lecellier V, Joost S (2020) Seascape genomics as a new tool to empower coral reef conservation strategies: an example on North-Western Pacific Acropora digitifera. Evol Appl 13:1923–1938. https://doi.org/10.1111/eva.12944

Shanks AL (2009) Pelagic larval duration and dispersal distance revisited. Biol Bull 216:373–385. https://doi.org/10.1086/BBLv216n3p373

Shinzato C, Narisoko H, Nishitsuji K, Nagata T, Satoh N, Inoue J (2021) Novel mitochondrial DNA markers for Scleractinian corals and generic-level environmental DNA Metabarcoding. Front Mar Sci 8

Siqueira AC, Bellwood DR, Cowman PF (2019) The evolution of traits and functions in herbivorous coral reef fishes through space and time. Proc R Soc B 286:20182672

Smallegange IM, Caswell H, Toorians ME, de Roos AM (2017) Mechanistic description of population dynamics using dynamic energy budget theory incorporated into integral projection models. Methods Ecol Evol 8:146–154

Smith S (1978) Coral-reef area and the contributions of reefs to processes and resources of the world's oceans. Nature 273:225–226

Souter D, Planes S, Wicquart J, Logan M, Obura D, Staub F (2021) Status of coral reefs of the world: 2020

Spalding M, Ravilious C, Green EP (2001) World atlas of coral reefs. University of California Press

Speers AE, Besedin EY, Palardy JE, Moore C (2016) Impacts of climate change and ocean acidification on coral reef fisheries: an integrated ecological–economic model. Ecol Econ 128:33–43

Stiger-Pouvreau V, Zubia M (2020) Macroalgal diversity for sustainable biotechnological development in French tropical overseas territories. Bot Mar 63:17–41. https://doi.org/10.1515/bot-2019-0032

Sully S, Burkepile DE, Donovan MK, Hodgson G, Van Woesik R (2019) A global analysis of coral bleaching over the past two decades. Nat Commun 10:1264

Swain TD, Lax S, Gilbert J, Backman V, Marcelino LA (2021) A phylogeny-informed analysis of the global coral-Symbiodiniaceae interaction network reveals that traits correlated with thermal bleaching are specific to symbiont transmission mode. mSystems 6. https://doi.org/10.1128/msystems.00266-21

Sweatman H (2008) No-take reserves protect coral reefs from predatory starfish. Curr Biol 18:R598–R599

Taglioni F, Guiltat S, Teurlai M, Delsaut M, Payet D (2019) A spatial and environmental analysis of shark attacks on Reunion Island (1980–2017). Mar Policy 101:51–62. https://doi.org/10.1016/j.marpol.2018.12.010

Tea Y-K, Xu X, DiBattista JD, Lo N, Cowman PF, Ho SY (2022) Phylogenomic analysis of concatenated ultraconserved elements reveals the recent evolutionary radiation of the fairy wrasses (Teleostei: Labridae: Cirrhilabrus). Syst Biol 71:1–12

Tekwa EW, Catalano KA, Bazzicalupo AL, O'Connor MI, Pinsky ML (2023) The sizes of life. PLoS One 18:e0283020. https://doi.org/10.1371/journal.pone.0283020

Thiault L, Gelcich S, Marshall N, Marshall P, Chlous F, Claudet J (2020a) Operationalizing vulnerability for social–ecological integration in conservation and natural resource management. Conserv Lett 13:e12677. https://doi.org/10.1111/conl.12677

Thiault L, Weekers D, Curnock M, Marshall N, Pert PL, Beeden R, Dyer M, Claudet J (2020b) Predicting poaching risk in marine protected areas for improved patrol efficiency. J Environ Manag 254:109808

Thirukanthan CS, Azra MN, Lananan F, Sara' G, Grinfelde I, Rudovica V, Vincevica-Gaile Z, Burlakovs J (2023) The evolution of coral reef under changing climate: a Scientometric review. Animals 13:949

Trebilco R, Baum JK, Salomon AK, Dulvy NK (2013) Ecosystem ecology: size-based constraints on the pyramids of life. Trends Ecol Evol 28:423–431

Udyawer V, Goiran C, Shine R (2021) Peaceful coexistence between people and deadly wildlife: why are recreational users of the ocean so rarely bitten by sea snakes? People Nat 3:335–346

Urban ER Jr, Ittekkot V (2022) Blue economy: an ocean science perspective. Springer Nature

Urbina-Barreto I, Chiroleu F, Pinel R, Fréchon L, Mahamadaly V, Elise S, Kulbicki M, Quod J-P, Dutrieux E, Garnier R (2021) Quantifying the shelter capacity of coral reefs using photogrammetric 3D modeling: from colonies to reefscapes. Ecol Indic 121:107151

Vajed Samiei J, Saleh A, Mehdinia A, Shirvani A, Kayal M (2015) Photosynthetic response of Persian Gulf acroporid corals to summer versus winter temperature deviations. PeerJ 3:e1062

Van Woesik R, Shlesinger T, Grottoli AG, Toonen RJ, Vega Thurber R, Warner ME, Marie Hulver A, Chapron L, McLachlan RH, Albright R (2022) Coral-bleaching responses to climate change across biological scales. Glob Change Biol 28:4229–4250

Vanhatalo J, Hosack GR, Sweatman H (2017) Spatiotemporal modelling of crown-of-thorns starfish outbreaks on the Great Barrier Reef to inform control strategies. J Appl Ecol 54:188–197. https://doi.org/10.1111/1365-2664.12710

Vercelloni J, Kayal M, Chancerelle Y, Planes S (2019) Exposure, vulnerability, and resiliency of French Polynesian coral reefs to environmental disturbances. Sci Rep 9:1027. https://doi.org/10.1038/s41598-018-38228-5

Veron JEN (2000) Corals of the World

Walker JA, Alfaro ME, Noble MM, Fulton CJ (2013) Body fineness ratio as a predictor of maximum prolonged-swimming speed in coral reef fishes. PLoS One 8:e75422

Wegley Kelly L, Nelson CE, Petras D, Koester I, Quinlan ZA, Arts MGI, Nothias L-F, Comstock J, White BM, Hopmans EC, van Duyl FC, Carlson CA, Aluwihare LI, Dorrestein PC, Haas AF (2022) Distinguishing the molecular diversity, nutrient content, and energetic potential of exometabolomes produced by macroalgae and reef-building corals. Proc Natl Acad Sci 119:e2110283119. https://doi.org/10.1073/pnas.2110283119

West TO, Marland G, Singh N, Bhaduri BL, Roddy AB (2009) The human carbon budget: an estimate of the spatial distribution of metabolic carbon consumption and release in the United States. Biogeochemistry 94:29–41

West KM, Stat M, Harvey ES, Skepper CL, DiBattista JD, Richards ZT, Travers MJ, Newman SJ, Bunce M (2020) eDNA metabarcoding survey reveals fine-scale coral reef community variation across a remote, tropical Island ecosystem. Mol Ecol 29:1069–1086. https://doi.org/10.1111/mec.15382

Westcott DA, Fletcher CS, Kroon FJ, Babcock RC, Plagányi EE, Pratchett MS, Bonin MC (2020) Relative efficacy of three approaches to mitigate Crown-of-Thorns Starfish outbreaks on Australia's Great Barrier Reef. Sci Rep 10:12594

Wiedenmann J, D'Angelo C, Mardones ML, Moore S, Benkwitt CE, Graham NA, Hambach B, Wilson PA, Vanstone J, Eyal G (2023) Reef-building corals farm and feed on their photosynthetic symbionts. Nature 1–7

Williams GJ, Graham NA, Jouffray J-B, Norström AV, Nyström M, Gove JM, Heenan A, Wedding LM (2019) Coral reef ecology in the Anthropocene. Funct Ecol 33:1014–1022

Wilmes JC, Hoey AS, Pratchett MS (2020) Contrasting size and fate of juvenile crown-of-thorns starfish linked to ontogenetic diet shifts. Proc R Soc B 287:20201052

Winters G, Beer S, Willette DA, Viana IG, Chiquillo KL, Beca-Carretero P, Villamayor B, Azcárate-García T, Shem-Tov R, Mwabvu B (2020) The tropical seagrass Halophila stipulacea: reviewing what we know from its native and invasive habitats, alongside identifying knowledge gaps. Front Mar Sci 7:300

Wolfe K, Desbiens A, Stella J, Mumby PJ (2020) Length–weight relationships to quantify biomass for motile coral reef cryptofauna. Coral Reefs 39:1649–1660

Wolfe K, Kenyon TM, Desbiens A, de la Motte K, Mumby PJ (2023) Hierarchical drivers of cryptic biodiversity on coral reefs. Ecol Monogr e1586

Wolff NH, Wong A, Vitolo R, Stolberg K, Anthony KRN, Mumby PJ (2016) Temporal clustering of tropical cyclones on the Great Barrier Reef and its ecological importance. Coral Reefs 35:613–623. https://doi.org/10.1007/s00338-016-1400-9

Woodhead AJ, Hicks CC, Norström AV, Williams GJ, Graham NA (2019) Coral reef ecosystem services in the Anthropocene. Funct Ecol 33:1023–1034

Wooldridge SA, Brodie JE (2015) Environmental triggers for primary outbreaks of crown-of-thorns starfish on the Great Barrier Reef, Australia. Mar Pollut Bull 101:805–815

Yuan T, Song H, Wood R, Wang C, Oreopoulos L, Platnick SE, von Hippel S, Meyer K, Light S, Wilcox E (2022) Global reduction in ship-tracks from sulfur regulations for shipping fuel. Sci Adv 8:eabn7988

Zhang T, Chen T, Liu S, Lin X, Li S, Yan W (2023) Coral reef resilience persisted for a millennium but has declined rapidly in recent decades. Front Mar Sci 10:1143728

11 Living Among the Reefs: Insights and Perspectives from Island Archaeology in the Pacific

Eric Conte, Gabrielle Traversat, Vahine Ahuura Rurua, Philippe Béarez, René Galzin, and Guillaume Molle

Abstract

Focusing on the Pacific Islands, we explore the long relationship between human communities and coral reefs through the lens of archaeology. Building upon multidisciplinary approaches as well as new and emerging methods, we present various ways of reconstructing complex marine socio-ecosystems in the past and highlight the future challenges posed by global climate change.

Keywords

Coastal archaeology · Marine socio-ecosystems · Zooarchaeology · Palaeoenvironmental reconstruction

E. Conte
UAR-2503 MSHP, Maison des Sciences de l'Homme du Pacifique, Université de la Polynésie française, CNRS, Tahiti, French Polynesia, France
e-mail: eric.conte@upf.pf

G. Traversat (✉)
UMR 8068 TEMPS (Technologie et Ethnologie des Mondes Préhistoriques), Centre International de Recherche Archéologique sur la Polynésie (CIRAP), MSH Mondes, Bâtiment René Ginouvès, 21 Allée de l'Université, Nanterre Cedex, France

V. A. Rurua · G. Molle
Centre International de Recherche Archéologique sur la Polynésie (CIRAP) Université de la Polynésie française, Tahiti, French Polynesia, France
e-mail: guillaume.molle@upf.pf

P. Béarez
UMR 7209 AASPE (Archéozoologie, archéobotanique: sociétés, pratiques et environnements), CNRS/MNHN, Muséum d'Histoire Naturelle, Paris, France
e-mail: philippe.bearez@mnhn.fr

R. Galzin
USR 3278 CRIOBE, PSL University, CNRS-EPHE, CRIOBE-Université de Perpignan, Moorea, French Polynesia, France

Laboratoire d'Excellence CORAIL, Moorea, French Polynesia
e-mail: galzin@univ-perp.fr

11.1 Between Constraints and Opportunities: Past Human Adaptation to Coastal Environments

Adaptation to coastal environments by modern humans has been a widely discussed topic among archaeologists for decades. Early evidence of marine exploitation in the African continent dates back to the Middle Pleistocene, circa 160 ka (Jerardino and Marean 2010), with increased evidence of coastal exploitation in the region during the Late Pleistocene, around 130 ka (Will et al. 2022). Coastal environments have long been thought to hold a marginal role in human dispersal, and debates among researchers had usually favored inland environments for resource availability (Erlandson 2001). More recently though, archaeologists have suggested that coasts may have played a significant role in the dispersal of human groups out of Africa and toward Asia and Australia (Erlandson and Braje 2015) as well as the Americas (Braje et al. 2019). Although little remains from early coastal occupation due to postglacial transgression and subsequent sea-level rising, this coastal migration theory (CMT) has gained traction and prompted more research in such areas (Davis and Madsen 2020). Coasts present a variety of morphologies making them more or less hospitable and prone to resource extraction and settlement. Steep cliffs of rocky substrates, strong currents, and exposition to climatic phenomena might deter groups from settling, while others offer refuge in bays, dune systems, and mangroves. Coastal dunes and beach ridges have long been considered an interface between the sea and the land and therefore a strategic location for many oceanic communities settling on a newly discovered island (Molle and Conte 2016).

Human groups reaching the tropical and subtropical waters have met with highly productive and complex coral reef ecosystems. While covering an incredibly small portion of coasts worldwide (around 0.1% of the Earth's surface), reefs have sustained human communities throughout history, as they still do today. Coasts, and by extension reefs, in archaeological settings represent the "living edge" of human

S. Planes (ed.), *The Future of Coral Reefs*, Coral Reefs of the World 21, https://doi.org/10.1007/978-3-031-98584-3_11

communities, a transitional zone between land and the open ocean (Ford 2012). Coral reefs offer a wide array of attractive faunal and floral resources, ranging from invertebrates, osseous and cartilaginous fishes, and marine mammals and reptiles to seaweed as well as plenty of crafting and building material. A holistic approach of reef socio-ecosystems is based on the "ridge to reef" concept to encompass both landscapes and seascapes into the study of long-term anthropogenic impacts (Fitzpatrick and Giovas 2021). Reef structures offer protection to coastal settlement, serving as a natural barrier and breakwater by dissipating wave energy along the reef flat (Ferrario et al. 2014). Reefs are key areas of human interactions, the starting and ending point of sea voyages, exchange routes, and often hold deep symbolic value to the people living in those islands (Breckwoldt et al. 2022). Archaeological research has revealed a deep history of interactions between humans and coral reefs in Africa (Walter et al. 2000), the Farasan islands in Saudi Arabia (Inglis et al. 2019), the Caribbean (Giovas 2016; Serrand and Bonnissent 2018), and Australia (McNiven et al. 2014). Similarly, in the Pacific, coral reefs have sustained human groups since their initial settlement in the region.

The human colonization of Oceania dates back to the late Pleistocene, with groups reaching the Sahul continent (now mainland Australia, Tasmania, the Aru Islands, and New Guinea) from Island Southeast Asia around 65 ka. Human presence is identified c. 45–40 ka in New Guinea and the Bismarck archipelago. Across the region, archaeological deposits in caves and rock-shelters attest to the early exploitation of shell and fish resources, together with the manufacture of marine technology (Langley et al. 2021; O'Connor et al. 2011). Much later, new migrations from people sharing East-Asian ancestry led to the exploration of Remote Oceania and the discovery of the West Pacific archipelagos. Departing from Taiwan the so-called Lapita groups brought with them major technological features, ceramics, and outrigger canoes (Kirch 2010). They engaged with long-distance seafaring around 3000 years ago and successfully settled the Reef/Santa Cruz Islands, Vanuatu, and New Caledonia, to finally reach the islands of Tonga and Samoa a couple of centuries later (Kirch 2017). Following an apparent long pause in voyaging, Polynesian navigators started again explorations eastward by the end of the first millennium AD, reaching islands of Central East Polynesia around the eleventh century (Conte and Molle 2014; Kirch and Molle 2024; Sear et al. 2020), and then Hawaii, Rapa Nui (Easter Island), and finally Aotearoa, New Zealand, by the thirteenth century (Bunbury et al. 2022).

From the massive landforms and continental islands in the West to the thousands of oceanic islands scattered eastward, coral reefs form a nearly ubiquitous feature of Pacific environments. Over time, Islanders have discovered, settled, exploited, and sometimes transformed these ecological areas whether fringing, barrier reefs or coralline atolls. Island environments offer both opportunities and constraints for human activities. The first Pacific settlers overcame those constraints by transporting and introducing plants and animals to newly discovered islands (Kirch 2017) and developing advanced sailing technology to maintain long distance voyaging and exchange routes (Hermann et al. 2023). If tropical high islands provided exploitable lush vegetation, freshwater, and a variety of lithic resources, life on low coral islands, however, was all the more precarious due to their inherent limitations, from difficult access to freshwater and lack of fertile soil to increased exposure to storm events. Atoll-dwelling people found innovative ways to adapt to these environments (Cramb and Thompson 2022; Thomas 2014). Maybe the most impressive one is the development of cultivation pits, a technique documented on most Pacific atolls from Micronesia to Eastern Polynesia. Taking advantage of the Ghyben-Herzberg freshwater lens, islanders dug large pits in which they grew a variety of cultigens (Chazine 2003; Burley et al. 2020).

Pacific Islanders' adaptation to these ecological milieus has been increasingly explored through the lens of human ecodynamics (Fitzhugh et al. 2019). This multidisciplinary approach aims at combining knowledge from the environmental disciplines (geology, biology, oceanography, geography) with social sciences (archaeology, anthropology, history, linguistics, etc.) to assess complex human-environment interactions (Kirch and Hunt 1997). This framework recognizes humans as integral parts of the ecosystems they inhabit as well as active managers of their surroundings, in a constant feedback loop of environmental alterations and cultural readjustments (Leppard and Fitzpatrick 2021). In short, it allows for the integration of localized human agency (in space and time) within a finer understanding of ecosystems' dynamic processes. Because it encompasses a wide array of complex, nonlinear, interdependent, and contingent histories, human ecodynamics welcomes a variety of theoretical standpoints and emerging methodologies for tackling research questions while avoiding the pitfalls of previous deterministic and reductive framings (Fitzhugh et al. 2019). Such an approach proves suitable for the long-term study of marine socio-ecosystems as demonstrated, for instance, by Aswani and Allen (2009). To assess human ecodynamics on the large Anaho coral reef, in the Marquesas, they integrated marine biology, traditional ecological knowledge, and archaeology and offered a historical perspective on changing marine environments shaped by human activities over nearly 700 years.

Here, we review the development of archaeological research on Pacific coral reef socio-ecosystems and emerging perspectives on both theoretical and methodological sides. We tackle two major themes in Pacific archaeology, bearing potential worldwide implications: the reconstruction of exploitation and uses of reef resources in the past and the reconstruction of past anthropogenic and climatic environmental impacts on coral reefs. Reflecting upon the success of multifocal and multidisciplinary engagement toward these critical questions, we call for further and closer collaborations between archaeology, anthropology, and the natural sciences, both on the field and in the lab. Now more than ever, a clearer understanding of long-term marine socio-ecosystems dynamics may open new avenues for present-day resource management and conservation strategies in the face of global climatic changes.

11.2 Exploitation and Uses of Coral Reef Resources Through Time

11.2.1 Reconstructing Uses and Exploitation of Coral Reef in the Past: A Multifocal Endeavor

Coral reef, nearshore, and oceanic environments have always been a formidable reservoir of resources for human communities. Pacific Islanders learned how to exploit a large variety of resources for multiple reasons which basically fall within two main categories: subsistence fishing (here understood as a collective term for all aquatic resources being harvested) and acquisition of raw material for its subsequent uses in material culture. While this relationship to the sea appears at first economical, it also conveys a symbolic dimension that is particularly relevant to many coastal communities in history.

Oceanic and nearshore resources have always been a major food source for island-dwelling people (Ono et al. 2013). Reliance on aquatic resources proved critical upon the arrival of human groups, and the exploitation of marine biotopes continued to be an important activity later on despite the development of inland agriculture. The vast amounts of faunal remains recovered in archaeological excavations attest to the intense, though changing, exploitation of a large variety of resources ranging from mollusks, echinoderms, crustaceans, osseous, and cartilaginous fishes to marine mammals, seabirds, and reptiles. Seaweed consumption was also important in many indigenous cultures but is mostly documented ethnographically as it leaves no archaeological trace (Abbott 1978; Conte and Payri 2002).

While the sea offers much foodstuffs, coastal communities further used raw material such as fish teeth, skin, bones, turtle shell, sea-urchin spines, shells, pearl shell, and coral to manufacture artifacts including weapons, tools (knives, scrapers, tattooing implements, files, fishing gears, etc.), and ornaments (Lavondès 1971; Emory 1975). Coral, limestone, and beachrock were also employed as construction material and incorporated into ceremonial monuments whether as gravel, raw blocks, or cut-and-dressed slabs. Consequently, marine products are found in various archaeological contexts ranging from domestic to community, religious, and/or funerary.

The archaeological record is by nature fragmented, and only a fraction of ancient activities has left material traces. In Island contexts, archaeologists seek to reconstruct ancient fishing activities in their whole complexity (Mauss' *total social fact*), that is, going beyond the description of faunal remains and artifacts but also documenting the targeted environments; the strategies; the social, technical, and even ritual organizations; etc. To do so, different approaches and methods for identifying archaeological fauna, microenvironments, and material culture are combined within an ethnoarchaeological process. Research conducted in French Polynesia over the past decade illustrates the implementation of various methods and some innovative ones currently under exploration (Fig. 11.1). Such endeavor relies on multidisciplinary collaborations between archaeology, anthropology, marine biology, geochemistry, and ecology, focusing on a range of questions and investing long-term efforts in targeted locations.

11.2.2 Zooarchaeology and the Challenges of Faunal Identifications

Archaeologists reconstruct past subsistence through a combination of different approaches, of which the first step consists in the identification of faunal remains recovered from excavations. This builds upon qualitative and quantitative assessments of the fauna assemblages. Taxonomic identification is key, and zooarchaeologists have been relying on comparative anatomy since the beginnings of the discipline. Comparison of archaeological elements to modern taxa for which precise identification to species, weights, lengths, and capture locations are known allows for identification of exploited diversity in the past. While simple in appearance, this process does not come without difficulties (Lambrides and Weisler 2016). First, preservation of archaeological remains varies from complete to highly fragmented or burnt, hampering identifications. Additionally, recovery techniques during excavations and sieving are not standardized among archaeologists and thus create biases (Nagaoka 2005; Allen 2014; Rebolledo et al. 2021). Second, for fish remains especially, archaeologists base on different sets of diagnostic skeletal elements with inherent limitations. For a long time, in Oceania, ichthyoarchaeologists have relied on a few cranial and limited postcranial elements to identify fish, relegating vast amounts of other skeletal parts to "unidentified" bones and therefore creating a statistical bias in the assemblage (Lambrides and Weisler 2013). Some researchers have then extended identifications to other parts, now using up to

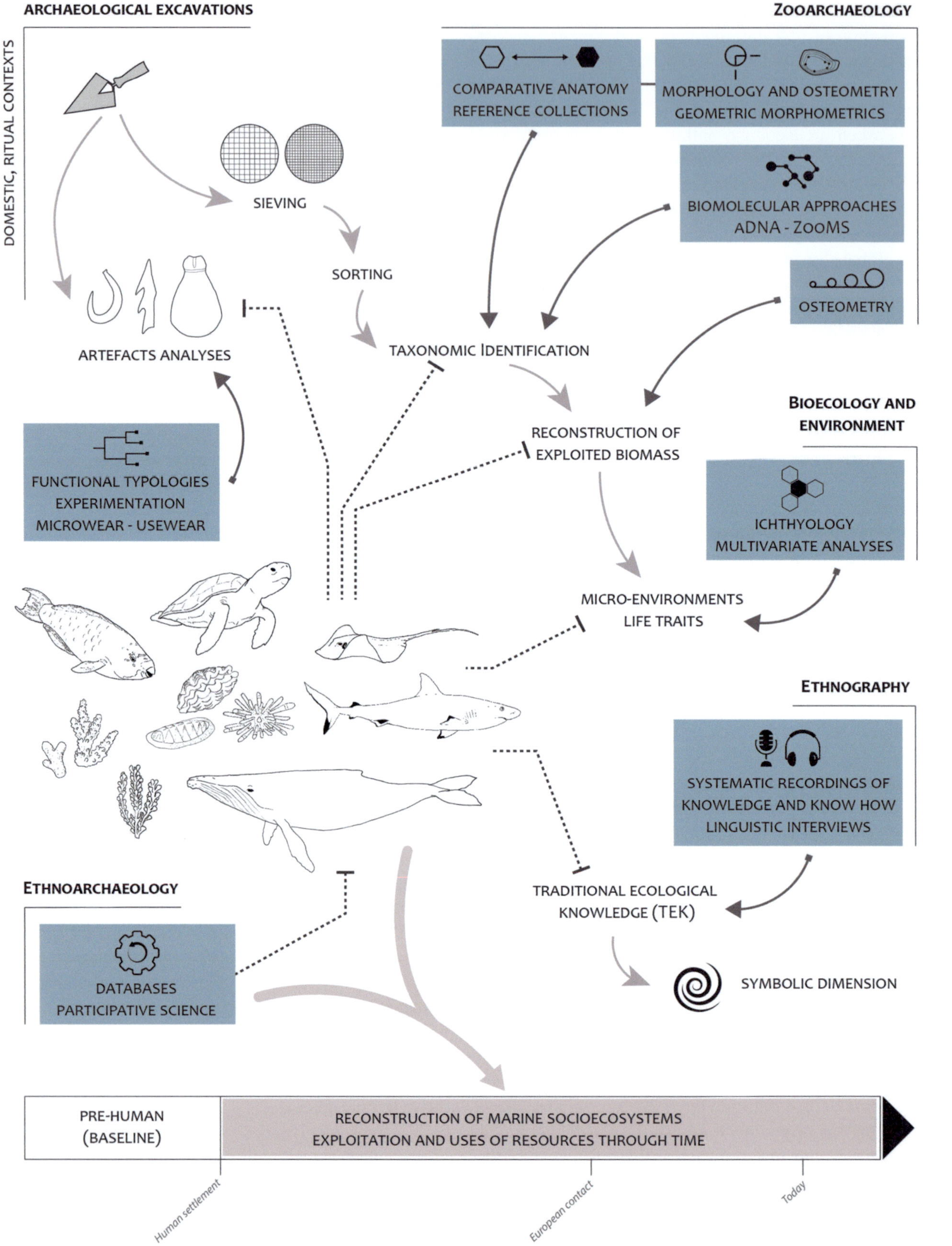

Fig. 11.1 Multifocal approach to reconstruct exploitation of marine resources in the past. Shaded boxes indicate current and future methods applied to each area of investigation

more than 40 diagnostic elements (Bouffandeau et al. 2018; Lambrides et al. 2018; Rurua 2020). Finally, the quality of identifications heavily depends on that of the reference collection that ideally should take into consideration intraspecies anatomical variations. For these reasons, analyses of ancient fauna often stay at the family or genera level and more rarely reach the species level, at least for fishes and marine turtles. Giovas et al. (2017) and Rurua (2020, 2022) have shown how detrimental the lack of level-species identification was, as it prevents any clear reconstruction of animal life traits and marine biotopes. Enhancing taxonomic identification of archaeological fauna down to the species-level thus proves a critical challenge (Lambrides and Weisler 2015).

In the case of the Arabian Peninsula, Yeomans and Beech (2021) demonstrated a strong correlation between the number of species in the reference collections and the number of fish species identifications. In the Pacific, researchers have constituted independent, nonlocalized collections, however hardly shared. Dye and Longenecker (2004) pioneered the accessibility of their Hawaiian modern and archaeological fish collections which became the only reference for tropical coral reefs. Regarding the high diversity of fish across the Pacific tropical zones, one of our biggest challenges lies in both the increase and the refinement of modern reference collections and the sharing of data at a regional scale. This could take the form of open publications of subregional anatomical atlases advocated, for instance, by Desse (1984) as a global initiative. Rurua et al. (2020) and Traversat et al. (2021) have constituted large and continuously expanding fish and shells reference collections for French Polynesia which will be made accessible online. The Indo-Pac Fishnet initiative will also offer access to digital reference material (Samper-Carro et al. 2024).

Collection practices must be standardized to overcome archipelagic endemicity biases. When difficult, morphological identifications of modern collection specimens should be supplemented by DNA barcoding (Hubert et al. 2010; http://www.mooreabiocode.org/). Because discrete characters can change their shape with the age or size of the fish, several individuals of the same species at different stages of ontogeny should be captured. Cartilaginous fishes and other protected species pose significant problems regarding capture, hence the need for further collaborations with museums of natural history (Moyer et al. 2015; Kozuch and Fitzgerald 1989). Depending on the studied areas, species-level identifications can be enhanced with morphometry, otolithometry, as well as new techniques including biomolecular approaches.

Morphological and osteometrical studies can overcome the limitations of comparative anatomy in the search for specific quantitative markers (Desse-Berset 2011). Taking advantage of scanning techniques such as X-ray computed tomography and recent advances in photogrammetry (Evin et al. 2016; Spyrou et al. 2022; Viviani et al. 2022), geometric morphometrics (GM) is used to measure and analyze the size and shape of bones using coordinates from specific, homologous locations—called landmarks—or along curves and surface (Adam et al. 2023). Samper-Carro et al. (2018) examined shape variations in vertebrae of Indo-Pacific reef and pelagic families of fish to differentiate habitats. GM studies should be developed with a focus on diagnostic elements used by zooarchaeologists. Morphometrics is also increasingly applied to otoliths, biomineral concretions present in the cranium of bony fishes, and used to discriminate between species (Duque-Velez et al. 2016). The analysis of otoliths growth rings further document age and season of capture. While seasonality is still poorly documented in the tropics, this approach bears potential for correlating age, weights, and lengths of the prey in the context of overpressure questioning. The geochemistry of coral reef fish otoliths could also segregate different habitats selected by the fish during the ontogenetic phase (Morat et al. 2015).

Biomolecular techniques offer new avenues for taxa identification and have seen major developments over the past years, along two directions. Improved DNA extractions methods and ancient genome-wide sequencing allow for exciting investigation in fish aDNA, addressing evolutionary, environmental, and taxonomic questions (Oosting et al. 2019). However, if genetic analyses of archaeological fish bones have proven successful in some regions, remains from tropical zones are more challenging. Nicholls et al. (2003) analyzed mitochondrial DNA of a Serranidae archaeological assemblage from Aitutaki, Cook Islands. Their pilot study showed that bone density (in maxilla, for instance) is important for preservation of genetic material, but DNA degradation caused by exposure and weathering of faunal remains greatly hampers the sequencing. Indeed, poor DNA preservation appears to be a common issue in the island contexts of the Pacific.

A more recent method is ZooMS or Zooarchaeology by Mass Spectrometry (Richter et al. 2022). Based on peptide mass fingerprinting (PMF), it uses preserved, robust collagen in bones to identify marker peptides specifically associated with taxonomic categories and therefore targeting samples in which decay has destroyed DNA information. This technique allows for minimally destructive sampling and is relatively "cost-effective" compared to other DNA-centered analyses. Currently available collagen markers are still biased toward European mammals but could be applied to other species through shared global databases. Harvey et al. (2019) have succeeded in turtle identification to species level with collagen. Buckley et al. (2021) combined microCT and ZooMS in application to scombrid remains from Tahuata in the Marquesas and confirmed identification of hypurals as skipjack with implications on day/night fishing strategies. An ongoing collaborative project between French Polynesia and Australia aims at enhancing collagen fingerprint references for Indo-Pacific fishes and will significantly improve archaeological applications.

Eventually, the combination of these different techniques allows for a fine-grained reconstruction of exploited biomass. At the moment, comparative anatomy informs sizes of captures but osteometry can achieve more robust reconstructions (Rurua 2020). The aim of osteometry is to build allometric regression models to determine reliable relations between the size of a particular fish bone and both the length and weight of the corresponding modern specimen. Despite an important application to archaeological material, very few studies have tackled central Pacific reef species (Marrast and Béarez 2019; Rurua et al. 2020). It thus matters to further develop such an approach with models focused on diagnostic elements often recovered in archaeological contexts (Desse 1984).

11.2.3 Exploited Diversity and Microenvironments

Reconstruction of marine resources exploitation must take into account the microenvironments that were targeted in the past. Archaeologists used to rely on biotopes to infer animal behaviors and fishing activities (Kirch and Dye 1979). In the Cook Islands, Butler (1994) highlighted the relationship between diet and chosen fishing gear. Rolett (1998) later integrated the mobility and nocturnal/diurnal activities of fish species in the case of Tahuata, in the Marquesas. On Rurutu, Austral Is., Weisler et al. (2010) interpreted anthropic activities based on fish size distribution between near- and offshore. These studies shed light on the significant variations across capture zones. Expanding on previous works, Rurua (2020, Rurua et al. in prep.) defined microenvironments as ecological milieu where species gather for reasons summarized in biological life traits, including diet, mobility, gregariousness, nycthemeral rhythm, type of reproduction, habitat distribution along the water column, and life stages (Galzin and Harmelin-Vivien 2002). Each trait comprises different attributes to score as presence/absence per identified archaeological remain (i.e., for mobility: sedentary, mobile, very mobile). Compiled in databases, information feeds into multivariate analyses to evaluate the Bray-Curtis dissimilarity/similarity index. Resulting graphic representations on MDS-plots show groupings characterized by a series of attributes that help us to interpret the microenvironments associated with archaeological specimens. Multidimensional scaling tools are flexible and allow future integration of various parameters such as highly adapted fishing techniques. Increased collaborations with marine biologists are critical to record life traits of species identified in archaeological assemblages. When data is lacking, dedicated multidisciplinary missions could be of great benefit to all disciplines (Planes et al. 1995).

11.2.4 Exploring Material Culture

Among material remains recovered in archaeological excavations are fishing gear including fish hooks, mostly made of pearl shell, some fishing weights, rare harpoon heads in bone or shell, etc. (Fig. 11.2). Associated elements such as fishing lines or nets were made of perishable material and therefore have not stood the test of time. Ethnography also informs us about traditional fishing techniques that did not make use of material implements (capture by hand, poisonous substances etc.). Consequently, archaeological artifacts only provide us with a truncated view of the ancient fishing technology deployed by islanders. Let's add that this equipment excavated from sites inland is disconnected from their areas of use, either on the reefs or offshore, which can complicate functional interpretation.

Aside from portable implements, ancient islanders also produced and managed seafood, thanks to fishing ponds and weirs. In Hawaii, large fishponds called *loko i'a* still function as estuaries providing habitats to juveniles (Summers 1964; Wyban 1992). On Pacific atolls, fish weirs are located in channels connecting the lagoon to the ocean and trap fishes in tidal waters.

Pacific archaeology has long focused on the study of fish hooks, by far the most common type of artifact recovered from coastal sites, made either of pearl shell, bone, trochus and turbo shell, sea turtle shell, or mussel. They were also made out of wood for large shark hooks and even in vegetal material such as candlenut endocarps in the case of Rapa iti (Szabó et al. 2012). The large diversity in shapes and sizes led to morphological classifications little informative about fishing techniques but rather employed as cultural markers and proxies for human migrations in the Pacific, within diffusionist perspectives (Emory et al. 1959). Yet, it remains critical to relate a type of fish hook to both the prey(s) and associated fishing strategy, which calls for a functional approach to these implements (Allen 1996). To do so, building standardized typologies is essential; however, a vast quantity of recovered remains are fragmented, and the complete hooks useful for analyses only represent a small fraction of the assemblages. Archaeologists have attempted to reconstruct complete hooks from fragments with different methods, but it is a difficult and time-consuming task. Building on a new measuring method (Carlier and Conte 2009), Chabrier et al. (2012) proposed an automatized protocol to measure hooks with high precision. Next step will consist in identifying types and sizes of the complete fish hook from a single fragment, thanks to machine-learning processes.

Fig. 11.2 Uses of marine resources in material culture (scale 2 cm). (**a**) Pearl shell trolling lure; (**b**) pearl shell fishhook; (**c**) urchin-spine file; (**d**) Cypraecassis rufa pu (wind instrument); (**e**) Tridacna shell adze (**a–d**) Marquesas. (Photos by G. Traversat, e. Tikopia, by I. Shaw 2022)

Of great interest is also the manufacture sequence of actions (*chaîne opératoire*) from the acquisition of raw material to the use and discard of the artifact. Natural stocks of *Pinctada margaritifera* pearl shell vary among islands, resulting in some material limitations that Polynesians overcame by replacing *Pinctada* by *Turbo setosus* hooks, for instance, in the Cook Is. (Carlier 2017). In the case of the Marquesas Is. where stocks are limited, part of the pearl shell was likely traded and imported from rich lagoons in the Tuamotu atolls. Preliminary analyses have shown microstructural variations between islands with different ecological parameters that may help discriminating atolls sources in the future. Sourcing analysis of mollusk shell compositions has already been successful elsewhere through stable isotopes (Grimstead et al. 2013) or more recently rare earth elements through laser-ablation inductively coupled plasma mass spectrometry (Mouchi et al. 2021). Future sourcing analyses of Marquesan fish hooks and pearl shell material can then shed light on regional exchanges routes and relationships between ancient Polynesian chiefdoms (Weisler et al. 2016). Documenting manufacturing sequences and uses would also benefit greatly from technological experimentation coupled with micro-wear and use-wear analyses, so far only applied to obsidian or bone material.

11.2.5 Ecological and Cultural Continuities: The Ethnoarchaeological Approach

To go over the technical dimension and to tackle the complexity of the human exploitation of coral reef resources, present-day "ethnographic" information can enrich and foster interpretations of archaeological data in the past. Polynesian Islands are ideal for such an ambitious ethnoarchaeological approach: first, the chronology of human occupation is relatively short, less than 1000 years between the first Polynesian settlements and the European contact. While major sociopolitical transformations took place in some places (for instance, in Hawai'i, Tonga, or Tahiti), many subsistence practices have on the other hand persisted in time. Cautious readings of early Westerners accounts can thus carry useful information even about the distant past, notably in terms of exploited species and targeted biotopes. Since initial human occupations in Central East Polynesia, nearshore and marine environments have retained most characteristics with regard to tidal cycles, currents, and species composition. Therefore, archaeological and ethnographic studies have highlighted some trends of continuous exploitation of marine resources (Rolett 1998; Rurua 2020; Traversat 2022; Traversat et al. 2021).

Despite some obvious changes in the fishing equipment since the nineteenth century, technical and ecological knowledge continued to be transmitted through generations. For example, Marquesan fishermen on Ua Huka still know very well the fishing spots, *toka*, located thanks to a combination of landmarks and of which a large part have been exploited for centuries. Therefore, the ecological and cultural continuity attested in some Pacific regions allow for an interpretive framework.

While potentially informative on technical, social, and even ritual aspects, reliance on ethnohistorical sources, either written or iconographic, must be moderated. Information cannot be directly applicable and is very heterogeneous between island groups. European accounts often insisted on the most spectacular techniques (capture of sharks or turtles) to the detriment of the daily, humble fishing practices. Reconstruction of ancient fishing is then achievable only through an integrated ethnological information that aims at documenting all dimensions of the exploitation activities and relating prey, implements, targeted environments, ecological conditions to fishing strategies. It further informs on social aspects (collective vs. individual, gendered activities, etc.) as well as symbolic and ritual dimensions (fishing lore, embodiment of spirits and ancestors into certain animals, propitiatory rites, sacred prohibitions, ritual management of the resources, etc.).

Such approach has been deployed to some extent, in Tokelau (Ono and Addison 2009), Niuatoputapu (Kirch and Dye 1979), the Marquesas (Rolett 1998), Hawaii (Kirch 1982; Jones 2004), the Lau Is. in Fiji (Jones 2009), and more thoroughly in the Tuamotus (Conte 1988). In that latter case, the exceptional preservation of many traditional practices related to marine exploitation on Napuka and Tepoto atolls enabled the recording of no less than 103 fishing techniques and associated ecological knowledge. This study also documented the treatment of the catches after the capture, from scaling to cooking and discarding. The constitution of such a body of information is a necessary step toward a fine-grained comprehension of human-coral reef relationships in the *longue durée*.

11.2.6 Beyond Ethnography: Developing a Tool to Understand the Logics of Adaptation

The overall ambition of the ethnoarchaeological approach is to deconstruct the mechanisms that determine the adoption of a fishing implement and technique depending on preys' behaviors and the ecological parameters in which they can be captured. In consequence, one can offer a body of credible hypotheses about fishing strategies targeting species whose remains are found in the archaeological assemblages. Like this, we can widen the interpretive potential regardless of time period or location.

We call for the implementation of a large-scale database to aggregate all data on recorded fishing strategies, divided into elemental analytical units, i.e., all criteria described above. Increased information on fishing strategies will also feed into fine-grained multivariate analyses of life traits as discussed supra. Starting with the Napuka-Tepoto corpus, databases of this kind will mobilize all sources including ethnohistorical observations, ethnographic studies, and future, focused investigations with the practitioners and indigenous communities. This project would also benefit from participative science as professional fishermen and elders from various islands can record themselves and share knowledge in their native language. Ongoing collaborations with linguists already demonstrate the value of topical interviews, which further help the preservation and protection of traditional knowledge. Therefore, the involvement of researchers from diverse areas of expertise is essential to such a multidisciplinary endeavor.

11.3 More Than Food: Accessing the Symbolic Dimension of Marine-Reef Resources

All around the world, coastal societies have always cultivated a deep symbolic connection with the sea. This is particularly true in Oceania where Islanders have integrated its conceptual and ritual dimensions in their daily lives (Bataille-Benguigui 1994).

In many Polynesian cultures, ancestors were thought to transform into or to penetrate an animal to watch over their kingroup. These protector spirits are called *tuputupua* in the Tuamotus (Prévost 2017), or `*aumakua* in Hawaii (Beckwith

1917). Ottino (1965) also discussed the term *tāura* referring to almost exclusively animal symbols attached to people, social groups, or geographical regions. Quite often, such incarnations are sharks (see Torrente et al. 2018) but also sperm whales, coconut crabs, moray eels, octopuses, etc. Turtles were the most sacred animals and considered as the messengers of the deities and the ancestors which seasonal migrations marked the traditional calendar and the beginning of the abundance season. First returning turtles were captured, cooked, and consumed collectively by the clan for strengthening social cohesion (Conte 1988). Similarly, marine animals were "manipulated" within various rituals of which only a small part is recorded in ethnography and oral traditions. Among them, fertility/propitiatory rites to favor the capture of certain species necessitated the deposition of a specimen on a sacred site, like the magic fish stones (*puna i'a*) on Polynesian *marae*. Other rites aimed at protecting the fishermen or the canoe-voyagers, or imbuing equipment with efficiency. Marine animals were also carved in coral, wood, or stone, either as portable objects or petroglyphs.

While sea-related ritual culture in Oceania is documented through artifacts and paraphernalia collected since the eighteenth century, they remain unfortunately out of context and often prevent archaeologists from accessing their original meaning. Larger discussions of ritual practices and Oceanic religious systems have been hampered by limited excavation data and an uncritical overreliance on ethnohistorical material. Consequently, efforts in developing an "archaeology of ritual framework" should be pursued. Evidence of ritual practices can be retrieved with appropriate excavating methodologies and targeted zooarchaeological approaches. Along this line, investigation of ritual features on ceremonial sites in Fakahina (Tuamotus) yielded faunal remains of turtles, fish, and shells which informed us on selected species but also aspects of preparation (butchery marks), cooking, and offering of marine resources manipulated in space and time (Claringbold 2021; Claringbold et al. 2024; Molle et al. 2021).

In the same vein, Molle et al. (2023) reviewed the ritual value of coral across central-east Polynesia, New Caledonia, and the Torres Strait Islands (see also Jimenez and Orejas 2016 for a worldwide survey). Islanders gathered alive and dead coral, the "stone from the sea," and used it in different forms: coral was transformed into coral lime to decorate the lapita pots 3000 years BP (Bedford 2006) and employed in the construction of catholic missionary villages in the nineteenth century (Lagarde et al. 2020); branches could serve as offerings on ceremonial sites; and cut-and-dressed blocks or slabs were integrated in the dry-masonry of temples (Fig. 11.3b). Coral material (both coral files, offerings, and construction material) prove particularly useful to archaeologists for 230Th/U uranium series high-resolution dating (Kirch and Sharp 2005; Montaggioni et al. 2019). This method continues to improve and offer new opportunities to refine chronological sequences in combination with other radiocarbon dating techniques.

Beyond a long and deep ecological knowledge of their marine environments, Pacific Islanders' seascapes also encompass specific locations (within the lagoons, on the reefs, etc.) subject to various prohibitions, taboos, and access rules. While the ocean is connecting close and distant communities together, it is also a way among others to reach out to the ancestors. On most Polynesian islands, the souls' jumping-off spot for the spirits of the dead to sail back to the mythical birthplace of Hawaiki usually consists of a point or a rock by the sea facing westward. Within the complex funerary systems documented in Oceania, burials at sea were probably a common custom in the past, notably for atoll-dwelling people (Spenneman 1999). Therefore, lagoons, reefs, and ocean were not just exploited areas but veritable parts of the territories continuously used and experienced by islanders since the first installations.

11.4 Modeling Anthropic and Natural Impacts on Coral Reefs Through Time

Whether for economic purposes or symbolic reasons, the exploitation of reef resources by human groups in the Pacific has always been subjected to various environmental constraints, hazards, and unpredictability that may drive changes in decision-making. Within these socio-ecological systems, we can examine adaptive cycles through which human communities play a transformative role in their environment and reciprocally respond to natural phenomena influencing their lifeways. Archaeology can help to better characterize those often-elusive processes to understand past human responses, adaptability, failures, and resilience. Furthermore, it adds to ever-growing palaeoenvironmental data sets and contributes to reconstruction of past climatic events.

11.4.1 Direct and Indirect Anthropic Impacts on Coral Reef Systems

Human societies have drastically modified island landscapes since their initial settlement. Whether conscious or involuntary, these processes may have consequences on their nearshore and marine areas in some ways. Direct impacts include the overexploitation of nearshore resources to the point of extirpation of species, while indirect impacts stem from either the introduction of invasive or predatory species triggering chain reactions in ecosystem structures, or terrestrial activities causing shoreline transformations, terrigenous sediment runoff, or habitat loss.

Fig. 11.3 (**a**) Niu maru, coral magic stone, Tuamotu (drawing by J.-L. Saquet after Emory, in Molle 2014); (**b**) coral slabs on the marae Apataki, Fakahina, Tuamotu. (Photo by G. Molle)

The anthropic impacts on the Pacific islands' flora and fauna have been widely identified in the archaeological record. Bird and seabird extinction is a well-documented pan-Pacific phenomenon and reflects both human overexploitation and rat predation on nests and juveniles during the first centuries of installation (Steadman 1995). In some cases, archaeological science can track the effects of such events on nutrient cycling and trophic ecology. The use of stable carbon and nitrogen isotopes on rat remains (*Rattus exulans* being a commensal species introduced by earlier colonists) permitted the detection of nutrient flow across time and derived landscape changes. Therefore, in the Gambier Is., the removal of seabirds and guano sources in islands naturally lacking nutrient-rich soils provoked a decline in phosphorus and nitrogen inputs, which later drove an impoverishment of the vegetal cover already disturbed by human deforestation (Kirch et al. 2015; Swift et al. 2018, 2021).

Such ecological changes prove more challenging to assess for marine environments. As Allen (2003) pointed out, nearshore environments were not impacted as dramatically as the land, or at least not on the same timescale. On one hand, humans have not translocated marine alien taxa that could have significantly disrupted marine ecosystems until fairly recently. On the other hand, anthropic activities such as cultivation or forest clearing led to slope erosion and shoreline expansion (Kahn et al. 2015a). Such processes undoubtedly had consequences on coastal water turbidity, the overall reef health, and its resources. In Fiji, quantity of filter feeder bivalves *Atactodea striata* and *Gafrarium tumidum* likely lessened because of the increase of clay and silt particles generated by terrigenous runoff (Morrison and Cochrane 2008).

Archaeologists use to infer overexploitation of resources from three documented outcomes: extirpation of species, change in size, and age structures and trophic alteration visible through change in diversity, richness, and abundance (Harris and Weisler 2017). Nonetheless, identifying anthropic pressure, and in the most extreme case overexploitation, remains difficult. Because many marine species are mobile and have rapid reproducing rate, they are fairly resilient from overexploitation. Extirpation cases have been documented for mollusks, especially Tridacna bivalves, in several archipelagos, and were partly interpreted through human pressure, because of its slow growth rate and clustered lifestyle (Morrison and Allen 2017).

Size and age structure change in molluscan remains have also been used to detect human pressure (Kirch et al. 1995; Masse et al. 2006; Swalding 1976, 1986; Rogers and Weisler 2019). In the Marquesas, at the Hane site, we observed a decrease in plate mean size coupled with an abrupt diminution of remains of the polyplacophoran *Acanthopleura gemmata*. This loss seems to be correlated with the quantitative increase in the much smaller yet plentiful *Cellana radiata* limpets (Traversat et al. 2021, 2024). However, size change cannot be solely attributed to human agency as molluscan size and size at age are complex and dynamic intraspecific processes. Moreover, a breadth of non-anthropic factors, such as climate change, might have been a play. In Palau, Micronesia, Giovas and colleagues proposed a series of eight nonexclusive hypotheses as to why *Strombus gibberulus*'s size increased while being consistently exploited from 3000 BP to present (Giovas et al. 2010).

On a larger ecological scale, archaeologists further look for signs of trophic alteration through change in diversity, richness, and abundance of assemblages (Morrison and Addison 2009). Removal of predators or key species from the environment may drive changes in the overall ecosystem structure. Among taxa deserving more attention are sea turtles. Turtles are a key species for reef ecosystems as they transfer nutrients from their foraging to nesting grounds (Bjorndal and Jackson 2003) and graze off seagrass. They have been a favored prey for Pacific communities for centuries, holding high symbolic value (Allen 2007). Because of their long gestation periods and low fecundity, they are highly susceptible to anthropic pressure and quick extirpation. This is attested by sometimes drastic reduction of turtle remains in faunal assemblage (ibid.). However, human/turtle interactions and the wider impact of their depletion in nearshore ecosystems are poorly documented. Trophic-level analyses of such taxa could not only inform archaeology on the role that turtles held in reef ecosystems but also benefit modern day fishery management in providing deep-time records of resource exploitation which too often rely on new baselines (Erlandson and Rick 2010).

Islanders have impacted their environment in multiple ways and to various degrees. Archaeology offers both past snapshots and long-term perspective to consider the sustainability of practices and subsequent resilience (or lack of) of targeted taxa (Swift et al. 2021). To tackle the diversity of factors impacting reef ecosystems, multifocal studies of coastal sites offering long-term sequences of occupation are favored. However, locating such sites can prove challenging as coastal transformations tend to mask early traces of human presence. Paleoenvironmental surveys through coring transects can facilitate the discovery of subdeposits, and this methodology was successfully applied in different places (Kahn et al. 2015b). Additionally, sequences spanning both pre- and post-European periods further document modern impacts of fishing on coral reef structures as demonstrated in Samoa (Craig et al. 2008) and New Guinea (Berzunza-Sanchez et al. 2013).

11.4.2 Integrating Past Climatic Changes and Events with Archaeological Records

Climatic phenomena, ranging from slow, gradual change (sea-level rise) and cyclical oscillations (ENSO) to paroxysmal events (cyclones, tsunami), have played a significant role, though long neglected, in the development of Pacific societies. They either hindered or facilitated long-distance voyaging and exploration, modified coral reef structures and their resources, and drove temporary or permanent displacement of populations. Consequently, archaeology increasingly acknowledges climatic phenomena in the comprehension of sociocultural processes. This materializes in two nonexclusive but rather complementary ways: first, archaeologists integrate paleoenvironmental data deriving from various disciplines (paleoclimatology, paleoecology, geomorphology) into their models; second, archaeological material offer innovative proxies thanks to recent advancements in analytical methods. The area of paleoclimate change thus appears a growing field of interest to many disciplines. Below, we discuss these two exciting avenues in relation to coral reef contexts in the Pacific.

Archaeology has long considered ENSO variability in the settlement models, as a driver for human dispersal in Oceania (Irwin 1992; Conte 2023). Using δ18O and δ15O shell records from archaeological sites in Vanuatu and New Caledonia, Duprey et al. (2014) demonstrated a strong climatic variability coupled with an intensification of El Niño activity around 3000 BP, which matches the first Lapita movements to Remote Oceania. A recent study combining archaeological data with indicators from lake coring reconstructed multiproxy hydroclimate sequences from Vanuatu, Samoa, and the Southern Cook. It especially shows a prolonged drought in the region which may have instigated initial exploring voyages eastward from AD 800, followed by progressive colonization a few centuries later, maybe supported by increased precipitations (Sear et al. 2020). Incremental human settlements of Pacific islands can thus greatly benefit from fine-grained palaeoclimatic reconstructions, complemented with geological estimations of atolls viability (Dickinson 2003; Montaggioni et al. 2019).

It is still a challenging endeavor as the effects of ENSO-related events are highly variable regionally (Allen 2006). Integrated studies remain scarce in the Pacific and would take advantage of more local data to apprehend the complexity of ENSO patterns in time. Future research should focus on establishing both global comparative data and high-resolution localized data of Holocene paleotemperature records for which archaeological assemblages can provide additional proxies.

Paleoclimatic reconstructions use oxygen and carbon stable isotope ratios to determine different variables like sea surface temperature (SST) or sea surface salinity (SSS). To do so, they rely on bicarbonates such as mollusk shells or coral skeletons. Likewise, climate-sensitive trace elements/calcium ratios can yield complementary data to stable isotopes (Lazareth 2017). Bivalves from the family Tridacnidae in particular are a relevant proxy for isotopic analysis (i.e., the measurement of stable isotopes ratios of chemical elements within organic or inorganic compounds) with high-resolution monthly records. As giant clams are common remains found on coastal archaeological sites, a large amount of shell material is available. Remains should be identified to the species level so that robust calibrations of δ18O are established from modern records. Archaeologists have already started to explore *Tridacna crocea* as a proxy for ambient water temperature records on Palau (Jew et al. 2019).

Traditional knowledge, in the form of oral traditions, has recorded climatic events of both low and high intensity. Nunn (2014) suggested that euhemeristic myths, that is, the interpretation of myths from an historical perspective, can translate observations and experiences of past events. I-Kiribati oral tradition from Makin Island informed researchers on a palaeotsunami occurring by the end of the sixteenth century, which was geologically identified (Terry et al. 2021). Similarly, in the Tuamotu, oral stories from Anaa and neighboring atolls recount catastrophic cyclones by the mid-seventeenth century that indisputably marked the cultural history of the western region (Torrente 2010; see also Ottino 1965 for sixteenth-century tsunami waves in Rangiroa). These events were recently confirmed thanks to geomorphological studies. The outer reef flats of atolls present mega-clasts, large carbonate boulders pulled out from the reef and transported meters away by waves of more than 10 to 15 m high during secular or millennial cyclones (Montaggioni et al. 2023). As they can be precisely dated through U/Th series, these geological features are valuable proxies for reconstructing major events in the history of the atoll-dwelling societies. While some may have forced population displacement, millennial cyclones of the kind mentioned above have eliminated all human life on the low coral islands, thus having wider regional repercussions on sociopolitical, economic, and religious structures.

On an even smaller scale, cyclones and tsunamis can cause variable mechanical damages to coral reefs, leading to habitats and biodiversity (Bozec et al. 2015). For instance, reefs might become more suitable to rubble specialists and grazing species (McClure et al. 2019). Hence, climatic disturbances likely impacted resource availability (Claar et al. 2018) and forced changes in diet and adaptive capacities. Hence, collaborative investigation of major marine hazards in the past must continue, especially in the contexts of low-coral islands.

Finally, and from a contemporary perspective, global climate change is actively endangering coastal and marine archaeological sites (Dawson et al. 2020). Sea-level rise, coastal erosion, human activities, and the increasing frequency of storm events constitute major threats to coastal sites. Archaeologists have called for increased monitoring and mapping of coastal sites which can be achieved through innovative, yet usually costly, geospatial tools. Remote sensing technology like LiDAR, nearshore maritime surveys for intertidal sites such as fish weirs, 3D modeling of at-risk features could help mitigate the progressive and inevitable loss of Pacific coastal heritage (McCoy 2018).

11.5 Lessons from the Past? Archaeology and the Future of Coral Reefs

In the last decades, archaeological sciences have seen major advancements in the development and application of analytical methods which are yielding promising results. As demonstrated in this chapter, coastal island archaeology greatly benefits from specialists' insights in biology, ecology, as well as palaeoenvironmental sciences, and the future of the field relies on the reinforcement and the emergence of new multidisciplinary collaborations designed around specific research questions.

Inversely, archaeology's capacity to reconstruct long-term socio-ecological dynamics, often inaccessible to other disciplines, can serve ecological sciences and inform conservation and management efforts. Historical ecology which articulates diverse sources of information proves to be fruitful in designing adapted local management strategies (Shackeroff et al. 2011), as recent assessments highlight the variability of climate change-related impacts between regions, urbanization levels, and island types (Duvat et al. 2021). Indeed, in the Pacific like elsewhere in the world, lessons from the past can help to better inform present decisions and future action (Boivin and Crowther 2021). Cooperation across the scientific fields, as well as increased sharing of knowledge between researchers and public actors (Terorotua et al. 2021), becomes critical to fully address questions about the future of coral reefs and marine ecosystems at large. Yet, social sciences and the overall human dimension of these pressing issues remain rarely included in debates (Trakadas et al. 2019).

Social sciences and archaeology have much to bring to today's debates about the preservation of cultural and natural heritage as well as traditional practices and restoration projects (such as *rahui* or fish weirs rehabilitation; Bambridge 2019), as revitalization of such practices can help compensate or replace "more productive but less sustainable practices" (Jeffrey 2013). This is crucial in order to better include indigenous communities in decision-making and to highlight and value their local long-lasting traditions and practices, providing more localized and adapted solutions to contemporary issues. As science moves forward to establish more ethical and inclusive ways of building and sharing knowledge, including communities in a collective way of science-making are critical necessities.

References

Abbott IA (1978) The uses of seaweed as food in Hawaii. Econ Bot 32:409–412

Adam AJ, Di Natale A, Addis P, Piattoni F, Onar V, Bernal-Casola D, Aniceti V et al (2023) Vertebrae reveal industrial-era increases in Atlantic bluefin tuna catch-at-size and juvenile growth. ICES J Mar Sci 80(4):836–847. https://doi.org/10.1093/icesjms/fsad013

Allen MS (1996) Style and function in East Polynesian fish-hooks. Antiquity 70(267):97–116

Allen MS (2003) Human impact on Pacific nearshore marine ecosystems. In: Sand C (ed) Pacific archaeology: assessments and prospects. Proceedings of the international Conference for the 50th anniversary of the first Lapita excavation. Les Cahiers de l'Archéologie en Nouvelle-Calédonie 15, Koné, Nouméa, pp 317–325

Allen MS (2006) New ideas about late Holocene climate variability in the Western and Central Pacific Ocean. Curr Anthropol 47(3):521–535

Allen MS (2007) Three millennia of human and sea turtle interactions in remote Oceania. Coral Reefs 26:959–970

Allen MS (2014) Variability is in the mesh-size of the sorter: Harataonga Beach and spatio-temporal patterning in Northern Māori fisheries. J Pac Archeol 5(1):21–38

Aswani S, Allen MS (2009) A Marquesan coral reef (French Polynesia) in historical context: an integrated socio-ecological approach. Aquat Conserv Mar Freshwat Ecosyst 19(6):614–625

Bambridge T (2019) Un commun foncier, maritime et terrestre en Polynésie : le rahui. In: Delmas B, Le Roy E (eds) Les communs, aujourd'hui ! enjeux planétaires d'une gestion locale des ressources renouvelables. Académie des sciences d'outre mer, Karthala, pp 159–163. https://doi.org/10.3917/kart.delma.2019.01.0159

Bataille-Benguigui M-C (1994) Le côté de la mer. Quotidien et imaginaire aux îles Tonga Polynésie Occidentale. Îles et Archipels 19. Centre de Recherche des Espaces Tropicaux de l'Université Michel de Montaigne, Bordeaux-Talence, 328 p

Beckwith MW (1917) Hawai'ian Shark Aumakua. Am Anthropol 19:503–517

Bedford S (2006) Pieces of the Vanuatu puzzle: archaeology of the North, South and Centre. Terra Australis 23. Pandanus Books, Canberra. 326 p

Berzunza-Sanchez MM, Gomez Cabrera MC, Pandolfi JM (2013) Historical patterns of resource exploitation and the status of Papua New Guinea Coral Reefs. Pac Sci 67(3):425–440. https://doi.org/10.2984/67.3.9

Bjorndal KA, Jackson JBC (2003) Roles of sea turtles in marine ecosystems: reconstructing the past. In: Lutz PL, Musick JA, Wyneken J (eds) The biology of sea turtles, vol 2. CRC Press, Boca Raton, pp 259–273

Boivin N, Crowther A (2021) Mobilizing the past to shape a better Anthropocene. Nat Ecol Evol 5(3):273–284. https://doi.org/10.1038/s41559-020-01361-4

Bouffandeau L, Béarez P, Bedford S, Valentin F, Spriggs S, Nolet E (2018) Fishing at Arapus-Mangaasi, Efate, Vanuatu (2800–2200 BP): new methodological approaches and results. J Archaeol Sci Rep 18:356–369. https://doi.org/10.1016/j.jasrep.2018.01.025

Bozec Y-M, Alvarez-Filip L, Mumby PJ (2015) The dynamics of architectural complexity on coral reefs under climate change. Glob Chang Biol 21(1):223–235

Braje TJ, Maloney JM, Gusick AE, Erlandson JM, Nyers A, Davis L, Gill KM, Reeder-Myers L, Ball D (2019) Working from the known to the unknown: linking the subaerial archaeology and the submerged landscapes of Santarosae Island, Alta California, USA. Open Quat 5(1):1–15. https://doi.org/10.5334/oq.66

Breckwoldt A, Dombal Y, Sabinot C, David G, Riera L, Ferse S, Fache E (2022) A social-ecological engagement with reef passages in New Caledonia: connectors between coastal and oceanic spaces and species. Ambio 51(13):2401–2413. https://doi.org/10.1007/s13280-022-01762-8

Buckley M, Pinsonneault M, Brassey C, Rolett BV (2021) High-throughput microCT and ZooMS collagen fingerprinting of Scombrid bone from the Marquesas Islands. J Archaeol Sci 136(337):105475. https://doi.org/10.1016/j.jas.2021.105475

Bunbury MME, Petchey F, Bickler SH (2022) A new chronology for the Maori settlement of Aotearoa (NZ) and the potential role of climate change in demographic development. PNAS 119(46):e2207609119. https://doi.org/10.1073/pnas.2207609119

Burley DV, Horrocks M, Weisler MI (2020) Earliest evidence for pit cultivation provides insight on the nature of first Polynesian settlement. J Isl Coast Archaeol 15(1):127–147. https://doi.org/10.1080/15564894.2018.1501441

Butler VL (1994) Fishfeeding behavior and fish capture: the case for variation in Lapita fishing strategies. Archaeol Ocean 29:81–90

Carlier C (2017) The Tangatatau fishhook assemblage: a typological analysis. In: Kirch PV (ed) Tangatatau Rockshelter: the evolution of an eastern Polynesian socio-ecosystem. Monumenta archaeological 40. Cotsen Institute of Archaeology Press, Los Angeles, pp 219–236

Carlier C, Conte E (2009) Proposition de nouvelles approches dans l'étude des hameçons océaniens. J Soc Ocean 1(128):133–146

Chabrier S, Molle G, Conte E, Carlier C (2012) Automation of fishhooks objective measures. Int J Hum Soc Sci 6(12):3670–3678

Chazine M (2003) Recherches archéologiques à Makemo – Tuamotu. Dossier d'Archéologie Polynésienne 2:117–136

Claar D, Szostek L, McDevitt-Irwin JM, Schanze JJ, Baum JK (2018) Global patterns and impacts of El Niño events on coral reefs: a meta-analysis. PLoS One 13(2):e01909577. https://doi.org/10.1371/journal.pone.0190957

Claringbold I (2021) The archaeology of Polynesian feasting: a zooarchaeological and taphonomic approach to turtle feasting deposits at Fakahina marae sites. Honours dissertation, Australian National University, Canberra, 192 p

Claringbold I, Samper-Carro SC, Molle G (2024) Turtles for the ancestors: a zooarchaeological study of ritual deposits on Fakahina, Tuamotu archipelago (French Polynesia). Archaeol Ocean 59(2):176–197

Conte E (1988) Les techniques de pêche pré-européennes et leurs survivances en Polynésie française : l'exploitation traditionnelle des ressources marines à Napuka (Tuamotu, Polynésie française). PhD dissertation: Université Paris I Panthéon-Sorbonne, Paris, 915 p

Conte E (2023) Sur le chemin des étoiles - Navigation traditionnelle et peuplement des îles du Pacifique. Au Vent des Îles, Papeete. 205 p

Conte E, Molle G (2014) Reinvestigating a key site for Polynesian prehistory: new results from the Hane dune site, Ua Huka (Marquesas). Archaeol Ocean 49(3):121–136. https://doi.org/10.1002/arco.5037

Conte E, Payri C (2002) La consommation des algues en Polynésie française. Premiers résultats d'une enquête. J Soc Ocean 114–115:165–172

Craig P, Green A, Tuilagi F (2008) Subsistence harvest of coral reef resources in the outer islands of American Samoa: modern, historic and prehistoric catches. Fish Res 89:230–240

Cramb J, Thompson VD (2022) Dynamic sustainability, resource management, and collective action on two atolls in the remote Pacific. Sustainability 14:5174

Davis LG, Madsen DB (2020) The coastal migration theory: formulation and testable hypotheses. Quat Sci Rev 249:106605. https://doi.org/10.1016/j.quascirev.2020.106605

Dawson T, Hambly J, Kelley A, Lees W, Miller S (2020) Coastal heritage, global climate change, public engagement, and citizen science. PNAS 117(15):8280–8286

Desse J (1984) Proposition pour une réalisation collective d'un corpus : fiches d'identification et d'exploitation métrique du squelette des poissons. In: Desse-Berset N (ed) 2eme Rencontres d'Archéo-Ichtyologie. Notes et monographies techniques 16, Editions du CNRS, Paris, pp 67–86

Desse-Berset N (2011) Discrimination of *Acipenser sturio*, *Acipenser oxyrinchus* and *Acipenser naccarii* by morphology of bones and osteometry. In: Williot P, Rochard E, Desse-Berset N, Kirschbaum F, Gessner J (eds) Biology and conservation of the European Sturgeon Acipenser sturio L. 1758. Springer, Berlin, Heidelberg, pp 23–51. https://doi.org/10.1007/978-3-642-20611-5_3

Dickinson WR (2003) Impact of mid-Holocene hydro-isostatic highstand in regional sea level on habitability of islands in Pacific Oceania. J Coast Res 19(3):489–502

Duprey N, Glipaud J-C, Cabioch G, Lazareth C (2014) Isotopic records from archaeological giant clams reveal a variable climate during the southwestern Pacific colonization ca. 3.0 ka BP. Palaeogeogr Palaeoclimatol Palaeoecol 404:97–108. https://doi.org/10.1016/j.palaeo.2014.04.002

Duque-Velez LM, Béarez P, Cornette R (2016) Otolith shape differences between *Micropogonias manni* and *M. altipinnis* (Sciaenidae). Société Française d'Ichtyologie 40(1):67–73

Duvat VKE, Magnan AK, Perry CT, Spencer T, Bell JD, Wabnitz CCC, Webb A, White I, McInnes KL, Gattuso J-P, Graham NAJ, Nunn PD, Le Cozannet G (2021) Risks to future atoll habitability from climate-driven environmental changes. Wiley Interdiscip Rev Clim Chang 12(3):e700. https://doi.org/10.1002/wcc.700

Dye TS, Longenecker KR (2004) Manual of Hawaiian fish remains identification based on the skeletal reference collection of Alan C. Ziegler and including Otoliths, vol 1. Society for Hawaiian Archaeology Special Publication, pp 1–134

Emory KP (1975) Material culture of the Tuamotu archipelago. Pacific Anthropological Records 22. Department of Anthropology. Bishop Museum Press, Honolulu. 253 p

Emory KP, Bonk WJ, Sinoto YH (1959) Hawaiian archaeology: fishhooks. Bernice P. Bishop museum special publication 47. Bishop Museum Press, Honolulu, p 45

Erlandson JM (2001) The archaeology of aquatic adaptations: paradigms for a new millennium. J Archaeol Res 9(4):287–350

Erlandson JM, Braje TJ (2015) Coasting out of Africa: the potential of mangrove forests and marine habitats to facilitate human coastal expansion via the Southern dispersal route. Quat Int 382:31–41

Erlandson JM, Rick TC (2010) Archaeology meets marine ecology: the antiquity of marine cultures and human impacts on marine fisheries and ecosystems. Annu Rev Mar Sci 2:165–185

Evin A, Souter T, Hulme-Beaman A, Ameen C, Allen R, Viacava P, Larson G, Cucchi T, Dobney K (2016) The use of close-range photogrammetry in zooarchaeology: creating accurate 3D models of wolf crania to study dog domestication. J Archaeol Sci Rep 9:87–93. https://doi.org/10.1016/j.jasrep.2016.06.028

Ferrario F, Beck MW, Storlazzi CD, Micheli F, Shepard CC, Airoldi L (2014) The effectiveness of coral reefs for coastal hazard risk reduction and adaptation. Nat Commun 5:3794. https://doi.org/10.1038/ncomms4794

Fitzhugh B, Butler VL, Bovy KM, Etnier MA (2019) Human ecodynamics: a perspective for the study of long-term change in socioecological systems. J Archaeol Sci Rep 23(1):1077–1094. https://doi.org/10.1016/j.jasrep.2018.03.016

Fitzpatrick SM, Giovas CM (2021) Tropical islands of the Anthropocene: deep histories of anthropogenic terrestrial-marine entanglement in the Pacific and Caribbean. PNAS 118(40):e2022209118

Ford B (2012) Coastal archaeology. In: Ford B, Hamilton DL, Catsambis A (eds) The Oxford handbook of maritime archaeology. Oxford University Press, New York, pp 763–785

Galzin R, Harmelin-Vivien M (2002) Ecologie des poissons des récifs coralliens. Océanis 26(3):465–495

Giovas CM (2016) Though she be but little: resource resilience, Amerindian foraging, and long-term adaptive strategies in the Grenadines, West Indies. J Isl Coast Archaeol 11:238–263. https://doi.org/10.1080/15564894.2016.1193572

Giovas CM, Fitzpatrick SM, Meagan C, Mira A (2010) Evidence for size increase in an exploited mollusc: humped conch (*Strombus gibberulus*) at Chelechol ra Orrak, Palau from ca. 3000-O BP. J Archaeol Sci 37:2788–2798. https://doi.org/10.1016/j.jas.2010.06.013

Giovas CM, Lambrides ABJ, Fitzpatrick S, Kataoka O (2017) Reconstructing prehistoric fishing zones in Palau, Micronesia using fish remains: a blind test of inter-analyst correspondence. Archaeol Ocean 52(1):45–61. https://doi.org/10.1002/arco.5119

Grimstead DN, Pailes MC, Dungan KA, Dettman DL, Martínez TN, Clark AE (2013) Identifying the origin of Southwestern shell: a geochemical application to Mogollon Rim archaeomolluscs. Am Antiq 78(4):1–22. https://doi.org/10.7183/0002-7316.78.4.640

Harris M, Weisler MI (2017) Prehistoric human impacts to marine mollusks and intertidal ecosystems in the Pacific Islands. J Isl Coast Archaeol 13(2):235–255. https://doi.org/10.1080/15564894.2016.1277810

Harvey VL, LeFebvre MJ, deFrance SD, Toftgaard C, Drosou K, Kitchener AC, Buckley M (2019) Preserved collagen reveals species identity in archaeological marine turtle bones from Caribbean and Florida sites. R Soc Open Sci 6(10):191137. https://doi.org/10.1098/rsos.191137

Hermann A, Gutiérrez P, Chauvel C, Maury R, Liorzou C, Willie E, Phillip I, Forkel R, Rzymski C, Bedford S (2023) Artefact geochemistry demonstrates long-distance voyaging in the Polynesian outliers. Sci Adv 9(16):1–12. https://doi.org/10.1126/sciadv.adf4487

Hubert N, Delrieu-Trottin E, Irisson JO, Meyer C, Planes S (2010) Identifying coral reef fish larvae through DNA barcoding: a test case with the families Acanthuridae and Holocentridae. Mol Phylogenet Evol 55(3):1195–1203. https://doi.org/10.1016/j.ympev.2010.02.023

Inglis RH, Bosworth W, Rasul NMA, Al-Saeedi AO, Bailey GN (2019) Investigating the Palaeoshorelines and coastal archaeology of the Southern Red Sea. In: Rasul NMA, Stewart ICF (eds) Geological setting, Palaeoenvironment and archaeology of the Red Sea. Springer, Cham, pp 533–581. https://doi.org/10.1007/978-3-319-99408-6_25

Irwin G (1992) The prehistoric exploration and colonization of the Pacific. Cambridge University Press Cambridge. 240 p

Jeffrey B (2013) Reviving community spirit: furthering the sustainable, historical and economic role of fish weirs and traps. J Marit Archaeol 8(1):29–57

Jerardino A, Marean CW (2010) Shellfish gathering, marine paleoecology and modern human behavior: perspectives from cave PP13B, Pinnacle Point, South Africa. J Hum Evol 59(3–4):412–424. https://doi.org/10.1016/j.jhevol.2010.07.003

Jew NP, Dodrill T, Fitzpatrick SM (2019) Evaluating the efficacy of the mollusc *Tridacna crocea* for reconstructing ancient sea-surface temperatures in the Rock Islands of Palau, Micronesia. Archaeol Oceania 54(3):107–119. https://doi.org/10.1002/arco.5182

Jimenez C, Orejas C (2016) The builders of the oceans – part II: corals from the past to the present (the stone from the sea). In: Rossi S, Bramanti L, Gori A, Orejas C (eds) Marine animal forests. Springer, Cham, pp 657–697. https://doi.org/10.1007/978-3-319-17001-5_56-1

Jones O'DS (2004) Marine resource exploitation and diversity in Kahikinui, Maui, Hawaii: bringing together multiple lines of evidence to interpret the past. Archaeofauna 13:97–108

Jones S (2009) A long-term perspective on biodiversity and marine resource exploitation in Fiji's Lau group. Pac Sci 63(4):617–648

Kahn JG, Dotte-Sarout E, Molle G, Conte E (2015a) Mid- to late prehistoric landscape change, settlement histories, and agricultural practices on Maupiti, Society Islands (Central Eastern Polynesia). J Isl Coast Archaeol 10(3):363–391. https://doi.org/10.1080/15564894.2014.1001922

Kahn JG, Nickelsen C, Stevenson J, Porch N, Dotte-Sarout E, Christensen CC, May L, Athens JS, Kirch PV (2015b) Mid- to late Holocene landscape change and anthropogenic transformations on Mo'orea, Society Islands: A multi-proxy approach. The Holocene 25(2):333–347. https://doi.org/10.1177/0959683614558649

Kirch PV (1982) The impact of the prehistoric Polynesians on the Hawaiian ecosystem. Pac Sci 36(1):1–14

Kirch PV (2010) Peopling of the Pacific: a holistic anthropological perspective. Annu Rev Anthropol 39:131–148

Kirch PV (2017) On the road of the winds: an archaeological history of the Pacific islands before European contact. University of California Press, Berkeley. 446 p

Kirch PV, Dye T (1979) Ethnoarchaeology and the development of Polynesian fishing strategies. J Polyn Soc 88(1):53–76

Kirch PV, Hunt TL (eds) (1997) Historical ecology in the Pacific Islands: prehistoric environmental and landscape change. Yale University Press, New Haven, p 331

Kirch PV, Molle G (2024) The archaeology of East Polynesia. In: Fitzpatrick SM, Erlandson J (eds) The Oxford handbook of Island and coastal archaeology. https://doi.org/10.1093/oxfordhb/9780197607770.013.28

Kirch PV, Sharp WD (2005) Coral 230Th dating of the imposition of a ritual control hierarchy in precontact Hawaii. Science 307(5706):102–104

Kirch PV, Steadman DW, Butler VL, Hather J, Weisler MI (1995) Prehistory and human ecology in eastern Polynesia: excavations at Tangatatau Rockshelter, Mangaia, Cook Islands. Archaeol Ocean 30(2):47–65

Kirch PV, Molle G, Nickelsen C, Mills P, Dotte-Sarout E, Swift J, Wolfe A, Horrocks M (2015) Human ecodynamics in the Mangareva Islands: a stratified sequence from Nenega-Iti Rock Shelter (site AGA-3, Agakauitai Island). Archaeol Ocean 50(1):23–42. https://doi.org/10.1002/arco.5050

Kozuch L, Fitzgerald C (1989) A guide to identifying shark centra from southeastern archaeological sites. Southeast Archaeol 8(2):146–157

Lagarde L, Nolet E, Molle G (2020) Hokikakika: history and archaeology of a catholic village in the Eastern Tuamotus. J Pac Archeol 11(2):74–89

Lambrides ABJ, Weisler MI (2013) Assessing protocols for identifying Pacific Island archaeological fish remains: the contribution of vertebrae. Int J Osteoarchaeol 25(6):838–848. https://doi.org/10.1002/oa.2354

Lambrides ABJ, Weisler MI (2015) Applications of vertebral morphometrics in Pacific Island archaeological fishing studies. Archaeol Ocean 50(2):53–70. https://doi.org/10.1002/arco.5059

Lambrides ABJ, Weisler MI (2016) Pacific Islands Ichthyoarchaeology: implications for the development of prehistoric fishing studies and global sustainability. J Archaeol Res 24(3):275–324. https://doi.org/10.1007/s10814-016-9090-y

Lambrides ABJ, Weisler MI, Charleux M (2018) New taxonomic records and regional trends for the Marquesan prehistoric marine fishery, Eiao Island, Polynesia. J Pac Archeol 9(1):44–62

Langley MC, O'Connor S, Kealy S, Mahirta (2021) Fishhooks, lures, and sinkers: intensive manufacture of marine technology from the terminal Pleistocene at Makpan Cave, Alor Island, Indonesia. J Isl Coast Archaeol 18(1):33–52. https://doi.org/10.1080/15564894.2020.1868631

Lavondès A (1971) Le Polynésien et la Mer, catalogue d'une exposition. Société des Etudes Océaniennes et ORSTOM, Papeete. 50 p

Lazareth CE (2017) High-resolution climatic reconstruction in the tropical Pacific based on biocarbonate archives. Mémoire HDR. Université Pierre et Marie Curie, p 109

Leppard TP, Fitzpatrick SM (2021) Theory beyond the calm ocean? The Pacific contribution to global Island archaeology. In: Thomas T (ed) Theory in the Pacific, the Pacific in theory: archaeological perspectives. Routledge, London, pp 37–57. https://doi.org/10.4324/9780203730973-2

Marrast A, Béarez P (2019) Osteometry and size reconstruction of the Indian and Pacific oceans' *Euthynnus* species, *E. affinis* and *E. lineatus* (Scombridae). Cybium 43(2):187–198. https://doi.org/10.26028/cybium/2019-423-007

Masse WB, Liston J, Carucci J, Athens JS (2006) Evaluating the effects of climate change on environment, resource depletion, and culture in the Palau Islands between AD 1200 and 1600. Quat Int 151(1):106–132. https://doi.org/10.1016/j.quaint.2006.01.017

McClure E, Richardson LE, Graba-Landry A, Loffler Z, Russ GR, Hoey AS (2019) Cross-shelf differences in the response of herbivorous fish assemblages to severe environmental disturbances. Diversity 11(2):23. https://doi.org/10.3390/d11020023

McCoy MD (2018) The race to document archaeological sites ahead of rising sea levels: recent applications of geospatial technologies in the archaeology of Polynesia. Sustainability 10(1):185. https://doi.org/10.3390/su10010185

McNiven IJ, De Maria N, Weisler M, Lewis T (2014) Darumbal voyaging: intensifying use of Central Queensland's Shoalwater Bay islands over the past 5000 years. Archaeol Ocean 49(1):2–42. https://doi.org/10.1002/arco.5016

Molle G (2014) Ancêtres-dieux et temples de corail. Approche ethnoarchéologique du complexe marae dans l'archipel des Tuamotu, Polynésie française. Les Cahiers du CIRAP 3(1):85

Molle G, Conte E (2016) Variabilité des espaces dunaires dans l'archipel des Marquises : approche comparative des séquences d'occupation littorale sur l'île de Ua Huka. In: Valentin F, Molle G (eds) La pratique de l'espace en Océanie : découverte, appropriation et émergence des systèmes sociaux traditionnels, vol 7. Séances de la Société Préhistorique française, Paris, pp 125–140

Molle G, Nolet E, Lagarde L, Valentin F, Claringbold I, Rurua V, Traversat G, Dotte-Sarout E, Conte E (2021) Recherches archéologiques et anthropologiques sur l'atoll de Fakahina (Tuamotu). In: Rapport final du projet 2018–2019. CIRAP- MSHP, Tahiti. 165 p

Molle G, Wadrawane J-M, Lagarde L, Wright D (2023) The sacred stone from the sea. Archaeological and ethnographic perspectives on the ritual value of coral across the Pacific. Archaeol Ocean 58:40–55. https://doi.org/10.1002/arco.5284

Montaggioni LF, Salvat B, Aubanel A, Pons-Branchu E, Martin-Garin B, Dapoigny A, Goeldner-Gianella L (2019) New insights into the Holocene development history of a Pacific, low-lying coral reef Island: Takapoto Atoll, French Polynesia. Quat Sci Rev 223:105947. https://doi.org/10.1016/j.quascirev.2019.105947

Montaggioni LF, Salvat B, Pons-Branchu E, Dapoigny A, Martin-Garin B, Poli G, Zanini J-M, Wan R (2023) Mid-late accretional history of low-lying, coral-reef rim islets, South-Marutea Atoll, Tuamotu, central South Pacific: the key role of marine hazard events. Nat Hazards Res 3(2):219–239. https://doi.org/10.1016/j.nhres.2023.02.004

Morat F, Blamart D, Lecomte-Finiger R, Planes S, Galzin R (2015) From the geochemistry of coral reef fish otoliths to climate: what can we learn from the example of Taiaro lagoon? Vie et Milieu 65(1):21–28

Morrison AE, Addison DJ (2009) Examining causes and trends in marine trophic level change: 1500 years of fish exploitation at Fatuma-Futi, Tutuila Island, American Samoa. J Isl Coast Archaeol 4(2):177–194. https://doi.org/10.1080/15564890903245264

Morrison AE, Allen MS (2017) Agent-based modelling, molluscan population dynamics, and archaeomalacology. Quat Int 427:170–183. https://doi.org/10.1016/j.quaint.2015.09.004

Morrison AE, Cochrane EE (2008) Investigating shellfish deposition and landscape history at the Natia Beach Site, Fiji. J Archaeol Sci 35(8):2387–2399. https://doi.org/10.1016/j.jas.2008.03.013

Mouchi V, Godbillot C, Dupont C, Vella M-A, Forest V, Ulianov A, Lartaud F, de Rafélis M, Emmanuel L, Verrecchia EP (2021) Provenance study of oyster shells by LA-ICP-MS. J Archaeol Sci 132(1):105418. https://doi.org/10.1016/j.jas.2021.105418

Moyer JK, Hamilton ND, Seeley RH, Riccio ML, Bemis WE (2015) Identification of Shark Teeth (Elasmobranchii: Lamnidae) from a historic Fishing Station on Smuttynose Island, Maine, using computed tomography imaging. Northeast Nat 22(3):585–597. https://doi.org/10.1656/045.022.0313

Nagaoka L (2005) Differential recovery of Pacific Islands fish remains. J Archaeol Sci 3:941–955

Nicholls A, Matisso-Smith E, Allen MS (2003) A novel application of molecular techniques to Pacific Archaeofish remains. Archaeometry 45(1):133–147

Nunn PD (2014) Lashed by sharks, pelted by demons, drowned for apostasy: the value of myths that explain geohazards in the Asia-Pacific region. Asian Geogr 31(1):59–82. https://doi.org/10.1080/10225706.2013.870080

O'Connor S, Ono R, Clarkson C (2011) Pelagic fishing at 42,000 years before the present and the maritime skills of modern humans. Science 334(6059):1117–1121. https://doi.org/10.1126/science.1207703

Ono R, Addison DJ (2009) Ethnoecology and Tokelauan fishing lore from Atafu Atoll, Tokelau. SPC TMRMKIB 26:3–22

Ono R, Morrison A, Addison DJ (eds) (2013) Prehistoric Marine Resource Use in the Indo-Pacific Regions. Terra Australis 39. ANU Press, Canberra. 204 p. https://doi.org/10.22459/TA39.12.2013

Oosting T, Star B, Barrett JH, Wellenreuther M, Ritchie PA, Rawlence NJ (2019) Unlocking the potential of ancient fish DNA in the genomic era. Evol Appl 12(8):1513–1522. https://doi.org/10.1111/eva.12811

Ottino P (1965) Ethno-histoire de Rangiroa (archipel des Tuamotu). Centre O.R.S.T.O.M, Papeete. 167 p

Planes S, Chancerelle Y, Galzin R (1995) Étude de l'environnement marin de deux baies à Ua Huka. Contrat Centre Polynésien des Sciences Humaines. Rapport Centre de Recherches Insulaires et Observatoire de l'Environnement, E.P.H.E., 1995RA63. 70 p

Prévost F (2017) Les *tuputupua* : ancêtres protecteurs et animaux tutélaires aux Tuamotu. J Soc Ocean 1–2(144–145):329–344. https://doi.org/10.4000/jso.7894

Rebolledo S, Béarez P, Zurro D (2021) Fishing during the early human occupations of the Atacama Desert coast: what if we standardize the data? Archaeol Anthropol Sci 13(9):152. https://doi.org/10.1007/s12520-021-01387-0

Richter KK, Codlin MC, Seabrook M, Warinner C (2022) A primer for ZooMS applications in archaeology. PNAS 119(20):e2109323119. https://doi.org/10.1073/pnas.2109323119

Rogers AJ, Weisler MI (2019) Assessing the efficacy of genus-level data in Archaeomalacology: a case study of the Hawaiian limpet (*Cellana* spp.), Moloka'i, Hawaiian Islands. J Isl Coast Archaeol 15(1):28–56

Rolett BV (1998) Hanamiai: prehistoric colonization and cultural change in the Marquesas Islands, East Polynesia. Yale University Press, New Haven. 277 p

Rurua VA (2020) Biodiversité et exploitation des ressources marines en Polynésie française sur la longue durée : étude comparative des archipels des Marquises et des Gambier. PhD dissertation, Université de la Polynésie française, Tahiti, 211 p

Rurua VA (2022) De la famille à l'espèce: aide à l'identification des vestiges osseux de poissons. Post-doctoral report, CIRAP-UPF, Tahiti, 105 p

Rurua VA, Béarez P, Hermann A, Conte E (2020) Length and weight reconstruction of *Chlorurus microrhinos* (Scaridae) from isolated cranial bones and vertebrae. Cymbium 44(1):61–68. https://doi.org/10.26028/cybium/2020-441-008

Samper-Carro SC, Louys J, O'Connor S (2018) Shape does matter: a geometric morphometric approach to shape variation in Indo-Pacific fish vertebrae for habitat identification. J Archaeol Sci 99:124–134. https://doi.org/10.1016/j.jas.2018.09.010

Samper-Carro SC, Rurua VA, Molle G (2024) IndoPac-fishnet workshop: current research and future directions in indo-Pacific ichthyoarchaeology. ICAZ Newsletter 24(1):24–25

Sear DA, Allen MS, Hassal JD, Maloney AE, Langdon PG, Morrison AE, Genderson ACG et al (2020) Human settlement of East Polynesia earlier, incremental, and coincident with prolonged South Pacific drought. PNAS 177(16):8813–8819. https://doi.org/10.1073/pnas.1920975117

Serrand N, Bonnissent D (2018) Interacting Pre-Columbian Amerindian societies and environments: insights from five millennia of archaeological invertebrate record on the Saint-Martin Island (French Lesser Antilles). Environ Archaeol 26(16):1–16. https://doi.org/10.1080/14614103.2018.1450463

Shackeroff JM, Campbell LM, Crowder LB (2011) Social-ecological guilds: putting people into marine historical ecology. Ecol Soc 16(1):52. https://doi.org/10.5751/ES-03410-160152

Shaw I (2022) From the sea floor, not far out: a life-history analysis of Tikopia shell blades from the Spillius collection, and the re-appreciation of unstratified assemblages in archaeology. Masters dissertation, Australian National University, Canberra, 91 p

Spenneman DHR (1999) No room for the dead. Practices in a constrained environment. Anthropos 94:35–56

Spyrou A, Nobles G, Hadjikoumis A, Evin A, Hulme-Beaman A, Çakirlar C, Ameen C, Loucas N, Nikita E, Hanot P, de Boer NM, Avgousti A, Zohar I, May H, Rehren TH (2022) Digital Zooarchaeology: state of the art, challenges, prospects and synergies. J Archaeol Sci Rep 45(1916):103588. https://doi.org/10.1016/j.jasrep.2022.103588

Steadman DW (1995) Prehistoric extinctions of Pacific Island birds: biodiversity meets zooarchaeology. Science 267(5201):1123–1131

Summers CC (1964) Hawaiian fishponds. Bernice P. Bishop museum special publication 52. Bishop Museum Press, Honolulu. 26 p

Swalding P (1976) Changes induced by human exploitation in prehistoric shellfish populations. Man 10(3):181–191

Swalding P (1986) Lapita shellfishing: evidence from sites in the Reef/Santa Cruz Group, southeast Solomons. In: Anderson AJ (ed) Pacific anthropological records 37: traditional fishing in the Pacific, Department of Anthropology. Bernice Pauahi Bishop Museum, Honolulu, pp 137–148

Swift AJ, Roberts P, Boivin N, Kirch PV (2018) Restructuring of nutrient flows in Island ecosystems following human colonization evidenced by isotopic analysis of commensal rats. PNAS 115(25):6392–6397. https://doi.org/10.1073/pnas.1805787115

Swift JA, Kirch PV, Ilgner J, Brown S, Lucas M, Marzo S, Roberts P (2021) Stable isotopic evidence for nutrient rejuvenation and long-term resilience on Tikopia Island (Southeast Solomon Islands). Sustainability 13(15):8567. https://doi.org/10.3390/su13158567

Szabó C, Vogel Y, Anderson A (2012) Marine resource exploitation on Rapa Island. Archaeology, material culture and ethnography. In: Anderson A, Kennett DJ (eds) Taking the high ground: the archaeology of Rapa, a fortified Island in remote Oceania. Terra Australis 37. ANU Press, Canberra, pp 146–166. https://doi.org/10.22459/TA37.11.2012.09

Terorotua H, Duvat VKE, Maspataud A, Ouriqua J (2021) Assessing perception of climate change by representatives of public authorities and designing coastal climate services: lessons learnt from French Polynesia. Front Mar Sci 7:160. https://doi.org/10.3389/fmars.2020.00160

Terry JP, Karoro R, Gienko GA, Wieczorek M, Lau AYA (2021) Giant palaeotsunami in Kiribati: converging evidence from geology and oral history. Island Archaeol 30(1):e12417. https://doi.org/10.1111/iar.12417

Thomas F (2014) 'Sustainability Archaeology' and landscape transformation in Eastern Micronesia: Kiribati case study. Paper presented at the Samoa Conference III: opportunities and challenges for a sustainable cultural and natural environment 25–29 August 2014. National University of Samoa, Apia

Torrente F (2010) Ethnohistoire de Anaa un atoll des Tuamotu. PhD dissertation, Université de Polynésie française, Tahiti, 564 p

Torrente F, Bambridge T, Planes S, Guiart J, Clua EG (2018) Sea swallowers and land devourers: can shark lore facilitate conservation? Hum Ecol 46(2):717–726. https://doi.org/10.1007/s10745-018-0018-7

Trakadas A, Firth A, Gregory D, Elkin D, Guerin U, Henderson J, Kimura J, Scott-Ireton D, Shashoua Y, Underwood C, Viduka A (2019) The ocean decade heritage network: integrating cultural heritage within the UN decade of ocean science 2021–2030. J Marit Archaeol 14(1):153–165. https://doi.org/10.1007/s11457-019-09241-0

Traversat G (2022) Exploitation, consommation et technologies coquillières à l'époque pré-européenne en Polynésie centrale. PhD dissertation, Université Paris I Panthéon-Sorbonne, Paris, 390 p

Traversat G, Conte E, Molle G, Salvat B (2021) An archaeomalacological investigation of chitons on the Hane Dune site, Ua Huka, Marquesas Islands. J Isl Coast Archaeol 18(3):475–497. https://doi.org/10.1080/15564894.2021.1986177

Traversat G, Conte E, Molle G (2024) The archaeology of shellfishing practices on Ua Huka, Marquesas archipelago (French Polynesia). Archaeol Ocean 59(2):251–274

Viviani J, LeBlanc A, Rurua V, Mou T, Liao V, Lecchini D, Galzin R, Viriot L (2022) Plicidentine in the oral fangs of parrotfish (Scarinae, Labriformes). J Anat 241(3):601–615. https://doi.org/10.1111/joa.13673

Walter RC, Buffler RT, Bruggemann JH, Guillaume MMM, Berhe SM, Begassi B, Libsekal Y, Cheng H, Edwards RL, von Cosel R, Néraudeau D, Gagnon M (2000) Early human occupation on the Red Sea coast of Eritrea during the last interglacial. Nature 405(6782):65–69. https://doi.org/10.1038/35011048

Weisler MI, Bollt R, Findlater A (2010) Prehistoric fishing strategies on the makatea Island of Rurutu. Archaeol Ocean 45(3):130–143. https://doi.org/10.1002/j.1834-4453.2010.tb00089.x

Weisler MI, Bolhar R, Ma J, Kirch PV (2016) Cook Island artifact geochemistry demonstrates spatial and temporal extent of pre-European interarchipelago voyaging in East Polynesia. PNAS 113(29):8150–8155. https://doi.org/10.1073/pnas.1608130113

Will M, Bader GD, Sommer C, Cooper A, Green A (2022) Coastal adaptations on the eastern seaboard of South Africa during the Pleistocene and Holocene? Current evidence and future perspectives from archaeology and marine geology. Front Earth Sci 10:964423. https://doi.org/10.3389/feart.2022.964423

Wyban CA (1992) Tide and current: fishponds of Hawai'i. University of Hawaii Press, Honolulu. 240 p

Yeomans L, Beech MJ (2021) An aid to the identification of fish bones from Southeast Arabia : the influence of reference collections on taxonomic diversity. Int J Osteoarchaeol 31(1):3–17. https://doi.org/10.1002/oa.2920

MIX
Papier aus verantwortungsvollen Quellen
Paper from responsible sources
FSC® C105338

If you have any concerns about our products,
you can contact us on
ProductSafety@springernature.com

In case Publisher is established outside the EU,
the EU authorized representative is:
Springer Nature Customer Service Center GmbH
Europaplatz 3, 69115 Heidelberg, Germany

Printed by Libri Plureos GmbH
in Hamburg, Germany